The Many-Body Problem in Quantum Mechanics

N. H. MARCH
Coulson Professor of Theoretical Chemistry, University of Oxford

W. H. YOUNG
Professorial Fellow in Physics, University of East Anglia

and

S. SAMPANTHAR
formerly at the University of Salford

DOVER PUBLICATIONS, INC.
New York

Copyright

Published in Canada by General Publishing Company, Ltd., 30 Lesmill Road, Don Mills, Toronto, Ontario.
Published in the United Kingdom by Constable and Company, Ltd., 3 The Lanchesters, 162–164 Fulham Palace Road, London W6 9ER.

Bibliographical Note

This Dover edition, first published in 1995, is an unaltered and unabridged republication of the work originally published by Cambridge University Press, London, in 1967 as part of the series *Cambridge Monographs on Physics.*

Library of Congress Cataloging-in-Publication Data

March, Norman H. (Norman Henry), 1927–
The many-body problem in quantum mechanics / N. H. March, W. H. Young and S. Sampanthar.
p. cm.
Originally published: London : Cambridge University Press, 1967, in series: Cambridge monographs on physics.
Includes bibliographical references and index.
ISBN 0-486-68754-6 (pbk.)
1. Many-body problem. 2. Quantum theory. I. Young, W. H. II. Sampanthar, S. III. Title.
QC174.17.P7M365 1995
530.1′44—dc20 95-22121
CIP

Manufactured in the United States of America
Dover Publications, Inc., 31 East 2nd Street, Mineola, N.Y. 11501

CONTENTS

CHAPTER 4

Many-body perturbation theory

CHAPTER 5

Fermi fluids

CHAPTER 6

Nuclear matter

CHAPTER 7

Superconductivity

CHAPTER 8

Many-Boson systems

CHAPTER 9

Grand partition functions

CHAPTER 10

Green functions

APPENDIXES

PREFACE

Our aim in writing this book was to provide, for the reader who wanted to learn about the many-body problem, an account both of the methods used and of the physics which emerges, within a single cover. By the standards of the literature on many-body theory, this is not an advanced book. We build largely on the assumption that the reader has a good basic understanding of elementary quantum mechanics.

We have not been able to cite all references in which important contributions to topics covered in the book have been made. Studies covering the latest developments in a particular field and the assigning of credit to the contributors were regarded as outside the scope of our programme, and more properly left to the writers of review articles.

Many people have helped us in our task, and to all these we express our sincere thanks. In particular Drs P. Carruthers, K. T. R. Davies, V. J. Emery, T. Gaskell, D. G. Graham, W. Jones, G. E. Kilby, A. Meyer, N. Ogg, E. R. Pike, H. C. Schweinler, G. L. Sewell and J. C. Stoddart made valuable comments on individual chapters. In addition, many former colleagues of one of us (W.H.Y.) at Oak Ridge National Laboratory contributed greatly by general discussion and we are particularly indebted to Drs J. H. Barrett, G. Czjzek, J. S. Faulkner and D. K. Holmes. The only person to see the whole book in manuscript was Professor J. M. Ziman, who made numerous suggestions which have considerably improved the layout and presentation.

Needless to say, the sole responsibility for errors which remain lies with the authors. It is too much to hope that the account can be free from misunderstandings of detail, and occasionally perhaps, because of the very broad field we have attempted to cover, even of fundamentals. If the book proves of interest to readers, we invite them to take the trouble to write to us about obscurities and mistakes of any kind.

Mrs Elaine Lycett typed the whole of the manuscript with truly outstanding skill and with great patience and we are most grateful to her. Finally, the officers of the Cambridge University Press are to be thanked for their help and advice through a long and difficult task.

N.H.M. W.H.Y. S.S.

CHAPTER 1

SINGLE-PARTICLE APPROXIMATIONS

1.1. Introduction

It is well known, even in classical mechanics, that the problem of interacting particles presents great difficulties when exact solutions are sought. Likewise, in quantum mechanics, there is hardly a single worthwhile problem with realistic interactions which we can solve precisely. We have learnt, therefore, to be content with approximate solutions, which either contain the essential features of the whole problem or, at least, contain within themselves a criterion of validity. With approximate solutions, it becomes crucial, for an understanding of the problem, to have a clearly delineated régime of applicability of the theory.

Indeed, it is often true that approximate physical theories, based on a combination of skilled intuition and powerful use of mathematical techniques, have been of more conceptual value than any exact numerical solutions of the appropriate equations might have been. As an outstanding example of this, we may mention the enormous pattern and clarification which emerged in atomic theory when Hartree proposed his famous self-consistent field method, in which every electron in an atom is given its own personal wave function and energy level. The whole construct of atomic theory has essentially been erected on this basis. Indeed, in molecular physics, solid-state theory, nuclear shell structure and so on, this idea—the reduction of the many-body problem to that of describing the motion of a single particle in some field, contributed to by the other particles—has been of the utmost utility.

Therefore, it is natural enough to begin our discussion from the standpoint of single-particle approximations. In this chapter, after a brief introduction to the basic many-body problem in section 1.2, we shall develop the single-particle theories using density matrices as well as more conventional treatments in terms of wave functions. Both perturbative and variational methods will be discussed. In particular, the central results of perturbation theory are given in section 1.7 in the density matrix framework, while in section 1.8 we discuss more general perturbation theory based directly on wave functions. The Hartree–Fock equations, the nub of single-particle approximations, are presented in section 1.10, while the semi-classical, but intuitively appealing, Thomas–Fermi theory is discussed in section 1.9.

1.2. Hamiltonian and symmetries of eigenfunctions

The Hamiltonian for a system of N non-relativistic particles interacting via two-body forces is

$$H = \sum_{i=1}^{N} -\frac{\hbar^2}{2m}\nabla_i^2 + \sum_{i<j} v(\mathbf{r}_i, \mathbf{r}_j). \tag{1.2.1}$$

If there are internal degrees of freedom, most usually spin, which affect the particle interaction, and if the particles are also subject to some external potential, we can effect a generalization of (1.2.1) to read

$$H = \sum_{i=1}^{N} U(\mathbf{r}_i\,\sigma_i) + \sum_{i<j} v(\mathbf{r}_i\,\sigma_i, \mathbf{r}_j\,\sigma_j). \tag{1.2.2}$$

Here the single-particle operator $U(\mathbf{r}_i\,\sigma_i)$ includes the kinetic energy operator and σ_i is the internal co-ordinate of the ith particle (usually the spin). The second summation involves $\frac{1}{2}[N(N-1)]$ terms, and may be rewritten in the form

$$H = \sum_{i=1}^{N} U(\mathbf{r}_i\,\sigma_i) + \frac{1}{2}\sum_{i=1}^{N}\sum_{j=1}^{N}{}' v(\mathbf{r}_i\,\sigma_i, \mathbf{r}_j\,\sigma_j), \tag{1.2.3}$$

where the prime indicates that terms with $i = j$ are omitted.

We wish to obtain the eigenvalues and eigenfunctions of this Hamiltonian. Explicitly we have to find E and Φ satisfying the Schrödinger equation

$$H\Phi = E\Phi. \tag{1.2.4}$$

As remarked in section 1.1, since we cannot solve (1.2.4) exactly in general, we either solve it approximately or, more frequently, we consider a model system which we choose such that it incorporates the central features of the actual system. If the Hamiltonian of the model system is H_0, then we solve for its eigenfunctions and eigenvalues exactly. We may improve the approximation later, should it prove necessary, by use of perturbation theory based on the model problem, or by some other suitable approximate method.

If we take as our model Hamiltonian

$$H_0 = \sum_1^N U(\mathbf{r}_i\,\sigma_i), \tag{1.2.5}$$

then its eigenvalues and eigenfunctions may be obtained in terms of the eigenvalues ϵ_l and the eigenfunctions $\phi_l(\mathbf{r}\sigma)$ of the one-body Hamiltonian $U(\mathbf{r}\sigma)$, satisfying

$$U(\mathbf{r}\sigma)\,\phi_l = \epsilon_l\,\phi_l. \tag{1.2.6}$$

From standard quantum mechanics, we know that the ϕ_l's may be made to form a complete set of orthonormal states, that is

$$\sum_\sigma \int \phi_l^*(\mathbf{r}\sigma)\,\phi_k(\mathbf{r}\sigma)\,d\mathbf{r} = \delta_{lk}, \tag{1.2.7}$$

where δ_{lk} is the Kronecker delta. In (1.2.7) we have used a notation appropriate to a discrete variable σ and discrete eigenvalues ϵ_l of (1.2.6). If ϵ_l is a continuous function of l, then we must replace the Kronecker delta of (1.2.7) by the Dirac delta function.

A typical eigenfunction of (1.2.5) is

$$\prod_{i=1}^N \phi_{l_i}(\mathbf{r}_i\,\sigma_i), \tag{1.2.8}$$

where ϕ_{l_i} are in general any N of the (not necessarily distinct) ϕ_l's of (1.2.6). The eigenvalue corresponding to the eigenfunction (1.2.8) is evidently $\sum_{i=1}^N \epsilon_{l_i}$. These statements take no account of the indistinguishability of the particles, and to proceed further we must now specify the statistics.

1.2.1. *Bosons*

If the particles are identical and are Bosons, then any acceptable many-body wave function must be symmetric with respect to the interchange of the co-ordinates of any two particles. Therefore we form symmetrized products Φ_L^B from (1.2.8) where L labels the N single-particle levels which occur in the symmetrized products. Then a general N-Boson wave function can be expanded in terms of these symmetrized products as

$$\Phi^B(\mathbf{r}_1\,\sigma_1 \ldots \mathbf{r}_N\,\sigma_N) = \sum_L C_L\,\Phi_L^B(\mathbf{r}_1\,\sigma_1 \ldots \mathbf{r}_N\,\sigma_N), \tag{1.2.9}$$

since the Φ_L^B's are (cf. (1.2.7)) orthonormal, and thus form a suitable basis for describing a general N-Boson wave function.

1.2.2. *Fermions*

For Fermions, the wave functions appropriate for describing the ground state and the excited states must be antisymmetrical in the interchange of the co-ordinates (space and spin) of any pair of particles. This is the generalization for many interacting particles of the famous Pauli Exclusion Principle, which states that with independent Fermi particles, no two can occupy the same state.

Thus, this time, we must form antisymmetrized products from (1.2.8), with no two single-particle wave functions the same. These antisymmetrized products, or determinants, Φ_L^F say, form a basis for N-Fermion wave functions through

$$\Phi^F(\mathbf{r}_1\,\sigma_1 \ldots \mathbf{r}_N\,\sigma_N) = \sum_L C_L \Phi_L^F(\mathbf{r}_1\,\sigma_1 \ldots \mathbf{r}_N\,\sigma_N). \quad (1.2.10)$$

In this equation, it is usually convenient to regard the Φ_L^F's as normalized determinants. In any determinant, the single-particle states ϕ_l which appear will be referred to as occupied states (or levels) and the other ϕ_l's making up the complete set as unoccupied states (or levels).

1.2.3. *Dirac notation*

We shall sometimes find it convenient to employ Dirac notation. Thus $\phi_l(\mathbf{r}\sigma)$ may be denoted by $\langle \mathbf{r}\sigma | l \rangle$, and the eigenvalue equation (1.2.6) may be written as

$$\langle \mathbf{r}\sigma | U | l \rangle = \epsilon_l \langle \mathbf{r}\sigma | l \rangle \quad (1.2.11)$$

or, without denoting the co-ordinate representation explicitly, in terms of the ket $|l\rangle$ as

$$U|l\rangle = \epsilon_l |l\rangle. \quad (1.2.12)$$

Then the orthonormality relation (1.2.7) may be written as

$$\sum_\sigma \int \langle l | \mathbf{r}\sigma \rangle \langle \mathbf{r}\sigma | k \rangle \, d\mathbf{r} = \delta_{lk}, \quad (1.2.13)$$

or, using the result $\sum_\sigma \int |\mathbf{r}\sigma\rangle\langle \mathbf{r}\sigma| \, d\mathbf{r} = 1,$

as

$$\langle l | k \rangle = \delta_{lk}. \quad (1.2.14)$$

The matrix element of a one-body operator between two single particle wave functions, namely

$$\sum_{\sigma}\int \phi_l^*(\mathbf{r}\sigma)\, U(\mathbf{r}\sigma)\, \phi_k(\mathbf{r}\sigma)\, d\mathbf{r}, \tag{1.2.15}$$

will be written $\langle l|U|k\rangle$. Similarly, the matrix element of a two-body operator between the products of two single-particle wave functions will be expressed as

$$\sum_{\sigma_1\sigma_2}\iint \phi_i^*(\mathbf{r}_1\,\sigma_1)\,\phi_j^*(\mathbf{r}_2\,\sigma_2)\, v(\mathbf{r}_1\,\sigma_1, \mathbf{r}_2\,\sigma_2)\, \phi_l(\mathbf{r}_1\,\sigma_1)\, \phi_k(\mathbf{r}_2\,\sigma_2)\, d\mathbf{r}_1\, d\mathbf{r}_2 = \langle ij|v|lk\rangle. \tag{1.2.16}$$

This latter form is clearly equivalent to $\langle ji|v|kl\rangle$ for a symmetrical interaction v.

1.3. Plane waves and Fermi sphere

In discussing free particles, it is almost always best to employ the eigenfunctions of momentum

$$\frac{1}{\Omega^{\frac{1}{2}}}e^{i\mathbf{k}.\mathbf{r}}, \tag{1.3.1}$$

where periodic boundary conditions are employed. Box boundary conditions and standing waves are generally less convenient, though they are occasionally employed (cf. problem P. 1 (iii)). With the volume Ω in (1.3.1) taken as a cube of side L, $\mathbf{k}$ has the allowed values

$$\mathbf{k} = \frac{2\pi}{L}(n_x, n_y, n_z), \tag{1.3.2}$$

with n_x, n_y, n_z taking all possible integral values, positive, negative and zero. The single-particle energies are $\epsilon_{\mathbf{k}} = \hbar^2k^2/2m$. For particles with spin, we must multiply (1.3.1) by the appropriate spin wave functions.

To avoid considering surface effects for a large system with finite density N/Ω, it is invariably convenient to take the limit $N\to\infty$, $\Omega\to\infty$.

Since, according to (1.3.2) there are $1/8\pi^3$ states per unit volume of $\mathbf{k}$ space, per unit volume of box in this limit, then the rule for converting summations over $\mathbf{k}$ into integrations is simply

$$\sum_{\mathbf{k}} \to \frac{\Omega}{(2\pi)^3}\int d\mathbf{k}. \tag{1.3.3}$$

1.3.1. *Matrix elements of one- and two-particle operators*

For a one-particle operator $U(\mathbf{r})$, the matrix element between free-particle states with wave vectors $\mathbf{k}$ and $\mathbf{m}$ is given by

$$\langle \mathbf{k}|U(\mathbf{r})|\mathbf{m}\rangle = \frac{1}{\Omega}\int e^{-i\mathbf{k}.\mathbf{r}}U(\mathbf{r})e^{i\mathbf{m}.\mathbf{r}}\,d\mathbf{r}$$

$$= \frac{1}{\Omega}\int U(\mathbf{r})e^{-i(\mathbf{k}-\mathbf{m}).\mathbf{r}}\,d\mathbf{r}$$

$$= \frac{1}{\Omega}U(|\mathbf{k}-\mathbf{m}|), \tag{1.3.4}$$

the last result defining the Fourier transform $U(|\mathbf{k}-\mathbf{m}|)$ of $U(\mathbf{r})$.

The matrix element of a two-particle operator, which depends only on the distance between the two particles, will often be required later. We may write

$$\langle \mathbf{kl}|v(|\mathbf{r}_1-\mathbf{r}_2|)\,|\mathbf{mn}\rangle$$

$$= \frac{1}{\Omega^2}\int e^{-i\mathbf{k}.\mathbf{r}_1}e^{-i\mathbf{l}.\mathbf{r}_2}v(|\mathbf{r}_1-\mathbf{r}_2|)e^{i\mathbf{m}.\mathbf{r}_1}e^{i\mathbf{n}.\mathbf{r}_2}\,d\mathbf{r}_1\,d\mathbf{r}_2. \tag{1.3.5}$$

We now change the variables to a centre-of-mass co-ordinate

$$\mathbf{R} = \frac{(\mathbf{r}_1+\mathbf{r}_2)}{2},$$

and a relative co-ordinate

$$\mathbf{r} = \mathbf{r}_1-\mathbf{r}_2.$$

Then, from (1.3.5), by integration over $\mathbf{R}$, we obtain

$$\langle \mathbf{kl}|v|\mathbf{mn}\rangle = \Omega^{-1}\delta_{\mathbf{k}+\mathbf{l},\,\mathbf{m}+\mathbf{n}}\int e^{-i(\mathbf{k}-\mathbf{m}).\mathbf{r}}v(\mathbf{r})\,d\mathbf{r}$$

$$= \Omega^{-1}\delta_{\mathbf{k}+\mathbf{l},\,\mathbf{m}+\mathbf{n}}v(|\mathbf{k}-\mathbf{m}|). \tag{1.3.6}$$

1.3.2. *Fermi surface and wave number*

For N non-interacting spinless Fermions, the ground-state wave function will be formed from the first N single-particle wave functions, corresponding to the lowest N eigenvalues $\epsilon_{\mathbf{k}}$. Thus the occupied states in $\mathbf{k}$ space lie inside a sphere, the Fermi sphere, described by a Fermi wave number k_f and a corresponding Fermi energy E_f. From (1.3.3), the number of particles N is

$$N = \sum_{k<k_f} 1 = \frac{\Omega}{(2\pi)^3}\int_{k<k_f} d\mathbf{k},$$

or

$$\frac{N}{\Omega} = \rho_0 = \frac{k_f^3}{6\pi^2}, \tag{1.3.7}$$

where ρ_0 is the mean particle density. Likewise, the ground-state energy is

$$E_0 = \sum_{k<k_f} \frac{\hbar^2 k^2}{2m} = \frac{\Omega}{(2\pi)^3}\int_{k<k_f} \frac{\hbar^2 k^2}{2m}\, d\mathbf{k}$$

$$= \frac{3}{10}\frac{\hbar^2 k_f^2}{m} N. \tag{1.3.8}$$

Excited state N-particle wave functions are formed when certain single-particle states with $k > k_f$ are included, while an equal number of states with $k < k_f$ are excluded from the determinantal N-particle wave function. The corresponding energies are readily calculated.

Finally, for Fermions with spin $\frac{1}{2}$, we have two degenerate wave functions

$$\phi_{\mathbf{k}}(\mathbf{r})\,\alpha \quad \text{and} \quad \phi_{\mathbf{k}}(\mathbf{r})\,\beta$$

for each state with wave vector $\mathbf{k}$, where α and β are the usual spin functions. Because of this spin degeneracy, (1.3.7) becomes

$$\rho_0 = \frac{k_f^3}{3\pi^2}, \tag{1.3.9}$$

while the energy/particle, E_0/N is $\frac{3}{5}E_f$ as in (1.3.8).

For Bosons, on the other hand, the ground state is trivially that in which all the particles are in states with $\mathbf{k} = 0$, and hence the total energy is zero.

1.4. Density matrices

The N-particle wave function corresponding to the Fermi sphere of section (1.3) contains a great deal of information which is almost wholly irrelevant in calculating many of the most important physical properties. More generally, the N-body wave function for interacting particles, when N is $\sim 10^{23}$, as in many problems of interest (cf. Chapters 5, 7 and 8), must be of such complexity that a full description is unthinkable.

Thus, it is very advantageous to introduce quantities, the so-called density matrices, defined from the many-body wave function Φ by integrating out much of the redundant information. These density matrices enable us to focus attention directly on such physically important quantities as the particle density ρ and the pair correlation function g (defined below).

The definitions of the density matrices can be conveniently divided into forms which enable us to calculate expectation values of single-particle operators, referred to as first-order matrices, and second-order forms which deal with two-particle dynamical variables. It is useful also to distinguish spin-dependent matrices and those in which summations over spin have been carried out. The latter forms are generally all that we require.

1.4.1. *First-order matrices and particle density*

The general first-order density matrix, $\gamma_\sigma(\mathbf{r}_1'\sigma_1', \mathbf{r}_1\sigma_1)$ is defined from the normalized many-body wave function $\Phi(\mathbf{r}_1\sigma_1, \ldots, \mathbf{r}_N\sigma_N)$ by

$$\gamma_\sigma(\mathbf{r}_1'\sigma_1', \mathbf{r}_1\sigma_1) = N \sum_{\sigma_2\ldots\sigma_N} \int \Phi^*(\mathbf{r}_1'\sigma_1', \mathbf{r}_2\sigma_2, \ldots, \mathbf{r}_N\sigma_N) \times \Phi(\mathbf{r}_1\sigma_1, \mathbf{r}_2\sigma_2, \ldots, \mathbf{r}_N\sigma_N)\, d\mathbf{r}_2\, d\mathbf{r}_3 \ldots d\mathbf{r}_N. \quad (1.4.1)$$

The very useful spinless form $\gamma(\mathbf{r}_1', \mathbf{r}_1)$ of (1.4.1) is simply

$$\gamma(\mathbf{r}_1', \mathbf{r}_1) = \sum_{\sigma_1} \gamma_\sigma(\mathbf{r}_1'\sigma_1, \mathbf{r}_1\sigma_1). \quad (1.4.2)$$

The normalization factor N in (1.4.1) has been chosen such that the diagonal element $\gamma(\mathbf{r}_1, \mathbf{r}_1)$ of (1.4.2) is the particle density $\rho(\mathbf{r}_1)$, which obviously must integrate to N. In section (1.5.1) we give an explicit calculation of γ for a plane-wave determinant, but before applying (1.4.2) we introduce the second-order forms.

1.4.2. *Second-order matrices and pair function*

In a similar way we define the second-order matrix $\Gamma_\sigma(\mathbf{r}_1'\sigma_1', \mathbf{r}_2'\sigma_2'; \mathbf{r}_1\sigma_1, \mathbf{r}_2\sigma_2)$ as

$$\Gamma_\sigma(\mathbf{r}_1'\sigma_1', \mathbf{r}_2'\sigma_2'; \mathbf{r}_1\sigma_1, \mathbf{r}_2\sigma_2) = \frac{N(N-1)}{2} \sum_{\sigma_3\ldots\sigma_N} \int \Phi^*(\mathbf{r}_1'\sigma_1', \mathbf{r}_2'\sigma_2', \mathbf{r}_3\sigma_3, \ldots, \mathbf{r}_N\sigma_N) \times \Phi(\mathbf{r}_1\sigma_1, \mathbf{r}_2\sigma_2, \ldots, \mathbf{r}_N\sigma_N)\, d\mathbf{r}_3 \ldots d\mathbf{r}_N \quad (1.4.3)$$

and its spinless form $\Gamma(\mathbf{r}_1', \mathbf{r}_2'; \mathbf{r}_1, \mathbf{r}_2)$ as

$$\Gamma(\mathbf{r}_1', \mathbf{r}_2'; \mathbf{r}_1, \mathbf{r}_2) = \sum_{\sigma_1 \sigma_2} \Gamma_\sigma(\mathbf{r}_1' \sigma_1, \mathbf{r}_2' \sigma_2; \mathbf{r}_1 \sigma_1, \mathbf{r}_2 \sigma_2). \quad (1.4.4)$$

As with the first-order matrix, the diagonal element of (1.4.4) has immediate physical significance, as it gives us the number of distinct pairs multiplied by the probability density of simultaneously finding one particle at $\mathbf{r}_1$ and another at $\mathbf{r}_2$. It is often useful to rewrite $\Gamma(\mathbf{r}_1, \mathbf{r}_2; \mathbf{r}_1, \mathbf{r}_2)$ as a constant times $g(\mathbf{r}_1, \mathbf{r}_2)$, where the constant is chosen such that $g \to 1$ for large separation $|\mathbf{r}_1 - \mathbf{r}_2|$. We shall then refer to $g(\mathbf{r}_1, \mathbf{r}_2)$ as the pair function.

1.4.3. *Relation between first- and second-order matrices, and total energy*

It is obvious from the definitions of Γ_σ and γ_σ that the former contains the latter, the explicit relation following immediately from (1.4.3) and (1.4.1) as

$$\gamma_\sigma(\mathbf{r}_1' \sigma_1', \mathbf{r}_1 \sigma_1) = \frac{2}{N-1} \sum_{\sigma_2} \int \Gamma_\sigma(\mathbf{r}_1' \sigma_1', \mathbf{r}_2 \sigma_2; \mathbf{r}_1 \sigma_1, \mathbf{r}_2 \sigma_2)\, d\mathbf{r}_2. \quad (1.4.5)$$

Similarly, for the spinless forms,

$$\gamma(\mathbf{r}_1', \mathbf{r}_1) = \frac{2}{N-1} \int \Gamma(\mathbf{r}_1', \mathbf{r}_2; \mathbf{r}_1, \mathbf{r}_2)\, d\mathbf{r}_2. \quad (1.4.6)$$

It will be valuable in what follows to express the total energy E of the system in terms of the density matrices. To do so, we note that the exact energy of a state defined by total wave function Φ is

$$E = \sum_{\text{spin}} \int \Phi^* H \Phi \, d\mathbf{r}_1 \ldots d\mathbf{r}_N. \quad (1.4.7)$$

Assuming a Hamiltonian of form (1.2.2), but without spin dependence, and writing the one-particle term $\sum_{i=1}^{N} U(\mathbf{r}_i)$ explicitly as

$$\sum_{i=1}^{N} \left(-\frac{\hbar^2}{2m} \nabla_i^2 + U_1(\mathbf{r}_i) \right), \quad (1.4.8)$$

it is clear that its expectation value can be expressed in terms of $\gamma(\mathbf{r}', \mathbf{r})$. Similarly, the average of the two-particle term

$$\sum_{i<j} v(\mathbf{r}_i, \mathbf{r}_j), \quad (1.4.9)$$

depends only on the diagonal element $\Gamma(\mathbf{r}_1, \mathbf{r}_2; \mathbf{r}_1, \mathbf{r}_2)$. The final form may be shown, after an elementary calculation, to be

$$E = -\frac{\hbar^2}{2m}\int [\nabla^2\gamma(\mathbf{r}', \mathbf{r})]_{\mathbf{r}'=\mathbf{r}}\, d\mathbf{r} + \int U_1(\mathbf{r})\rho(\mathbf{r})\, d\mathbf{r} + \int v(\mathbf{r}_1, \mathbf{r}_2)\,\Gamma(\mathbf{r}_1, \mathbf{r}_2; \mathbf{r}_1, \mathbf{r}_2)\, d\mathbf{r}_1\, d\mathbf{r}_2. \quad (1.4.10)$$

1.5. Dirac density matrix

The density matrices, as defined above, are to be calculated using the exact many-body wave functions. However, in the case when we form the total wave function as a determinant of one-particle states, the density matrices take a particularly simple form, as first emphasized by Dirac (1930). Thus, if the wave function Φ is

$$\Phi = \frac{1}{(N!)^{\frac{1}{2}}} \det \phi_i(\mathbf{r}_j\, \sigma_j), \quad (1.5.1)$$

then, in this case, it is readily shown that

$$\gamma_\sigma(\mathbf{r}'\sigma', \mathbf{r}\sigma) = \sum_1^N \phi_i^*(\mathbf{r}'\sigma')\,\phi_i(\mathbf{r}\sigma), \quad (1.5.2)$$

where the ϕ_i's are the one-particle states occurring in the determinant. Similarly, the second-order matrix is

$$\Gamma_\sigma(\mathbf{r}_1'\,\sigma_1', \mathbf{r}_2'\,\sigma_2'; \mathbf{r}_1\,\sigma_1, \mathbf{r}_2\,\sigma_2) = \frac{1}{2}\begin{vmatrix} \gamma_\sigma(\mathbf{r}_1'\,\sigma_1', \mathbf{r}_1\,\sigma_1) & \gamma_\sigma(\mathbf{r}_1'\,\sigma_1', \mathbf{r}_2\,\sigma_2) \\ \gamma_\sigma(\mathbf{r}_2'\,\sigma_2', \mathbf{r}_1\,\sigma_1) & \gamma_\sigma(\mathbf{r}_2'\,\sigma_2', \mathbf{r}_2\,\sigma_2) \end{vmatrix}. \quad (1.5.3)$$

When we consider levels occupied by two Fermions with opposed spin, then the states $\phi_i(\mathbf{r}_j\,\sigma_j)$ are built up as simple products of $\frac{1}{2}N$ space orbitals $\phi_i(\mathbf{r}_j)$ with the spin functions α and β. It then follows that

$$\gamma(\mathbf{r}_1, \mathbf{r}_2) = 2\sum_1^{\frac{1}{2}N} \phi_i^*(\mathbf{r}_1)\,\phi_i(\mathbf{r}_2) \quad (1.5.4)$$

and

$$2\Gamma(\mathbf{r}_1, \mathbf{r}_2; \mathbf{r}_1, \mathbf{r}_2) = \rho(\mathbf{r}_1)\rho(\mathbf{r}_2) - \tfrac{1}{2}[\gamma(\mathbf{r}_1, \mathbf{r}_2)]^2. \quad (1.5.5)$$

Equations (1.5.2) and (1.5.4) define the Dirac density matrix, the latter, of course, being the spinless form. Furthermore, with a single determinant, we see from (1.5.5) that the Dirac matrix also determines the second-order density matrix. This is not generally true. Combining the results (1.5.4) and (1.5.5) with the energy expression (1.4.10), it is clear that we could obtain the best possible orbitals ϕ_i, in the sense of the variational method, by minimization. This is the Hartree–Fock theory, which we describe in section (1.10).

In many circumstances, we may ask for the one-particle orbitals ϕ_i which are not the best possible, but which satisfy a one-particle wave equation

$$\left[-\frac{\hbar^2}{2m}\nabla^2 + V(\mathbf{r})\right]\phi_i(\mathbf{r}) = \epsilon_i\,\phi_i(\mathbf{r}), \tag{1.5.6}$$

where $V(\mathbf{r})$ is a common potential in which all the particles move, and which should eventually be calculated by a 'self-consistent' procedure. We illustrate this in some detail in Chapter 2, section 2.3, for atoms, and for the present we regard $V(\mathbf{r})$ as given.

In this simplified case, the problem is reduced to finding Dirac's density matrix as a functional of $V(\mathbf{r})$. Unfortunately, no closed solution of this problem has, as yet, been found for an arbitrary potential $V(\mathbf{r})$. However, for the most important case when the N lowest single-particle states are occupied, that is for the ground state, an interesting perturbation solution may be obtained, as has been shown by March & Murray (1960, 1961) and we shall derive their results below. Since their expansion is based on plane waves, it will be convenient at this point to derive explicit forms for the density matrices in this free-particle case.

1.5.1. *Plane-wave results*

As a simple, but nevertheless important, example of the calculation of the density matrices γ and Γ in the approximation of a single determinant, let us consider the ground state for the case of free particles, discussed in section 1.3. The single-particle wave functions

$$\frac{1}{\Omega^{\frac{1}{2}}}e^{i\mathbf{k}.\mathbf{r}},$$

when substituted in (1.5.4) yield

$$\gamma(\mathbf{r}', \mathbf{r}) = \frac{2}{\Omega}\sum_{k<k_f} e^{-i\mathbf{k}.\mathbf{r}'}e^{i\mathbf{k}.\mathbf{r}}, \tag{1.5.7}$$

and passing from a summation to an integration according to (1.3.3) we have

$$\gamma(\mathbf{r}',\mathbf{r}) = \frac{2}{8\pi^3}\int_{k<k_f} e^{i\mathbf{k}.(\mathbf{r}-\mathbf{r}')}\,d\mathbf{k}. \tag{1.5.8}$$

The integration over angles simply means replacing the plane wave $e^{i\mathbf{k}.\mathbf{r}}$ by the s wave, $(\sin kr)/kr$, of its expansion in spherical waves, and hence

$$\gamma(\mathbf{r}',\mathbf{r}) = \frac{1}{\pi^2}\int_0^{k_f} k\,\frac{\sin k|\mathbf{r}'-\mathbf{r}|}{|\mathbf{r}'-\mathbf{r}|}\,dk$$

$$= \frac{k_f^3}{\pi^2}\frac{j_1(k_f|\mathbf{r}'-\mathbf{r}|)}{k_f|\mathbf{r}'-\mathbf{r}|}, \tag{1.5.9}$$

where $j_1(x)$ is the first-order spherical Bessel function defined explicitly by

$$j_1(x) = \frac{\sin x - x\cos x}{x^2}. \tag{1.5.10}$$

Taking the limit $\mathbf{r}'\to\mathbf{r}$ in (1.5.9) and noting that, for small x, $j_1(x) \sim \frac{1}{3}x$, we immediately obtain

$$\gamma(\mathbf{r},\mathbf{r}) = \frac{k_f^3}{3\pi^2}, \tag{1.5.11}$$

which is simply the constant particle density ρ_0 of (1.3.9).

The second-order density matrix is readily obtained from (1.5.3), and, in particular, its diagonal element $\Gamma(\mathbf{r}_1,\mathbf{r}_2;\,\mathbf{r}_1,\mathbf{r}_2)$ is given from (1.5.5) and (1.5.9) as

$$\Gamma(\mathbf{r}_1,\mathbf{r}_2;\,\mathbf{r}_1,\mathbf{r}_2) = \frac{k_f^6}{18\pi^4} - \frac{k_f^6}{4\pi^4}\left\{\frac{j_1(k_f|\mathbf{r}_1-\mathbf{r}_2|)}{k_f|\mathbf{r}_1-\mathbf{r}_2|}\right\}^2. \tag{1.5.12}$$

This gives us the pair correlation function $g(r)$ as

$$g(r) = 1 - \frac{9}{2}\left\{\frac{j_1(k_f r)}{k_f r}\right\}^2. \tag{1.5.13}$$

These explicit forms for $\gamma(\mathbf{r}',\mathbf{r})$ and $g(r)$ are central to the considerations on interacting electrons in Chapter 5, as well as to the plane-wave perturbation theory of section 1.7.

1.6. Bloch density matrix

To derive the Dirac density matrix, it is convenient first to introduce a related quantity, which we shall refer to as the Bloch density matrix. This is defined in terms of the wave functions $\phi_i(\mathbf{r})$ and energy levels ϵ_i of (1.5.6), as

$$C(\mathbf{r}'\mathbf{r}\beta) = \sum_i \phi_i^*(\mathbf{r}')\,\phi_i(\mathbf{r})\,e^{-\epsilon_i\beta}, \tag{1.6.1}$$

where $\beta = (k_B T)^{-1}$, k_B being Boltzmann's constant and T the absolute temperature. It will be recognized that $C(\mathbf{r}'\mathbf{r}\beta)$ constitutes a 'generalized partition function', in the sense that if we integrate along the diagonal, we obtain the ordinary partition function of statistical mechanics. We shall have a good deal to say in Chapter 9 on the statistical mechanics of an assembly of interacting particles, but for the moment our purpose in introducing (1.6.1) is to allow us to calculate γ. It should be noted that (1.6.1) deals only with singly occupied levels.

1.6.1. *Bloch equation*

Before deriving the relation between C and γ, we note that if we operate with the one-particle Hamiltonian (we use units in which $\hbar = 1$, $m = 1$ here, to simplify the notation)

$$H_s = -\tfrac{1}{2}\nabla_{\mathbf{r}}^2 + V(\mathbf{r}) \tag{1.6.2}$$

on C, and compare the result with that obtained by differentiating C with respect to β, then we find the so-called Bloch equation

$$H_s C = -\frac{\partial C}{\partial \beta}, \tag{1.6.3}$$

which has the form of the time-dependent Schrödinger equation, with β playing the role of it. In Chapter 9 we exploit this analogy. The boundary condition required to define the solution of (1.6.3) follows from the completeness theorem for eigenfunctions, namely

$$C(\mathbf{r}'\mathbf{r}0) = \delta(\mathbf{r}' - \mathbf{r}) \tag{1.6.4}$$

and hence we can calculate C, at least in principle, from (1.6.3) and (1.6.4). In fact, such a calculation, starting from a free-particle problem, and using an iterative approach, was carried out by Green (1952).

While Green's calculation was very general, the derivation we give below, due to March & Murray (1961), has the merit of simplicity and is in a very suitable form for the purpose of calculating

Dirac's matrix. Before embarking on the perturbation development for C, we first demonstrate the relation between the Bloch and the Dirac matrices.

1.6.2. *Relation to Dirac matrix*

We introduce now the function

$$G(\mathbf{r}'\mathbf{r}\epsilon) = \sum_i \phi_i^*(\mathbf{r}')\,\phi_i(\mathbf{r})\,\delta(\epsilon-\epsilon_i), \tag{1.6.5}$$

and then, if we define the Fermi level ζ such that it lies just above the highest occupied level, we may write

$$\gamma(\mathbf{r}'\mathbf{r}\zeta) = \int_0^\zeta G(\mathbf{r}'\mathbf{r}\epsilon)\,d\epsilon. \tag{1.6.6}$$

Here, we have assumed that all the energy levels of (1.5.6) are positive, which can always be arranged by a suitable choice of V. From (1.6.1) and (1.6.5) we may write

$$\begin{aligned} C(\mathbf{r}'\mathbf{r}\beta) &= \int_0^\infty G(\mathbf{r}'\mathbf{r}\epsilon)\,e^{-\beta\epsilon}\,d\epsilon \\ &= \beta\int_0^\infty \left\{\int_0^E G(\mathbf{r}'\mathbf{r}E)\,dE\right\} e^{-\beta E}\,dE \\ &= \beta\int_0^\infty \gamma(\mathbf{r}'\mathbf{r}\zeta)\,e^{-\beta\zeta}\,d\zeta. \end{aligned} \tag{1.6.7}$$

Thus we see that if $C(\mathbf{r}'\mathbf{r}\beta)$ can be determined, then $\gamma(\mathbf{r}'\mathbf{r}\zeta)$ may be obtained as the inverse Laplace transform of C/β. It is also worth noting that G as defined by (1.6.5) is a Green function, and as such will be discussed in Chapter 10.

1.7. Plane-wave perturbation theory

As Sondheimer & Wilson (1951) have shown, the Bloch equation (1.6.3), subject to the condition (1.6.4), is readily solved for an assembly of free electrons, and the result may be written in the form

$$C_0(\mathbf{r}'\mathbf{r}\beta) = (2\pi\beta)^{-\frac{3}{2}}\exp\left(-|\mathbf{r}-\mathbf{r}'|^2/2\beta\right), \tag{1.7.1}$$

as is easily verified. We note that, by putting $\mathbf{r}' = \mathbf{r}$ and integrating over a volume Ω, we obtain the partition function $\sum_i e^{-\beta\epsilon_i}$ as

$$\frac{\Omega(k_B T)^{\frac{3}{2}}}{(2\pi)^{\frac{3}{2}}}, \tag{1.7.2}$$

which, in the units we are employing, is the well-known result for a perfect gas.

1.7.1. *Bloch density matrix*

Equation (1.6.3) may now be solved by treating $V(\mathbf{r})$ as a perturbation, and to do so it is convenient to convert the differential equation to an integral form into which the condition (1.6.4) has been inserted. The result we require is

$$C(\mathbf{r}\mathbf{r}_0\beta) = C_0(\mathbf{r}\mathbf{r}_0\beta) - \int d\mathbf{r}_1 \int_0^\beta d\beta_1\, C_0(\mathbf{r}\mathbf{r}_1\,\beta-\beta_1)\, V(\mathbf{r}_1)\, C(\mathbf{r}_1\mathbf{r}_0\beta_1) \tag{1.7.3}$$

as may be verified by direct substitution in the Bloch equation. We proceed iteratively, at this point inserting C_0 for C in the integral in the first step. The problem of performing the integral over β_1

$$\int_0^\beta d\beta_1\, C_0(\mathbf{r}\mathbf{r}_1\,\beta-\beta_1)\, C_0(\mathbf{r}_1\mathbf{r}_0\beta_1) \tag{1.7.4}$$

may be solved by making use of a theorem on the Laplace transform of a convolution (see March & Murray, 1961), and the final result of continuing this procedure may be written

$$C(\mathbf{r}\mathbf{r}_0\beta) = \sum_{j=0}^{\infty} C_j(\mathbf{r}\mathbf{r}_0\beta), \tag{1.7.5}$$

where

$$C_j(\mathbf{r}\mathbf{r}_0\beta) = (2\pi\beta)^{-\frac{3}{2}} \int^j \prod_{l=1}^{j} \left\{-\frac{d\mathbf{r}_l\, V(\mathbf{r}_l)}{2\pi}\right\} \left(\sum_{l=1}^{j+1} s_l\right) \times \exp\left\{-\frac{1}{2\beta}\left(\sum_{l=1}^{j+1} s_l\right)^2\right\} \div \prod_{l=1}^{j+1} s_l, \tag{1.7.6}$$

and $\quad s_l = |\mathbf{r}_l - \mathbf{r}_{l-1}|,\ \mathbf{r}_{j+1} = \mathbf{r}.$

1.7.2. *Dirac density matrix*

At this stage, we can use the Laplace transform relation (1.6.7) between γ and C/β, and hence Dirac's density matrix may be obtained. The final result may be written, with $2E = k^2$, as

$$\gamma(\mathbf{r}\mathbf{r}_0k) = \sum_{j=0}^{\infty} \gamma_j(\mathbf{r}\mathbf{r}_0k), \tag{1.7.7}$$

where

$$\gamma_j(\mathbf{r}\mathbf{r}_0k) = \frac{k^2}{2\pi^2} \int^j \prod_{l=1}^{j} \left\{\frac{-d\mathbf{r}_l\, V(\mathbf{r}_l)}{2\pi}\right\} j_1\left(k\sum_{l=1}^{j+1} s_l\right) \div \prod_{l=1}^{j+1} s_l. \tag{1.7.8}$$

The subscript j in (1.7.7) distinguishes the different orders of perturbation theory and has nothing to do with the subscript σ of the spin-dependent density matrices of section 1.4. Writing down the low-order terms of (1.7.7), we have first

$$\gamma_0(\mathbf{r}\mathbf{r}_0 k) = \frac{k^2}{2\pi^2} \frac{j_1(k|\mathbf{r}-\mathbf{r}_0|)}{|\mathbf{r}-\mathbf{r}_0|}, \tag{1.7.9}$$

which is the free-particle result of (1.5.9) apart from the fact that the present argument has not included spin degeneracy. Furthermore the first-order term is given by

$$\gamma_1(\mathbf{r}\mathbf{r}_0 k) = -\frac{k^2}{2\pi^2}\int d\mathbf{r}_1 \frac{V(\mathbf{r}_1)}{2\pi} \frac{j_1(k|\mathbf{r}-\mathbf{r}_1|+k|\mathbf{r}_1-\mathbf{r}_0|)}{|\mathbf{r}-\mathbf{r}_1|\,|\mathbf{r}_1-\mathbf{r}_0|}. \tag{1.7.10}$$

We shall use these results in section 1.9 to derive the Thomas–Fermi theory, and also (1.7.10) is important in Chapter 5 when we discuss screening in an electron gas. For the moment, we turn to perturbation theory formulated in terms of wave functions.

1.8. Perturbation theory from general unperturbed states

In the previous section, we developed density matrix perturbation theory, with the unperturbed state formed from eigenfunctions of the kinetic energy operator. The theory could then be carried through to all orders in the perturbation because of the simplicity of the zeroth-order density matrices.

It is often advantageous to make a more careful choice of unperturbed solution. In such cases we cannot form the density matrices explicitly in zeroth-order, and the advantages of the method are largely lost. Thus, we shall present here the formal results of perturbation theory in terms of wave functions.

But there is another reason for considering wave function perturbation theory at this point. The two forms of perturbation theory discussed below, the Brillouin–Wigner and Rayleigh–Schrödinger formulations, will be analysed in some detail in Chapter 4 and shown to lead to unphysical features in the energy series as the number of particles N becomes indefinitely large. While the difficulty in the Brillouin–Wigner theory seems basic, we shall see that there is a way of adapting the Rayleigh–Schrödinger theory to deal with a many-particle system. Thus, it is important to acquire some familiarity with the structure of these theories at this point, for we can then proceed fairly quickly to the heart of the matter in applying perturbation theory to the many-body case in Chapter 4.

1.8.1. *Level shift*

Consider any quantum mechanical system with Hamiltonian

$$H = H_0 + V. \tag{1.8.1}$$

H_0 is the unperturbed or model Hamiltonian and V is the perturbation. The eigenvalues and normalized eigenkets of H_0 given by (1.8.2) below are supposed known:

$$(H_0 - E_i)|\xi_i\rangle = 0 \quad (i = 0, 1, 2, \ldots). \tag{1.8.2}$$

E_0 is the eigenvalue corresponding to the ground state of the unperturbed system $|\xi_0\rangle$ which is assumed to be non-degenerate.

We wish to calculate, in particular, the ground-state eigenvalue E and eigenket $|\psi\rangle$ of the Hamiltonian H, given by

$$(E - H_0 - V)|\psi\rangle = 0. \tag{1.8.3}$$

Let $|\psi\rangle$ be normalized such that

$$\langle\xi_0|\psi\rangle = 1, \tag{1.8.4}$$

and then, from (1.8.2) and (1.8.3), we obtain

$$\langle\psi|H_0|\xi_0\rangle = E_0\langle\psi|\xi_0\rangle \tag{1.8.5}$$

and

$$\langle\xi_0|H_0|\psi\rangle + \langle\xi_0|V|\psi\rangle = E\langle\xi_0|\psi\rangle. \tag{1.8.6}$$

Subtracting (1.8.5) from (1.8.6), using the Hermiticity of H_0 and the normalization condition, we obtain

$$E - E_0 = \langle\xi_0|V|\psi\rangle, \tag{1.8.7}$$

which gives us an exact formula for the level shift, $E - E_0$. As it stands, of course, (1.8.7) is not of practical value, as it involves the wave function $|\psi\rangle$, which is as yet unknown.

1.8.2. *Perturbed wave function*

Introduce now a projection operator P, which, when applied to an arbitrary ket $|\rangle$ gives its component $\langle\xi_0|\rangle$ multiplied by the ket $|\xi_0\rangle$. Thus

$$P|\psi\rangle = |\xi_0\rangle \tag{1.8.8}$$

and hence

$$\begin{aligned} |\psi\rangle &= |\xi_0\rangle + (1-P)|\psi\rangle \\ &= |\xi_0\rangle + Q|\psi\rangle, \end{aligned} \tag{1.8.9}$$

where (1.8.9) defines a new operator Q. Since $P^2 = P$, it follows from the definition that $Q^2 = Q$, and hence Q is also a projection operator. We have also, from (1.8.3) for an arbitrary ϵ,

$$(\epsilon - H_0)|\psi\rangle = (\epsilon - E + V)|\psi\rangle, \tag{1.8.10}$$

which may be written equivalently as

$$|\psi\rangle = \frac{1}{\epsilon - H_0}(\epsilon - E + V)|\psi\rangle. \tag{1.8.11}$$

Using (1.8.11) in (1.8.9) we have

$$|\psi\rangle = |\xi_0\rangle + Q\frac{1}{\epsilon - H_0}(\epsilon - E + V)|\psi\rangle, \tag{1.8.12}$$

and iterating (1.8.12) we obtain

$$|\psi\rangle = \sum_{n=0}^{\infty}\left\{\frac{Q}{\epsilon - H_0}(\epsilon - E + V)\right\}^n|\xi_0\rangle. \tag{1.8.13}$$

We return at this point to the level shift formula (1.8.7), and substituting for $|\psi\rangle$ from (1.8.13) we find

$$E - E_0 = \sum_0^{\infty}\langle\xi_0|V\left\{\frac{Q}{\epsilon - H_0}(\epsilon - E + V)\right\}^n|\xi_0\rangle. \tag{1.8.14}$$

1.8.3. *Brillouin–Wigner theory*

From the general results (1.8.13) and (1.8.14), the Brillouin–Wigner formulae are obtained immediately by putting $\epsilon = E$. We may then write $|\psi\rangle$ in the form

$$|\psi\rangle = \sum_0^{\infty}\left\{\frac{Q}{E - H_0}V\right\}^n|\xi_0\rangle \tag{1.8.15}$$

and

$$E - E_0 = \sum_0^{\infty}\langle\xi_0|V\left\{\frac{Q}{E - H_0}V\right\}^n|\xi_0\rangle. \tag{1.8.16}$$

It will be seen that a characteristic feature of the Brillouin–Wigner formulation for the energy E is that we have the unknown energy appearing on both sides of (1.8.16). This is in contrast with the Rayleigh–Schrödinger method which we shall now discuss.

1.8.4. *Rayleigh–Schrödinger theory*

In this theory, we simply take ϵ in (1.8.13) and (1.8.14) as E_0, and these equations now become

$$|\psi\rangle = \sum_0^\infty \left\{\frac{Q}{E_0 - H_0}(E_0 - E + V)\right\}^n |\xi_0\rangle \qquad (1.8.17)$$

and

$$E - E_0 = \sum_0^\infty \langle \xi_0 | V \left\{\frac{Q}{E_0 - H_0}(E_0 - E + V)\right\}^n |\xi_0\rangle, \qquad (1.8.18)$$

respectively. From these equations, $|\psi\rangle$ and $E - E_0$ can be obtained to any desired order in V. We shall leave the theory at this point, taking up the structure of the low-order terms again in Chapter 4.

1.9. Thomas–Fermi theory

Having discussed general wave function perturbation theory, we return now to the perturbation series (1.7.7) for the Dirac matrix in order to derive the Thomas–Fermi theory. This method is based on the single-particle wave equation (1.5.6), but, in addition, the potential $V(\mathbf{r})$ is considered to vary only slowly over a characteristic wavelength of the Fermions (for example, in an electron gas, over a de Broglie wavelength for an electron at the Fermi surface, that is $2\pi/k_f$). If we take the diagonal element of (1.7.10) and write

$$\gamma_1(\mathbf{rr}k) = -\frac{k^2}{2\pi^2}\int d\mathbf{r}_1 \frac{V(\mathbf{r}_1)}{2\pi}\frac{j_1(2k|\mathbf{r}-\mathbf{r}_1|)}{|\mathbf{r}-\mathbf{r}_1|^2}, \qquad (1.9.1)$$

then the above assumption implies that $V(\mathbf{r}_1)$ in (1.9.1) may be approximated by $V(\mathbf{r})$. The integration may then be completed, with the result

$$\gamma_1(\mathbf{rr}k) = -kV(\mathbf{r})/2\pi^2. \qquad (1.9.2)$$

In a similar way, the higher terms of (1.7.6) may be calculated and we obtain

$$\gamma(\mathbf{rr}k) = \frac{k^3}{6\pi^2}\left(1 - \frac{3V}{k^2} + \frac{3V^2}{2k^4} + \frac{V^3}{2k^6} + \dots\right)$$

$$= \frac{1}{6\pi^2}(k^2 - 2V(\mathbf{r}))^{\frac{3}{2}},$$

or

$$\gamma(\mathbf{rr}E) = \frac{2^{\frac{3}{2}}}{6\pi^2}[E - V(\mathbf{r})]^{\frac{3}{2}}. \tag{1.9.3}$$

It is this relation between density and potential which characterizes the Thomas–Fermi approximation. In general, in applying (1.9.3) to a physical problem, γ is required at $E = E_f$, the Fermi energy. As usual, E_f must then be determined such that $\int \gamma \, d\mathbf{r} = N$, the number of particles in the system.

It is worth giving the elementary physical derivation of (1.9.3) as an alternative to the more formal calculation from the density matrix perturbation theory.

Essentially we divide the Fermi gas into small regions, within each of which we then assume that the density $\gamma(\mathbf{rr})$ varies only slowly with position. Then $\gamma(\mathbf{rr})$ is related to the maximum momentum $p_f(\mathbf{r})$ by the usual result (cf. (1.5.11)) for a uniform Fermi gas, namely

$$\gamma(\mathbf{rr}) = \frac{8\pi}{3h^3} p_f^3(\mathbf{r}). \tag{1.9.4}$$

Now $p_f(\mathbf{r})$ must be expressed in terms of the potential $V(\mathbf{r})$ in which the Fermions move, and this is achieved by writing down the classical energy equation for the fastest Fermion, namely that the Fermi energy E_f is given by

$$E_f = \frac{p_f^2(\mathbf{r})}{2m} + V(\mathbf{r}). \tag{1.9.5}$$

Combining (1.9.4) and (1.9.5), we recover (1.9.3) immediately when we use units in which $\hbar = 1$, $m = 1$ and remember to doubly fill the energy levels.

The relation (1.9.3) will be used explicitly to calculate atomic properties and, in particular, binding energies of heavy atoms, in Chapter 2.

1.10. Hartree–Fock equations

As we saw in section 1.5, the best one-particle description of the many-body system is to be obtained by minimizing the energy (1.4.10) with respect to the orbitals ϕ_i used in the determinantal wave function. So far, we have been content with less, and in particular with ϕ_i's derivable from a common potential $V(\mathbf{r})$ in which all particles move. The Thomas–Fermi theory discussed in the previous section gives us a method of finding the self-consistent field $V(\mathbf{r})$, as explained fully later in Chaper 2. A better method would still use a common potential field, but would make no additional approximations regarding the gradient of the potential. This is, in essence, the method of section 1.7, though unfortunately the answer there is a perturbation series which it has not, so far, proved possible to sum.

It is natural that we turn finally to the equations defining the best choice of ϕ_i. This gives us the Hartree–Fock theory. To derive this theory, we consider N spinless Fermions (spin is readily included and we omit it to avoid complicating the notation) with Hamiltonian

$$H = \sum_{i=1}^{N} U(\mathbf{r}_i) + \sum_{i<j} v(\mathbf{r}_i - \mathbf{r}_j). \tag{1.10.1}$$

At this stage we can choose a model Hamiltonian

$$H_0 = \sum_{i=1}^{N} H_s(\mathbf{r}_i), \tag{1.10.2}$$

and then the ground-state wave function of the model system, when used as a variational wave function, will give an upper bound for the ground-state energy of the real system.

It is immediately clear from (1.10.2) that the ground-state wave function of the model system will take the form (cf. equation 1.5.1)

$$\Phi_m = \frac{1}{(N!)^{\frac{1}{2}}} \det \psi_i(\mathbf{r}_j), \tag{1.10.3}$$

where the $\psi_i(\mathbf{r})$ ($i = 1 \ldots N$) will be the N eigenfunctions with the lowest eigenvalues of the single-particle Hamiltonian H_s. This Hamiltonian, of course, is as yet unspecified. The single-particle states ψ_i will be chosen to be orthonormal; that is

$$\int \psi_i^*(\mathbf{r})\, \psi_j(\mathbf{r})\, d\mathbf{r} = \delta_{ij}. \tag{1.10.4}$$

1.10.1. *Single-particle self-consistent Hamiltonian*

The expectation value of H with respect to the wave function (1.10.3) is

$$\langle \Phi_m | H | \Phi_m \rangle = \sum_{i=1}^{N} \langle \psi_i | U | \psi_i \rangle + \frac{1}{2} \sum_{ij} \langle \psi_i \psi_j | v | \psi_i \psi_j \rangle - \frac{1}{2} \sum_{ij} \langle \psi_i \psi_j | v | \psi_j \psi_i \rangle. \tag{1.10.5}$$

For (1.10.5) to be a minimum with respect to variations in the ψ_i's which preserve (1.10.4), we must have

$$\sum_{i=1}^{N} \langle \delta\psi_i | U | \psi_i \rangle + \sum_{ij} \langle \delta\psi_i \psi_j | v | \psi_i \psi_j \rangle - \sum_{ij} \langle \delta\psi_i \psi_j | v | \psi_j \psi_i \rangle + \sum_{ij} \lambda_{ij} \langle \delta\psi_i | \psi_j \rangle = 0, \quad (1.10.6)$$

where λ_{ij} are Lagrange multipliers.

Since (1.10.6) must be true for arbitrary $\langle \delta\psi_i |$, we obtain

$$U\psi_i(\mathbf{r}_1) + \sum_j \int \psi_j^*(\mathbf{r}_2)\, v(\mathbf{r}_1 - \mathbf{r}_2)\, \psi_j(\mathbf{r}_2)\, d\mathbf{r}_2\, \psi_i(\mathbf{r}_1)$$
$$- \sum_j \int \psi_j^*(\mathbf{r}_2)\, v(\mathbf{r}_1 - \mathbf{r}_2)\, \psi_i(\mathbf{r}_2)\, d\mathbf{r}_2\, \psi_j(\mathbf{r}_1) - \sum_j \lambda_{ij}\, \psi_j(\mathbf{r}_1) = 0. \quad (1.10.7)$$

The λ_{ij}'s and the $\psi_i(\mathbf{r})$'s which satisfy (1.10.4) and (1.10.7) help to define a single-particle Hamiltonian. Thus, multiplying by $\psi_k(\mathbf{r}_1)$ and integrating, we obtain matrix elements of some single-particle operator H_s, in the representation given by $\psi_1 \ldots \psi_r \ldots$; namely

$$\langle \psi_k | H_s | \psi_i \rangle = \langle \psi_k | U | \psi_i \rangle + \langle \psi_k \psi_j | v | \psi_i \psi_j \rangle - \langle \psi_k \psi_j | v | \psi_j \psi_i \rangle = \lambda_{ik}. \quad (1.10.8)$$

This Hamiltonian can be diagonalized, such that

$$\langle k | H_s | i \rangle = \epsilon_i\, \delta_{ik}, \quad (1.10.9)$$

by taking suitable linear combinations of $\psi_1 \ldots \psi_N$ to form new single-particle wave functions $\phi_1 \ldots \phi_N$. In (1.10.9), we have written $\langle k | H_s | i \rangle$ for $\langle \phi_k | H_s | \phi_i \rangle$, that is for the matrix elements in the ϕ representation. In terms of these new wave functions, (1.10.7) becomes

$$U\phi_i(\mathbf{r}_1) + \sum_j \int \phi_j^*(\mathbf{r}_2)\, v(\mathbf{r}_1 - \mathbf{r}_2)\, \phi_j(\mathbf{r}_2)\, d\mathbf{r}_2\, \phi_i(\mathbf{r}_1)$$
$$- \sum_j \int \phi_j^*(\mathbf{r}_2)\, v(\mathbf{r}_1 - \mathbf{r}_2)\, \phi_i(\mathbf{r}_2)\, d\mathbf{r}_2\, \phi_j(\mathbf{r}_1) = \epsilon_i\, \phi_i(\mathbf{r}_1). \quad (1.10.10)$$

These equations (1.10.10) form the basis of the Hartree–Fock theory.

1.10.2. *Physical interpretation of Hartree–Fock equations*

To see the meaning of the single-particle equations (1.10.10) it will be convenient to specialize the interaction and to deal with Coulombic forces. This will also serve as a convenient introduction to the problems of atomic and molecular structure considered in Chapter 2. Then, the second term on the left-hand side of (1.10.10) obviously is the electrostatic potential energy at $\mathbf{r}_1$ due to the electronic distribution $\sum_j \phi_j^* \phi_j$, multiplied by the one-electron wave function $\phi_i(\mathbf{r}_1)$ which we are calculating.

Let us now suppose that we retain only one term, that for $i = j$, in the summation over j in the last contribution to the left-hand side of (1.10.10). This term is evidently subtracting off the potential created by the electron under consideration. This is then the Hartree approximation, and in the cruder common potential method of sections 1.5 and 1.9, we neglect even this correction. This might be expected naïvely to be good when the number of electrons is very large.

This Hartree correction, which clearly involves a deficit of one electronic charge in calculating the potential acting on the electron under consideration is only a crude representation of the third (exchange) term on the left-hand side of equation (1.10.10). This is physically again correcting for the fact that the electron does not act on itself, and the charge deficiency is once more one unit. Furthermore, this charge is generally rather localized around the electron we are considering, and so the correction is often important.

We shall see more fully this effect of an electron 'digging' a hole around itself in Chapter 2. We should stress, however, that some simple modifications in these arguments are called for when we have equal numbers of particles with $\uparrow$ and $\downarrow$ spin, the case we have been considering being essentially that when all the spins are parallel. An illuminating physical discussion which goes beyond the brief remarks above may be found in a paper by Slater (1951).

1.10.3. *Expectation value of Hamiltonian*

The Hamiltonian H_s defined above involves its own eigenfunctions. Thus, both H_s and its eigenfunctions must be obtained by an iterative procedure, in which a first approximation is chosen for the eigenfunctions, the Hamiltonian is then calculated, and solved for new eigenfunctions. The iteration must clearly be continued until the new eigenfunctions reproduce the Hamiltonian from which they were derived. Self-consistency has then been achieved.

Once we know the self-consistent Hamiltonian we can readily calculate the total energy in the Hartree–Fock approximation. Thus, the expectation value of H is still given by (1.10.5) with the ϕ_i's replacing the ψ_i's. Using (1.10.8) and (1.10.9) we find

$$\langle \Phi_m | H | \Phi_m \rangle = \frac{1}{2} \sum_i \langle i | H_s | i \rangle + \frac{1}{2} \sum \langle i | U | i \rangle$$

$$= \frac{1}{2} \sum_{i=1}^{N} \{\langle i | U | i \rangle + \epsilon_i\}. \tag{1.10.11}$$

H_s is, of course, the Hartree–Fock Hamiltonian, which we encounter again in Chapter 3.

1.11. Quasi-particles and elementary excitations. Qualitative remarks

The main emphasis so far has been in calculating the one-particle orbitals in a single determinant designed to approximate the many-body wave function. Both common potential methods, which can be referred to as symmetrized Hartree theory, and the full Hartree–Fock method, have been derived.

In view of the central theme of the book, namely the treatment of systems of strongly interacting particles, it seems important at this stage to describe qualitatively the way something akin to single-particle behaviour can remain, even though the wave functions are not adequately described by symmetrized or antisymmetrized products.

The key to many physical properties is, of course, the energy level spectrum, and it turns out in practice that the low-lying excitations can often be neatly and conveniently classified. The origin of this classification is the rather general result (whatever the statistics) that, without necessarily assuming weak interactions between the original (bare) particles, the Hamiltonian can be reduced to the form

$$H = E + H_{\text{q.p.}} + H_{\text{q.p. int.}}, \tag{1.11.1}$$

where E is a constant, $H_{\text{q.p.}}$ describes a set of independent 'particles' of prescribed momenta and energies and $H_{\text{q.p. int.}}$ corresponds to a weak interaction between these 'particles'. In general, the independent 'particles' described by the Hamiltonian $H_{\text{q.p.}}$ are different from the original particles, and are referred to as quasi-particles (hence the subscript q.p. in the Hamiltonian). For this reduction, in first approximation, to independent quasi-particles to be meaningful and useful, it is, of course, necessary that the interaction term $H_{\text{q.p. int.}}$ be small.

Important differences arise between the quasi-particles and the bare particles. Thus, in contrast with the original system we can immediately note that: (i) the number of quasi-particles need not be fixed, and (ii) the statistics of the quasi-particles can be different from those obeyed by the bare particles.

With regard to (i), in the ground state we can say that there are no quasi-particles present at all, and the ground-state energy is simply E from (1.11.1). We can now usefully describe the excited states by starting from an elementary excitation, defined as one in which only a single quasi-particle is present. Its momentum and energy as specified by $H_{\text{q.p.}}$ are the total momentum and excitation energy of the whole system.‡ Now suppose that we have two quasi-particles with momenta $\hbar\mathbf{k}_1$ and $\hbar\mathbf{k}_2$, and energies $\epsilon(\mathbf{k}_1)$ and $\epsilon(\mathbf{k}_2)$. In so far as $H_{\text{q.p. int.}}$ can be neglected, the total momentum of the system is $\hbar(\mathbf{k}_1+\mathbf{k}_2)$ and the excitation energy is $\epsilon(\mathbf{k}_1)+\epsilon(\mathbf{k}_2)$. In this way, it might appear that we can systematically construct all the excited states of the many-body system. However, this argument begins to break down as the number of quasi-particles becomes large, since, even though $H_{\text{q.p. int.}}$ is small, the effect of many pair interactions is eventually to build up to destroy the simple picture. The most important point which emerges from this discussion is that the low-lying excitations are specified by describing the quasi-particles which are present.

‡ We have in mind here homogeneous systems specifically.

Point (ii) above is best introduced at this stage by specific examples. As we shall see in detail in Chapter 5, the uniform electron gas, where the bare particles are, of course, Fermions, is described in terms of elementary excitations of both Bose (plasmons) and Fermi type, a fact that was known experimentally for some years before a good theoretical basis was provided for these concepts. Secondly, in Chapter 8, where we discuss briefly the elastic vibrations of a crystalline solid, we shall see that the quasi-particles, the well-known phonons, obey Bose statistics. It should be noted that this is true even if the atoms or ions making up the solid are themselves Fermions (for example, such as in solid ^{3}He).

Finally we remark that the single-particle methods derived in this chapter are quite appropriate for the discussion of atoms and molecules taken up in Chapter 2. As we have seen, however, when the 'independent' particles are not the original bare particles, the conservation of particle number may have to be relaxed and this is customarily effected through the method of second quantization which we therefore derive in Chapter 3.

Problems

P.1 (i). Derive the electron density

$$\rho(\mathbf{r}) = \sum_{\mathbf{k}}^{k_f} \psi_{\mathbf{k}}^* \psi_{\mathbf{k}},$$

to first order in the potential energy $V(\mathbf{r})$, starting from the integral formulation of the wave equation

$$\psi_{\mathbf{k}}(\mathbf{r}) = e^{i\mathbf{k}.\mathbf{r}} - \frac{1}{2\pi} \int G(\mathbf{r}, \mathbf{r}')\, V(\mathbf{r}')\, \psi_{\mathbf{k}}(\mathbf{r}')\, d\mathbf{r}',$$

where the free particle Green function $G(\mathbf{r}, \mathbf{r}')$ is given by

$$G(\mathbf{r}, \mathbf{r}') = \frac{e^{ik|\mathbf{r}-\mathbf{r}'|}}{|\mathbf{r}-\mathbf{r}'|}.$$

Hence regain (1.9.1).

P.1 (ii). For bound states, the wave equation may be written alternatively as

$$\psi(\mathbf{r}) = -\frac{1}{2\pi} \int \frac{e^{i\sqrt{(2E)}|\mathbf{r}-\mathbf{r}'|}}{|\mathbf{r}-\mathbf{r}'|}\, V(\mathbf{r}')\, \psi(\mathbf{r}')\, d\mathbf{r}'.$$

Verify that, with the ground-state energy $E = -\frac{1}{2}Z^2$ in atomic units, the hydrogen wave function $\psi = e^{-Zr}$ is a solution when $V = -Z/r$. Why will the Rayleigh–Schrödinger perturbation theory, based on plane waves as the unperturbed problem, give an incorrect result in this case?

P.1 (iii). Show that, for N spinless Fermi particles confined to move on the x axis in a one-dimensional box of length l, the number density $\rho_0(x)$ is given by

$$\rho_0(x) = \frac{N+\frac{1}{2}}{l} - \frac{\sin\left\{\dfrac{\pi(2N+1)x}{l}\right\}}{2\sin\left(\dfrac{\pi x}{l}\right)}.$$

Hence show that the Dirac density matrix is given by

$$\gamma_0(x', x) = \rho_0\left(\frac{x'+x}{2}\right) - \rho_0\left(\frac{x'-x}{2}\right).$$

P.1 (iv). Show that the Dirac density matrix of section 1.5 is a projection operator satisfying $\gamma^2 = \gamma$. [In co-ordinate representation, this implies

$$\int \gamma(\mathbf{rr}')\,\gamma(\mathbf{r}'\mathbf{r}_0)\,d\mathbf{r}' = \gamma(\mathbf{rr}_0).]$$

P.1 (v). Using the definition of G given in (1.6.5), convert the integral equation (1.7.3) into

$$G(\mathbf{rr}_0\,\epsilon) = G_0(\mathbf{rr}_0\,\epsilon) - \int d\mathbf{r}_1\, G_0(\mathbf{rr}_1\,\epsilon)\, V(\mathbf{r}_1)\, G(\mathbf{r}_1\,\mathbf{r}_0\,\epsilon).$$

P.1 (vi). Show that the diagonal element of the Bloch density matrix for a three-dimensional isotropic harmonic oscillator is given by

$$C(\mathbf{r}\mathbf{r}\beta) = \left(\frac{m}{2\pi\hbar}\right)^{\frac{3}{2}} \left(\frac{\omega}{\sinh\hbar\omega\beta}\right)^{\frac{3}{2}} \exp\left(-\frac{m}{\hbar}\,\omega r^2 \tanh\tfrac{1}{2}\hbar\omega\beta\right),$$

the potential energy $V(r)$ being written as $\frac{1}{2}m\omega^2 r^2$, and $\beta = (k_B\,T)^{-1}$.

CHAPTER 2

ATOMS AND MOLECULES

2.1. Introduction

While, in this book, we shall generally be concerned with a many-body problem of N particles in a volume Ω, where N is so large that we can proceed to the limit $N \to \infty$, $\Omega \to \infty$, $N/\Omega \to$ finite value, in this chapter we wish to consider some results which may be obtained for 'small' systems, that is for isolated atoms and molecules.

We shall first consider briefly the consequences of the one-particle approximations of Chapter 1, with particular reference to the binding energies of atoms. Though it should be said from the outset that a full quantitative study of the consequences of the Hartree–Fock theory in atomic physics is of the greatest interest, complete accounts of this already exist (Hartree, 1957; Herman & Skillman, 1965). Thus, after a brief discussion of, at best, semi-quantitative aspects of atomic theory, in the one-electron picture, we turn to a somewhat more detailed discussion of two-electron systems, the He atom and more general two-electron ions, and the H_2 molecule. This affords us the opportunity of introducing and illustrating some of the basic concepts essential to a description of correlated motions of interacting particles in the very simplest systems.

We include a brief description of correlations in H_2, for the reason that, although molecules are not again referred to in the book, we shall use the concept which arises there of the 'strong-correlation' or 'strong-coupling' limit directly in Chapter 5, when we discuss the electron gas in metals.

2.2. Hydrogen-like atoms, without electron-electron interactions

Since the hydrogen-like atom is the prototype for all atomic problems, we shall briefly summarize the elementary results for this case, and stress their qualitative consequences.

As is well known, the eigenfunctions of any central field problem, and hence in particular for atomic hydrogen, are

$$\psi_{nlm}(\mathbf{r}) = R_{nl}(r)\, Y_l^m(\theta, \phi), \tag{2.2.1}$$

where the spherical harmonics $Y_l^m(\theta, \phi)$ are eigenfunctions of the square of the orbital angular momentum, L^2, with eigenvalues $l(l+1)\hbar^2$ $(l=0, 1, 2, \ldots)$ and its projection L_z on the z axis, the corresponding eigenvalues of L_z being $m\hbar$ $(-l \leqslant m \leqslant l)$. If we include the case of an arbitrary nucleus, with charge Ze, then the radial wave functions $R_{nl}(r)$ for the hydrogen-like atom have the form (unnormalized)

$$R_{nl}(r) = e^{-\frac{1}{2}\rho}\rho^l L_{n+l}^{2l+1}(\rho). \tag{2.2.2}$$

The functions of the right-hand side of (2.2.2) are the associated Laguerre functions while

$$n = 1, 2, 3\ldots \; (l=0, 1, 2, \ldots, n-1) \quad \text{and} \quad \rho = 2Zr/na_0.$$

We shall use these wave functions in section 2.4, but, for the present, the important point to note is that the corresponding energy levels are given in terms of the principal or total quantum number n by

$$E_n = -\frac{Z^2}{2n^2}\frac{e^2}{a_0}, \tag{2.2.3}$$

where $a_0 = \hbar^2/me^2$ is the first Bohr radius for hydrogen.

2.2.1. *Binding energies of hydrogen-like atoms*

Let us now consider briefly the model in which we regard the Z electrons in an atom as interacting with the nucleus, but not with one another. Somehow, then, we 'switch off' the electron-electron interactions. Furthermore, let us assume that N closed shells are occupied by electrons. It is clear that the total energy will be given by the sum of the one-particle energy levels (2.2.3) and, since each shell of total quantum number n contains $2n^2$ particles, it follows from (2.2.3) that the contribution of this shell to the energy is $-Z^2e^2/a_0$. Thus, the sum ϵ of the eigenvalues for these N closed shells is simply

$$\epsilon = -Z^2N\frac{e^2}{a_0}, \tag{2.2.4}$$

and for a neutral atom we must have that the total number of electrons Z is related to N through

$$Z = \sum_{n=1}^{N} 2n^2 \tag{2.2.5}$$

$$= \frac{N(N+1)(2N+1)}{3}. \tag{2.2.6}$$

If we now assume that the number of electrons Z is very large, so that N is also large (i.e. $N \gg 1$) then (2.2.6) yields

$$N \sim (\tfrac{3}{2})^{\frac{1}{3}} Z^{\frac{1}{3}}. \tag{2.2.7}$$

Thus, from (2.2.5) we find

$$\epsilon = -(\tfrac{3}{2})^{\frac{1}{3}} Z^{\frac{7}{3}} \frac{e^2}{a_0}, \tag{2.2.8}$$

and we see the way in which the total binding energy of a heavy atom would vary with atomic number if it were valid to employ a non-relativistic wave equation. Actually, when Z is very large, the inner electrons move very rapidly, as we shall see when we calculate the momentum distribution of electrons in atoms in section 2.4 (cf. equation (2.4.7)), and it is strictly necessary to use a relativistic wave equation. Most of our considerations in this book use, however, non-relativistic theory, and we shall be content to state the limits of validity of our treatment, without giving the correct theory based on the Dirac equation, in detail.

In Chapters 5 and 6 we shall contrast the situation in an electron gas and in nuclear matter with that in an atom, as described by (2.2.8). In both the electron gas and in nuclear matter we obtain expressions for the energy which are proportional to the number of particles, in contrast to (2.2.8).

2.3. Approximate treatment of binding energies in heavy atoms

The treatment which led to the eigenvalue sum (2.2.8) for large atomic number must be modified to make it quantitative, for it neglected electron-electron interactions. To illustrate the way in which these interactions are included in a single-particle framework, we shall employ the approximate density matrix theory developed in section 1.9.

Clearly, we wish to replace the Coulomb potential energy $-Ze^2/r$ by a screened field which we can write in the form $V(\mathbf{r}) = -Z(r)\,e^2/r$ where $Z(r) \to Z$ as $r \to 0$ and tends to zero as $r \to \infty$. To approximate to $Z(r)$, let us consider the symmetrized Hartree method of Chapter 1. This method assumes that each electron moves in the same potential field $V(\mathbf{r})$. In Hartree's original work on the self-consistent field for atoms, he took cognizance of the fact that the electron whose motion is under discussion does not act on itself, and this is overlooked in the symmetrized Hartree method (though not in the Hartree–Fock method of section 1.10). The effect of allowing an electron to act on itself clearly becomes less important as Z becomes large.

If we could then sum the density matrix perturbation theory of section 1.7 to infinite order, we would obtain for the electron density $\rho(\mathbf{r}E)$ a result having the functional form

$$\rho(\mathbf{r}E) = F(V(\mathbf{r}), E). \tag{2.3.1}$$

As remarked in Chapter 1, this summation has unfortunately not proved tractable to date, and while we can circumvent these difficulties by direct numerical calculation of the Hartree wave functions (see Herman & Skillman, 1965), we shall content ourselves with the approximate Thomas–Fermi result that F is to be replaced, apart from a constant, by $[E - V(\mathbf{r})]^{\frac{3}{2}}$ for $E \geqslant V$, and otherwise by zero.

The condition of self-consistency discussed in Chapter 1, section 1.10, implies that the charge density $\rho(\mathbf{r})$ must be related to $V(\mathbf{r})$ by the Poisson equation

$$\nabla^2 V = -4\pi e^2 \rho(\mathbf{r}), \tag{2.3.2}$$

where
$$\left.\begin{aligned} \rho(r) &= \frac{8\pi}{3h^3}[2m(E_f - V)]^{\frac{3}{2}} \quad (E_f > V) \\ &= 0 \qquad\qquad\qquad\qquad\quad \text{otherwise.} \end{aligned}\right\} \tag{2.3.3}$$

If we write
$$E_f - V = \frac{Ze^2}{r}\chi \quad (r = bx), \tag{2.3.4}$$

where
$$b = \left(\frac{3}{32\pi^2}\right)^{\frac{2}{3}} \frac{h^2}{2me^2 Z^{\frac{1}{3}}} = \frac{0{\cdot}88534}{Z^{\frac{1}{3}}} a_0,$$

then x and χ are clearly dimensionless. The non-linear equation

$$\frac{d^2\chi}{dx^2} = \frac{\chi^{\frac{3}{2}}}{x^{\frac{1}{2}}} \tag{2.3.5}$$

follows from (2.3.2) to (2.3.4), and for neutral atoms the desired solution $\chi(x)$ must satisfy the boundary conditions

$$\chi(0) = 1, \quad \chi(\infty) = 0. \tag{2.3.6}$$

Actually, it is readily shown from (2.3.5) that the solution tending to zero at infinity behaves asymptotically as

$$\chi(x) \sim 144x^{-3}, \tag{2.3.7}$$

but the solution for all x has to be found numerically. Gombàs (1949) gives a useful tabulation of χ, and its general form is shown in Fig. 2.1. An approximate analytic solution of (2.3.5), obtained by Sommerfeld (1932) on the basis of the asymptotic form (2.3.7), namely

$$\chi = \{1+(x/a)^d\}^{-c}, \tag{2.3.8}$$

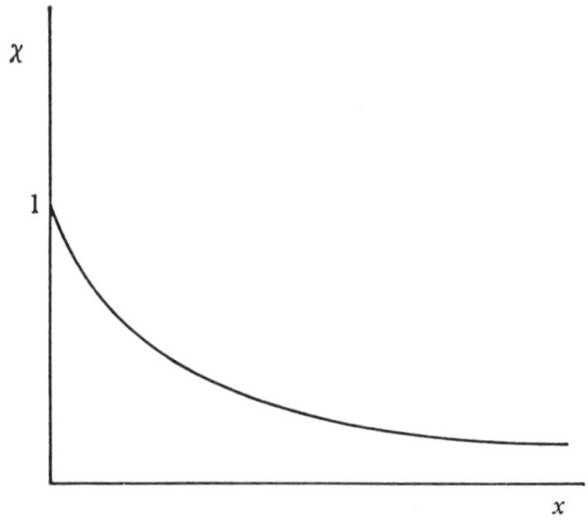

Fig. 2.1. Solution $\chi(x)$ of dimensionless Thomas–Fermi equation (2.3.5), with boundary conditions (2.3.6).

with $a = 12^{\frac{2}{3}}$, $d = 0{\cdot}772$, $c = 3/d = 3{\cdot}886$, is sometimes useful for rough work, though it has the defect that it leads to an infinite slope for χ at the origin (cf. 2.3.13 below).

The important point to make is that, according to (2.3.4), the Coulomb field is now screened and the total binding energy $\mathcal{E}$ for the electrons is correspondingly smaller. Actually, in any screened field such as (2.3.4) we can calculate the eigenvalue sum, ϵ_{sum}, analogous to the Coulomb field result (2.2.8). The Thomas–Fermi result for this is found by noting that the average potential energy associated with a one-electron wave function ϕ_i is $\int \phi_i^* V \phi_i \, d\mathbf{r}$, and summing over i up to the Fermi level we clearly obtain $\int \rho(\mathbf{r}) V(\mathbf{r}) \, d\mathbf{r}$. To this we must add the kinetic energy, and in this model we construct it either by evaluating

$$\int -\frac{\hbar^2}{2m} [\nabla^2 \gamma(\mathbf{r}'\mathbf{r})]_{\mathbf{r}'=\mathbf{r}} \, d\mathbf{r}$$

from the density matrix perturbation theory of Chapter 1, section 1.7, as we did for the density $\rho(\mathbf{r})$, or, more simply, we note from (1.3.7) and (1.3.8) that the kinetic energy/unit volume for free electrons is proportional to the five-thirds power of the density. Hence, we find almost immediately

$$\epsilon_{\text{sum}} = \frac{3h^2}{10m} \left(\frac{3}{8\pi}\right)^{\frac{2}{3}} \int \rho^{\frac{5}{3}} \, d\mathbf{r} + \int \rho V \, d\mathbf{r}. \tag{2.3.9}$$

For the Thomas–Fermi potential defined through (2.3.4) and (2.3.5), this yields

$$\epsilon_{\text{sum}} = -0{\cdot}5125 Z^{\frac{7}{3}} \frac{e^2}{a_0}. \tag{2.3.10}$$

This is not the binding energy though, for in the Hartree eigenvalue sum we have counted electron-electron interactions twice. Correcting (2.3.9) for this, we find

$$\epsilon_{\text{binding energy}} = -0{\cdot}7687 Z^{\frac{7}{3}} (e^2/a_0), \tag{2.3.11}$$

whereas from (2.2.8) we had $-1{\cdot}2 Z^{\frac{7}{3}} (e^2/a_0)$. The self-consistent field is seen then to be important quantitatively, though it does not change the qualitative picture.

Actually, if we analyse the theory a little further, it may be shown that the potential energy, U_{eN} say, of the electron-nuclear interaction is simply related to the potential energy of the electron-electron interaction. Hence, combining this result with the virial theorem for a system in equilibrium under the action of solely Coulomb forces:

$$2\,(\text{kinetic energy}) + \text{potential energy} = 0, \tag{2.3.12}$$

we find a relation between the binding energy ϵ_b and U_{eN} which turns out to be simply

$$\epsilon_b = \tfrac{3}{7} U_{eN}. \qquad (2.3.13)$$

U_{eN} of (2.3.13) is obviously the same as the potential $V_e(0)$ due to the electron cloud at the nucleus, times the nuclear charge Ze.

2.3.1. *Hartree results for potential due to electron cloud at nucleus*

Figure 2.2 now shows a plot of results obtained for $V_e(0)$ from the Thomas–Fermi model and from Hartree calculations of Dickinson (1950). It will be seen that the Hartree points from He to Hg fall on a straight line to remarkable accuracy, the equation of the straight line being

$$eV_e(0) = (\tfrac{6}{5})\, Z^{\frac{7}{5}}(e^2/a_0). \qquad (2.3.14)$$

By an approximate argument, based on a theorem due to Feynman (1939), Foldy (1951) showed how (2.3.14) may be used to obtain the Hartree atomic binding energies. We shall not give the argument in detail, for the essential point can be seen from (2.3.13) and (2.3.14); namely that, for heavy atoms, we would expect that, since U_{eN} of (2.3.13) is proportional to $ZV_e(0)$, the binding energies, using (2.3.14) would go as $Z^{\frac{12}{5}}$. Thus, whereas the exponent of Z in the Thomas–Fermi theory is $\frac{7}{3} = 2{\cdot}33$, that in the Foldy treatment is 2·40. The agreement in this respect is good, but such differences as exist are, no doubt, largely masked experimentally by relativistic effects, for Scott (1952) has made some rough estimates which indicate that

$$\epsilon_{\text{relativistic}} - \epsilon_{\text{non-relativistic}} \sim 4 \times 10^{-6} Z^{\frac{9}{2}}(e^2/a_0). \qquad (2.3.15)$$

Having discussed the qualitative forms of the space wave functions and total energies in atoms, we turn now to a brief consideration of wave functions in momentum space.

2.4. Momentum distribution of electrons in atoms

In the case of an atom with N electrons, given the space wave function $\Psi(\mathbf{r}_1 \mathbf{r}_2 \ldots \mathbf{r}_N)$ we may obtain the momentum wave function $\chi(\mathbf{p}_1 \mathbf{p}_2 \ldots \mathbf{p}_N)$ by means of the Fourier transform

$$\chi(\mathbf{p}_1 \mathbf{p}_2 \ldots \mathbf{p}_N) = \frac{1}{(2\pi)^{\frac{3}{2}N}} \int \exp\{-i(\mathbf{p}_1 . \mathbf{r}_1 + \ldots + \mathbf{p}_N . \mathbf{r}_N)\} \times \Psi(\mathbf{r}_1 \mathbf{r}_2 \ldots \mathbf{r}_N)\, d\mathbf{r}_1\, d\mathbf{r}_2 \ldots d\mathbf{r}_N. \qquad (2.4.1)$$

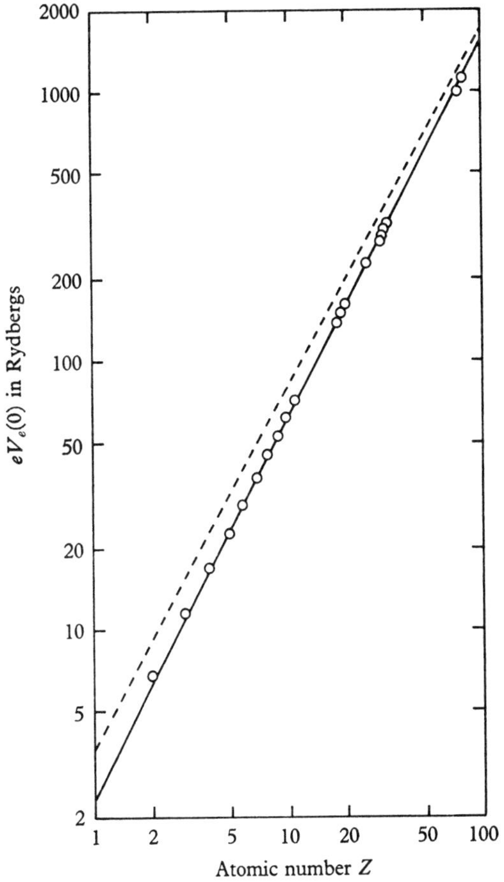

Fig. 2.2. Potential at nucleus of atom due to electronic charge cloud. ----, Thomas–Fermi result; ——, result of equation (2.3.14); Circles, individual Hartree calculations.

Particularly in problems with many electrons, the momentum distribution is of great physical importance, and it is therefore useful at this stage to give briefly an elementary discussion of momenta in atoms.

The simplest case is then the ground state of a hydrogen-like atom of charge Z. From (2.2.1) and (2.2.2) we have for the normalized wave function $\psi_{100}(\mathbf{r})$ the result

$$\psi_{100}(\mathbf{r}) = \left(\frac{Z^3}{\pi}\right)^{\frac{1}{2}} e^{-Zr}, \tag{2.4.2}$$

with units in which $a_0 = 1$, and, using the Fourier transform relation

$$\int e^{-\alpha r + i(\mathbf{p}.\mathbf{r})}\, d\mathbf{r} = \frac{8\pi\alpha}{(p^2+\alpha^2)^2}, \tag{2.4.3}$$

we have for the momentum wave function

$$\chi(\mathbf{p}) = \frac{(8Z^5)^{\frac{1}{2}}}{\pi(p^2+Z^2)^2}. \tag{2.4.4}$$

The probability of finding an electron with momentum between p and $p+dp$ is clearly given by

$$I(p)\, dp = |\chi(p)|^2\, 4\pi p^2\, dp \tag{2.4.5}$$

$$= \frac{32p^2Z^5}{\pi(p^2+Z^2)^4}. \tag{2.4.6}$$

The mean momentum $\bar{p}$ given by

$$\bar{p} = \int_0^\infty p\, I(p)\, dp$$

$$= \frac{8Z}{3\pi}. \tag{2.4.7}$$

The momentum wave functions for the hydrogen-like atom in the general case were first given by Pauling & Podolsky (1929), who demonstrated in particular that the angular dependence in the momentum wave function is the same as in the space wave function.

We shall return to a discussion of the momentum distribution $I(p)$ in section 2.7, and shall show its connection with the first-order spinless density matrix $\gamma(\mathbf{rr}')$ when we have discussed the role of electron correlations, to which we now turn.

2.5. Fermi and correlation holes in atoms

We have taken no account, so far, of the detailed correlations induced in the electronic motions by the Coulomb repulsions. It is quite true that the self-consistent field takes these into account in an average way, but we must now consider the correlated motion.

It might be argued, with some justification, that the problem of atoms is hardly a many-body problem (cf. the introductory remarks in section 2.1). Nevertheless, some of the features we shall be continually encountering in general many-body theory come up even in light atoms, and we can readily introduce some of the basic concepts here.

2.5.1. *Fermi hole*

As we stressed in Chapter 1, the total wave function Φ of any system of Fermions must be antisymmetrical in the interchange of co-ordinates, both space and spin, of any pair of particles. As the most elementary example, a starting point for the discussion of the ground-state wave function Φ_0 of the helium atom would be

$$\Phi_0(\mathbf{r}_1\,\sigma_1, \mathbf{r}_2\,\sigma_2) = \phi_0(\mathbf{r}_1\,\mathbf{r}_2)\,[\alpha(\sigma_1)\,\beta(\sigma_2) - \alpha(\sigma_2)\,\beta(\sigma_1)], \quad (2.5.1)$$

where the space wave function $\phi_0(\mathbf{r}_1\,\mathbf{r}_2)$ must be symmetric. In the simplest approximation to ϕ_0, we form it as a product of hydrogen-like 1*s* functions (2.4.2), with effective nuclear charge Z' say, and thus we write

$$\phi_0(\mathbf{r}_1\,\mathbf{r}_2) = \frac{Z'^3}{\pi}\,e^{-Z'r_1}\,e^{-Z'r_2}, \quad (2.5.2)$$

which already has the required symmetry. Clearly there is no correlation between electrons with such a product wave function.

On the other hand, let us form a wave function for a state (excited level) of helium in which the two electrons have parallel spins. This we might write, if $\mathbf{x}$ denotes both space and spin co-ordinates, as

$$\Psi_1(\mathbf{x}_1\,\mathbf{x}_2) = \psi_1(\mathbf{r}_1\,\mathbf{r}_2)\,\alpha(1)\,\alpha(2) \quad (2.5.3)$$

and we must now have an antisymmetric space part ψ_1. We might form this from an antisymmetrized product of a 1*s* and a 2*s* space wave function, that is

$$\psi_1(\mathbf{r}_1\,\mathbf{r}_2) = \psi_{1s}(\mathbf{r}_1)\,\psi_{2s}(\mathbf{r}_2) - \psi_{1s}(\mathbf{r}_2)\,\psi_{2s}(\mathbf{r}_1). \quad (2.5.4)$$

But we see now that if we ask for the probability that electron 1 is at $\mathbf{r}_1$ and electron 2 is simultaneously at $\mathbf{r}_1$, which is clearly proportional to $\{\psi_1(\mathbf{r}_1\,\mathbf{r}_2)^2\}_{\mathbf{r}_2=\mathbf{r}_1}$, then this is identically zero from (2.5.4), in contrast to the non-zero result given by (2.5.2).

This result, proved in this approximate and elementary way, is a perfectly general consequence of the antisymmetry of the many-body wave function for Fermions. There is zero probability of finding two Fermions with parallel spins at the same point in space. In other words, irrespective of the presence of interactions, a Fermi particle 'digs a hole' round itself, the hole being a region deficient in parallel spin particles. This is the so-called 'Fermi hole' effect, and clearly is already present in an 'orbital' or one-electron theory, when there are electrons with parallel spins.

In the language of Chapter 1, we could say that the diagonal element of the second-order density matrix is zero for parallel spins when their separation is zero, or the pair function vanishes in this case.

This is a fortunate circumstance, because it means that although we neglect, in one-particle approximations, the correlations arising from Coulomb repulsions, the statistics already keep parallel spin electrons apart. This is not true when the spins are antiparallel, as in the ground state of helium, and this leads now to the concept of the correlation hole.

2.5.2. *Correlation hole*

Because electrons repel one another, there is less chance of finding antiparallel spins close together than at large separations. This feature is lost in the approximate ground-state wave function given by (2.5.1) and (2.5.2), and we must refine the calculation to include it.

Put another way, if we denote the distance between the electrons in the ground state by r_{12}, then by introducing electron repulsion into an uncorrelated wave function, we will expect that the mean value $\langle r_{12} \rangle$ is increased. It is worth considering this in a little detail, to see just how this increase comes about.

To this end, introduce a distribution function $f(r_{12})$, which we normalize such that

$$\int_0^\infty f(r_{12})\, dr_{12} = 1. \tag{2.5.5}$$

$f(r_{12})$ will now show us the effect of correlations rather directly if we compute it from both accurate correlated wave functions, and from the best wave function which includes no correlation, namely the Hartree–Fock wave function. The difference between f as derived from the exact wave function and the Hartree–Fock wave function yields a convenient description of the correlation hole.

(*a*) *Distribution in r_{12} for helium atom.* For helium, the simplest two-electron problem, the space wave function $\phi_0(\mathbf{r}_1\mathbf{r}_2)$ of (2.5.1) is readily shown to depend only on the distances r_1 and r_2 of electrons 1 and 2 from the nucleus, and on the interelectronic distance r_{12}. Then we have the result that

$$f(r_{12})\,dr_{12} = \int \phi_0^2(\mathbf{r}_1\mathbf{r}_2)\,d\mathbf{r}_1\,d\mathbf{r}_2, \tag{2.5.6}$$

where, however, the integrations are performed over all positions of the two electrons such that the interelectronic distance lies between r_{12} and $r_{12}+dr_{12}$.

As an example, we choose first the simple uncorrelated function (2.5.2). The result for $f(r_{12})$ is given by

$$f(x) = \frac{Z'^3}{6}(3x^2+6Z'x^3+4Z'^2x^4)\,e^{-2Z'x}. \tag{2.5.7}$$

This is plotted in Fig. 2.3 for the 'best' value $Z' = 2-\frac{5}{16} = \frac{27}{16}$, where we also show $f(x)$ as derived by Coulson & Neilson (1961), from the Hylleraas space wave function

$$\phi_0(\mathbf{r}_1\mathbf{r}_2) = \Sigma C_{lmn}(r_1+r_2)^l\,(r_1-r_2)^{2m}\,r_{12}^n\,e^{-Z'(r_1+r_2)}, \tag{2.5.8}$$

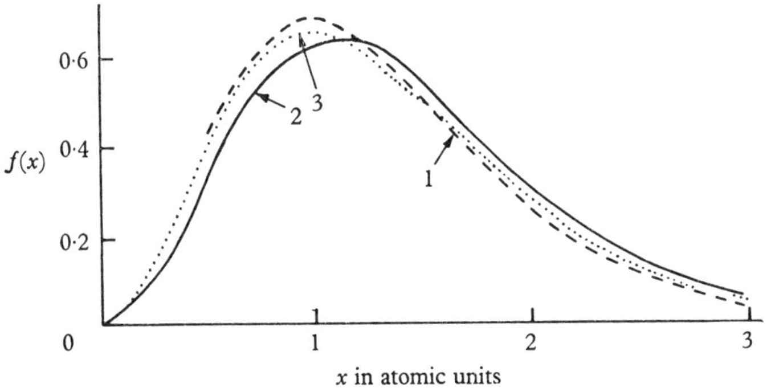

Fig. 2.3. Distribution function $f(x)$ for ground state of helium atom. Curve 1: result of equation (2.5.7). Curve 2: 'exact' result from Hylleraas wave function (2.5.8). Curve 3: self-consistent result from analytic wave function of Roothaan, Sachs & Weiss (1960).

with inclusion of six terms, the coefficients of which are determined variationally. This wave function will essentially give the exact $f(x)$ to graphical accuracy. It will be seen that there is a quite marked broadening out of the distribution function, and the mean value $\langle r_{12}\rangle$, in atomic units, is increased from 1·296 to 1·420.

The corresponding change in the interelectronic repulsion energy $\langle 1/r_{12}\rangle$ is from 1·055 to 0·946, that is 0·11 atomic unit, or 3 electron volts. This is a very substantial reduction, but the kinetic energy is thereby raised, and, from the virial theorem (2.3.12), only a half of this energy represents a gain in total energy. This is a general consequence of correlating electronic motions; we lose on the kinetic energy but regain more potential energy.

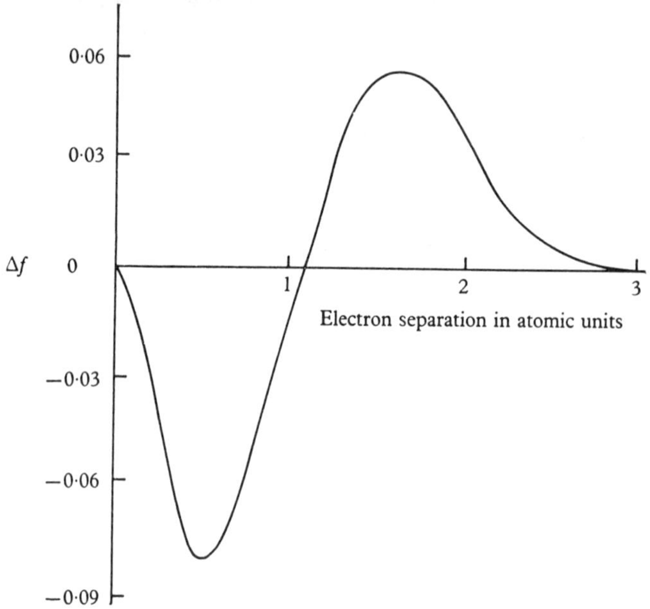

Fig. 2.4. Correlation hole $\Delta f = f_{\text{exact}} - f_{\text{Hartree-Fock}}$.

If we plot the correlation hole $\Delta f = f_{\text{exact}} - f_{\text{H-F.}}$, then it has the form shown in Fig. 2.4, and reveals that, in He, the probability of the two electrons lying anywhere within a distance $\sim$ 1 atomic unit of each other is less than it would be without correlation and, correspondingly, the probability that the electrons are separated by more than $\sim a_0$ is greater. If, following Coulson & Neilson, we adopt the radius of $\sim a_0$ as a rough measure of the size of the correlation hole (these workers term it the Coulomb hole), then the total charge moved is $\sim \frac{1}{20}$ of an electronic charge, in marked contrast to the Fermi hole, where the charge involved may be shown to correspond precisely to one electron (see Chapter 5).

(*b*) *Correlation holes for other two-electron ions.* Calculation of wave functions having the accuracy of (2.5.8) is a tedious business, and one object of such studies must be to search for systematic properties to be associated with Coulomb correlations in atoms and ions.

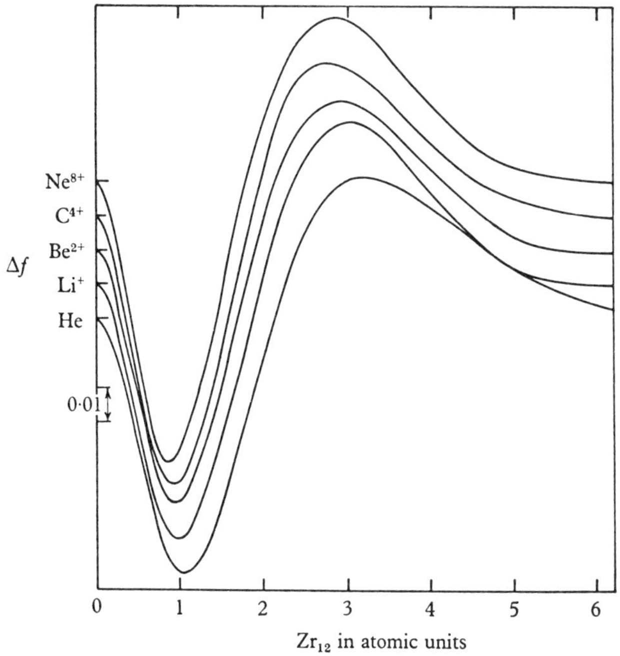

Fig. 2.5. Correlation holes for isoelectronic sequence He, Li^+, ..., Ne^{8+}.

It is therefore of interest to remark briefly that the form of the correlation hole turns out to be remarkably constant through the isoelectronic sequence He, Li^+, ..., Ne^{8+}. Indeed, if Δf is plotted against Zr_{12}, where Z is the atomic number, then the curves only vary slowly with Z, and hardly change shape at all (Lester & Krauss, 1964) as shown in Fig. 2.5.

Obviously, when electrons with parallel spins are involved, the Fermi and correlation holes must be considered together, and some calculations on the Ne atom have been reported by Maslen (1956), to which the reader should refer for further details. Indeed, no completely clear-cut separation of statistics and forces can be made in general, in systems of interacting particles.

To summarize this very brief discussion on two-electron systems, we have seen that relatively small changes in the mean interelectronic separation ($\sim 10\%$) can result in energy gains of easily 1 eV. It will be obvious then, that when one is asking the theory to yield energies useful for comparison with measurement, inclusion of Coulomb repulsions in a careful way is essential. With parallel spin electrons, it is fortunate that the Exclusion Principle (or more strictly, antisymmetry) already forbids two electrons to be close together.

2.6. Perturbation calculation of first-order density matrix for helium isoelectronic sequence

The results given in section 2.5.2 for helium and other two-electron ions have been obtained by the variational method. The interesting possibility of calculating the electron density in an atom by perturbation theory, without knowing the wave function, was discovered by Schwartz (1959) and this result has recently been generalized by Hall, Jones & Rees (1965) to yield the off-diagonal element of the spinless first-order density matrix defined in Chapter 1, section 1.4.1. We shall briefly consider their calculation here, for the ground state of the helium isoelectronic sequence, though we must refer the reader to the original papers for full details.

Essentially, the idea is to split up the Hamiltonian into one-electron terms

$$-\tfrac{1}{2}\nabla^2 - \left(\frac{Z-\alpha}{r}\right) \tag{2.6.1}$$

for each electron, plus a perturbation which is evidently

$$\frac{1}{r_{12}} - \frac{\alpha}{r_1} - \frac{\alpha}{r_2}. \tag{2.6.2}$$

α is clearly a screening constant, to be chosen later, and by a change of scale it is easy to show that if (2.6.1) is used to define the unperturbed problem, then the expansion parameter is the inverse of the effective nuclear charge, that is $(Z-\alpha)^{-1}$. Since the wave function obtained for a sum of one-electron Hamiltonians of form (2.6.1) is simply (2.5.2), with $Z' = Z-\alpha$, the zeroth-order density matrix is readily found as

$$\gamma_0(\mathbf{r}'\mathbf{r}) = \frac{(Z-\alpha)^3}{\pi} \exp\{-(Z-\alpha)(r'+r)\}. \tag{2.6.3}$$

As explained below, the perturbation (2.6.2) may be handled by writing

$$\gamma_1(\mathbf{r}'\mathbf{r}) = \gamma_0(\mathbf{r}'\mathbf{r})\,[F(r')+F(r)-2\langle F\rangle], \tag{2.6.4}$$

and by a generalization of the wave-function perturbation theory of Chapter 1, section 1.8, $F(r)$ may be obtained. The result is

$$(Z-\alpha)\,F(r) = \tfrac{1}{8}(-8\alpha+5)\,y-\tfrac{1}{8}-\tfrac{1}{4}\,e^{-2y}-\frac{3}{16}\frac{(e^{-2y}-1)}{y} + \frac{3}{8}\int_0^y \frac{e^{-2x}-1}{x}\,dx, \tag{2.6.5}$$

where $y = (Z-\alpha)\,r$. The constant term in (2.6.4) is readily found to be given by

$$\langle F\rangle = \frac{-24\alpha+15}{16(Z-\alpha)}-\frac{11}{32(Z-\alpha)}-\frac{3}{8(Z-\alpha)}\ln 2. \tag{2.6.6}$$

This result generalizes the electron density calculated by Schwartz (1959) for the ground state of the helium isoelectronic sequence.

α has so far not been specified. According to Hall (1961), in calculating the mean value of any one-electron operator with the density matrix (2.6.4), the 'best' value of α is found by making the mean value stationary with respect to α. The result thus obtained for α ensures that the contribution to the mean value from the first-order term vanishes. In general, this value of α is different from that which minimizes the energy (cf. P.2 (v)).

Having summarized the main results of the calculation for the density matrix in (2.6.3) and (2.6.4), let us briefly look at the basic reasons why a closed result may be obtained in this case, in spite of the presence of the interelectronic interaction r_{12}^{-1} in the perturbation (2.6.2). To do so, we can, without essential loss of generality, restrict ourselves to the case $\alpha = 0$. Then consider the wave function ϕ expanded as

$$\phi = \phi_0+\phi_1+\ldots, \tag{2.6.7}$$

where ϕ_0 is given by (2.5.2) with $Z' = Z$. The one-electron density $\rho(\mathbf{r}_1)$ can be obtained from ϕ by integrating over $\mathbf{r}_2$ (cf. also problem P.2 (i)), and clearly, to first order, this becomes

$$\rho(\mathbf{r}_1) = \int \phi_0^2(\mathbf{r}_1\mathbf{r}_2)\,d\mathbf{r}_2+2\int \phi_0(\mathbf{r}_1\mathbf{r}_2)\,\phi_1(\mathbf{r}_1\mathbf{r}_2)\,d\mathbf{r}_2. \tag{2.6.8}$$

From the Rayleigh–Schrödinger perturbation theory of Chapter 1, section 1.8.4, we may now write

$$\phi_1(\mathbf{r}_1\mathbf{r}_2) = \sum_n{}' |n\rangle \frac{\langle n|1/r_{12}|0\rangle}{E_0-E_n}, \tag{2.6.9}$$

where $|n\rangle$ are the hydrogenic states for two electrons, which are symmetric for constructing the ground (singlet) states.

Now the crucial point in Schwartz's argument is to observe that, because of the integration over $\mathbf{r}_2$ in (2.6.8), only those products $|n\rangle$ in (2.6.9) which have one electron in a $1s$ state contribute. It will then be seen that the matrix element $\langle n|1/r_{12}|0\rangle$ in (2.6.9), for terms which contribute to ρ, will involve

$$\int \phi_{1s}(r_2)\frac{1}{r_{12}}\phi_{1s}(r_2)\,d\mathbf{r}_2 = \frac{1}{r_1}[1-e^{-2r_1}(1+r_1)], \tag{2.6.10}$$

and, effectively, we can reduce the problem to perturbing solely s states with the 'potential' defined in (2.6.10). The matrix form (2.6.9) then is equivalent to a one-dimensional differential equation in perturbation theory, and (cf. Young & March, 1958) this can always be solved by quadrature. This is the basic reason why the result for $F(r)$ given in (2.6.5) can be obtained.

The exact result (2.6.5) was used in an interesting way by Schwartz to check features of the approximate wave function (2.5.8) used variationally by Hylleraas, in addition to its application to calculate mean values.

It will be clear that the method is in an early stage of development, and the case of helium is, of course, exceptionally simple. It would be of great interest if it could be extended to deal with the expansion of the second-order density matrix for more general atomic systems.

2.7. Discussion of electron correlation in hydrogen molecule

To conclude this chapter, we turn briefly to a consideration of electron correlation effects in the simplest molecule H_2. This has been a testing ground for all sorts of approximate theories, but, in the last resort, the only really accurate wave functions have come from complicated trial forms employed in the variational principle, with explicit inclusion of the interelectronic distance r_{12}. We shall consider the results obtained from one such wave function, due to James & Coolidge (1939), in some detail later.

For the moment, however, the point we stress here is that either we can begin to build up the total wave function from atomic orbitals, localized on their own nuclei, or from molecular orbitals, which spread over the whole nuclear framework. The first approach, exemplified by the work of Heitler & London, is valid when the atomic orbits have radii very small compared with the actual internuclear distance R, while the realm of validity of the molecular orbital approach is that in which there is large overlap between the orbitals. In the H_2 molecule, as indeed in wider aspects of both molecular physics and solid-state physics, the actual situation with which we must deal is intermediate between these two extremes. For some properties, though, as we shall see, the two different philosophies lead to rather similar final results.

Let us write down the approximate ground-state wave functions in terms of the same hydrogen-like 1s functions. Then we have, in an obvious notation in which a and b label the nuclei:

$$\Psi_{HL}(\mathbf{x}_1\mathbf{x}_2) = [\psi_a(1)\,\psi_b(2)+\psi_a(2)\,\psi_b(1)]\,[\alpha(1)\,\beta(2)-\alpha(2)\,\beta(1)], \tag{2.7.1}$$

where $$\psi(r) = \frac{Z^{\frac{3}{2}}\,e^{-Zr}}{\pi^{\frac{1}{2}}}.$$

In contrast, the molecular orbital wave function takes the form

$$\Psi_{mo}(\mathbf{x}_1\,\mathbf{x}_2) = [\psi_a(1)+\psi_b(1)]\,[\psi_a(2)+\psi_b(2)]\,[\alpha(1)\,\beta(2)-\alpha(2)\,\beta(1)], \tag{2.7.2}$$

where, in (2.7.2), we have made the further approximation of forming the molecular orbitals as a linear combination of atomic orbitals.

But, while (2.7.1) and (2.7.2) are too crude to be taken very seriously nowadays, they demonstrate one point of importance immediately. Multiplying out (2.7.2) we see that it already contains the Heitler–London terms, but in addition, terms like $\psi_a(1)\,\psi_a(2)$ and $\psi_b(1)\,\psi_b(2)$, which clearly describe situations in which both electrons can be found on nuclei a or b respectively. Obviously, at large distances, this is very unfavourable energetically because of Coulomb repulsion.

We can say, in general, that the Heitler–London method overestimates electron correlation and is a 'strong coupling' approximation, while the molecular orbital method, for antiparallel spins, fails to include it. This will be important later, when we come to discuss the electron gas in a metal, and its behaviour as a function of density, in Chapter 5.

Finally we will summarize briefly results for three physical quantities of interest in the H_2 molecule: the charge density, which can be compared with X-ray scattering results, the momentum density, which, for example, determines the shape of the Compton modified line, and the correlations or the probability of electron separation r_{12}.

2.7.1. *Charge density in* H_2

In terms of the space part of the wave function, $\Psi(\mathbf{r}_1\mathbf{r}_2)$, the charge density $\rho(\mathbf{r})$ is given by

$$\rho(\mathbf{r}_1) = \int |\Psi(\mathbf{r}_1\mathbf{r}_2)|^2\,d\mathbf{r}_2. \tag{2.7.3}$$

Since this is axially symmetric, we can expand it in Legendre polynomials and write

$$\rho(\mathbf{r}) = \Sigma\rho_l(r)\,P_l(\cos\theta). \tag{2.7.4}$$

We shall confine ourselves to a discussion of $\rho_0(r)$, the spherically averaged charge density about the mid-point of the H_2 bond, θ measuring the angle between the bond and $\mathbf{r}$ in (2.7.4).

We show in Fig. 2.6, $\rho_0(r)$ for various wave functions:

Curve 1. The accurate variational wave function of James & Coolidge (1939).

Curve 2. The self-consistent wave function (Coulson, 1938).

Curve 3. A Heitler–London function of the form (2.7.1), with the 1*s* orbitals allowed to 'float' away from the protons (Gurnee & Magee, 1950). This function, somewhat surprisingly, gives a good binding energy ($\sim$ 4·2 eV; cf. experimental value of 4·7 eV).

The features we wish to stress are

(*a*) All three functions used lead to a very substantial increase (a factor of $\sim \frac{3}{2}$) in the charge density at the midpoint of the bond

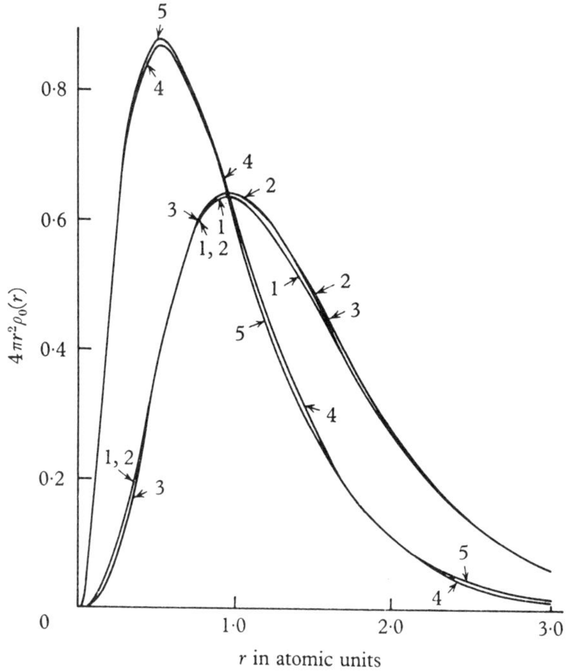

Fig. 2.6. Spherical average $4\pi r^2\rho_0(r)$ of charge distribution in H_2 and He.

over the atomic superposition density. Here is, of course, the reason why all the methods lead to a stable bond: charge is moved from the peripheral regions into the bond.

(*b*) The effect of correlations on the charge density is not large, or in the language of Chapter 1, section 1.4, we may say that the diagonal element of the first-order density matrix is one of the quantities which is relatively insensitive to electron correlation.

(*c*) The self-consistent field picture gives a density somewhat too diffuse, correlations tending to contract the charge cloud slightly.

2.7.2. *Momentum density*

Given the space wave function of a system, we can construct the momentum wave function as shown in section 2.4. We have said earlier that the charge density determines the diagonal element of the first-order density matrix. The dynamics, on the other hand, is reflected in the off-diagonal elements and hence we expect to obtain rather different information about the role of electron correlations by studying the momentum distribution of the electrons.

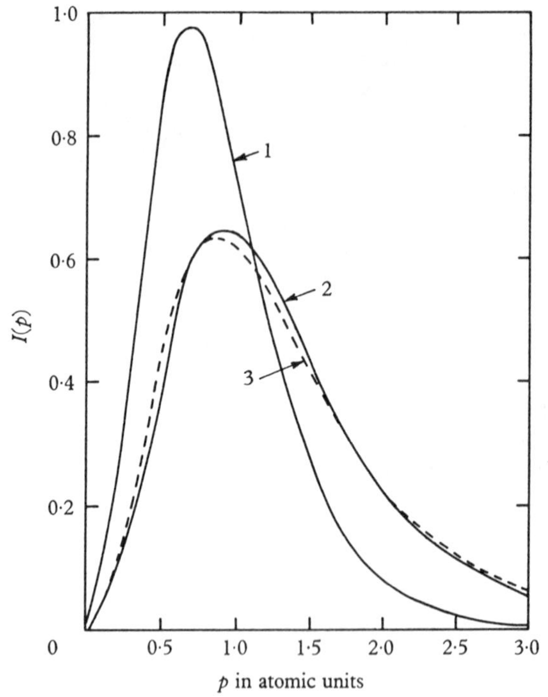

Fig. 2.7. Momentum distribution function $I(p)$ for He and H_2.

If, as in section 2.4 for atoms, we define the momentum distribution $I(p)$ such that $I(p)\,dp$ is the probability of finding an electron with momentum of magnitude between p and $p+dp$, then, in terms of the first-order spinless density matrix $\gamma(\mathbf{r}\mathbf{r}')$, we have

$$I(p) = \frac{p^2}{2\pi^2}\int\int \gamma(\mathbf{r}\mathbf{r}')\frac{\sin p|\mathbf{r}-\mathbf{r}'|}{p|\mathbf{r}-\mathbf{r}'|}\,d\mathbf{r}\,d\mathbf{r}'. \qquad (2.7.5)$$

Unfortunately, $I(p)$ does not seem to have been calculated for other than the 'best' self-consistent wave function (Kilby, 1961), this result being shown in curve 1 of Fig. 2.7.

Therefore, to enable us to draw some conclusions on the role of electron correlations in affecting $I(p)$, we show, in the same figure, results for He for (*a*) the self-consistent field wave function (curve 2), (*b*) a correlated wave function due to Eckart & Hylleraas (curve 3).‡

Qualitatively, these results are like the function $I(p)$ of equation (2.4.6) which we derived for the ground state of the hydrogen atom. However, it may be seen by comparing the two curves for He that the effect of correlations is to make $I(p)$ more diffuse, and hence to increase the mean momentum $\bar{p}$ and the kinetic energy $\overline{p^2}/2m$. This is, of course, in keeping with our earlier statement that introducing correlations into an independent-particle wave function increases the kinetic energy. We turn finally to discuss the average value of r_{12}, which is the source of the gain in potential energy.

2.7.3. *Probability of electron separation*

Barnett, Birss & Coulson (1958) have discussed the way in which the mean value of r_{12} varies according to different degrees of correlation in the wave function for the ground state of H_2, and we shall briefly report their results. For the molecular orbital function (2.7.2), with hydrogenic orbitals without screening constants (i.e. $Z = 1$), the mean value $\langle r_{12} \rangle$ is $2{\cdot}37a_0$ when the internuclear distance is $1{\cdot}4a_0$. This is changed, if we employ the 'best' variational value $Z = 1{\cdot}193$, to $2{\cdot}04a_0$. On the other hand, the Heitler–London values for the two screening constants are $2{\cdot}49a_0$ and $2{\cdot}20a_0$ respectively. The electrons are, naturally enough, further apart in the 'strong-correlation' approximation. As stressed above, the potential energy, depending on the average value of the inverse of r_{12}, is substantially lowered by keeping the electrons further apart.

In summary, we have seen in this chapter, that, in atoms, an approximate estimate of the role of the self-consistent field leads us to the (non-relativistic) results for heavy atoms: (i) the total binding energy is proportional to $Z^{\frac{7}{3}}$; (ii) the mean momentum/electron

‡ This wave function has a space part given by

$$\psi(\mathbf{r}_1\mathbf{r}_2) = N\{e^{-ar_1}e^{-br_2}+e^{-br_1}e^{-ar_2}\},$$

where N is a normalizing constant, $a = 2{\cdot}1832$ and $b = 1{\cdot}1886$. It is a wave function in which different orbits are allocated to electrons with ↑ and ↓ spins. The resulting charge distribution is shown in curve 5 of Fig. 2.6 along with the self-consistent field density (curve 4).

is proportional to $Z^{\frac{2}{3}}$; (iii) the mean radius of the electronic cloud is proportional to $Z^{-\frac{1}{3}}$ (cf. 2.3.4). We wish to reiterate that this $Z^{\frac{7}{3}}$ dependence of the total energy in an atom on the number of electrons Z is in marked contrast to the situation we shall encounter in the truly many-body problems described in the ensuing chapters, where the number of particles is allowed to become indefinitely large, and only the energy/particle has meaning.

Secondly, we have stressed especially the importance of dynamic correlations, as well as statistical correlations for Fermi particles arising from the Exclusion Principle. Two quantities, in particular, which we have introduced and discussed for two-electron systems, will be of central importance throughout our discussion; the momentum distribution, and the probability density of particle separation.

Finally, we have made no attempt, of course, to discuss the detailed shell structure which is very characteristic of atomic theory. This is because, so far, the vast majority of authors have discussed electronic structure of atoms by direct numerical solution of the Hartree–Fock equations, and these results are described extensively elsewhere. However, the methods of many-body perturbation theory, discussed in Chapter 4, hold out promise that they can be applied to atoms with some success, and further progress in atomic theory may be expected along such lines in the next decade.

Problems

P.2 (i). Calculate the charge density in the H_2 molecule from the wave functions (2.7.1) and (2.7.2), using hydrogen 1*s* wave functions for the orbitals. Evaluate the overlap integral

$$S = \int \psi_a(1)\,\psi_b(1)\,d\mathbf{r}_1,$$

for the internuclear spacing $R = 1{\cdot}4a_0$. Hence find the value of $\rho(\mathbf{r})$ at the mid-point of the H_2 bond and compare with the superposition density for the same value of R.

P.2 (ii). Derive the result given in (2.5.7) for the distribution in r_{12} of electrons in the He atom.

P.2 (iii). Show that the momentum distribution function $I(p)$ for the Eckart–Hylleraas function for He is explicitly given by

$$I(p) = \frac{p^2}{4\pi\left[\dfrac{a^{-3}b^{-3}}{64}+(a+b)^{-6}\right]}\left\{\frac{a^2}{b^3(a^2+p^2)^4}\right.$$
$$\left.+\frac{16ab}{(a+b)^3(a^2+p^2)^2(b^2+p^2)^2}+\frac{b^2}{a^3(b^2+p^2)^4}\right\}.$$

P.2 (iv). Calculate the correction terms of order Z^2 and $Z^{\frac{5}{3}}$ to the binding energy formula (2.2.8).

$$\left[\text{Result: } \epsilon = \left\{-\left(\frac{3}{2}\right)^{\frac{1}{3}} Z^{\frac{7}{3}} + \tfrac{1}{2}Z^2 - \frac{1}{18}\left(\frac{3}{2}\right)^{\frac{2}{3}} Z^{\frac{5}{3}}\ldots\right\}\frac{e^2}{a_0}.\right]$$

Estimate very crudely the correction from the exchange energy to the Thomas–Fermi result (2.3.11), given that the exchange energy density for N free electrons in a volume Ω is

$$-\frac{3e^2}{4}\left(\frac{3}{\pi}\right)^{\frac{1}{3}}\left(\frac{N}{\Omega}\right)^{\frac{4}{3}}.$$

In particular, how does it depend on atomic number Z for large Z?

P.2 (v)*. Use the first-order density matrix defined by (2.6.4) to calculate, for the ground state of the helium atom: (a) the mean kinetic energy of the electrons; (b) the mean value of r.

In each case discuss the choice of the screening constant α, and compare with that value of α which minimizes the total energy.

P.2 (vi). Illustrate further the concept of the Fermi hole by showing that the pair function $g(r)$ vanishes for $r = 0$ for free particles with ↑ spin occupying the states inside the Fermi sphere. Why is $g(0)$ equal to $\frac{1}{2}$, and not zero, from (1.5.13)?

CHAPTER 3

SECOND QUANTIZATION

3.1. Introduction

Previously we have worked in terms of the Schrödinger wave function corresponding to the Hamiltonian (1.2.2), or the density matrices derived from it. But in Chapter 1, while we wrote down the many-body wave function for Fermions, for example, as an expansion in determinants, in practice we employed only a single determinant.

When we attempt to generalize this approximation, to include interactions between configurations, then the language so far employed becomes rather inadequate for practical application. We are therefore led to introduce the idea of an occupation number representation, as we discuss below.

3.2. Occupation number representation

In forming many-body Boson (Fermion) wave functions, the symmetrization (antisymmetrization) which had to be carried out to avoid the possibility of distinguishing between particles leads to a clumsy notation, and this suggests that we employ a new procedure, where we merely specify which of the possible single-particle levels are occupied and the occupation numbers of each. This is the maximum information compatible with indistinguishability. Then we can represent the basis functions in the expansions of the many-body wave functions, given in (1.2.9) and (1.2.10) for Bosons and Fermions respectively, by

$$|n_1 \ldots n_N \ldots\rangle, \tag{3.2.1}$$

where n_k is the number of particles in the level ϕ_k. For Fermions, of course, we must allow n_k to take only the values 0 or 1, whereas there are no restrictions for the Boson case. In the form (3.2.1) we must satisfy $\sum_i n_i = N$ for the systems with a fixed number of particles N. A further advantage of this notation, as we shall see below, is that it permits consideration of systems with variable particle number.

We can view (3.2.1) as a state vector in Fock space. The coordinate representation of such a vector is then either Φ_L^B of (1.2.9), if the particles are Bosons, or Φ_L^F of (1.2.10) if the particles are Fermions. To use (3.2.1), we must clearly be able to write the Hamiltonian in terms of operators which can act on such state vectors.

3.3. Creation and annihilation operators

To achieve this end, we introduce operators a_k called annihilation operators, and their Hermitian conjugates $a_k^\dagger$ called creation operators.

3.3.1. *Bosons*

We consider Bosons first, and in this case we define a_k and $a_k^\dagger$ by

$$a_k|\ldots, n_k, \ldots\rangle = \sqrt{n_k}|\ldots, n_k - 1, \ldots\rangle \tag{3.3.1}$$

and

$$a_k^\dagger|\ldots, n_k, \ldots\rangle = \sqrt{(1+n_k)}|\ldots, n_k + 1, \ldots\rangle. \tag{3.3.2}$$

Equation (3.3.1) evidently implies that a_k, acting on a state vector in Fock space in which there are n_k particles in the level ϕ_k, yields a state with $n_k - 1$ particles in this level, while all other occupation numbers remain unchanged. Similarly, $a_k^\dagger$ increases by unity the occupation number of the level ϕ_k. The factor multiplying the state vector on the right-hand side of (3.3.1) must be included to ensure normalization of the N- and the $(N-1)$-particle wave functions and to make the definitions consistent with the assertion that a_k and $a_k^\dagger$ are Hermitian conjugates, the factor $\sqrt{(1+n_k)}$ must then be included in (3.3.2), as may be checked by evaluating $\langle n_{k'}|a_k|n_k\rangle$ and comparing it with $\langle n_k|a_k^\dagger|n_{k'}\rangle$. In these expressions, we have written $|n_k\rangle$ for $|\ldots, n_k, \ldots\rangle$ above, a procedure which we shall often employ later.

From the definitions (3.3.1) and (3.3.2) it follows readily that

$$a_k a_k^\dagger|n_k\rangle = (n_k + 1)\,|n_k\rangle \tag{3.3.3}$$

and

$$a_k^\dagger a_k|n_k\rangle = n_k|n_k\rangle. \tag{3.3.4}$$

Thus, by subtracting (3.3.4) from (3.3.3), we have

$$(a_k a_k^\dagger - a_k^\dagger a_k)\,|n_k\rangle = |n_k\rangle. \tag{3.3.5}$$

Since in Boson problems we are only concerned with operands which can be expressed as linear combinations of symmetrized products, we may write, for this class of functions

$$a_k a_k^\dagger - a_k^\dagger a_k \equiv [a_k, a_k^\dagger] = 1. \tag{3.3.6}$$

In fact, by a similar technique, one can easily prove a more general result than (3.3.6), namely

$$[a_k, a_{k'}^\dagger] = \delta_{kk'} \tag{3.3.7}$$

and, furthermore, that

$$[a_k, a_{k'}] = 0 = [a_k^\dagger, a_{k'}^\dagger]. \tag{3.3.8}$$

For most purposes, the basic commutation relations (3.3.7) and (3.3.8) represent all that we need to know about the creation and annihilation operators.

3.3.2. *Fermions*

Turning now to the Fermion case, we suppose all single-particle levels to be arranged in some definite order. Since n_k is either 0 or 1 for Fermions as we have seen above, we need only show in the state vector the k values of the occupied levels. We can then write $|\ldots k \ldots\rangle$ for a state vector where the N levels k which appear are ordered as above, that is for an N-particle determinant. If we write $|\ldots \not{k} \ldots\rangle$, we imply a many-particle determinant in which the state ϕ_k is absent. Since it is unimportant what the spatial variables are, one obtains the determinant $|\ldots \not{k} \ldots\rangle$ from $|\ldots k \ldots\rangle$ by striking out the column (row) appropriate to k, removing at the same time any one row (column) and revising the normalization. A destruction or annihilation operator for this Fermion case will now be defined as follows:

$$a_k|\ldots k \ldots\rangle = (-1)^m|\ldots \not{k} \ldots\rangle, \tag{3.3.9}$$

where there are m levels immediately preceding k in the state vector, and

$$a_k|\ldots \not{k} \ldots\rangle = 0. \tag{3.3.10}$$

Equations (3.3.9) and (3.3.10) define the operation of a_k on any arbitrary determinant and thus on any arbitrary N-particle wave function which can be expanded as the sum of such determinants.

An analogous creation operator $a_k^\dagger$ can be defined as follows:

$$a_k^\dagger|\ldots \not{k} \ldots\rangle = (-1)^m\,|\ldots k \ldots\rangle, \tag{3.3.11}$$

where there are m states strictly preceding k in the state vector, and

$$a_k^\dagger|\ldots k\ldots\rangle = 0. \tag{3.3.12}$$

As a consequence of these definitions, we are now able to deduce the analogous relations to (3.3.6)–(3.3.8) already discussed for Bosons.

Consider first the effect on an arbitrary N-particle determinant of a creation-destruction pair.

Case $k' \neq k$. We may assume k' precedes k in the ordering, a similar argument holding in the other case. Thus we may write

$$a_{k'}^\dagger a_k|\ldots \bar{k}'\ldots k\ldots\rangle = (-1)^m|\ldots k'\ldots \bar{k}\ldots\rangle,$$

where there are m levels between k' and k.

Obviously, if k' and k are both absent, a_k gives zero and, if both are present, $a_{k'}^\dagger$ gives zero.

Case $k' = k$. Here if k' $(= k)$ is present

$$\begin{aligned} a_k^\dagger a_k|\ldots k\ldots\rangle &= a_k^\dagger(-1)^m|\ldots \bar{k}\ldots\rangle \\ &= (-1)^m(-1)^m|\ldots k\ldots\rangle \\ &= 1|\ldots k\ldots\rangle, \end{aligned}$$

where there are m levels preceding k.

If k' $(= k)$ is absent, then

$$a_k^\dagger a_k|\ldots \bar{k}\ldots\rangle = 0.$$

Thus again, as for Bosons (cf. equation 3.3.4), $a_k^\dagger a_k$ tells us the occupation number of state k.

Turning now to the effect of a destruction-creation pair we have:

Case $k' \neq k$. As before, suppose k' precedes k. Then

$$a_k a_{k'}^\dagger|\ldots \bar{k}'\ldots k\ldots\rangle = (-1)^{m-1}|\ldots k'\ldots \bar{k}\ldots\rangle,$$

where there are m states between k' and k. Otherwise

$$a_k a_{k'}^\dagger|\ldots\rangle = 0.$$

Case $k' = k$. Here, much as before

$$a_k a_k^\dagger|\ldots k\ldots\rangle = 0$$

and

$$a_k a_k^\dagger|\ldots \bar{k}\ldots\rangle = |\ldots \bar{k}\ldots\rangle.$$

The important point that emerges from the detail is that for a quite arbitrary determinant $|\rangle$

$$(a_{k'}^{\dagger} a_k + a_k a_{k'}^{\dagger}) | \rangle = \delta_{k'k} | \rangle,$$

or, if we always understand the a's and the $a^{\dagger}$'s to operate on wave functions expressible as linear combinations of determinants, we may write (cf. the discussion preceding (3.3.7))

$$a_{k'}^{\dagger} a_k + a_k a_{k'}^{\dagger} = \delta_{k'k}. \tag{3.3.13}$$

Other simple relations arise from considering annihilation-annihilation pairs and creation-creation pairs and these are

$$a_{k'} a_k + a_k a_{k'} = 0, \tag{3.3.14}$$

$$a_{k'}^{\dagger} a_k^{\dagger} + a_k^{\dagger} a_{k'}^{\dagger} = 0. \tag{3.3.15}$$

For most purposes, all we need to remember are these anticommutation (to be contrasted with commutation for Bosons) relations.

3.4. Number operator

As anticipated above, for both types of statistics the number operator N_k for the single-particle state ϕ_k is given by

$$N_k = a_k^{\dagger} a_k \tag{3.4.1}$$

and our basis kets $|...n_k...\rangle$ are eigenkets of this operator with eigenvalue n_k. For Fermions, it follows readily from the anticommutation relations (3.3.13)–(3.3.15) that $N_k^2 = N_k$. Hence the eigenvalues of N_k are zero or unity (see also the case $k' = k$ following (3.3.12)). For Bosons the eigenvalues are either zero or any positive integer (see (3.3.4)).

We can define the total number operator N as

$$N = \sum_k N_k, \tag{3.4.2}$$

and since it may be very trivially shown that

$$[N_k, N_{k'}] = 0, \tag{3.4.3}$$

it follows in a straightforward way that our basis kets are also eigenkets of N, with eigenvalues

$$\sum_k n_k.$$

Indeed, as remarked above, one advantage of any treatment using occupation number representation is that we are not restricted to a fixed number of particles.

We may deduce further that for either type of statistics

$$[N_k, a_k] = -a_k \tag{3.4.4}$$

and

$$[N_k, a_k^\dagger] = a_k^\dagger. \tag{3.4.5}$$

3.5. Vacuum state

It should be noted at this point that if we successively extract creation operators from an N particle determinant; for example we write

$$|...k...\rangle = (-1)^m a_k^\dagger|...\not{k}...\rangle,$$

where m is the number of levels preceding k, we shall reach a point where a product of such operators is acting on a single-particle state. If we halt the process of extracting these operators at this point, we place this single-particle level on a different footing from the others. This is undesirable both physically and mathematically and we complete the process by defining a vacuum state $|0\rangle$, a property of which is

$$a_k^\dagger|0\rangle = |k\rangle \quad \text{(all } k\text{)}. \tag{3.5.1}$$

This defines one property of the vacuum state, and, in principle, any other (non-contradictory) properties can also be assigned to it, but certain special ones turn out to be desirable for the following reason. In the derivation of the anticommutation relations for Fermions, it was tacitly assumed that the set of one-particle states in the determinantal wave functions was sufficiently large for all operations to be defined. Thus, for example, if the set consisted of only one element, it makes no sense so far to talk of a_k operating on any single-particle state $|k'\rangle$ and, in particular, the relation (3.3.13) is only proved for two- or more-particle states. However, since we would like to apply our operator rules indiscriminately without regard to how many (or rather how few!) states are contained in the operand, we ascribe to $|0\rangle$ the following simple properties augmenting (3.5.1):

$$a_k|k\rangle = |0\rangle \quad \text{(all } k\text{)}, \tag{3.5.2}$$

$$a_k|0\rangle = 0 \quad \text{(all } k\text{)}, \tag{3.5.3}$$

and

$$\langle 0|0\rangle = 1. \tag{3.5.4}$$

One may verify by inspection of each case that the anti-commutation rules continue to hold for all 0, 1 and 2 particle states provided (3.5.1), (3.5.2) and (3.5.3) are observed. The property (3.5.4), satisfactorily normalizing the vacuum to unity as in the case of the determinantal wave functions, is necessary to ensure that a_α, $a_\alpha^\dagger$ are Hermitian conjugates. Thus, for example,

$$\langle 0|a_\alpha|\beta\rangle = \langle 0|0\rangle\,\delta_{\alpha\beta} = \delta_{\alpha\beta} = \langle\beta|\alpha\rangle = \langle\beta|a_\alpha^\dagger|0\rangle.$$

3.6. Operators in second quantized form

We are now in a position to return to our basic task, that of writing operators, and in particular the Hamiltonian of (1.2.2) in terms of a_k's. Let us define, for convenience, separate kinetic and potential energy parts by

$$T = -\frac{\hbar^2}{2m}\sum_{i=1}^{N}\nabla_i^2 \tag{3.6.1}$$

and

$$V = \sum_{i<j}^{N} v(\mathbf{r}_i, \mathbf{r}_j). \tag{3.6.2}$$

The second quantized form of a general one-particle operator is derived in Appendix 3A. Using this result, we can immediately show that T is given by

$$T = \sum_{\text{all }\mathbf{k}} \epsilon_{\mathbf{k}}\, a_{\mathbf{k}}^\dagger a_{\mathbf{k}}, \tag{3.6.3}$$

where $a_{\mathbf{k}}^\dagger$ and $a_{\mathbf{k}}$ are referred to a set of plane wave states and $\epsilon_{\mathbf{k}} = \hbar^2k^2/2m$. The subscript may, in general, refer to momentum and spin.

The form of V can be obtained by a straightforward, if somewhat lengthy, extension of the arguments of Appendix 3A, the final result being

$$V = \tfrac{1}{2}\sum_{\mathbf{k}_1\mathbf{k}_2\mathbf{k}_1'\mathbf{k}_2'} \langle \mathbf{k}_1\mathbf{k}_2|v|\mathbf{k}_2'\mathbf{k}_1'\rangle\, a_{\mathbf{k}_1}^\dagger a_{\mathbf{k}_2}^\dagger a_{\mathbf{k}_1'} a_{\mathbf{k}_2'}. \tag{3.6.4}$$

The ordering of the operators in (3.6.4) should be stressed, and must be preserved when we are dealing with Fermions, since the Fermion operators anticommute. For Bosons, the operators commute and the ordering is therefore unimportant.

If we had chosen a more general orthonormal set of single-particle states ϕ_α and the corresponding creation and annihilation operators $c_\alpha^\dagger$ and c_α, the kinetic energy operator would naturally involve off-diagonal terms, and the Hamiltonian may then be written

$$H = \Sigma \langle \alpha | T | \beta \rangle c_\alpha^\dagger c_\beta + \tfrac{1}{2}\Sigma \langle \alpha\beta | v | \delta\gamma \rangle c_\alpha^\dagger c_\beta^\dagger c_\gamma c_\delta. \qquad (3.6.5)$$

We shall have occasion to use (3.6.5) below, but generally in this book we shall use the forms (3.6.3) and (3.6.4). Our object of expressing the Hamiltonian in a form suitable for acting on state vectors in Fock space has now been achieved.

3.7. Field operators

There is an alternative description to that in terms of a_k and $a_k^\dagger$, which it will be convenient to introduce at this point. Thus, let us write

$$\psi(\mathbf{x}) = \sum_k a_k \phi_k(\mathbf{x}). \qquad (3.7.1)$$

Then

$$\psi^\dagger(\mathbf{x}) = \sum_k a_k^\dagger \phi_k^*(\mathbf{x}), \qquad (3.7.2)$$

and from the basic rules for the creation and annihilation operators it may readily be proved that

$$[\psi(\mathbf{x}), \psi^\dagger(\mathbf{x}')] = \delta(\mathbf{x}-\mathbf{x}'), \qquad (3.7.3)$$

$$[\psi(\mathbf{x}), \psi(\mathbf{x}')] = 0 = [\psi^\dagger(\mathbf{x}), \psi^\dagger(\mathbf{x}')], \qquad (3.7.4)$$

for Bosons, the Fermion results being obtained by replacing the commutator brackets by anticommutators. $\psi(\mathbf{x})$ can be interpreted (see, for example, Schweber, 1961, pp. 133–4) as an operator annihilating a particle at the point $\mathbf{x}$, while $\psi^\dagger(\mathbf{x})$ is a creation operator of the same kind. Then, for example, the total number operator may be written in terms of the ψ's as

$$N = \sum_k N_k = \sum_k a_k^\dagger a_k = \int \sum_{kk'} a_k^\dagger \phi_k^*(\mathbf{x})\, a_{k'} \phi_{k'}(\mathbf{x})\, d\mathbf{x}$$

$$= \int \psi^\dagger(\mathbf{x})\, \psi(\mathbf{x})\, d\mathbf{x}. \qquad (3.7.5)$$

Finally, it is often useful to express H as given by (3.6.3) and (3.6.4) in terms of the operators ψ and $\psi^\dagger$ defined in (3.7.1) and (3.7.2). Considering the potential energy operator V given by (3.6.2) and (3.6.4), we can rewrite it as

$$\frac{1}{2} \sum_{k_1 k_2 k_1' k_2'} \iint \phi_{k_1}^*(\mathbf{x}_1)\, \phi_{k_2}^*(\mathbf{x}_2)\, v(\mathbf{x}_1, \mathbf{x}_2)\, \phi_{k_1'}(\mathbf{x}_1)\, \phi_{k_2'}(\mathbf{x}_2)$$

$$\times a_{k_2}^\dagger a_{k_1}^\dagger a_{k_1'} a_{k_2'}\, d\mathbf{x}_1\, d\mathbf{x}_2$$

$$= \frac{1}{2} \iint \psi^\dagger(\mathbf{x}_2)\, \psi^\dagger(\mathbf{x}_1)\, v(\mathbf{x}_1, \mathbf{x}_2)\, \psi(\mathbf{x}_1)\, \psi(\mathbf{x}_2)\, d\mathbf{x}_1\, d\mathbf{x}_2. \qquad (3.7.6)$$

Similarly, if we include a one-particle contribution U_1 in the potential energy, this would lead to the result

$$\int \psi^{\dagger}(\mathbf{x})\, U_1(\mathbf{x})\, \psi(\mathbf{x})\, d\mathbf{x}, \tag{3.7.7}$$

in terms of the field operators. Equations (3.7.6) and (3.7.7) supplemented by the kinetic energy term, give us the desired form for the total Hamiltonian.

3.8. Time-independent Hartree–Fock theory

It is very instructive to write down the Hartree–Fock equations in terms of the second quantized creation and annihilation operators for Fermions.

If $a_k^{\dagger}$ and a_k are creation and annihilation operators corresponding to some single-particle level $|k\rangle$, let there be a set of these which are orthonormal. Then a suitable N-body ket $|N\rangle$ is given by $a_{k_1}^{\dagger} \ldots a_{k_N}^{\dagger}|0\rangle$, where $|k_1\rangle \ldots |k_N\rangle$ are the occupied levels of this ket.

3.8.1. *Non-diagonal form of Hamiltonian*

It will be convenient to consider a new set of creation and annihilation operators, defined as linear combination of the a's through

$$\left.\begin{aligned} c_k &= \sum_{s=1}^{N} \alpha_{ks} a_s \qquad (k = 1 \text{ to } N), \\ c_k &= \sum_{s=N+1}^{\infty} \alpha_{ks} a_s \quad (k = N+1 \text{ to } \infty), \end{aligned}\right\} \tag{3.8.1}$$

where α_{ks} are some constants, (the $c_k^{\dagger}$'s being given by the Hermitian conjugates of (3.8.1)). The Hamiltonian then takes the form

$$H = \Sigma \langle k|T|m\rangle c_k^{\dagger} c_m + \tfrac{1}{2}\Sigma \langle kl|v|nm\rangle c_k^{\dagger} c_l^{\dagger} c_m c_n. \tag{3.8.2}$$

The expectation value of H in a general state $|N\rangle$ is

$$\langle N|H|N\rangle = \sum_{km} \langle k|T|m\rangle \langle N|c_k^{\dagger} c_m|N\rangle + \tfrac{1}{2}\Sigma \langle kl|v|nm\rangle \langle N|c_k^{\dagger} c_l^{\dagger} c_m c_n|N\rangle.$$

Making use of the completeness property of basis wave functions, it may readily be shown that

$$\langle N|c_k^{\dagger} c_l^{\dagger} c_m c_n|N\rangle = \langle N|c_k^{\dagger} c_n|N\rangle \langle N|c_l^{\dagger} c_m|N\rangle - \langle N|c_k^{\dagger} c_m|N\rangle \langle N|c_l^{\dagger} c_n|N\rangle. \tag{3.8.3}$$

Hence $\langle N|H|N\rangle$ can be expressed as

$$\langle N|H|N\rangle = \sum_{km} \langle k|T|m\rangle\langle m|\gamma|k\rangle + \frac{1}{2}\sum_{klmn} \langle kl|v|nm\rangle\langle n|\gamma|k\rangle\langle m|\gamma|l\rangle$$

$$-\frac{1}{2}\sum_{klmn} \langle kl|v|nm\rangle\langle m|\gamma|k\rangle\langle n|\gamma|l\rangle$$

$$= \sum_{km} \langle k|T|m\rangle\langle m|\gamma|k\rangle + \frac{1}{2}\sum_{klmn} (\langle kl|v|nm\rangle - \langle kl|v|mn\rangle) \times \langle m|\gamma|l\rangle\langle n|\gamma|k\rangle, \quad (3.8.4)$$

where
$$\langle m|\gamma|k\rangle = \langle N|c_k^\dagger c_m|N\rangle. \quad (3.8.5)$$

Two properties of γ may be deduced at once from (3.8.5) if we choose the c operators to be the same as the a's. Then

$$\langle m|\gamma|k\rangle = \langle N|a_k^\dagger a_m|N\rangle = 0 \quad \text{if} \quad m \neq k$$

or if $m > N$; and $\langle m|\gamma|k\rangle = 1$ if $m = k < N$. Hence we deduce that

$$\operatorname{Tr}\gamma = N \quad (3.8.6)$$

and
$$\gamma^2 = \gamma. \quad (3.8.7)$$

These results, proved for a very special choice of α_{ks} in (3.8.1), can be shown to be true in general.

The Hartree–Fock equations are obtained when we minimize $\langle N|H|N\rangle$ subject to the conditions (3.8.6) and (3.8.7) on γ (cf. section 1.10 of Chapter 1).

Formally we have $\delta\langle N|H|N\rangle = 0$, that is

$$\langle \delta N|H|N\rangle = 0, \quad (3.8.8)$$

where $|\delta N\rangle$ is a state that is different from $|N\rangle$. We restrict ourselves to states which are still expressible in terms of antisymmetrized products of single-particle levels, that is we take

$$|\delta N\rangle = \epsilon c_p^\dagger c_q|N\rangle, \quad (3.8.9)$$

where $p > N$, $q < N$, and ϵ is a small quantity. Substituting (3.8.9) into (3.8.8), we obtain

$$\left.\begin{aligned} &\sum_{km} \langle N|c_q^\dagger c_p \langle k|T|m\rangle c_k^\dagger c_m|N\rangle + \frac{1}{2}\sum_{klmn} \langle N|c_q^\dagger c_p \langle kl|v|nm\rangle \\ &\qquad\qquad\qquad\qquad \times c_k^\dagger c_l^\dagger c_m c_n|N\rangle \\ &= 0 \quad \text{for any } p > N \\ &\qquad\quad \text{and } q < N. \end{aligned}\right\} \quad (3.8.10)$$

All terms of the summation over k, m vanish except those for $k = p$ and $m = q$.

Similarly, of the terms of summation over k, l, m, n, only the terms with one of the following values for k, l, m, n survive:

(i) $k = p, n = l, m = q$; (ii) $k = m, l = p, n = q$;
(iii) $k = p, m = l, n = q$; (iv) $k = n, l = p, m = q$.

Equation (3.8.10) therefore becomes

$$\langle p|T|q\rangle + \frac{1}{2}\sum_{k=1}^{N} \langle kp|v|kq\rangle - \langle kp|v|qk\rangle + \frac{1}{2}\sum_{l=1}^{N} (\langle pl|v|ql\rangle - \langle pl|v|lq\rangle) = 0,$$

that is
$$\langle p|T|q\rangle + \sum_{l=1}^{N} (\langle pl|v|ql\rangle - \langle pl|v|lq\rangle) = 0, \qquad (3.8.11)$$

all $p > N$, $q \leqslant N$. We now introduce a single-particle Hamiltonian

$$H_s = \left[\sum_p \sum_q \left\{\langle p|T|q\rangle + \sum_{l=1}^{N} (\langle pl|v|ql\rangle - \langle pl|v|lq\rangle)\right\}\right] c_p^\dagger c_q, \qquad (3.8.12)$$

with no restriction on p, q.

The important property of this Hamiltonian is that

$$\langle p|H_s|q\rangle = 0, \qquad (3.8.13)$$

where $|q\rangle$ is an occupied level and $|p\rangle$ is an unoccupied level.

We have already emphasized that we could have different complete sets of single-particle states. It is useful to choose a different set of single-particle levels by suitable linear combinations of occupied single-particle levels only and of unoccupied single-particle levels only.

This transformation to new single-particle levels leaves (3.8.12) unaltered. Now we choose the new single-particle levels in this way to make (3.8.12) diagonal. With this choice we have from (3.8.12),

$$\langle p|T|p\rangle + \sum_{l=1}^{N} (\langle pl|v|pl\rangle - \langle pl|v|lp\rangle) = \epsilon_p, \qquad (3.8.14)$$

where ϵ_p are the eigenvalues of the single-particle levels p. ϵ_p are termed the self-consistent energies of the levels and they are good approximations to the single-particle excitations of the system for p near to the Fermi surface.

The expectation value of H with the trial ground state thus obtained is

$$E_0 = \sum_{p=1}^{N} \langle p|T|p\rangle + \frac{1}{2}\sum_{p}\sum_{l}(\langle pl|v|pl\rangle - \langle pl|v|lp\rangle). \quad (3.8.15)$$

As shown by direct arguments based on wave functions in Chapter 1 (3.8.14) are the Hartree–Fock equations, which lead to the total energy of the system given by (3.8.15).

3.9. Particle-hole description for Fermions

For independent Fermions, as we saw in Chapter 1, the lowest states in $\mathbf{k}$ space are occupied out to a Fermi wave number k_f with corresponding wave functions $e^{i\mathbf{k}.\mathbf{r}}$. The effect of interactions, as discussed in general terms in Chapter 2, is to promote some particles into higher momentum states. It is useful then to focus attention only on the change from the original Fermi distribution, and re-describe the situation in terms of particles above the Fermi surface, and the holes remaining inside. To this end, let us define new operators by

$$b_{\mathbf{k}} = \begin{cases} a_{\mathbf{k}} \ (k > k_f), \\ a_{\mathbf{k}}^{\dagger} \ (k < k_f), \end{cases} \quad b_{\mathbf{k}}^{\dagger} = \begin{cases} a_{\mathbf{k}}^{\dagger} \ (k > k_f), \\ a_{\mathbf{k}} \ (k < k_f). \end{cases} \quad (3.9.1)$$

The b's and $b^{\dagger}$'s thus defined are easily seen to satisfy anticommutation relations of type (3.3.13), (3.3.14) and (3.3.15). The ground-state wave function for non-interacting spinless Fermions, namely

$$|g\rangle = \prod_{k<k_f} a_{\mathbf{k}}^{\dagger}|0\rangle \quad (3.9.2)$$

is such that, from (3.5.1),

$$b_{\mathbf{k}}|g\rangle = 0 \quad (\text{all } k). \quad (3.9.3)$$

It is readily shown from (3.5.3) and the anticommutation relations that $\langle g|g\rangle = 1$. Thus, the occupied Fermi sphere is playing the role of the vacuum state in the new description.

Physically, we are generally concerned with creation of particle-hole pairs, a hole being an absent particle inside the Fermi sphere. Thus, in the new description, we are always concerned with $2n$-particle states, a single-particle-hole pair corresponding to

$n = 1$. For example, the two- and four-particle states specified by Figs. 3.1 and 3.2 are respectively

$$b_k^\dagger b_{k'}^\dagger |g\rangle \quad \text{and} \quad b_k^\dagger b_{k'}^\dagger b_{k''}^\dagger b_{k'''}^\dagger |g\rangle. \tag{3.9.4}$$

A general $2n$-particle state is readily constructed in a similar way.

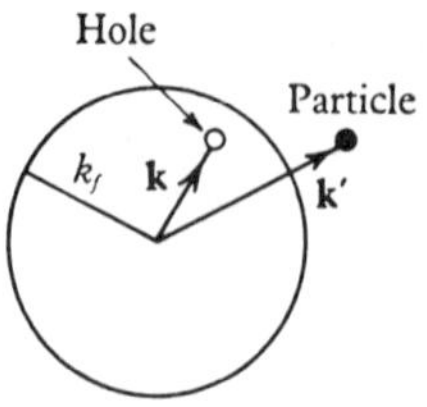

Fig. 3.1

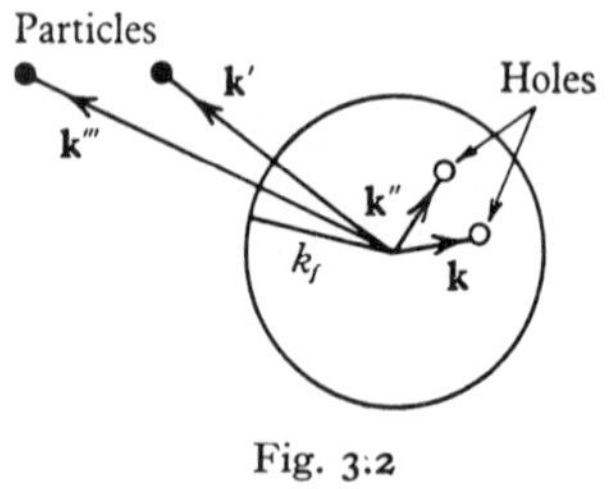

Fig. 3.2

A point to note is that this mathematical description is not unique (though it is perhaps the most obvious one). For example, if the b's and $b^\dagger$'s had been defined by

$$b_{\mathbf{k}} = \begin{cases} a_{\mathbf{k}} & (k > k_f), \\ -a_{-\mathbf{k}}^\dagger & (k < k_f), \end{cases} \quad b_{\mathbf{k}}^\dagger = \begin{cases} a_{\mathbf{k}}^\dagger & (k > k_f), \\ -a_{-\mathbf{k}} & (k < k_f), \end{cases} \tag{3.9.5}$$

the anticommutation relations would still hold. We shall have occasion to recall this alternative description later (see Chapter 7).

The different way in which we have viewed the concept of 'particle' in this section should be emphasized. In (3.9.4), for example, the particles consist of the original particles above the Fermi sea and the holes. More precisely we could use the term 'quasi-particle', this being the simplest illustration of this concept, which we introduced in Chapter 1, section 1.11.

We should note that this description in terms of particles and holes does not imply that we can neglect the particles that occupy the states with $k < k_f$ to form the vacuum. These particles, termed passive particles, do in fact play an essential role in any theory of interacting Fermions. Indeed, it is the interactions between the passive particles, and the interaction between these and any particle that may propagate through the medium, which mainly determine the physical properties of the system.

Finally, it should be remarked that it is sometimes convenient to analyse the ψ's of section 3.7, using the above particle-hole formalism. We can conveniently generalize this to include spin if we denote by K a state with a definite momentum and a definite spin s. Equation (3.7.1) then gives

$$\begin{aligned}\psi(\mathbf{x}) &= \sum_{k<k_f} \phi_K(\mathbf{x})\, a_K + \sum_{k>k_f} \phi_K(\mathbf{x})\, a_K \\ &= \sum_{k<k_f} \phi_K(\mathbf{x})\, b_K^\dagger + \sum_{k>k_f} \phi_K(\mathbf{x})\, b_K \\ &= \psi_+(\mathbf{x}) + \psi_-(\mathbf{x}),\end{aligned} \tag{3.9.6}$$

while (3.7.2) becomes

$$\begin{aligned}\psi^\dagger(\mathbf{x}) &= \sum_{k<k_f} \phi_K^*(\mathbf{x})\, a_K^\dagger + \sum_{k>k_f} \phi_K^*(\mathbf{x})\, a_K^\dagger \\ &= \sum_{k<k_f} \phi_K^*(\mathbf{x})\, b_K + \sum_{k>k_f} \phi_K^*(\mathbf{x})\, b_K^\dagger \\ &= \psi_-^\dagger(\mathbf{x}) + \psi_+^\dagger(\mathbf{x}).\end{aligned} \tag{3.9.7}$$

We should note the convention $\psi_-^\dagger = (\psi^\dagger)_-$, $\psi_+^\dagger = (\psi^\dagger)_+$, which reflects the more fundamental status of the superscript, and also that $(\psi_-)^\dagger \neq \psi_-^\dagger$, etc. Because of (3.9.3), we have

$$\psi_-|g\rangle = 0 = \psi_-^\dagger|g\rangle. \tag{3.9.8}$$

These results will be of use in the perturbation theory of Chapter 4.

Problems

P.3 (i). As explained in section 3.9, for an N-Fermion system, we can define creation and annihilation operators for particles and holes.

Show that the creation and annihilation operators for a particle-hole pair

$$A_\mathbf{k}^\dagger = a_{\mathbf{k}+\mathbf{q}}^\dagger b_\mathbf{k}^\dagger \quad \text{and} \quad A_\mathbf{k} = b_\mathbf{k} a_{\mathbf{k}+\mathbf{q}},$$

where

$$|\mathbf{k}| < k_f \quad \text{and} \quad |\mathbf{k}+\mathbf{q}| > k_f,$$

satisfy the commutation relations appropriate to Bosons.

P.3 (ii). Assume that a complete subset of single Fermion states $|m\rangle$ of equal energy can be specified by operators a_m (and $a_m^\dagger$), where m characterizes an additive quantum number. Assume further that

$$|m\rangle^* = |-m\rangle.$$

Consider

$$H' = -\lambda \sum_{\substack{m_1>0 \\ m_2>0}} a_{m_1}^\dagger a_{-m_1}^\dagger a_{m_2} a_{-m_2}$$

as a perturbation.

Let n be the number of different positive values for m_1 and m_2 in the summation. Two particles, prior to the application of this perturbation, can occupy n different degenerate states $(m, -m)$. Show that this perturbation splits this degenerate level into one with energy $n\lambda$ and $n-1$ with energy zero.

[*Hint*. Define

$$A = \sum_{m>0}^{n} a_m^\dagger a_{-m}^\dagger.$$

Calculate $[H', A]$ and obtain an eigenvalue equation.]

P.3 (iii). Interpret the state vector

$$|\psi\rangle = \prod_{m>0} (u_m + v_m a_m^\dagger a_{-m}^\dagger)|0\rangle,$$

where u_m and v_m are constants such that

$$u_m^2 + v_m^2 = 1.$$

P.3 (iv). Define new operators

$$\left.\begin{aligned} \alpha_m &= u_m a_m - v_m a_{-m}^\dagger \\ \alpha_{-m} &= u_m a_{-m} + v_m a_m^\dagger \end{aligned}\right\} \quad (m > 0).$$

Show that

$$[\alpha_m^\dagger, \alpha_{m'}]_+ = \delta_{mm'}, \quad [\alpha_m, \alpha_{m'}]_+ = [\alpha_m^\dagger, \alpha_{m'}^\dagger]_+ = 0,$$

and that $\alpha_m|\psi\rangle = 0$, all m, where $|\psi\rangle$ is as defined in Problem (iii).

Interpret these results. ($[\alpha, \beta]_+$ is defined as $\alpha\beta + \beta\alpha$.)

CHAPTER 4

MANY-BODY PERTURBATION THEORY

4.1. Introduction

For small systems, such as atoms, we saw in Chapter 2 that the dependence of the total energy on the number of particles N was not, in general, such that the energy/particle was independent of N. When this latter condition is satisfied, we speak of a saturating system, and it is with such systems that we are concerned in the remainder of the book. For calculating the bulk properties (as opposed to surface properties) of the system, we consider the limit in which N and Ω both become very large, but the density N/Ω remains finite. To be definite, we consider Fermions, interacting via given two-body forces, the corresponding potential having a finite Fourier transform. The modifications required for Bosons are not major and will not be discussed in this chapter (cf., however, Chapter 8).

The point of view adopted in this chapter is that we have complete knowledge of the system when no forces are acting and so it ought, in general, to be possible to apply perturbation theory in the (coupling constant premultiplying the) potential to calculate ground-state properties.

The perturbation theory developed in Chapter 1 will be applied below in sections 4.2 and 4.3, to the case of N spinless distinguishable particles, in order, with the minimum detail, to expose the difficulties in a direct application of this perturbation theory. Indeed, we shall see in section 4.2 that the Brillouin–Wigner form of the theory is completely inappropriate, for reasons considered in detail below. On the other hand, the Rayleigh–Schrödinger theory of section 1.8.4, if carefully and systematically used, can yield an energy/particle proportional to N as required, in spite of the appearance of spurious terms proportional to N^2, etc., in any given order. In fact, Brueckner (1955) showed that the non-physical terms cancelled up to fourth order, and the generalization to all orders was effected by Goldstone.

The formulation given below is largely due to Goldstone (1957) and Hubbard (1957), and automatically ensures that only terms having direct physical significance arise. This treatment assumes, of course, that the interacting ground state can be generated by perturbation theory from the free-particle ground state. This is certainly not always true, as we shall see when we discuss the superconducting state in Chapter 7. In that case, the presence of a certain attractive potential can never be correctly dealt with by perturbation theory. In addition, there are even suspicions about repulsive potentials (Van Hove, 1960). Roughly, such difficulties could be attributed to the fact that by turning on the forces, we may have induced 'level-crossing'.

Quite apart from the sign, other features of the potential can give concern. For example, our formalism specifically excludes hard cores (as these have no Fourier transforms). One way round this has been pointed by Brueckner and others (see Chapter 6), who effectively describe two-body encounters using phase-shift analysis rather than Born approximation. A second example is the Coulomb interaction (see Chapter 5). There, the Fourier transform exists, but is of such a form that there are divergencies in each order of the energy-shift series beyond the first. This problem has been fully solved by Gell-Mann & Brueckner (1957) following earlier work by Macke (1950). Also we should mention the case of non-spherically symmetrical potentials. Kohn & Luttinger (1960) have used grand canonical ensemble perturbation theory to carry out elevated temperature calculations. Because of the diffuseness of the Fermi surface, more scattering processes are allowable than at zero temperature. These give rise to contributions which have only been proved to vanish in the zero-temperature limit for the special case of spherically symmetrical potentials. Thus, while the realm of validity of the many-body perturbation theory has not been rigorously settled, we shall go on to develop the formalism under the assumption that it has rather wide applicability.

There have been two general lines of attack on the problem, the time-dependent and time-independent approaches. The latter is obviously more natural for a time-independent problem, but the former has mathematical advantages and for this reason is preferred below. The language of second quantization developed in Chapter 3 will prove very appropriate. In this connection we should point out that whereas some authors use field operators (the ψ's), others use the a's appropriate to a set of plane wave states. The two descriptions are, of course, equivalent, although each has its own individual advantages. In general, one might say that the ψ's lead to a more compact and elegant development of the general theory, whereas the a's are likely to be more useful for calculational purposes. The choice is rather a matter of taste. In this book, we shall place stronger emphasis on the latter than the former, but because the reader will require a knowledge of both in order to read the literature satisfactorily, the two techniques will be simultaneously developed in this chapter.

We are now in a position to begin the explicit discussion for the Hamiltonian (1.2.1), where we reiterate that the two-body interaction $v(\mathbf{r}_i - \mathbf{r}_j)$ is supposed to have a finite Fourier transform $v(\mathbf{k})$ defined by

$$\int e^{i\mathbf{k}.\mathbf{r}}\, v(\mathbf{r})\, d\mathbf{r}.$$

The following two sections are concerned with the application of the standard perturbation theories of Chapter 1. As hinted above, the conclusion reached is that they are not satisfactory, as they stand, for dealing with very many particles, the peculiar nature of the difficulties being spelled out in detail. This leads us, beginning in section 4.4, to take a different and successful approach, namely that of Goldstone & Hubbard. Thus, attention to the fine detail of sections 4.2 and 4.3 is, perhaps, not as important as in the paragraphs subsequent to them.

4.2. Terms in Brillouin–Wigner series

4.2.1. *Energy*

In section 1.8.3, we wrote down the Brillouin–Wigner perturbation series for the ground-state energy E as

$$E = E_0 + \langle \xi_0 | V | \xi_0 \rangle + \langle \xi_0 | V \frac{Q}{E-H_0} V | \xi_0 \rangle + \langle \xi_0 | V \frac{Q}{E-H_0} V \frac{Q}{E-H_0} V | \xi_0 \rangle + \dots \quad (4.2.1)$$

We want to examine the dependence on N of successive terms in this series in the limit $N \to \infty$, $\Omega \to \infty$, with the density $N/\Omega = \rho$ finite. The unperturbed energy E_0 is given by (1.3.8) and is of course proportional to N.

Proceeding to the first-order term we have explicitly

$$\langle \xi_0 | V | \xi_0 \rangle = \frac{1}{2} \sum_{i,j} \langle \xi_0 | v(\mathbf{r}_i - \mathbf{r}_j) | \xi_0 \rangle.$$

Obviously, the matrix element of v between the N-particle wave function ξ_0 reduces, by integration over $N-2$ particles, to a two-particle matrix element of the form (1.3.6), with the result

$$\langle \xi_0 | V | \xi_0 \rangle = \frac{1}{2} \sum_{\mathbf{ij}} \langle \mathbf{ij} | v | \mathbf{ij} \rangle,$$

where $\langle \mathbf{ij} | v | \mathbf{ij} \rangle$ from (1.3.6) is proportional to $1/\Omega$. Also $\sum_{ij}$ gives a factor $\propto N^2$, and thus $\langle \xi_0 | V | \xi_0 \rangle$ is proportional to $N\rho$. The next term in (4.2.1) may be written

$$\sum_{\alpha \neq 0} \frac{\langle \xi_0 | V | \xi_\alpha \rangle \langle \xi_\alpha | V | \xi_0 \rangle}{E - E_\alpha}, \tag{4.2.2}$$

where E_α is the energy of the state $|\xi_\alpha\rangle$ and is such that $E_0 \neq E_\alpha$ since $|\xi_0\rangle$ was assumed non-degenerate.

Expression (4.2.2) may be written in terms of two-particle matrix elements as

$$\frac{1}{4} \sum_{\mathbf{ij}\,\mathbf{i'j'}} \frac{\langle \mathbf{ij} | v | \mathbf{i'j'} \rangle \langle \mathbf{i'j'} | v | \mathbf{ij} \rangle}{E - E_\alpha},$$

since $|\xi_\alpha\rangle$ can only differ from $|\xi_0\rangle$ in two single-particle states. Since v conserves momentum from (1.3.6), $\sum_{\mathbf{i'j'}}$ involves only one summation over momentum giving a factor $\propto \Omega$ from (1.3.3). Furthermore, $\sum_{\mathbf{ij}}$ gives the usual factor $\propto N^2$, the two matrix elements each give factors $1/\Omega$, and $1/(E-E_\alpha)$ gives a factor $\propto 1/N$, since E_α differs from E_0 by $o(1)$ for terms with non-vanishing matrix elements. Combining these results we obtain immediately

$$\langle \xi_0 | V \frac{Q}{E - H_0} V | \xi_0 \rangle \sim 1.$$

By similar arguments we can show that the higher-order terms of the perturbation series give contributions which are either independent of N or vanish for large N.

Thus we arrive at the remarkable result that the many-body perturbation theory gives the first term $\propto N$ and all succeeding terms individually negligible compared to the first term, whatever the strength of the perturbation. However, it would be wrong to conclude that the above argument proves that the Brillouin–Wigner series is convergent and that only the first term need be considered. Among other reasons, we know that $\langle\xi_0|V|\xi_0\rangle$ does not include dynamical correlations between particles and these must be important (on physical grounds) for a strongly interacting system. The situation therefore must be that many ($\sim N$) of the small terms beyond the first are of roughly comparable size, and add up to change the energy/particle by a finite amount. Thus it will be completely misleading to apply many-body perturbation theory in the Brillouin–Wigner form, short of considering an infinite number of terms in the limit of large N.

4.2.2. *Wave function*

We next consider whether the perturbed wave functions as given by (1.8.15) yield a better representation of the true many-body wave function than the unperturbed wave function $|\xi_0\rangle$.

Consider the first-order wave function

$$|\psi^{(1)}\rangle = |\xi_0\rangle + \left(\frac{Q}{E-H_0}V\right)|\xi_0\rangle. \tag{4.2.3}$$

Equation (4.2.3) may also be written, using the matrix element representation,

$$\begin{aligned}|\psi^{(1)}\rangle &= |\xi_0\rangle + \sum_{\alpha\neq 0}\frac{\langle\xi_\alpha|V|\xi_0\rangle}{E-E_\alpha}|\xi_\alpha\rangle \\ &= |\xi_0\rangle + \sum_{\mathbf{ij\,i'j'}}\frac{\langle\mathbf{i'j'}|v|\mathbf{ij}\rangle}{E-E_\alpha}|\xi_\alpha\rangle.\end{aligned} \tag{4.2.4}$$

Equation (4.2.4) states that $|\psi^{(1)}\rangle$ consists of a superposition of a state $|\xi_0\rangle$ in which no particles are excited and states in which one pair of particles are excited (and no others), the pair being selected in all possible ways. However, in the actual physical state for large N, it is clear that the number of particles excited must $\to\infty$ with N, and hence the probability of finding the system with only two particles excited must be essentially zero. Clearly therefore we cannot hope to describe the true physical situation by considering $\psi^{(1)}$, or for that matter $\psi^{(r)}$ for any finite r.

We are thus led to conclude that the essentially non-convergent nature of the Brillouin–Wigner form of the perturbation theory for the energy, together with the unphysical character of the perturbed wave function to any finite order, makes it unprofitable to use this form of perturbation theory to study many-body systems.

4.3. Terms in Rayleigh–Schrödinger series

4.3.1. *Energy*

The Rayleigh–Schrödinger form of perturbation theory was discussed in Chapter 1, section 1.8.4. In particular, we may write from (1.8.18)

$$E = E_0 + E^{(1)} + E^{(2)} + E^{(3)} + \ldots,$$

where

$$E^{(1)} = \langle \xi_0 | V | \xi_0 \rangle,$$

$$E^{(2)} = \langle \xi_0 | V \frac{Q}{E_0 - H_0} V | \xi_0 \rangle,$$

$$E^{(3)} = \langle \xi_0 | V \frac{Q}{E_0 - H_0} V \frac{Q}{E_0 - H_0} V | \xi_0 \rangle - E^{(1)} \langle \xi_0 | V \frac{Q}{(E_0 - H_0)^2} V | \xi_0 \rangle$$

$$= \langle \xi_0 | V \frac{Q}{E_0 - H_0} (V - \langle \xi_0 | V | \xi_0 \rangle) \frac{Q}{E_0 - H_0} V | \xi_0 \rangle. \qquad (4.3.1)$$

A similar analysis to that employed for the Brillouin–Wigner series yields immediately $E_0 \sim N$ and $E^{(1)} \sim N\rho$ as before. However

$$E^{(2)} = \sum_\alpha \langle \xi_0 | V | \xi_\alpha \rangle \langle \xi_\alpha | V | \xi_0 \rangle / (E_0 - E_\alpha)$$

$$= \frac{1}{4} \sum_{\mathbf{ij i'j'}} \frac{\langle \mathbf{ij} | v | \mathbf{i'j'} \rangle \langle \mathbf{i'j'} | v | \mathbf{ij} \rangle}{E_0 - E_\alpha},$$

since, as before, non-vanishing matrix elements occur between N-particle states which only differ in the levels occupied by two particles. Thus, we could write explicitly for the denominator

$$E_0 - E_\alpha = \frac{k_i^2}{2m} + \frac{k_j^2}{2m} - \frac{k_{i'}^2}{2m} - \frac{k_{j'}^2}{2m}$$

and hence $E_0 - E_\alpha$ is independent of N, in marked contrast to the Brillouin–Wigner case. Therefore, it follows that

$$E^{(2)} \sim N^2 \Omega \frac{1}{\Omega^2} = N\rho.$$

The first term of $E^{(3)}$ in (4.3.1) may be written

$$\sum_{\alpha\beta} \frac{\langle\xi_0|V|\xi_\alpha\rangle\langle\xi_\alpha|V|\xi_\beta\rangle\langle\xi_\beta|V|\xi_0\rangle}{(E_0-E_\alpha)(E_0-E_\beta)}. \tag{4.3.2}$$

This differs from $E^{(2)}$ in that it involves an additional factor

$$\sum_\beta \frac{\langle\xi_\alpha|V|\xi_\beta\rangle}{E_0-E_\beta},$$

and as before the matrix element is of order Ω^{-1}, E_0-E_β is $o(1)$, and the summation gives a factor Ω. Thus the additional factor is of order unity if

$$|\xi_\alpha\rangle \neq |\xi_\beta\rangle.$$

This contribution to $E^{(3)}$ is thus proportional to $N\rho$.

In every term of the form

$$\langle\xi_0|V\left(\frac{QV}{E_0-H_0}\right)^r|\xi_0\rangle \quad (r \geqslant 3),$$

it follows by similar arguments that there are always terms proportional to $N\rho$.

But we must still consider the case

$$|\xi_\alpha\rangle = |\xi_\beta\rangle.$$

Then (4.3.2) includes the term

$$\sum_{\mathbf{ij\,i'j'\,kl}} \frac{\langle\mathbf{ij}|v|\mathbf{i'j'}\rangle\langle\mathbf{i'j'}|v|\mathbf{ij}\rangle\langle\mathbf{kl}|v|\mathbf{kl}\rangle}{(E_0-E_\alpha)^2} \tag{4.3.3}$$

and, exactly as before, this is found to be proportional to $(N\rho)^2$. This term is clearly non-physical, but a closer examination reveals that the second contribution to $E^{(3)}$ in (4.3.1), namely

$$-E^{(1)}\langle\xi_0|V\frac{Q}{(E_0-H_0)^2}V|\xi_0\rangle,$$

exactly cancels the non-physical N^2 dependence. The proof that this term is indeed proportional to N^2 is trivial, but detailed calculation is required to show exact cancellation.

To third order (and to any higher order) one can in fact verify that the terms of the Rayleigh–Schrödinger perturbation expansion either vanish or increase as N in the limit $N\to\infty$, the non-physical terms proportional to N^2, N^3, etc., all neatly cancelling each other. We shall not go further into the details as we shall later prove a perturbation expansion which explicitly gives terms which are proportional to N only.

Assuming the validity of this cancellation for the present, we notice at once that the Rayleigh–Schrödinger series is more useful than the Brillouin–Wigner form in that it yields an expansion leading to the energy per particle as independent of the size of the system for large N.

4.3.2. *Wave function*

The first-order Rayleigh–Schrödinger wave function is

$$|\psi^{(1)}\rangle = |\xi_0\rangle + \frac{Q}{E_0 - H_0} V|\xi_0\rangle, \tag{4.3.4}$$

and we now consider the variational approximation to the energy which it affords. This is given by

$$\frac{\langle\psi^{(1)}|H|\psi^{(1)}\rangle}{\langle\psi^{(1)}|\psi^{(1)}\rangle},$$

which, from (4.3.4), is equal to

$$\frac{\langle\xi_0|H|\xi_0\rangle + 2\langle\xi_0|V\frac{Q}{E_0-H_0}V|\xi_0\rangle + \langle\xi_0|V\frac{Q}{E_0-H_0}H\frac{Q}{E_0-H_0}V|\xi_0\rangle}{1+\langle\xi_0|V\frac{Q}{(E_0-H_0)^2}V|\xi_0\rangle}$$

$$= E_0 + E^{(1)} + \left\{\frac{-E_0\langle\xi_0|V\frac{Q}{(E_0-H_0)^2}V|\xi_0\rangle}{1+\langle\xi_0|V\frac{Q}{(E_0-H_0)^2}V|\xi_0\rangle} + \frac{\langle\xi_0|V\frac{Q}{E_0-H_0}V|\xi_0\rangle + E^{(2)} + E^{(3)}}{1+\langle\xi_0|V\frac{Q}{(E_0-H_0)^2}V|\xi_0\rangle}\right\}$$

$$+ \frac{\langle\xi_0|V\frac{Q}{E_0-H_0}H_0\frac{Q}{E_0-H_0}V|\xi_0\rangle}{1+\langle\xi_0|V\frac{Q}{(E_0-H_0)^2}V|\xi_0\rangle}$$

$$= E_0 + E^{(1)} + \frac{E^{(2)} + E^{(3)}}{1+\langle\xi_0|V\frac{Q}{(E_0-H_0)^2}V|\xi_0\rangle}.$$

Now as before

$$\langle \xi_0 | V \frac{Q}{(E_0 - H_0)^2} V | \xi_0 \rangle = \sum_{\alpha} \frac{\langle \xi_0 | V | \xi_\alpha \rangle \langle \xi_\alpha | V | \xi_0 \rangle}{(E_0 - E_\alpha)^2},$$

$\sim N\rho$ for large N. Thus, the term additional to $E_0 + E^{(1)}$ is of order unity, and we regain the result of a variational calculation with the unperturbed wave function.

A similar result can be established for any specific higher-order wave function.

Thus the Rayleigh–Schrödinger expansion for the wave function is no more useful than the Brillouin–Wigner form. The reason in each case is that any finite-order wave function is only capable of describing a finite number of simultaneous pair excitations while a true many-body wave function must be capable of describing the simultaneous excitation of an infinite number of particles for $N \to \infty$.

In view of the above, one might be tempted to ask how it is that one is able to get a possibly useful (convergent) energy series, though never a useful approximation for a wave function. The answer to the paradox may be given along the following lines. To describe an extended system accurately we must know the wave function, not only locally, but over the entire region with equal accuracy and this effectively requires us to go to infinite order. But to determine the expectation value of a local operator (and calculation of the energy is that of finding the expectation value of a finite-range two-body potential) we need to know accurately only the wave function in the immediate neighbourhood ($\sim$ (range)^3 of the potential). This, the Rayleigh–Schrödinger (or Brillouin–Wigner) wave function is capable of giving.

Having seen then the difficulties of the conventional formulation of perturbation theory, we are in a position to turn to the Goldstone–Hubbard treatment. However, as a convenient preliminary, we shall give a simple way of representing the Rayleigh–Schrödinger energy series by graphs. These will require generalization eventually to take account of time dependence and particle indistinguishability but will serve as an introduction to some important ideas needed later.

4·3·3. *Graphs*

We represent Fermions by solid 'vertical' lines and interactions by 'horizontal' dotted lines. Thus Fig. 4.1 (i) represents the interaction (scattering) of two Fermions.

If we label the solid lines by indices i, j, i' and j' we can say that two Fermions in states i and j interact and are scattered into states i' and j'. Usually these states are momentum states and the interaction conserves total momentum. It is customary to regard Fig. 4.1 (i) as representing the matrix element $\langle i'j'|v|ij\rangle$ and also

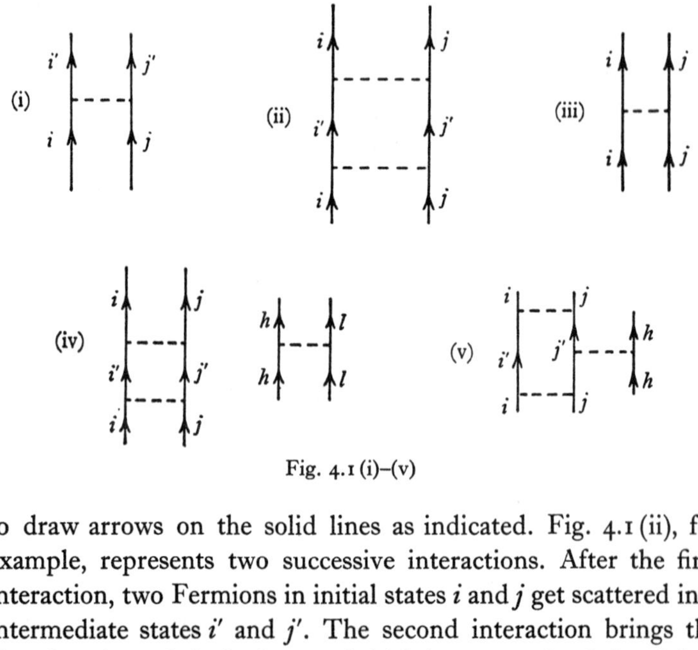

Fig. 4.1 (i)–(v)

to draw arrows on the solid lines as indicated. Fig. 4.1 (ii), for example, represents two successive interactions. After the first interaction, two Fermions in initial states i and j get scattered into intermediate states i' and j'. The second interaction brings the Fermions into their final states (which happen to be their original states in this example). The corresponding matrix element is

$$\langle ij|v|i'j'\rangle \langle i'j'|v|ij\rangle.$$

According to the above convention, Fig. 4.1 (iii), which represents the matrix element $\langle ij|v|ij\rangle$, may also be construed to represent $E^{(1)}$, if we incorporate the summation over i and j (with the right factor) which $E^{(1)}$ implies.

Similarly, Fig. 4.1 (ii) may also be construed to represent $E^{(2)}$, if we incorporate the summation convention (over i, j, i' and j') and the provision of an energy denominator for the intermediate state. This energy denominator is simply the difference between the (kinetic) energies of the intermediate and initial states.

No new conventions arise when we consider the representation of higher-order terms by means of diagrams.

Corresponding to a given $E^{(r)}$, there will be several diagrams, each diagram having r interaction lines. There may be anything from two to $2r$ solid lines in the diagram. These diagrams may be connected (linked) or not. Fig. 4.1 (iv), which represents (4.3.3), is an example of a third-order diagram which is unlinked (disconnected).

Fig. 4.1 (v), however, is a third-order connected diagram. We have already seen that the term corresponding to Fig. 4.1 (iv) gave a contribution proportional to N^2. However, the contribution of Fig. 4.1 (v) is simply proportional to the number of particles.

Indeed, the major burden of the rest of the chapter is to show that, in a suitable formulation of the perturbation theory, terms giving unphysical N dependence do not arise at all. Since it is easy to see that all unlinked diagrams give such unphysical contributions, the end result will be to confine oneself to connected diagrams only. Thus we may anticipate the rest of the chapter and write down the Goldstone formula, which is simply (4.3.1) with the proviso that only contributions from linked diagrams should be taken into account. Thus we write

$$E = \langle\xi_0|V+V\frac{Q}{E_0-H_0}V+V\frac{Q}{E_0-H_0}V\frac{Q}{E_0-H_0}V+\ldots|\xi_0\rangle_L, \tag{4.3.5}$$

where L indicates that only linked diagrams should be considered.

Another result is worthy of note at this stage. With the restriction to connected diagrams, every contribution is proportional to Ω, the volume of the system, multiplied by a power of the density ρ. There is a simple connection between the number of Fermion lines appearing in the diagram and the power of the density; every diagram involving only two Fermion lines giving a contribution proportional to $\rho\Omega$. Similarly, each diagram involving only three Fermion lines will give a contribution proportional to $\Omega\rho^2$ and so on (see P. 4 (iv)).

4.4. Time-dependent perturbation theory

As we remarked in section 4.1, we shall base our treatment of many-body perturbation theory on the time-dependent approach, in contrast to the discussion above. Thus, in units with $\hbar = 1$, we wish to solve the wave equation

$$H\Phi = i\frac{\partial \Phi}{\partial t}, \tag{4.4.1}$$

where
$$H = H_0 + V. \tag{4.4.2}$$

H_0 and V need not be specified at this stage.

We work in the interaction picture which involves the following substitutions

$$\Psi = e^{iH_0t}\Phi, \quad V(t) = e^{iH_0t}V\,e^{-iH_0t}. \tag{4.4.3}$$

These change (4.4.1) to the form

$$i\frac{\partial}{\partial t}\Psi = V(t)\Psi. \tag{4.4.4}$$

This can be integrated at once to give

$$\Psi(t) = \Psi(t_0) - i\int_{t_0}^{t} V(t')\Psi(t')\,dt', \tag{4.4.5}$$

and iteration then yields

$$\Psi(t) = U(t, t_0)\Psi(t_0), \tag{4.4.6}$$

where

$$U(t, t_0) = 1 + \sum_{n=1}^{\infty}(-i)^n\int_{t_0}^{t} dt_n\int_{t_0}^{t_n} dt_{n-1}\ldots\int_{t_0}^{t_2} dt_1 V(t_n)\ldots V(t_1). \tag{4.4.7}$$

Following Dyson, we now rewrite the latter in the more convenient form

$$\begin{aligned} U(t, t_0) &= 1 + \sum_{n=1}^{\infty}\frac{(-i)^n}{n!}\int_{t_0}^{t} dt_n\int_{t_0}^{t} dt_{n-1}\ldots\int_{t_0}^{t} dt_1\, P\{V(t_1)\ldots V(t_n)\} \\ &= 1 + \sum_{n=1}^{\infty} U_n(t, t_0), \end{aligned} \tag{4.4.8}$$

where P is defined with respect to any operators $O_i(t_i)$ by

$$P\{O_1(t_1)\ldots O_n(t_n)\} = O_{\alpha_1}(t_{\alpha_1})\ldots O_{\alpha_n}(t_{\alpha_n}) \quad (t_{\alpha_1} > t_{\alpha_2} > \ldots > t_{\alpha_n}), \tag{4.4.9}$$

and is often called the chronological ordering operator. The idea of writing (4.4.7) as (4.4.8) is to introduce a common upper limit t in each integral, at the expense of complicating the integrand somewhat. The formal proof of the equivalence of (4.4.8) and (4.4.7) can be constructed by induction.

The U matrix introduced in equation (4.4.7) is central to the present theory. Let us, therefore, note the following basic properties:

$$U(t, t_0) = e^{iH_0 t} e^{-iH(t-t_0)} e^{-iH_0 t_0} \tag{4.4.10}$$

$$U(t_1, t_3)\, U(t_3, t_2) = U(t_1, t_2) \quad \text{(Composition)}, \tag{4.4.11}$$

$$U^{\dagger}(t_1, t_2) = U(t_2, t_1) \quad \text{(Hermiticity)}, \tag{4.4.12}$$

$$U(t_1, t_2)\, U^{\dagger}(t_1, t_2) = 1 \quad \text{(Unitarity)}. \tag{4.4.13}$$

Equation (4.4.10) follows directly by first checking that

$$\Phi(t) = e^{-iH(t-t_0)}\,\Phi(t_0)$$

satisfies the Schrödinger equation and is internally consistent at $t = t_0$. Then, (4.4.3) gives

$$\Psi(t) = e^{iH_0 t} e^{-iH(t-t_0)}\,\Phi(t_0) = e^{iH_0 t} e^{-iH(t-t_0)} e^{-iH_0 t_0}\,\Psi(t_0),$$

which may now be compared with (4.4.6). The exponential product does not simplify further because the operators in the exponents are in general non-commuting. The remaining three identities follow rather trivially from (4.4.10).

4.5. Adiabatic hypothesis

The basic idea in the time-dependent approach is to start off with a non-interacting system, 'switch on' the interactions and wait for an infinite time until the new interacting stationary state has been achieved. Mathematically, this means that if $\Psi(-\infty)$, say, is the non-interacting ground state in (4.4.6), then the state we want is

$$\Psi(0) = U(0, -\infty)\,\Psi(-\infty). \tag{4.5.1}$$

This is not quite a proper statement of the position. If we were to go ahead and use (4.5.1) to find $\Psi(0)$, we would eventually get into difficulties associated with divergent integrals and a more careful analysis of the situation becomes necessary. An acceptable way out has been indicated by Gell-Mann & Low (1951). This may be expressed physically by saying that when interactions are introduced, this should be done by adiabatic switching. If the 'turning on' is done sufficiently slowly, it may be plausibly assumed that the true perturbed ground state is generated.

The mathematical expression of this is that $\Psi(0)$, as given by (4.5.1), is an eigenfunction of H, provided we understand it to mean

$$\Psi(0) = \lim_{\alpha \to +0} U_\alpha(0, -\infty)\Psi(-\infty), \qquad (4.5.2)$$

where

$$U_\alpha(t, t_0) = 1 + \sum_{n=1}^{\infty} \frac{(-i)^n}{n!} \int_{t_0}^{t} dt_n \int_{t_0}^{t} dt_{n-1} \ldots \int_{t_0}^{t} dt_1 \, P\{V_\alpha(t_1) \ldots V_\alpha(t_n)\} \qquad (4.5.3)$$

and

$$V_\alpha(t) = V(t)\, e^{-\alpha|t|}. \qquad (4.5.4)$$

The proof of this assertion is straightforward in principle. One takes $H_0 - E_0$ and operates on (4.5.2). On suitably rearranging and taking the limit $\alpha \to 0$, all that remains is a constant (which is the energy shift ΔE, below) times (4.5.2). Some skilful manipulation is required in practice, however, and for details the reader is referred to Gell-Mann and Low's Appendix or the amplified discussion of Nozières (1964, pp. 176–8). Usually, we will not introduce α, but in cases of difficulty it should be remembered that (4.5.2) rather than (4.5.1) is the precise formulation.

It should be noted that for positive t, (4.5.4) represents adiabatic switching-off of the interactions, until at time $t = +\infty$, the non-interacting state is once more reached. There is no need, in the present chapter, to consider positive times, but the doubly infinite range is useful in the Green function theory of Chapter 10.

Using the above expression for the perturbed wave function, the energy shift ΔE is specified by a result essentially similar to the level shift formula of Chapter 1, (1.8.4) and (1.8.7), namely

$$\Delta E = \frac{\langle \Psi(-\infty) | V | \Psi(0) \rangle}{\langle \Psi(-\infty) | \Psi(0) \rangle} = \frac{\langle \Psi(-\infty) | V U(0, -\infty) | \Psi(-\infty) \rangle}{\langle \Psi(-\infty) | U(0, -\infty) | \Psi(-\infty) \rangle}. \qquad (4.5.5)$$

An alternative form of (4.5.5) which is very convenient is

$$\Delta E = \left[i \frac{\partial}{\partial t} \ln \langle \Psi(-\infty) | U(t, -\infty) | \Psi(-\infty) \rangle \right]_{t=0}. \qquad (4.5.6)$$

The equivalence of these two expressions follows since

$$\begin{aligned}\langle\Psi(-\infty)|U(t,-\infty)|\Psi(-\infty)\rangle\, i\frac{\partial}{\partial t}\ln\langle\Psi(-\infty)|U(t,-\infty)|\Psi(-\infty)\rangle\\ = i\frac{\partial}{\partial t}\langle\Psi(-\infty)|U(t,-\infty)|\Psi(-\infty)\rangle\\ = \langle\Psi(-\infty)|i\frac{\partial}{\partial t}U(t,-\infty)|\Psi(-\infty)\rangle\\ = \langle\Psi(-\infty)|i\frac{\partial}{\partial t}\Psi(t)\rangle,\end{aligned}$$

the last step following from (4.4.6). But employing (4.4.4), we may write immediately that

$$\langle\Psi(-\infty)|i\frac{\partial}{\partial t}\Psi(t)\rangle = \langle\Psi(-\infty)|V(t)|\Psi(t)\rangle,$$

which is equivalent to

$$\langle\Psi(-\infty)|V(t)\,U(t,-\infty)|\Psi(-\infty)\rangle,$$

when we again use (4.4.6). On putting $t = 0$, we obtain the desired result.

Our aim is to derive, using (4.5.6), a series for ΔE in powers of the coupling constant. Thus we see why it is so important to study U, and, in particular, its average with respect to $\Psi(-\infty)$.

4.6. Preliminary discussion of graphs for *U* matrix

At this point, we need to specify H_0 and V in (4.4.2). We adopt the choice discussed in Chapter 3, (recall convention preceding 3.9.6)

$$H_0 = \Sigma\epsilon_K a_K^\dagger a_K = -\frac{1}{2}\int\psi^\dagger(\mathbf{x})\nabla_\mathbf{r}^2\psi(\mathbf{x})\,d\mathbf{x} \qquad (4.6.1)$$

and

$$\begin{aligned}V &= \frac{1}{2}\sum_{KLMN}\langle KL|v|NM\rangle a_K^\dagger a_L^\dagger a_M a_N\\ &= \frac{1}{2}\iint d\mathbf{x}_1\,d\mathbf{x}_2\,\psi^\dagger(\mathbf{x}_1)\,\psi^\dagger(\mathbf{x}_2)\,v(\mathbf{r}_1\mathbf{r}_2)\,\psi(\mathbf{x}_2)\,\psi(\mathbf{x}_1). \qquad (4.6.2)\end{aligned}$$

From (4.4.8), the U matrix is obtained from products of V's (actually, $V(t)$'s in the interaction picture) and it is clear, using

(4.6.2) that a highly complex mathematical form results. We have to discuss this in full detail, and indeed, this analysis of the U matrix, together with the development of suitable tools by which it may be carried through, takes effectively the rest of this chapter. But it seems worthwhile here to outline the way progress can be made, in analogy with the simple graphical description of the terms in the Rayleigh–Schrödinger energy series given in section 4.3.3.

Roughly speaking, we wish to represent a term in (4.6.2) like

$$\langle KL|v|NM\rangle\, a_K^\dagger\, a_L^\dagger\, a_M\, a_N \tag{4.6.3}$$

by a suitable graph. Strictly, we have to convert this to the interaction representation (see section 4.7) and then the graphical method used in section 4.3.3 must be generalized to include the time dependence. The rules for such graphs are set out in detail in section 4.7.1, but it may be helpful to note that, without specifying further whether $KLMN$ represent particle or hole states, a set of diagrams could be drawn, one of which has the form of Fig. 4.1 (i) of section 4.3.3. We shall see later that, in fact, the operator in (4.6.3), when acting on $|\Psi(-\infty)\rangle$ as in (4.5.1), leads to the graph shown in

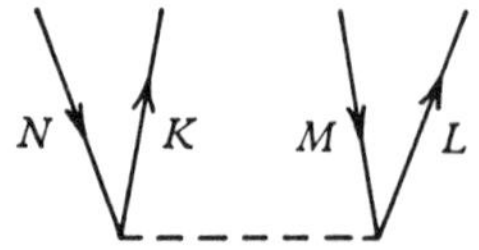

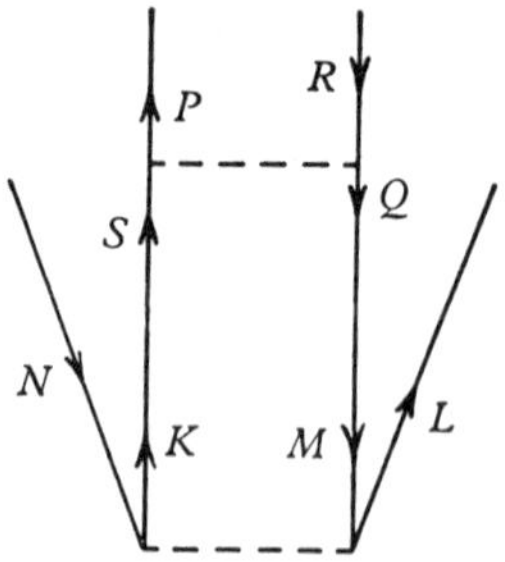

Fig. 4.1(vi) and (vii)

Fig. 4.1 (vi). Anticipating the convention of section 4.7.1 that time flows upwards, we can say that (4.6.3), or better, its time-dependent form, represents two holes N and M and two particles K and L created out of the vacuum state $|\Psi(-\infty)\rangle$. Clearly, from (4.4.8), products of terms like (4.6.3) will enter the analysis of U, and it is worthwhile to indicate one further simple example of the graphical representation of a second-order term. Thus, consider the graph shown in Fig. 4.1 (vii). This is clearly a second-order term, because of the two interaction lines. The lower part of this graph is the same as in Fig. 4.1 (vi), and hence represents again a term of the form (4.6.3). However, the upper part is arising from

$$\langle PQ|v|SR\rangle a_P^\dagger a_Q^\dagger a_R a_S, \tag{4.6.4}$$

the total graph therefore coming from the product of (4.6.3) and (4.6.4). These arguments may be generalized to arbitrary order, but the general term in U is extremely complex and we shall need further calculational aids, in addition to these Feynman graphs.

Thus, the programme for the rest of the chapter is as follows. The nth term in U is explicitly displayed in section 4.7 and this immediately focuses attention on the need for more powerful methods of dealing with the time ordering. A theorem supplying the desired result (Wick's theorem) is given in a limited form in section 4.10 and in more general terms in section 4.14. The perturbation terms in U can then be dealt with, the first-order theory being given in section 4.11, and the general theory in section 4.15. Finally the time integrations appearing in (4.4.8) must be carried out, and this is done in section 4.17. The end product is the linked cluster theorem, in the form anticipated from the Rayleigh–Schrödinger series in section 4.3.3.

4.7. Time-dependent particle-hole formalism

The essential ingredient of the U matrix to be studied is $V(t)$ as defined by (4.4.3). For this reason, we examine this interaction operator more closely. Using (4.6.2) we can write it in the forms

$$\begin{aligned} V(t_{12}) &= \frac{1}{2}\sum_{KLMN} \langle KL|v|NM\rangle a_K^\dagger(t_{12})\, a_L^\dagger(t_{12})\, a_M(t_{12})\, a_N(t_{12}) \\ &= \frac{1}{2}\iint d\mathbf{x}_1\, d\mathbf{x}_2\, \psi^\dagger(1)\, \psi^\dagger(2)\, v(\mathbf{r}_1\mathbf{r}_2)\, \psi(2)\, \psi(1) \quad (t_1 = t_2 = t_{12}), \end{aligned} \tag{4.7.1}$$

where $\psi(j) = e^{iH_0 t_j}\psi(\mathbf{x}_j)\, e^{-iH_0 t_j}, \quad a_K(t) = e^{iH_0 t} a_K\, e^{-iH_0 t}.$ (4.7.2)

The time-labelling convention used above should be noted. It is one we shall employ frequently in future.

The formal expressions shown in (4.7.2) are often very useful as they stand. But they can also be written in alternative forms. For example, we have

$$\psi(j) = \Sigma\phi_K(\mathbf{x}_j)\, a_K(t_j), \quad a_K(t) = e^{-i\epsilon_K t} a_K. \tag{4.7.3}$$

The latter follows if we consider the effect of either side on an arbitrary determinant of plane-wave states. The former is obtained by replacing $\psi(\mathbf{x}_j)$ in (4.7.2) using (3.7.1).

Now let us explicitly substitute (4.7.1) into (4.4.8). The nth order term in the U matrix becomes

$$\begin{aligned}
U_n(t, t_0) &= \frac{(-i)^n}{2^n n!}\int_{t_0}^{t} dt_{2n-1\,2n}\int_{t_0}^{t}\cdots\int_{t_0}^{t} dt_{12} \sum_{\substack{KLMN \\ XYZT}} \\
&\quad \times \langle KL|v|NM\rangle\langle PQ|v|SR\rangle\ldots\langle XY|v|TZ\rangle \\
&\quad \times P[a_K^\dagger(t_{12})\, a_L^\dagger(t_{12})\, a_M(t_{12})\, a_N(t_{12})] \\
&\quad \times [a_P^\dagger(t_{34})\, a_Q^\dagger(t_{34})\, a_R(t_{34})\, a_S(t_{34})] \\
&\quad \times \ldots[a_X^\dagger(t_{2n-1\,2n})\, a_Y^\dagger(t_{2n-1\,2n})\, a_Z(t_{2n-1\,2n})\, a_T(t_{2n-1\,2n})] \\
&= \frac{(-i)^n}{2^n n!}\int_{t_0}^{t} dt_{2n-1\,2n}\cdots\int_{t_0}^{t} dt_{12}\int d\mathbf{x}_1\, d\mathbf{x}_2\, v(\mathbf{r}_1\,\mathbf{r}_2)\int d\mathbf{x}_3\, d\mathbf{x}_4\ldots \\
&\quad \times \int d\mathbf{x}_{2n-1}\, d\mathbf{x}_{2n}\, v(\mathbf{r}_{2n-1}\,\mathbf{r}_{2n})\, P[\psi^\dagger(1)\,\psi^\dagger(2)\,\psi(2)\,\psi(1)] \\
&\quad \times [\psi^\dagger(3)\,\psi^\dagger(4)\,\psi(4)\,\psi(3)]\ldots \\
&\quad \times [\psi^\dagger(2n-1)\,\psi^\dagger(2n)\,\psi(2n)\,\psi(2n-1)],
\end{aligned} \tag{4.7.4}$$

where P is defined, once more through (4.4.9), with respect to products of four operators at equal times. For example, if $t_{34} > t_{12}$, then

$$\begin{aligned}
&P[\psi^\dagger(1)\,\psi^\dagger(2)\,\psi(2)\,\psi(1)]\,[\psi^\dagger(3)\,\psi^\dagger(4)\,\psi(4)\,\psi(3)] \\
&\qquad = [\psi^\dagger(3)\,\psi^\dagger(4)\,\psi(4)\,\psi(3)]\,[\psi^\dagger(1)\,\psi^\dagger(2)\,\psi(2)\,\psi(1)].
\end{aligned}$$

The expressions (4.7.4) emphasize the important part which the time-dependent ψ's and their components will play in the U-matrix analysis. Thus, we shall examine these quantities a little more before concluding this section. Our main concern is to introduce a time-dependent generalization of the particle-hole formalism discussed in section 3.9, which will later be used for studying (4.7.4).

Let us use (3.9.1) to replace a's by b's in ψ as given by (4.7.3). We have (cf. (3.9.6) and (3.9.7))

$$\psi(1) = \sum_{k<k_f} \phi_K(\mathbf{x}_1)\, a_K(t_1) + \sum_{k>k_f} \phi_K(\mathbf{x}_1)\, a_K(t_1) \tag{4.7.5}$$

$$= \sum_{k<k_f} \phi_K(\mathbf{x}_1)\, b_K^\dagger(t_1) + \sum_{k>k_f} \phi_K(\mathbf{x}_1)\, b_K(t_1) \tag{4.7.6}$$

$$= \psi_+(1) + \psi_-(1) \tag{4.7.7}$$

and

$$\psi^\dagger(1) = \sum_{k<k_f} \phi_K^*(\mathbf{x}_1)\, a_K^\dagger(t_1) + \sum_{k>k_f} \phi_K^*(\mathbf{x}_1)\, a_K^\dagger(t_1) \tag{4.7.8}$$

$$= \sum_{k<k_f} \phi_K^*(\mathbf{x}_1)\, b_K(t_1) + \sum_{k>k_f} \phi_K^*(\mathbf{x}_1)\, b_K^\dagger(t_1) \tag{4.7.9}$$

$$= \psi_-^\dagger(1) + \psi_+^\dagger(1). \tag{4.7.10}$$

We have defined $b_K(t)$ so that

$$b_K(t) = \begin{cases} a_K^\dagger(t) = e^{i\epsilon_K t} a_K^\dagger & (k < k_f) \\ a_K(t) = e^{-i\epsilon_K t} a_K & (k > k_f) \end{cases} \tag{4.7.11}$$

the latter step following from (4.7.3). These revert to (3.9.1), (3.9.6) and (3.9.7) on putting $t = 0$.

4.7.1. *Rules for Feynman graphs*

This is the appropriate stage at which to deal precisely with the rules for the Feynman diagrams we had introduced somewhat loosely by analogy with graphs from time-independent perturbation theory. While we had considered earlier graphs with at least four Fermion lines, these had arisen via the a's and it is worthwhile to stress at this point that likewise the ψ operators can be represented in a similar way. The composite graphs we used earlier were, of course, built up from the elementary structures shown in Fig. 4.2, the equivalent descriptions in terms of a's, b's or ψ's being noted.

We should emphasize that while such graphs are not absolutely essential to the theory, they provide great insight into the physics and are, in fact, part of the very language we shall use to express the results of this chapter.

The following points are important in understanding the graphical representation:

(i) As always (see (4.7.2) and (4.7.3)), the integers 1, 2, ..., etc., represent space-time points. In the diagrams, the time axis runs up the page from south to north. In this way, one may talk (as Feynman does) of 'hole lines' ($k < k_f$) running backwards in time and 'particle lines' ($k > k_f$) propagating forwards in time. The east–west axis has to represent three-dimensional co-ordinate space. The north-west, etc., directions of the lines shown are not significant —only the time directions are.

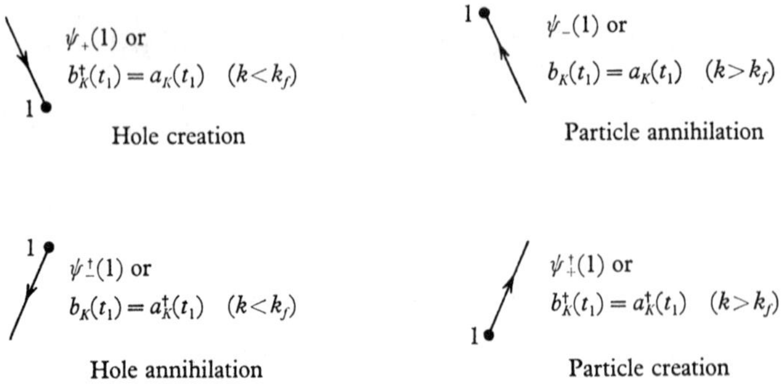

Fig. 4.2

(ii) Lines corresponding to $\psi^\dagger(1)$ and $a_K^\dagger(t_1)$ are directed away from the space-time point 1; $\psi(1)$ and $a_K(t_1)$ towards 1.

Creation lines (i.e. $\psi_+^\dagger(1)$, $\psi_+(1)$ and $b_K^\dagger(t_1)$ are located north of 1; destruction lines (i.e. $\psi_-^\dagger(1)$, $\psi_-(1)$ and $b_K(t_1)$) south.

If we bear this in mind, we can label Fig. 4.2 much more economically. For example, if we are using field operators, the first diagram of Fig. 4.2 is specified by  . If we are using components we need only add an index K alongside the arrow.

4.7.2. *Properties of time-dependent operators*

(i) Once more (see remarks following (3.9.7)) we have the convention

$$\psi_+^\dagger \equiv (\psi^\dagger)_+, \quad \psi_-^\dagger \equiv (\psi^\dagger)_-. \tag{4.7.12}$$

(ii) A number of other properties carry over from the time-independent formalism with obvious modification. These include

$$\left.\begin{aligned} a_K(t)\,a_{K'}^\dagger(t') + a_{K'}^\dagger(t')\,a_K(t) &= \delta_{KK'}\,e^{-i\epsilon_K(t-t')}, \\ b_K(t)\,b_{K'}^\dagger(t') + b_{K'}^\dagger(t')\,b_K(t) &= \delta_{KK'}\,e^{\mp i\epsilon_K(t-t')} \quad (k \gtrless k_f), \end{aligned}\right\} \tag{4.7.13}$$

all other anticommutators (of all a or all b types) vanishing. Also the vacuum property

$$\psi_-|\Psi(-\infty)\rangle = 0 = \psi_-^\dagger|\Psi(-\infty)\rangle; \quad b_K|\Psi(-\infty)\rangle = 0 \quad (4.7.14)$$

carries over for time-dependent ψ_-, $\psi_-^\dagger$ and b_K.

We have reached the stage at which the P product of the $V(t)$'s must now be tackled. In order to keep in mind our main objective, we reiterate that the energy shift ΔE, from (4.5.6), is given simply by the matrix element of U with respect to the non-interacting ground state. To calculate the nth-order term

$$\langle\Psi(-\infty)|U_n(t, -\infty)|\Psi(-\infty)\rangle,$$

we must be able to evaluate such P-products as are shown in (4.7.4) and this leads us to introduce the idea of a normal or N-product.

4.8. Normal products

Let us begin by considering the quantity $a_{K_1}(t_1)\,a_{K_2}(t_2)\,a_{K_3}^\dagger(t_3)$ as an example. Replacing the a's by b's from (4.7.11), and then using the anticommutation rules specified in section 4.7.2, the identity

$$a_{K_1}(t_1)\,a_{K_2}(t_2)\,a_{K_3}^\dagger(t_3)$$

$$= \begin{cases} b_{K_3}^\dagger(t_3)\,b_{K_1}(t_1)\,b_{K_2}(t_2) & \\ \quad -\delta_{K_1K_3}\,e^{-i\epsilon_{K_1}(t_1-t_3)}b_{K_2}(t_2) & \\ \quad +\delta_{K_2K_3}\,e^{-i\epsilon_{K_2}(t_2-t_3)}\,b_{K_1}(t_1) & (k_1, k_2, k_3 > k_f), \\ b_{K_1}(t_1)\,b_{K_2}(t_2)\,b_{K_3}(t_3) & (k_1, k_2 > k_f;\ k_3 < k_f), \\ b_{K_2}^\dagger(t_2)\,b_{K_3}^\dagger(t_3)\,b_{K_1}(t_1) & \\ \quad -\delta_{K_1K_3}\,e^{-i\epsilon_{K_1}(t_1-t_3)}b_{K_2}^\dagger(t_2) & (k_1, k_3 > k_f;\ k_2 < k_f), \\ -b_{K_2}^\dagger(t_2)\,b_{K_1}(t_1)\,b_{K_3}(t_3) & (k_1 > k_f;\ k_2, k_3 < k_f), \\ -b_{K_1}^\dagger(t_1)\,b_{K_3}^\dagger(t_3)\,b_{K_2}(t_2) & \\ \quad +\delta_{K_2K_3}\,e^{-i\epsilon_{K_2}(t_2-t_3)}b_{K_1}(t_1) & (k_1 < k_f;\ k_2, k_3 > k_f), \\ b_{K_1}^\dagger(t_1)\,b_{K_2}(t_2)\,b_{K_3}(t_3) & (k_1, k_3 < k_f;\ k_2 > k_f), \\ b_{K_1}^\dagger\,b_{K_2}^\dagger\,b_{K_3}^\dagger & (k_1, k_2 < k_f;\ k_3 > k_f), \\ b_{K_1}^\dagger\,b_{K_2}^\dagger\,b_{K_3} & (k_1, k_2,\ k_3 < k_f) \end{cases} \quad (4.8.1)$$

is readily verified. The properties of the various matrix elements of the original product now follow readily, if we remember that $b_K|\Psi(-\infty)\rangle = 0 = \langle\Psi(-\infty)|b_K^\dagger$ for all K. For example, since every term in (4.8.1) involves an odd number of operators, the average value with respect to $|\Psi(-\infty)\rangle$ is zero. However, the second term of line 1 of (4.8.1) shows that the matrix element with bra $\langle\Psi(-\infty)|b_{K_2}^\dagger(t_2)$ and ket $|\Psi(-\infty)\rangle$ has the value

$$-\delta_{K_1 K_3} \exp\{-i\epsilon_{K_1}(t_1 - t_3)\}$$

provided the states K_1, K_2 and K_3 all lie above the Fermi sea. In a similar fashion the other terms specify which matrix elements are non-vanishing and what their values are.

We now turn to consider matrix elements of U with respect to $|\Psi(-\infty)\rangle$. At high order, obviously it will be impracticable to proceed as in (4.8.1), and a more powerful technique must be used. This is the basic reason why we must deal with normal products, which we now define.

Suppose we have a simple product of time-dependent $a^\dagger$'s and a's. It may be described in particle-hole terms by replacing all these a's and $a^\dagger$'s by b's and $b^\dagger$'s according to (4.7.11). The normal product of this expression is then defined to be the product rearranged so that all the $b^\dagger$'s are on the left and the b's on the right, an appropriate signature being taken. The signature is $(-1)^p$, where p is the number of interchanges of factors required to change the old expression into the new.

As a simple example, if $k_1, k_3 > k_f$ and $k_2 < k_f$, we have

$$\begin{aligned} N(a_{K_1}(t_1)\, a_{K_2}(t_2)\, a_{K_3}^\dagger(t_3)) &= N(b_{K_1}(t_1)\, b_{K_2}^\dagger(t_2)\, b_{K_3}^\dagger(t_3)) \\ &= (-1)^2\, b_{K_2}^\dagger(t_2)\, b_{K_3}^\dagger(t_3)\, b_{K_1}(t_1) \\ &\equiv (-1)^2\, a_{K_2}(t_2)\, a_{K_3}^\dagger(t_3)\, a_{K_1}(t_1). \end{aligned} \tag{4.8.2}$$

The following additional points concerned with such normal products may be noted:

(i) The calculation of a normal product is done as though every pair of Fermion operators anticommutes. Among other things, this means $N(XY) = -N(YX)$. In (4.8.1), the normal product is obtained by dropping the δ function terms.

(ii) The definition is internally consistent, i.e. no matter in what order the separation is made, one always obtains the same final result. This depends on the fact that the $b^\dagger$'s anticommute among themselves, as do the b's. Thus, in the example (4.8.2), one might have conceivably written $(-1)^p\, b^\dagger_{K_3}(t_3)\, b^\dagger_{K_2}(t_2)\, b_{K_1}(t_1)$ and this is quite correct provided one bears in mind that, in this case, $p = 3$.

(iii) This formalism applies *a fortiori* to the time-independent a's, etc., since the time is an irrelevant variable in defining normal products.

(iv) Finally, and of prime importance, the expectation value of any normal product in the vacuum state is zero (see, however, a qualification to be made in section 4.9 below).

For the field operator (i.e. ψ) development of the U matrix, it is convenient to define normal products involving linear combinations of a's and $a^\dagger$'s. Thus, let $A, B, C, \ldots$ denote simple products of a's and $a^\dagger$'s and $\lambda, \mu, \ldots$ scalars (c-numbers). Then the normal product of a linear combination is defined by

$$N(\lambda A+\mu B) = \lambda N(A)+\mu N(B). \tag{4.8.3}$$

In this way, the normal product of more complicated expressions can be calculated. Thus, for example,

$$N\{A(B+C)\} = N(AB+AC) = N(AB)+N(AC) \tag{4.8.4}$$

and, more generally,

$$N\{\sum_i A_i \sum_j B_j\} = \sum_{ij} N(A_i B_j). \tag{4.8.5}$$

Using (4.8.5) and point (i) following (4.8.2), we have

$$N(XY) = -N(YX), \tag{4.8.6}$$

where now X and Y represent linear combinations of simple operators.

The definition (4.8.3) applies, in particular, to the operators $\psi, \psi^\dagger, \psi_\pm, \psi^\dagger_\pm$. Such expressions as

$$N\{\psi^\dagger(1)\,\psi(2)\,\psi(3)\,\psi^\dagger(4)\,\psi^\dagger(5)\ldots\}$$

can be evaluated in complete detail in terms of the a's, or split up in terms of the $\psi_\pm$'s. Thus, for example,

$$\begin{aligned} N\{\psi^\dagger(1)\,\psi^\dagger(2)\,\psi(2)\,\psi(1)\} &= N\{(\psi^\dagger_+(1)+\psi^\dagger_-(1))\,\psi^\dagger(2)\,\psi(2)\,\psi(1)\} \\ &= N\{\psi^\dagger_+(1)\,\psi^\dagger(2)\,\psi(2)\,\psi(1)\} \\ &\quad +N\{\psi^\dagger_-(1)\,\psi^\dagger(2)\,\psi(2)\,\psi(1)\}, \end{aligned} \tag{4.8.7}$$

and so on.

When performing calculations with the ψ's and $\psi^\dagger$'s, it is useful to note that when dealing with any single product of subscripted ψ's and $\psi^\dagger$'s, provided we treat the $\psi^\dagger_+$'s and ψ_+'s as though they were $b^\dagger$'s, and the $\psi^\dagger_-$'s and ψ_-'s as though they were b's, our original definition of normal product and the four subsequently enumerated points continue to apply. For example, we have

$$\begin{aligned} N\{\psi^\dagger(2)\,\psi(1)\} &= N\{\psi^\dagger_+(2)\,\psi_+(1)+\psi^\dagger_+(2)\,\psi_-(1) \\ &\qquad +\psi^\dagger_-(2)\,\psi_+(1)+\psi^\dagger_-(2)\,\psi_-(1)\} \\ &= \psi^\dagger_+(2)\,\psi_+(1)+\psi^\dagger_+(2)\,\psi_-(1)-\psi_+(1)\,\psi^\dagger_-(2) \\ &\qquad +\psi^\dagger_-(2)\,\psi_-(1). \end{aligned} \tag{4.8.8}$$

4.9. Pairings

We have seen above that the expectation value of a normal product with respect to $|\Psi(-\infty)\rangle$ vanishes. This implies that it is convenient to define a quantity $\underbracket{A\,B}$ through

$$AB = N(AB)+\underbracket{A\,B}, \tag{4.9.1}$$

where A, B are any operators for which a normal product has been defined. Then, taking matrix elements of (4.9.1) with respect to the vacuum state $|\Psi(-\infty)\rangle$, we find

$$\langle AB\rangle = \langle \underbracket{A\,B}\rangle. \tag{4.9.2}$$

$\underbracket{A\,B}$ is called the pairing of A and B (see, for example, Brout & Carruthers, 1963). In all applications of interest below, $\underbracket{A\,B}$ turns out to be a c-number and thus the brackets on the right-hand side of (4.9.2) can be removed. As an example, it is readily verified using (4.7.13) and the definition of normal products that

$$\left.\begin{aligned} \underbracket{a_K(t)\,a^\dagger_{K'}}(t') &= \delta_{KK'}\,e^{-i\epsilon_K(t-t')} \quad (k,k'>k_f), \\ \underbracket{a^\dagger_K(t)\,a_{K'}}(t') &= \delta_{KK'}\,e^{\,i\epsilon_K(t-t')} \quad (k,k'<k_f), \end{aligned}\right\} \tag{4.9.3}$$

all other pairings among the a's and $a^\dagger$'s being zero. In this example, the pairings $\underbracket{A\,B}$ are c-numbers and thus, by (4.9.2) are equal to $\langle AB\rangle$.

An obvious generalization of (4.9.1) occurs when one of the

operators is formed as a linear combination of others. Then we may write, for example,

$$\underline{A(B+C)} = A(B+C) - N\{A(B+C)\},$$

and using (4.8.4) this becomes

$$\begin{aligned} AB+AC-N(AB)-N(AC) &= \{AB-N(AB)\}+\{AC-N(AC)\} \\ &= \underline{AB}+\underline{AC}. \end{aligned} \tag{4.9.4}$$

The generalization of this result is clearly

$$\underline{(\Sigma A_i)(\Sigma B_j)} = \Sigma \underline{A_i B_j}. \tag{4.9.5}$$

4.9.1. *Examples*

We now consider two examples of these results.

(i) There are sixteen possible pairings which can be constructed from $\psi_\pm$, $\psi^\dagger_\pm$, of which only the two discussed below are non-zero. Introducing the notation $\phi_K(1) = \phi_K(\mathbf{x}_1)\, e^{-i\epsilon_K t}$, we have

$$\begin{aligned} \underline{\psi_-(1)\,\psi^\dagger_+(2)} &= \sum_{k_1, k_2 > k_f} \phi_{K_1}(\mathbf{x}_1)\,\phi^*_{K_2}(\mathbf{x}_2)\,\underline{a_{K_1}(t_1)\,a^\dagger_{K_2}(t_2)} \\ &= \sum_{k > k_f} \phi_K(1)\,\phi^*_K(2) \end{aligned} \tag{4.9.6}$$

and

$$\begin{aligned} \underline{\psi^\dagger_-(1)\,\psi_+(2)} &= \sum_{k_1, k_2 < k_f} \phi^*_{K_1}(\mathbf{x}_1)\,\phi_{K_2}(\mathbf{x}_2)\,\underline{a^\dagger_{K_1}(t_1)\,a_{K_2}(t_2)} \\ &= \sum_{k < k_f} \phi^*_K(1)\,\phi_K(2). \end{aligned} \tag{4.9.7}$$

(ii) There are four possible pairings which can be constructed from ψ and $\psi^\dagger$ of which only two are non-zero. We have, by example (i) above,

$$\begin{aligned} \underline{\psi(1)\,\psi^\dagger(2)} &= \underline{\psi_+(1)\,\psi^\dagger_+(2)} + \underline{\psi_+(1)\,\psi^\dagger_-(2)} + \underline{\psi_-(1)\,\psi^\dagger_+(2)} + \underline{\psi_-(1)\,\psi^\dagger_-(2)} \\ &= 0+0+\sum_{k>k_f} \phi_K(1)\,\phi^*_K(2) + 0. \end{aligned}$$

Similarly, the pairing $\underline{\psi^\dagger(1)\,\psi(2)}$ is given by

$$\underline{\psi^\dagger(1)\,\psi(2)} = \sum_{k < k_f} \phi^*_K(1)\,\phi_K(2). \tag{4.9.8}$$

Once again the pairings $\underline{AB}$ are c-numbers, and thus, by (4.9.2) are equal to $\langle AB \rangle$.

4.9.2. *Normal products which include pairings*

We define these as follows:

$$N(P\underbracket{Q\underbracket{R S}T\ldots X Y}Z) = \pm\underbracket{QS}\,\underbracket{RY}\ldots N(PT\ldots XZ). \qquad (4.9.9)$$

The pairings are removed from the normal product and the appropriate signature taken—plus for an even number of operators which are traversed in removing the pairing from the normal product, and minus for an odd number. (The sign is unaffected by the order in which pairings are removed.) For fully paired normal products, one takes $N(\underbracket{AB}) = \underbracket{AB}$. This is the qualification referred to in point (iv) following (4.8.2), since the expectation value of this normal product does not vanish.

As an example we have

$$N(\underbracket{ABC}\,\underbracket{DE}) = -N(\underbracket{AC}B\underbracket{DE}) = -\underbracket{AC}N(B\underbracket{DE}) = +\underbracket{AC}N(\underbracket{DBE})$$
$$= -\underbracket{AC}N(\underbracket{DE}B) = -\underbracket{AC}\,\underbracket{DE}N(B).$$

We now have the following result, which is useful for the development in terms of the field operators, although not necessary for that in terms of components:

$$N(\sum_k A_1^k \sum_l \underbracket{A_2^l \sum_m A_3^m \ldots \sum_p A_r^p}) = \sum_{klm\ldots p} N(A_1^k \underbracket{A_2^l A_3^m \ldots A_r^p}). \qquad (4.9.10)$$

The proof of (4.9.10) goes as follows. By the definition (4.9.9), the left-hand side of (4.9.10) can be written

$$\pm(\sum_l \underbracket{A_2^l \sum_p A_r^p})(\underbracket{\ })\ldots N(\sum_k A_1^k \sum_m A_3^m \ldots).$$

But (4.9.5) enables us to deal with pairings of sums, and (4.8.5) with normal products of sums. Hence the above expression becomes

$$\pm\sum_{lp} \underbracket{A_2^l A_r^p} \sum_{\underbracket{\ }} \sum_{\underbracket{\ }} \cdots \sum_{km\ldots} N(A_1^k A_3^m \ldots)$$
$$= \pm\sum \underbracket{A_2^l A_r^p}\,\underbracket{\ }\ldots N(A_1^k A_3^m \ldots),$$

the sum to be taken over all superscripts. We may now re-insert all pairings into the N-product using (4.9.9) to obtain the right-hand side of (4.9.10).

As an example of this formalism we consider the normal product of $\underbrace{\psi^\dagger(1)\,\psi^\dagger(2)\,\psi(2)}\,\psi(1)$. This may be written

$$
\begin{aligned}
&N(\underbrace{\psi^\dagger(1)\,\psi^\dagger(2)\,\psi(2)}\,\psi(1))\\
&\quad = N(\underbrace{\psi^\dagger(1)\,[\psi^\dagger_+(2)+\psi^\dagger_-(2)]\,\psi(2)}\,[\psi_+(1)+\psi_-(1)])\\
&\quad = N(\underbrace{\psi^\dagger(1)\,\psi^\dagger_+(2)\,\psi(2)}\,\psi_+(1)) + N(\underbrace{\psi^\dagger(1)\,\psi^\dagger_+(2)\,\psi(2)}\,\psi_-(1))\\
&\qquad + N(\underbrace{\psi^\dagger(1)\,\psi^\dagger_-(2)\,\psi(2)}\,\psi_+(1))\\
&\qquad + N(\underbrace{\psi^\dagger(1)\,\psi^\dagger_-(2)\,\psi(2)}\,\psi_-(1)).
\end{aligned}
\tag{4.9.11}
$$

This example is typical of a kind of analysis we find useful later.

4.10. Wick's theorem for ordinary products

Very often one wishes to write a given expression in normal product form.‡ The purpose of Wick's theorem is to do this as systematically and concisely as possible. We know the solution already for any simple product of a pair of a or ψ operators. Use of (4.9.1) solves the problem. In fact (4.9.1) is the content of Wick's theorem for this simple case. The generalization of this result to the case of the product of many operators must now be effected.

The theorem states that if $A_1, A_2, \ldots, A_n$ represent a set of a's b's, $a^\dagger$'s and $b^\dagger$'s, or linear combinations of such operators, then

$A_1 A_2 \ldots A_n = N(A_1 A_2 \ldots A_n)$ (All terms with no pairings—there is only one)

$+N(\underbrace{A_1 A_2} \ldots A_n)$

$+N(\underbrace{A_1 A_2 A_3} \ldots A_n) + \ldots$ (All terms with one pairing—there are ${}^nC_2$)

$+N(\underbrace{A_1 A_2}\, \underbrace{\ldots}\, A_n) + \ldots$ (All terms with two pairings—there are $\frac{1}{2}{}^nC_2{}^{n-2}C_2$)

$+\ldots$

$+$ all fully paired terms. (These occur if and only if n is even) (4.10.1)

‡ I.e. as a linear combination of operator products such that in each one, when expressed in terms of the b's and $b^\dagger$'s, the $b^\dagger$'s stand to the left and the b's to the right. An example is (4.8.1).

Since the expectation value of any N-product with respect to the non-interacting ground state is zero, (4.10.1) says in particular that this expectation value of a product is given by the sum of all fully paired terms. The right-hand side of (4.10.1) represents $A_1, A_2, \ldots, A_n$ in normal product form. The proof of this important theorem is given in Appendix (4A.1).

At this point we see how our illustrative example (4.8.1) may be written more succinctly. Using (4.10.1), we have

$$
\begin{aligned}
& a_{K_1}(t_1)\, a_{K_2}(t_2)\, a^{\dagger}_{K_3}(t_3) \\
&\quad = N\{a_{K_1}(t_1)\, a_{K_2}(t_2)\, a^{\dagger}_{K_3}(t_3)\} + N\{\underbrace{a_{K_1}(t_1)\, a_{K_2}}(t_2)\, a^{\dagger}_{K_3}(t_3)\} \\
&\qquad + N\{\underbrace{a_{K_1}(t_1)\, a_{K_2}(t_2)\, a^{\dagger}_{K_3}}(t_3)\} + N\{a_{K_1}(t_1)\, \underbrace{a_{K_2}(t_2)\, a^{\dagger}_{K_3}}(t_3)\} \\
&\quad = N\{a_{K_1}(t_1)\, a_{K_2}(t_2)\, a^{\dagger}_{K_3}(t_3)\} + 0 \\
&\qquad - \underbrace{a_{K_1}(t_1)\, a^{\dagger}_{K_3}}(t_3)\, a_{K_2}(t_2) + \underbrace{a_{K_2}(t_2)\, a^{\dagger}_{K_3}}(t_3)\, a_{K_1}(t_1). \qquad (4.10.2)
\end{aligned}
$$

The pairings are removed from the N-products using (4.9.9) and the second term is zero because of the remark following (4.9.3). For any specified set of K's, of course, we can use (4.7.11) and (4.9.3) to obtain the appropriate line of (4.8.1), but usually the formalism of (4.10.2) is employed to summarize the information in such unwieldy expressions as (4.8.1).

4.11. First-order perturbation theory

The object of this section is to analyse, with the help of Wick's theorem, the first non-trivial term of the U-matrix expansion (4.4.8). By (4.7.4), on putting $n = 1$, we have

$$
\begin{aligned}
U_1(t, t_0) &= \frac{-i}{2} \int_{t_0}^{t} dt_{12} \sum_{KLMN} \langle KL|v|NM\rangle\, a^{\dagger}_K(t_{12})\, a^{\dagger}_L(t_{12})\, a_M(t_{12})\, a_N(t_{12}) \\
&= \frac{-i}{2} \int_{t_0}^{t} dt_{12} \iint d\mathbf{x}_1\, d\mathbf{x}_2\, v(\mathbf{r}_1 \mathbf{r}_2)\, \psi^{\dagger}(1)\, \psi^{\dagger}(2)\, \psi(2)\, \psi(1).
\end{aligned}
\qquad (4.11.1)
$$

Our immediate task is to write the operators in the integrands in normal product form.

Using Wick's theorem, we have (suppressing the t_{12} variable for convenience),

$$\begin{aligned}
a_K^\dagger\, a_L^\dagger\, a_M\, a_N = {} & N(a_K^\dagger\, a_L^\dagger\, a_M\, a_N) \\
& + N(\underbrace{a_K^\dagger\, a_L^\dagger\, a_M\, a_N}) + N(a_K^\dagger\, \underbrace{a_L^\dagger\, a_M}\, a_N) \\
& + N(\underbrace{a_K^\dagger\, a_L^\dagger\, a_M}\, a_N) + N(a_K^\dagger\, \underbrace{a_L^\dagger\, a_M\, a_N}) \\
& + N(\underbrace{a_K^\dagger\, \underbrace{a_L^\dagger\, a_M}\, a_N}) \\
& + N(\underbrace{a_K^\dagger\, \rlap{\overbrace{}}a_L^\dagger\, a_M}\, a_N),
\end{aligned} \tag{4.11.2}$$

and, in a precisely similar manner,

$$\begin{aligned}
& \psi^\dagger(1)\, \psi^\dagger(2)\, \psi(2)\, \psi(1) = N(\psi^\dagger(1)\, \psi^\dagger(2)\, \psi(2)\, \psi(1)) \\
& \quad + N(\underbrace{\psi^\dagger(1)\, \psi^\dagger(2)\, \psi(2)\, \psi(1)}) + N(\psi^\dagger(1)\, \underbrace{\psi^\dagger(2)\, \psi(2)}\, \psi(1)) \\
& \quad + N(\underbrace{\psi^\dagger(1)\, \psi^\dagger(2)\, \psi(2)}\, \psi(1)) + N(\psi^\dagger(1)\, \underbrace{\psi^\dagger(2)\, \psi(2)\, \psi(1)}) \\
& \quad + N(\underbrace{\psi^\dagger(1)\, \underbrace{\psi^\dagger(2)\, \psi(2)}\, \psi(1)}) + N(\underbrace{\psi^\dagger(1)\, \rlap{\overbrace{}}\psi^\dagger(2)\, \psi(2)}\, \psi(1)).
\end{aligned} \tag{4.11.3}$$

It should be noted that such terms as $N(a_K^\dagger\, a_L^\dagger\, \underbrace{a_M\, a_N})$ which would occur in (4.11.2) by straightforward application of the theorem, are, in fact identically zero, since (see concluding part of the sentence containing (4.9.3)) $\underbrace{a^\dagger a^\dagger} = 0 = \underbrace{a\, a}$. A similar statement holds for (4.11.3) (see example (ii) of section 4.9.1).

We now proceed to a graphical representation of the various terms of (4.11.2) and (4.11.3) and we will do this in full detail because the general nth-order graphs are systematically built up in terms of first-order ones. The routine, in (4.11.3), is to split up each unpaired ψ and $\psi^\dagger$ into its $\psi_\pm$, $\psi_\pm^\dagger$ parts in the manner of (4.9.11). The analogue in (4.11.2) is to distinguish whether the K, L, M and N lie above or below the Fermi surface.

4.11.1. *Graphical representation*

We consider the diagrams appropriate to the terms on the right-hand side of (4.11.2) and (4.11.3).

(i) *No pairings.* Here, we consider the first terms in (4.11.2) and (4.11.3), when 16 cases arise. This is so, in $N(a_K^\dagger a_L^\dagger a_M a_N)$, because each index may correspond to a state either inside or outside the Fermi sphere. One can display this quite explicitly in the case of (4.11.3), for

$$N(\psi^\dagger(1)\psi^\dagger(2)\psi(2)\psi(1)) = N\{(\psi_+^\dagger(1)+\psi_-^\dagger(1))(\psi_+^\dagger(2)+\psi_-^\dagger(2)) \times(\psi_+(2)+\psi_-(2))(\psi_+(1)+\psi_-(1))\}. \quad (4.11.4)$$

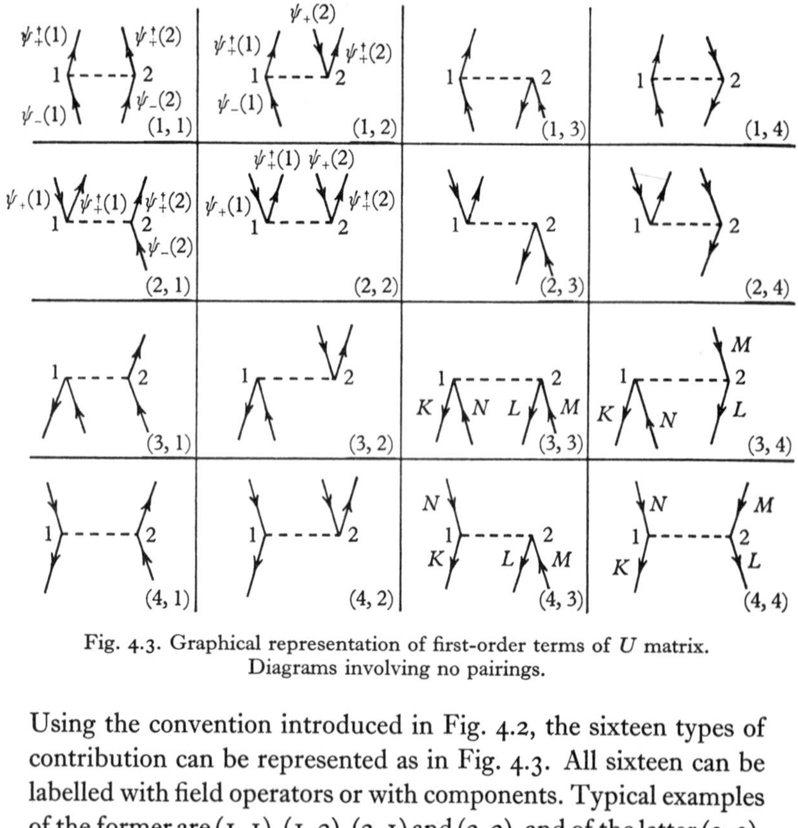

Fig. 4.3. Graphical representation of first-order terms of U matrix. Diagrams involving no pairings.

Using the convention introduced in Fig. 4.2, the sixteen types of contribution can be represented as in Fig. 4.3. All sixteen can be labelled with field operators or with components. Typical examples of the former are (1, 1), (1, 2), (2, 1) and (2, 2), and of the latter (3, 3), (3, 4), (4, 3) and (4, 4). The following comments should be made:

(*a*) The dotted line represents the interaction in (4.11.1) and is often referred to as a vertex. It is horizontal because $t_1 = t_2$; the interaction is said to be instantaneous. One talks of the scattering of particles due to the interaction. Thus, say, the (3, 4)th diagram shows two holes (K and L) and one particle (N) at time $t < t_{12}$. At time t_{12}, the hole K and the particle N annihilate each other, while the hole L becomes the hole M for subsequent times $t > t_{12}$.

(*b*) Certain characteristic features of these graphs follow from the special form of the product of four operators involved in (4.11.3) and from the conventions of Fig. 4.2 (see also the comments pertaining thereto). Since each $\psi^\dagger$ is directed away from the vertex and each ψ towards it, it follows from the form of (4.11.4) that one line enters and one leaves each end of the vertex. In the component formalism, K and N are associated with the outgoing and ingoing lines respectively at 1, while L and M play the same role at 2. Whether these indices represent particles or holes depends on the time directions of the respective lines. Particles are directed forward in time, holes backwards.

(ii) *One pairing*. For convenience, we divide these into two groups, the direct and exchange terms.

The direct part is given by the second and third terms of (4.11.2) and (4.11.3); to be specific, let us consider the second term of (4.11.3). (Here is an example of the convenience of the ψ's for general theoretical purposes, referred to at the end of section 4.1.)

Thus, the term under discussion is

$$N(\underbrace{\psi^\dagger(1)\,\psi^\dagger(2)\,\psi(2)\,\psi(1)}_{}) = \underbrace{\psi^\dagger(1)\,\psi(1)}_{}\,N(\psi^\dagger(2)\,\psi(2)), \tag{4.11.5}$$

where

$$\begin{aligned} N(\psi^\dagger(2)\,\psi(2)) &= N\{(\psi^\dagger_+(2)+\psi^\dagger_-(2))\,(\psi_+(2)+\psi_-(2))\} \\ &= N(\psi^\dagger_+(2)\,\psi_+(2)+\psi^\dagger_+(2)\,\psi_-(2)+\psi^\dagger_-(2)\,\psi_+(2) \\ &\qquad +\psi^\dagger_-(2)\,\psi_-(2)) \\ &= N(\psi^\dagger_+(2)\,\psi_+(2))+\underline{N(\psi^\dagger_+(2)\,\psi_-(2))} \\ &\qquad +\underline{N(\psi^\dagger_-(2)\,\psi_+(2))}+N(\psi^\dagger_-(2)\,\psi_-(2)) \\ &= \psi^\dagger_+(2)\,\psi_+(2)+\underline{\psi^\dagger_+(2)\,\psi_-(2)}-\underline{\psi_+(2)\,\psi^\dagger_-(2)} \\ &\qquad +\psi^\dagger_-(2)\,\psi_-(2). \end{aligned} \tag{4.11.6}$$

There are two non-zero contributions only, arising from the terms underlined. The graphs specifying these are shown in Fig. 4.4 (i). The two remaining terms of (4.11.6), which would be represented as in Fig. 4.4 (iii), give zero contributions to U_1 and can therefore be ignored, as we shall see shortly.

If we had studied the second term of (4.11.2), we would have had to consider explicitly whether the various momenta were above or below the Fermi surface. The basic development, however, must parallel the above, of course, and we arrive at terms represented by parts (ii) and (iv) of Fig. 4.4. Because of (4.9.3) and the remarks following it, for a non-zero contribution to U_1, the momentum-spin indices are necessarily equal and lie inside the Fermi sphere.

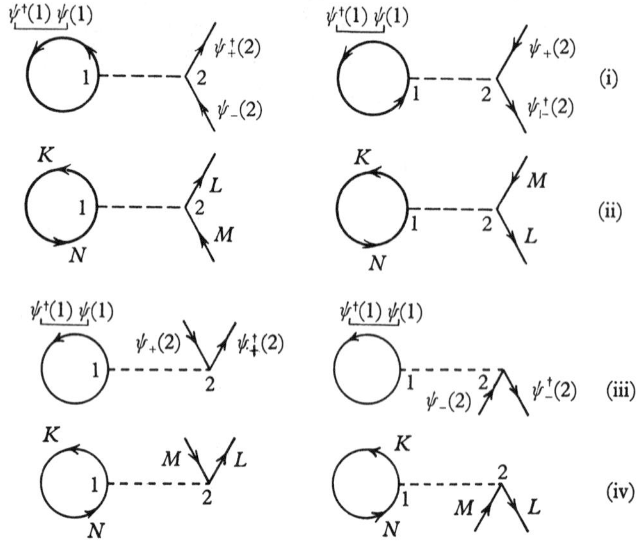

Fig. 4.4

To see that the contributions from Fig. 4.4 (iii) and (iv) vanish, all we need do is note that these graphs give rise, in (4.11.1), only to matrix elements $\langle KL|v|KM\rangle$ with $l > k_f$ and $m < k_f$ or vice versa. As (1.3.6) tells us, all such terms are zero for momentum conserving potentials.

Now a similar development is possible for the third terms of (4.11.2) and (4.11.3). The only difference from the above is that the roles of 1 and 2 are interchanged. We can give the following interpretation to the diagrams:

(*a*) Fig. 4.4 (i) or (ii) represents the scattering of a particle (hole) against a passive particle. The use of a circle to describe the pairing is meant to be suggestive of the physical description of the preceding sentence. It is also in conformity with the representation of (unequal time) contractions to be introduced later. The direction indicated by the arrow has no significance, as Figs. 4.5 and 4.6 indicate.

Fig. 4.5

The vanishing of the contribution due to Fig. 4.4 (iii) and (iv) is interpreted in the physical way indicated above to conform to the fact that it is impossible to create (or destroy) a single-particle hole pair and simultaneously conserve momentum.

(*b*) At each end of every vertex, there is still a line entering and a line leaving. The labelling of Fig. 4.4 is as for the no pairing case.

We turn now to the so-called 'exchange' contribution. This arises from the fourth and fifth terms of (4.11.2) and (4.11.3). The fourth term of (4.11.3) is

$$N(\underbracket{\psi^\dagger(1)\,\psi^\dagger(2)\,\psi(2)}\,\psi(1)) = -\underbracket{\psi^\dagger(1)\,\psi(2)}\,N(\psi^\dagger(2)\,\psi(1)), \quad (4.11.7)$$

where (cf. (4.11.6))

$$N(\psi^\dagger(2)\,\psi(1)) = \psi^\dagger_+(2)\,\psi_+(1) + \underline{\psi^\dagger_+(2)\,\psi_-(1)} - \underline{\psi_+(1)\,\psi^\dagger_-(2)} + \psi^\dagger_-(2)\,\psi_-(1). \quad (4.11.8)$$

There are two non-zero contributions only, arising from the terms underlined. The graphs specifying these are shown in Fig. 4.6 (i).

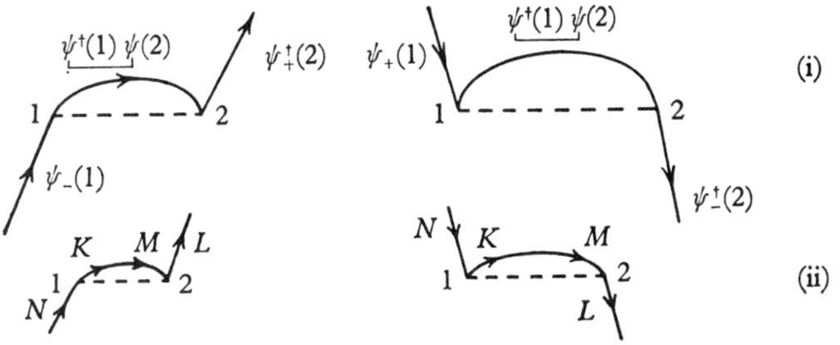

Fig. 4.6

The two remaining terms of (4.11.3), which would be represented as in Fig. 4.7, give zero contributions to U_1 and can, therefore, be ignored. The argument runs just as in the direct case where it was shown that the graphs of Fig. 4.4 did not contribute. The graphs shown in Fig. 4.7 correspond to cases where momentum is not conserved.

The analysis of the fourth term of (4.11.2) parallels the above. The graphs contributing to U_1 are as shown in Fig. 4.6 (ii), those corresponding to Fig. 4.7 once more giving zero contributions for momentum conserving potentials.

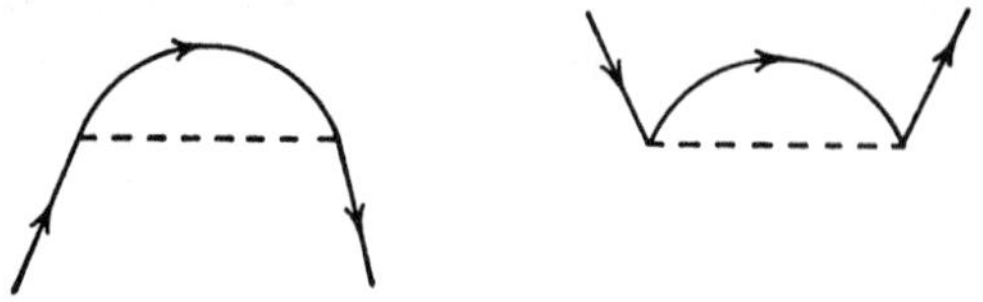

Fig. 4.7

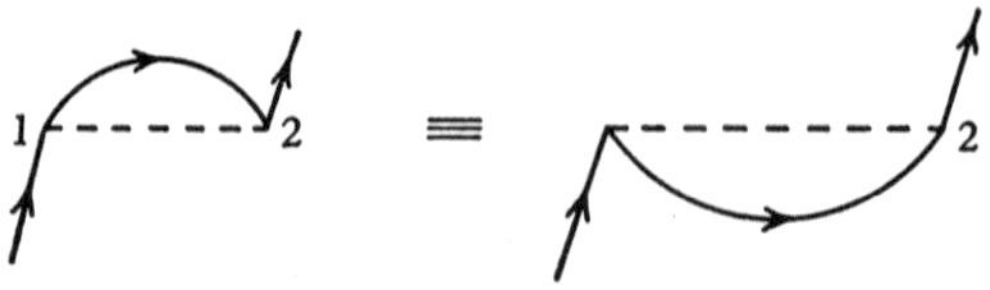

Fig. 4.8

The fifth terms of (4.11.2) and (4.11.3) may be analysed as above, the only difference being that the roles of 1 and 2 are interchanged. It should be remarked that Fig. 4.6 represents the destruction of a particle (hole) at 1 (2), followed instantaneously by the creation of a particle (hole) at 2 (1). The bend in the pairing line has no special significance as Fig. 4.8 indicates. Thus the time-dependence of both is more properly represented by a horizontal pairing line overlapping the dotted vertex line. For reasons of clarity, the latter is not used.

(iii) *Fully paired terms.* These are the remaining sixth and seventh terms of (4.11.2) and (4.11.3). Once more we can conveniently classify the terms as direct and exchange contributions.

The direct contribution is given by the sixth terms of (4.11.2) and (4.11.3) and the corresponding graphs are shown in Fig. 4.9.

The exchange contributions are the final terms in (4.11.2) and (4.11.3), the corresponding graphs being those shown in Fig. 4.10.

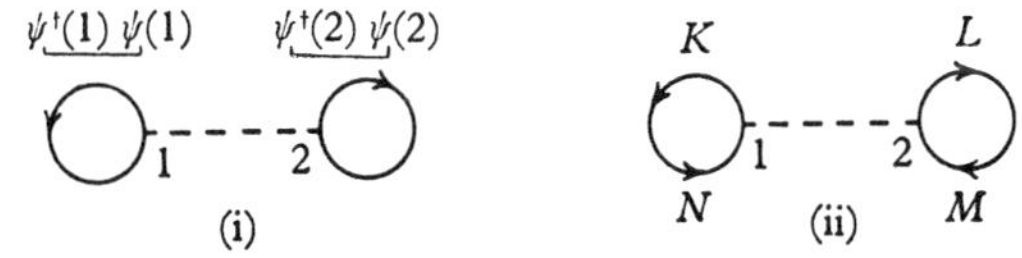

Fig. 4.9

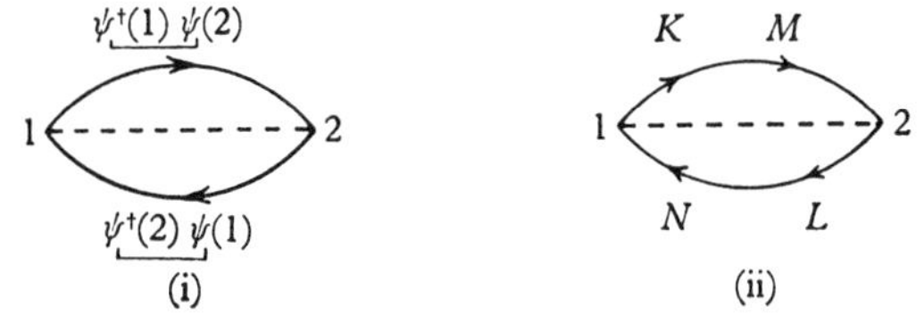

Fig. 4.10

We should note that Fig. 4.9 corresponds to the static interaction and Fig. 4.10 to the exchange of two passive particles.

To proceed beyond this first-order theory, which was based on Wick's theorem for ordinary products, it will prove highly convenient to generalize this theorem somewhat, to deal with time ordering. This is not essential, however, and, indeed, Goldstone in his original work circumvented the need for introducing time-ordered products. But the idea has proved of wide applicability in many-body theory, and in addition leads to the further important concept of contractions (see also the connection with Green functions in Chapter 10) introduced in section 4.13.

4.12. Time-ordered products

The central difficulty in dealing with higher-order terms in the U matrix lies in the reduction of the operator part of the integrand to normal product form. In view of our first-order analysis, we expect to do this with the help of Wick's theorem and this leads us to consider the form of this theorem for P-products. The simplest expression of Wick's theorem for ordinary products is that given by (4.9.1); let us then ask what the analogue of this result is for the P-product of two such operators. We have, using (4.8.6) and (4.9.1),

$$A(t_1)\,B(t_2) = N(A(t_1)\,B(t_2)) + \underbrace{A(t_1)\,B(t_2)}_{} \tag{4.12.1}$$

and

$$\begin{aligned} B(t_2)\,A(t_1) &= N(B(t_2)\,A(t_1)) + \underbrace{B(t_2)\,A(t_1)}_{} \\ &= -N(A(t_1)\,B(t_2)) + \underbrace{B(t_2)\,A(t_1)}_{}. \end{aligned} \tag{4.12.2}$$

Thus, the desired analogue of (4.9.1) is

$$P(A(t_1)\,B(t_2)) = \pm N(A(t_1)\,B(t_2)) + \begin{cases} \underbrace{A(t_1)\,B(t_2)} \\ \underbrace{B(t_2)\,A(t_1)} \end{cases} \quad (t_1 \gtrless t_2). \tag{4.12.3}$$

We saw that for simple products, (4.9.1) generalized to (4.10.1). It is clear, therefore, that if we wish to generalize (4.12.3) similarly, the alternating sign of the N-product is going to be a nuisance. This leads us, below, to define a T-product operator which can replace P in (4.7.4), eliminates the alternating sign feature, and leads to a convenient Wick's theorem. The full details are given below; suffice it to say at the moment that in the special case (4.12.3), it turns out that $P = \pm T\ (t_1 \gtrless t_2)$ and so

$$T(A(t_1)\,B(t_2)) = N(A(t_1)\,B(t_2)) + \overbrace{A(t_1)\,B(t_2)}, \tag{4.12.4}$$

where

$$\overbrace{A(t_1)\,B(t_2)} = \left\{\begin{matrix} \underbrace{A(t_1)\,B(t_2)} & (t_1 > t_2) \\ -\underbrace{B(t_2)\,A(t_1)} & (t_2 > t_1) \end{matrix}\right\}. \tag{4.12.5}$$

With this general motivation, we proceed to the more formal exposition. The development resembles closely that of the theory of normal products and pairings, so somewhat less time will be spent on the finer details as these should be apparent from our previous work.

Let $A_1(t_1)$, $A_2(t_2)$, ... denote a product of time-dependent a's, $a^\dagger$'s, b's and $b^\dagger$'s. Then we define the time-ordered product of this by

$$T(A_1(t_1)\,A_2(t_2)\ldots) = (-1)^P\,A_{\alpha_1}(t_{\alpha_1})\,A_{\alpha_2}(t_{\alpha_2})\ldots \quad (t_{\alpha_1} > t_{\alpha_2} > \ldots), \tag{4.12.6}$$

where P is the number of interchanges required to convert

$$A_1(t_1)\,A_2(t_2)\ldots \quad \text{into} \quad A_{\alpha_1}(t_{\alpha_1})\,A_{\alpha_2}(t_{\alpha_2})\ldots.$$

More generally, one can extend the definition to linear combinations of products as follows. Let A, B, C, ... denote simple products of the above kind and λ, μ, ν, ... c-numbers. Then we write

$$T(\lambda A + \mu B) = \lambda T(A) + \mu T(B), \tag{4.12.7}$$

and so, for example,

$$T\{A(B+C)\} = T(AB+AC) = T(AB) + T(AC). \tag{4.12.8}$$

4.13. Contractions

Let $A(t_A)$, $B(t_B)$ be operators, the normal and time-ordered products of which have been defined. Then, $\overbracket{AB}$, called the contraction of A and B, is defined by the equation

$$T(AB) = N(AB) + \overbracket{AB}. \qquad (4.13.1)$$

Once more, remembering that the expectation value of a normal product is zero, we have

$$\langle \overbracket{AB} \rangle = \langle T(AB) \rangle. \qquad (4.13.2)$$

All our contractions will turn out to be c-numbers and so the brackets on the left can be removed as before. It should be noted that there is a simple relationship between contraction and pairing. For, if $t_A > t_B$, then $T(AB) = AB$ and so

$$\overbracket{AB} = \underbracket{AB} \quad (t_A > t_B). \qquad (4.13.3)$$

If $t_B > t_A$, then $T(AB) = -BA = -N(BA) - \underbracket{BA} = N(AB) - \underbracket{BA}$, and so from the definition (4.13.1), we have

$$\overbracket{AB} = -\underbracket{BA} \quad (t_B > t_A). \qquad (4.13.4)$$

These equations are, of course, just (4.12.5), but now in a more general context. It will be noted also that (4.13.3) and (4.13.4) imply

$$\overbracket{AB} = -\overbracket{BA} \quad \text{(either time ordering).} \qquad (4.13.5)$$

A law analogous to (4.9.4) also holds for contractions, for

$$\begin{aligned} \overbracket{A(B+C)} &= T\{A(B+C)\} - N\{A(B+C)\} \\ &= T(AB) + T(AC) - N(AB) - N(AC) \\ &= \{T(AB) - N(AB)\} + \{T(AC) - N(AC)\} \\ &= \overbracket{AB} + \overbracket{AC}. \end{aligned} \qquad (4.13.6)$$

More generally,
$$\overbracket{(\sum_i A_i)(\sum_j B_j)} = \sum_{ij} \overbracket{A_i B_j}. \qquad (4.13.7)$$

Normal products, which include contractions, are defined by a relation similar to (4.9.9), namely

$$N(P\,Q\,R\,S\,T \ldots X\,Y\,Z) = \pm \overline{Q\,S}\,\overline{R\,Y} \ldots N(PT \ldots XZ), \quad (4.13.8)$$

the sign being decided as before. Analogous to (4.9.10) we have the theorem:

$$N(\sum_k A_1^k \sum_l A_2^l \sum_m A_3^m \ldots \sum_p A_r^p) = \sum_{k,l,m,\ldots,p} N(A_1^k A_2^l A_3^m \ldots A_r^p). \quad (4.13.9)$$

Using this result, we can make decompositions similar to those indicated in (4.9.11).

4.13.1. *Examples of contractions of operators*

Let us now illustrate the above formulae with some examples:

(i) Equations (4.13.3) and (4.13.4) show us how to write contractions as pairings. Hence, we can use the example following (4.9.2) to write down all contractions among the a's and $a^\dagger$'s. We have

$$\overline{a_K(t)\, a_{K'}^\dagger}(t') = \begin{cases} \delta_{KK'} e^{-i\epsilon_K(t-t')} & \text{if } t > t' \text{ and } k, k' > k_f \\ -\delta_{KK'} e^{-i\epsilon_K(t-t')} & \text{if } t' > t \text{ and } k, k' < k_f \end{cases}. \quad (4.13.10)$$

All other contractions of the form $\overline{a\, a^\dagger}$ are zero. The $\overline{a^\dagger\, a}$ contractions are immediately deduced from the $\overline{a\, a^\dagger}$ results using (4.13.5). In addition (see remarks following (4.9.3)),

$$\overline{a_K^\dagger(t)\, a_{K'}^\dagger}(t') = 0 = \overline{a_K(t)\, a_{K'}}(t'). \quad (4.13.11)$$

(ii) Use of (4.13.3), (4.13.4), and the results of section 4.9.1 give

$$\overline{\psi(1)\,\psi^\dagger(2)} = \begin{cases} \sum\limits_{k>k_f} \phi_K(1)\, \phi_K^*(2) & (t_1 > t_2) \\ -\sum\limits_{k<k_f} \phi_K(1)\, \phi_K^*(2) & (t_2 > t_1) \end{cases} \quad (4.13.12)$$

$$\overline{\psi^\dagger(1)\,\psi(2)} = \begin{cases} \sum\limits_{k<k_f} \phi_K^*(1)\, \phi_K(2) & (t_1 > t_2) \\ -\sum\limits_{k>k_f} \phi_K^*(1)\, \phi_K(2) & (t_2 > t_1) \end{cases} \quad (4.13.13)$$

and

$$\overline{\psi^\dagger(1)\,\psi^\dagger(2)} = 0 = \overline{\psi(1)\,\psi(2)}. \quad (4.13.14)$$

It should be noted that for any pair of field operators, A and B, their contraction is a c-number, and so, from (4.13.2),

$$\overbrace{AB} = \langle T(AB)\rangle. \qquad (4.13.15)$$

For reasons which will be discussed in Chapter 10, our contractions may be referred to as unperturbed Green functions or unperturbed propagators. Specifically, in view of (4.13.10), we write

$$\overbrace{a_K(t)\, a_K^\dagger}(t') = \langle T(a_K(t)\, a_{K'}^\dagger(t'))\rangle = -iG_0(K, t-t')\,\delta_{KK'}. \qquad (4.13.16)$$

Since G_0 will be such an important function, let us write it out in full. We have

$$G_0(K, t-t') = \begin{cases} \pm i \exp\{-i\epsilon_K(t-t')\} & (+\text{ if } t > t' \text{ and } k > k_f; \\ & \;-\text{ if } t < t' \text{ and } k < k_f) \\ 0 & (\text{otherwise}). \end{cases} \qquad (4.13.17)$$

4.14. Wick's theorem for time-ordered products

We are now in a position to state Wick's theorem for time-ordered products. The theorem states that if $A_1, A_2, \ldots, A_n$ represent a set of time-dependent operators for which a T-product is defined, then

$$T(A_1 A_2 \ldots A_n) = N(A_1 A_2 \ldots A_n) \qquad \text{(All terms with no contractions—there is only one)}$$

$$+N(\overbrace{A_1 A_2} \ldots A_n) + N(\overbrace{A_1 A_2 A_3} \ldots A_n) + \ldots \qquad \text{(All terms with one contraction—there are } {}^nC_2)$$

$$+\ldots \qquad \text{(All terms with two contractions—there are } \tfrac{1}{2}{}^nC_2\,{}^{n-2}C_2)$$

$$+\ldots$$

$$+\text{all fully contracted terms.} \qquad \text{(These occur if and only if } n \text{ is even).} \qquad (4.14.1)$$

Since the expectation value of any N-product with respect to the non-interacting ground state is zero, (4.14.1) implies, in particular, that this expectation value of a T-product is given by the sum of all fully contracted terms. The right-hand side of (4.14.1) is said to represent $T(A_1 A_2 \ldots A_n)$ in normal product form.

Equation (4.13.1) corresponds to (4.14.1) for two operators only. The proof of the theorem is given in Appendix 4A.1; it is a quite straightforward consequence of Wick's theorem for ordinary products as given by (4.10.1). With the aid of (4.14.1), we may now consider second- and higher-order perturbation theory.

4.15. Second- and higher-order perturbation theory

At this point, we are going to restrict our discussion in two ways. First, we will consider only vacuum-vacuum terms of the U matrix. The study of the non-vacuum terms constitute, in a sense, the theory of Green functions, but in this chapter at least, as (4.5.6) indicates, we are interested only in averages with respect to the unperturbed ground state.

Secondly, we will confine ourselves to the momentum space operators only. This is to avoid a great deal of repetition. Since the two developments are so similar, it is expected that the reader will be able to fill in the field operator development for himself. The following, then, based on the first of the two expressions (4.7.4) for U_n, is rather similar in spirit to the Goldstone formalism (see, however, the final paragraph of section 4.11).

Let us begin this discussion by considering the operator part

$$P[a_K^\dagger(t_{12})\, a_L^\dagger(t_{12})\, a_M(t_{12})\, a_N(t_{12})]\ldots[a_X^\dagger(t_{2n-1\,2n}) \times a_Y^\dagger(t_{2n-1\,2n})\, a_Z(t_{2n-1\,2n})\, a_T(t_{2n-1\,2n})] \quad (4.15.1)$$

of (4.7.4). We have stated that we expect to use Wick's theorem (4.14.1) for rewriting this in normal product form. But, it will be seen there are two immediate obstacles in the way of doing this. One is that Wick's theorem, as stated, is for T-products and not P-products. Secondly, the results of section 4.14 were derived on the tacit assumption that equal time operators do not arise. Such sets of operators do occur in (4.15.1), however (in groups of four) and, for example,

$$\overbrace{a_K^\dagger(t_{12})\, a_M}(t_{12})$$

is not yet defined.

In order to define the T-product and contraction of any pair of components in (4.15.1), we use the artifice of letting the equal-time operators appearing in the products have slightly different time arguments, the limit being taken at the end of the calculation. The standard way of doing this is to assume that, within any product of four a operators at equal times (referred to subsequently as a 4-product), the times go from later to earlier as one reads from left to right. Thus, for example, $a_K^\dagger(t_{12})\,a_L^\dagger(t_{12})\,a_M(t_{12})\,a_N(t_{12})$ is replaced by $a_K^\dagger(t_{12}+\alpha)\,a_L^\dagger(t_{12}+\beta)\,a_M(t_{12}+\gamma)\,a_N(t_{12})$, where $\alpha > \beta > \gamma > 0$ are infinitesimals. With this convention, for operations on these 4-products, we have

$$P = T. \tag{4.15.2}$$

This is so, since, within any 4-product, the operators are already ordered by our introduction of the infinitesimals. Thus, we need only permute in blocks of four. One such permutation gives a signature of $+1$, as usual, in the P case, while in the T case, it gives $(-1)^{16} = +1$.

On replacing P by T in (4.15.1), we may now use Wick's theorem. Once (4.14.1) has been applied, one can simplify a little further, reducing the number of limiting parameters α, β, γ within each 4-product from three to one. This may be seen by examining the right-hand side of (4.14.1) and observing that N-products are unaffected by time conventions. Thus, this leaves us only with contractions to worry about. But we know that contractions between two $a^\dagger$'s (or two a's) are zero. It follows, therefore, that it is never necessary to distinguish between the time orderings for two $a^\dagger$'s (or two a's). Thus we arrive at the rule that we can suppose (4.13.10) applies to equal time contractions and retain $a(t)$ in its original form but suppose $a^\dagger(t)$ is replaced by $a^\dagger(t+0)$. Thus, for example, we have

$$a_K^\dagger(t_{12})\,a_L^\dagger(t_{12})\,a_M(t_{12})\,a_N(t_{12}) = a_K^\dagger(t_{12}+0)\,a_L^\dagger(t_{12}+0)\,a_M(t_{12})\,a_N(t_{12}), \tag{4.15.3}$$

while (see (4.13.10) and subsequent remarks)

$$\overbracket{a_K^\dagger(t)\,a_{K'}}(t) \equiv \overbracket{a_K^\dagger(t+0)\,a_{K'}}(t) = \begin{Bmatrix} \delta_{KK'}\,e^{i\epsilon_K 0} & (k < k_f) \\ 0 & (k > k_f) \end{Bmatrix}. \tag{4.15.4}$$

Contractions of the kind $\overbracket{a_{K'}(t)\, a_K^\dagger}(t)$ do not naturally arise in (4.7.4), but it is sometimes convenient to express every contraction in terms of G_0's using (4.13.5) and (4.13.16). In that case, we have

$$-\overbracket{a_K^\dagger(t)\, a_{K'}}(t) = \overbracket{a_{K'}(t)\, a_K^\dagger}(t) = -iG_0(K, -0)\,\delta_{KK'}. \qquad (4.15.5)$$

It should be pointed out that because of our time convention and (4.13.3), we have

$$\overbracket{a_K^\dagger(t)\, a_{K'}}(t) = \underbracket{a_K^\dagger(t)\, a_{K'}}(t). \qquad (4.15.6)$$

Thus, in section 4.11, all pairings can be replaced by contractions. This is normally done for the sake of uniformity of notation, the concept of contraction being needed, of course, for the higher-order terms.

With these preliminaries, we may now begin a systematic study of the vacuum-vacuum matrix elements of (4.15.1), and their contributions to $\langle U(t, t_0)\rangle$.

The first-order terms have already been explored at the end of section 4.11. Let us briefly recapitulate. From equation (4.11.1), we had

$$\langle U_1(t, -\infty)\rangle = -\frac{i}{2}\int_{-\infty}^{t} dt_{12} \sum_{KLMN} \langle KL|v|NM\rangle$$
$$\times \langle a_K^\dagger(t_{12})\, a_L^\dagger(t_{12})\, a_M(t_{12})\, a_N(t_{12})\rangle, \qquad (4.15.7)$$

where (exercising our option of writing contractions instead of pairings)

$$\langle a_K^\dagger(t_{12})\, a_L^\dagger(t_{12})\, a_M(t_{12})\, a_N(t_{12})\rangle$$
$$= \overbracket{a_K^\dagger(t_{12})\, \overbracket{a_L^\dagger(t_{12})\, a_M}(t_{12})\, a_N}(t_{12}) + \overbracket{a_K^\dagger(t_{12})\, a_L^\dagger(t_{12})\, a_M}(t_{12})\, a_N(t_{12}). \qquad (4.15.8)$$

An alternative way of writing the terms on the right is

$$\overbracket{a_K^\dagger(t_{12})\, \overbracket{a_L^\dagger(t_{12})\, a_M}(t_{12})\, a_N}(t_{12}) = (-i)^2\,\delta_{KN}\,\delta_{LM}\,G_0(K, -0)\,G_0(L, -0), \qquad (4.15.9)$$

$$\overbracket{a_K^\dagger(t_{12})\, a_L^\dagger(t_{12})\, a_M}(t_{12})\, a_N(t_{12})$$
$$= -(-i)^2\,\delta_{KM}\,\delta_{LN}\,G_0(K, -0)\,G_0(L, -0), \qquad (4.15.10)$$

these expressions being obtained by unravelling the left-hand sides according to (4.13.8) (see also (4.9.9)), and then applying (4.15.5).

The graphs representing the respective contributions to (4.15.7) are as drawn in Figs. 4.9 (ii) and 4.10 (ii). Let us now give a rule for drawing such graphs and, conversely, when presented with a graph, of recovering the contribution to (4.15.7).

4.15.1. *Rules for first-order graphs*

(i) *To draw a graph from a fully contracted contribution.* Draw a vertex with end-points 1 and 2; associate K and N with 1 and L and M with 2, as shown in Fig. 4.11. Now look at the individual contractions $\overline{a^\dagger_\alpha a_\beta} = -\overline{a_\beta a^\dagger_\alpha}$ which compose the term. (Necessarily $\alpha = K$ or L, $\beta = M$ or N.) For each such contraction, draw a directed line from (the vertex end associated with) α to (that associated with) β. Transfer the labels α and β to the lines thus drawn. The result is to draw Figs. 4.9 (ii) and 4.10 (ii).

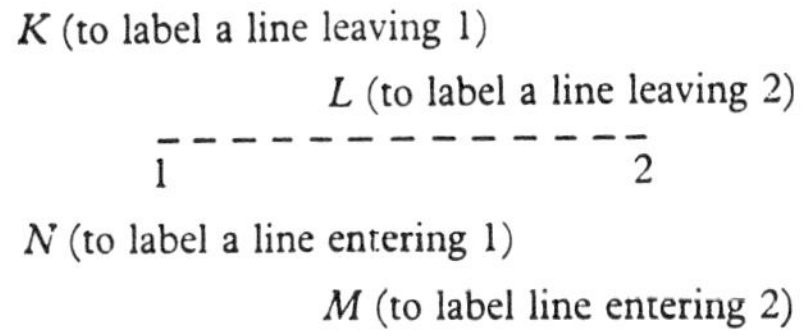

Fig. 4.11

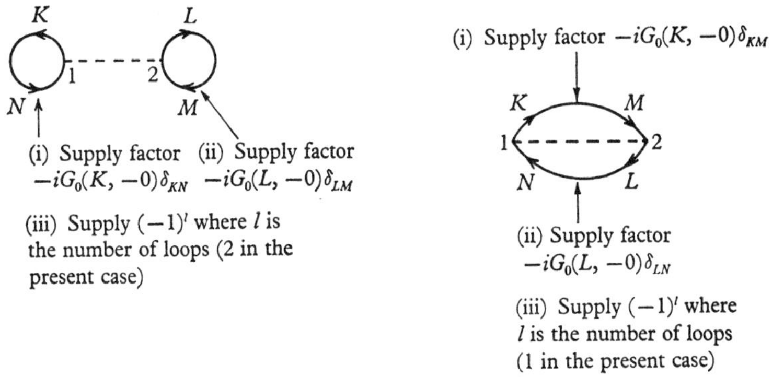

Fig. 4.12

(ii) *To write down a contribution from a graph.* One can, of course, simply write down the product $a^\dagger_K a^\dagger_L a_M a_N$ and supply the appropriate contractions as indicated by the diagram. There is another method, however, which is more convenient, though this will be more apparent when we reach higher orders. Though a little artificial in the present first-order theory, we will, nevertheless use it. The method is to write down the right-hand sides of (4.15.9) and (4.15.10) by using the prescriptions shown in Fig. 4.12.

4.15.2. *Second-order graphs*

Here, we have

$$\langle U_2(t, -\infty)\rangle = \frac{(-i)^2}{2^2 2!}\int_{-\infty}^{t} dt_{34}\int_{-\infty}^{t} dt_{12} \sum_{\substack{KLMN\\PQRS}} \times \langle KL|v|NM\rangle\langle PQ|v|SR\rangle$$
$$\times\langle Ta_K^\dagger(t_{12})\, a_L^\dagger(t_{12})\, a_M(t_{12})\, a_N(t_{12})\, a_P^\dagger(t_{34})\, a_Q^\dagger(t_{34})\, a_R(t_{34})\, a_S(t_{34})\rangle. \tag{4.15.11}$$

There are $4! = 24$ terms in the expansion of the operator average and it is clearly inconvenient to write them all down. Instead, we will examine three typical members in detail, show how to draw their graphs and then present all 24 graphs, from which, with his accumulated experience, the reader can, if he so wishes, write down the analytical equivalents.

(1) $a_K^\dagger(t_{12})\, a_L^\dagger(t_{12})\, a_M(t_{12})\, a_N(t_{12})\, a_P^\dagger(t_{34})\, a_Q^\dagger(t_{34})\, a_R(t_{34})\, a_S(t_{34})$

$$= -(-i)^4\,\delta_{KM}\,\delta_{NL}\,\delta_{PS}\,\delta_{QR}\,G_0(K, -0)\,G_0(L, -0) \times G_0(P, -0)\,G_0(Q, -0). \tag{4.15.12}$$

(2) $a_K^\dagger(t_{12})\, a_L^\dagger(t_{12})\, a_M(t_{12})\, a_N(t_{12})\, a_P^\dagger(t_{34})\, a_Q^\dagger(t_{34})\, a_R(t_{34})\, a_S(t_{34})$

$$=(-i)^4\,\delta_{KM}\,\delta_{LS}\,\delta_{NP}\,\delta_{QR}\,G_0(K, -0)\,G_0(L, t_{34}-t_{12}) \times G_0(N, t_{12}-t_{34})\,G_0(Q, -0). \tag{4.15.13}$$

(3) $a_K^\dagger(t_{12})\, a_L^\dagger(t_{12})\, a_M(t_{12})\, a_N(t_{12})\, a_P^\dagger(t_{34})\, a_Q^\dagger(t_{34})\, a_R(t_{34})\, a_S(t_{34})$

$$=(-i)^4\,\delta_{KS}\,\delta_{LR}\,\delta_{MQ}\,\delta_{NP}\,G_0(K, t_{34}-t_{12})\,G_0(L, t_{34}-t_{12}) \times G_0(M, t_{12}-t_{34})\,G_0(N, t_{12}-t_{34}). \tag{4.15.14}$$

The corresponding graphs are shown in Fig. 4.13. Once, more, there are rules for drawing such graphs and for the recovery of the corresponding analytical expression for a given graph.

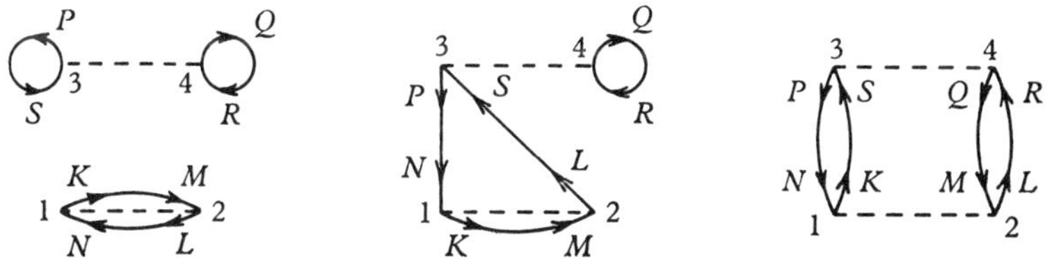

Fig. 4.13. Three typical examples from 24 second-order graphs.

We have now reached the point where we can summarize the rules for second-order graphs:

(i) *To draw a graph from a fully contracted contribution.* Draw two vertices with end-points 1, 2 and 3, 4; associate K, N with 1, L, M with 2, P, S with 3 and Q, R with 4 (see Fig. 4.14). Now look at the individual contractions $\overline{a^{\dagger}_{\alpha} a_{\beta}} = -\overline{a_{\beta} a^{\dagger}_{\alpha}}$. For each, draw a directed line from α to β and label. (Thus we see that the diagrams of Fig. 4.13 can be constructed from (4.15.12)–(4.15.14) using the above prescription.)

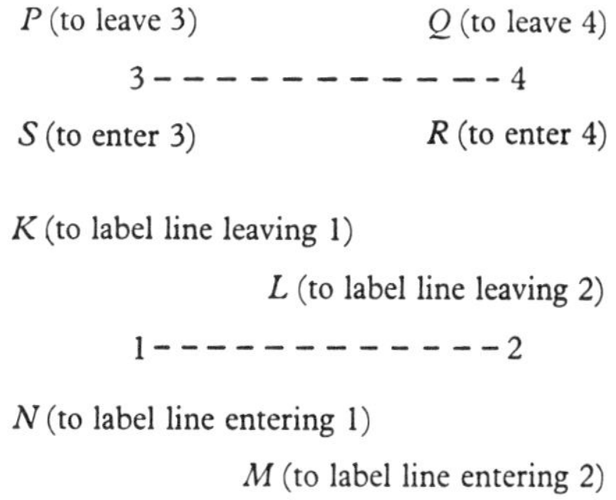

Fig. 4.14

As t_{12} and t_{34} vary, we imagine the vertices (1, 2), (3, 4) to move relative to each other. If $t_{34} > t_{12}$, the relative positions of the vertices are as in Fig. 4.14; if $t_{12} > t_{34}$, the positions are reversed. But, always, the topological structure of the graph remains the same because of our above means of construction. It should be noted that, because of this construction and remembering (4.13.16) and (4.13.17), when a diagram is constructed with due regard to the time order, contraction lines directed in the positive time direction contribute only for $k > k_f$ and those directed backwards in time, only for $k < k_f$. In short, one might say that particles travel forward in time, and holes backwards. Usually, for convenience we take the diagram with the time sequence shown in Fig. 4.14 to be the representative of both possible time orders.

(ii) *To write down a contribution from a graph.* Again, as in first order, one can simply work backwards from rule (i). But once more we will emphasize the alternative way of writing down a product of G_0's and δ-functions with appropriate signature. For every contraction line from A at time $t_{\lambda\mu}$ to B at time $t_{\nu\sigma}$ (not necessarily distinct from $t_{\lambda\mu}$), supply a factor $-iG_0(A, t_{\nu\sigma}-t_{\lambda\mu})\,\delta_{AB}$. Supply also a factor $(-1)^l$, where l is the number of loops in the diagram.

Using the above rules, one can easily write down the right-hand sides of (4.15.12), (4.15.13) and (4.15.14) from the appropriate diagrams in Fig. 4.13. It should be noted that there are three loops in (i) and two in (ii) and (iii). Thus (4.15.12) carries an extra minus sign.

The totality of second-order diagrams are as shown in Fig. 4.15. (Labelling has been left implicit in most of the graphs.) As stated in section 4.11, the higher-order terms are systematically built up from first-order ones. We see this illustrated now in second order. For example, clearly the (6, 3)th term of Fig. 4.15 is obtained by tying together, appropriately, the ends of two graphs of the kind (2, 2), (3, 3), shown in Fig. 4.3. The precise way in which this occurs is easily traced if we note that one way of writing out the contributions to (4.15.11) is to substitute for each of the equal-time 4-products using (4.11.2). Thus, denoting the right-hand side of (4.11.2) in an obvious way by

$$\sum_{i=1}^{7} n_i(K, L, M, N;\, t_{12}),$$

we have

$$\begin{aligned}
&\langle Ta_K^\dagger(t_{12})\,a_L^\dagger(t_{12})\,a_M(t_{12})\,a_N(t_{12})\,a_P^\dagger(t_{34})\,a_Q^\dagger(t_{34})\,a_R(t_{34})\,a_S(t_{34})\rangle \\
&\quad = \left\langle T\sum_{i=1}^{7} n_i(K, L, M, N;\, t_{12}) \sum_{j=1}^{7} n_j(P, Q, R, S;\, t_{34})\right\rangle \\
&\quad = \sum_{i,j=1}^{7} \langle Tn_i(K, L, M, N;\, t_{12})\,n_j(P, Q, R, \mathrm{S};\, (t_{34})\rangle. \qquad (4.15.15)
\end{aligned}$$

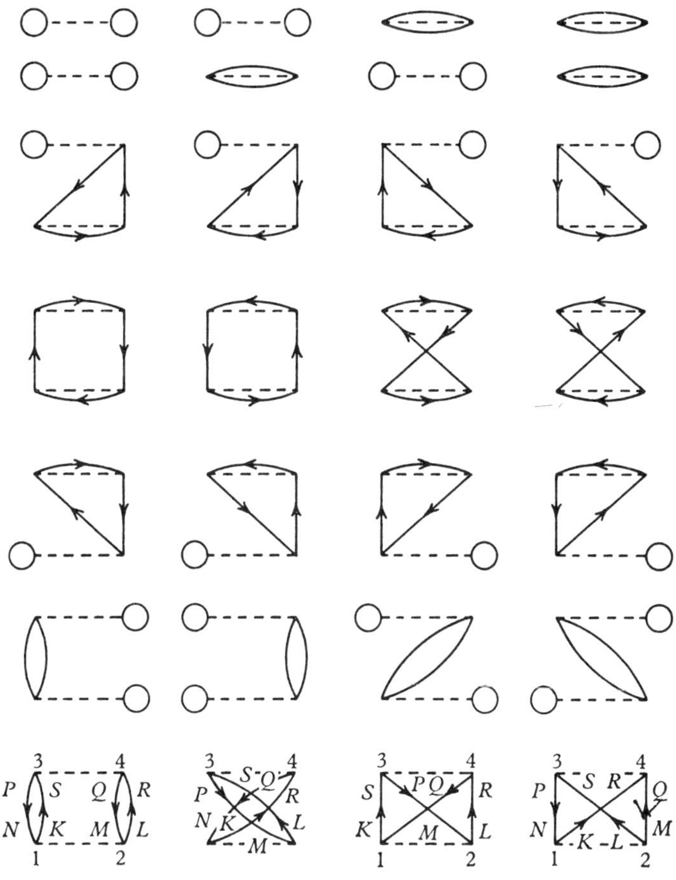

Fig. 4.15. All second-order diagrams.

The n_i and n_j already contain the pairings and the exhibited T accounts for the unequal time contractions. The latter, of course, are the common strings which tie together first-order diagrams.

It follows as a corollary of the above analysis (though one can also see it directly) that any composite diagram constructed from one or more vanishing first-order diagrams (see Figs. 4.4(iii) and (iv) and 4.7) is itself vanishing. Thus, we find that only the first and the sixth rows of Fig. 4.15 give non-zero contributions to (4.15.11).

Finally, to complete our second-order discussion, let us show how to write down the contribution to the (6, 3)th graph of Fig. 4.15

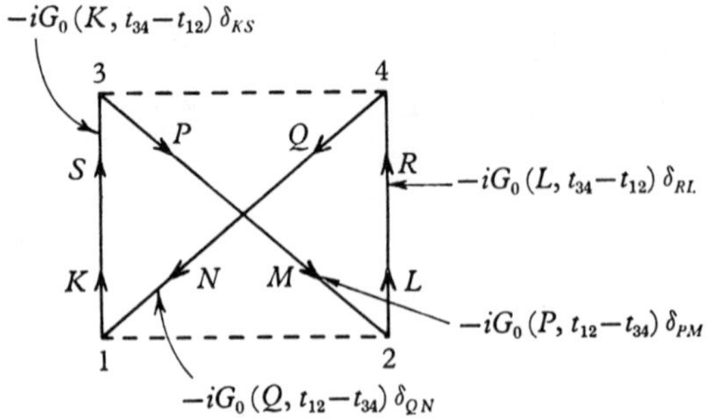

Fig. 4.16

using rule (ii) above. This is done in Fig. 4.16 where the contributing factors are indicated. In addition there is a $(-1)^1$ factor because we have a single loop. Thus this graph corresponds to the expression

$$-(-i)^4\,\delta_{KS}\,\delta_{RL}\,\delta_{PM}\,\delta_{QN}\,G_0(K, t_{34}-t_{12})\,G_0(L, t_{34}-t_{12}) \times G_0(P, t_{12}-t_{34})\,G_0(Q, t_{12}-t_{34}), \quad (4.15.16)$$

contributing to (4.15.11).

4.15.3. *nth-order theory*

There are no new essential difficulties beyond the second order, and thus we may proceed at once to the general term. The main thing is to keep one's head as the symbolism proliferates!

From (4.7.4), we have

$$\langle U_n(t, -\infty)\rangle = \frac{(-i)^n}{2^n n!}\int_{-\infty}^{t} dt_{2n-1\,2n}\cdots\int_{-\infty}^{t} dt_{12} \sum_{\substack{KLMN\\ \cdots\\ XYZT}} \times \langle KL|v|NM\rangle\ldots\langle XY|v|TZ\rangle \times \langle T a_K^\dagger(t_{12})\,a_L^\dagger(t_{12})\ldots a_Z(t_{2n-1\,2n})\,a_T(t_{2n-1\,2n})\rangle. \quad (4.15.17)$$

There are $n!$ component parts in (4.15.17) many of which will be vanishing. We will now state the general graphical rules for such terms:

(i) *To draw a graph from a fully contracted contribution.* Draw n vertices with end points $(1, 2), (3, 4), \ldots, (2n-1, 2n)$ (odd numbers on the left, even on the right). Associate K, N with 1, L, M with

2, ..., X, T with $2n-1$, and Y, Z with $2n$. Now look at the individual contractions $\overline{a_\alpha^\dagger a_\beta} = -\overline{a_\beta a_\alpha^\dagger}$. For each, draw a directed line from α to β and label.

For a specified time sequence we order the vertices accordingly. Thus, in Fig. 4.17 we have $t_{12} < t_{34} < \ldots < t_{2n-1\,2n}$. But no matter what the time sequence (cf. rule (i) in second order) the topological structure remains invariant and always particles travel forwards in time and holes backwards. Usually, for convenience, we allow the diagram with the time sequence indicated in Fig. 4.17 to represent the family of $n!$ members appropriate to all possible sequences.

(ii) *To write down a contribution from a graph.* For every contraction line from A at time $t_{\lambda\mu}$ to B at time $t_{\nu\sigma}$ (not necessarily distinct from $t_{\mu\lambda}$), supply a factor $-iG_0(A, t_{\nu\sigma}-t_{\lambda\mu})\,\delta_{AB}$. By construction, the particle (i.e. solid) lines of the graph necessarily constitute a number l of closed loops. Supply a factor $(-1)^l$.

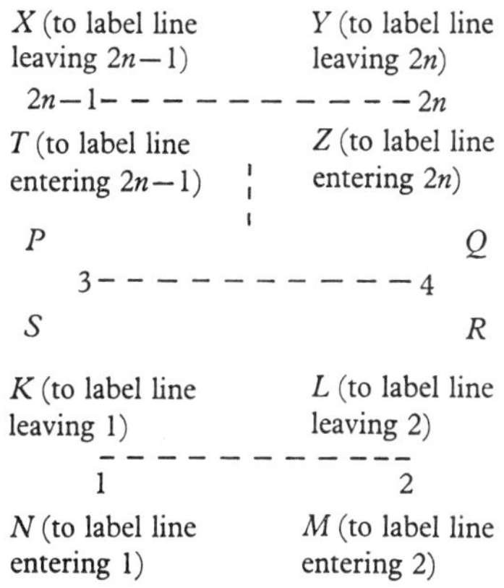

Fig. 4.17

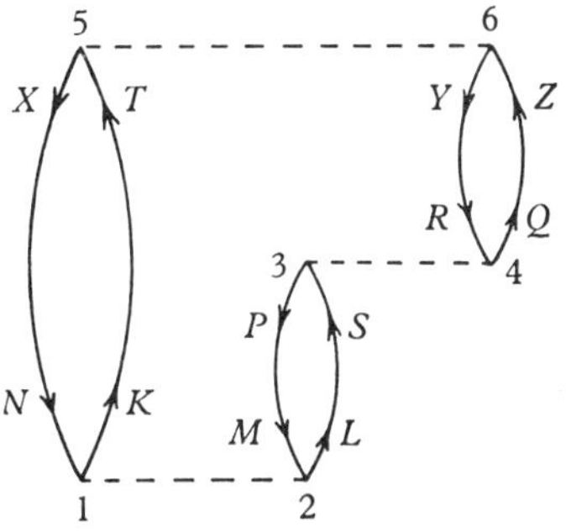

Fig. 4.18

That the latter prescription does indeed recover our original formulae is obvious, apart from the question of sign. The proof that the sign is as above is given in Appendix 4A.2.

As an example, consider the third-order graph shown in Fig. 4.18. This is of the type indicated in Fig. 4.17 ($n = 3$). The displacing and stretching and shortening of the vertex lines has been done for reasons of aesthetics and clarity only! The contribution to $\langle U_3(t, -\infty)\rangle$ is

$$\frac{(-i)^3}{2^3 3!}\int_{-\infty}^{t} dt_{56}\int_{-\infty}^{t} dt_{34}\int_{-\infty}^{t} dt_{12} \sum_{\substack{KLMN\\PQRS\\XYZT}} \langle KL|v|NM\rangle\langle PQ|v|SR\rangle$$

$$\times\langle XY|v|TZ\rangle(-i)^6\,\delta_{KT}\,\delta_{LS}\,\delta_{MP}\,\delta_{NX}\,\delta_{QZ}\,\delta_{RY}\,G_0(K, t_{56}-t_{12})$$
$$\times G_0(L, t_{34}-t_{12})\,G_0(M, t_{12}-t_{34})\,G_0(N, t_{12}-t_{56})\,G_0(Q, t_{56}-t_{34})$$
$$\times G_0(R, t_{34}-t_{56}). \quad (4.15.18)$$

Once again, any nth-order graph containing a first-order part of vanishing kind, itself vanishes.

4.16. Linked cluster theorem

The theory as developed so far, being based on (4.5.6), and therefore (4.5.5), is not really basically different from the Rayleigh–Schrödinger method. Inherent in it, therefore, are the unphysical terms (i.e. terms not proportional to the number of particles) in each order in the energy series, as emphasized in section 4.3. The advantage of the present approach, however, is that it is very amenable to a modification of such a form that only terms proportional to N appear.

To give us a clue to the correct procedure, it is instructive to consider how the present formalism works for the trivial case of a constant perturbation. In that case, of course, the normalized wave function of the system is the usual Fermi sphere solution. On the other hand we can inquire what (4.5.2) gives. We know that, in this simple case, V is independent of t and so we may write, using (4.4.7),

$$U_\alpha(0, -\infty) = 1+(-i)\int_{-\infty}^{0} dt_1\, V e^{\alpha t_1}$$
$$+(-i)^2\int_{-\infty}^{0} dt_2\int_{-\infty}^{t_2} dt_1\, V^2 e^{\alpha t_1} e^{\alpha t_2}+\dots. \quad (4.16.1)$$

The necessity of introducing the adiabatic factor $e^{\alpha t}$ should be noted. The evaluation of (4.16.1) is straightforward. One finds

$$U_\alpha(0, -\infty) = 1-\frac{iV}{\alpha}+\frac{(-i)^2}{2!}\frac{V^2}{\alpha^2}\dots = e^{-iV/\alpha}, \quad (4.16.2)$$

and thus, by (4.5.2),

$$\Psi_\alpha(0) = U_\alpha(0, -\infty)\Psi(-\infty) = e^{-iV/\alpha}\Psi(-\infty) = C(\alpha)\Psi(-\infty), \tag{4.16.3}$$

where $C(\alpha)$ is a c-number under present simple circumstances.

The form (4.16.3)—the correct form, since in summing (4.16.2) we are only rectifying the original 'mistake' of having expanded in the form (4.16.1)—is now trivially dealt with, since we are really interested in $\Psi(0)$, normalized to unity. All that has to be done is to drop $C(\alpha)$ from the right-hand side and with it all our difficulties.

We now remind the reader that our development of the general theory corresponds to the series stage of (4.16.2) above. (It turns out (see (4.17.3) and (4.17.4) below) that the time integration in (4.15.7) diverges, for example.) Thus, it seems desirable to sum our perturbation series for $\langle U(t, -\infty)\rangle$ in closed form if this is possible and in this way regain the correct analytic behaviour. This summation turns out to be not only possible, but also, like our simple example (4.16.2) above, to be of exponential form. We now see why the form (4.5.6) turns out to be so useful. All we have to do is identify the exponent and differentiate it.

In preparation for the theorem, let us recall a definition in section 4.3.3. A graph may or may not fall into two or more separate parts. If it does not, it is said to be linked or connected. Otherwise, it is unlinked or disconnected. For example, in Fig. 4.15, the graphs of the first row are unlinked, the others are linked.

The above is a geometrical-topological definition; there is an alternative analytical one, based on the observation that the mathematical contribution to $\langle U(t, -\infty)\rangle$ of an unlinked graph, factorizes in the following manner. Figure 4.19 shows symbolically three graphs, representing contributions to $\langle U(t, -\infty)\rangle$: (i) is meant to represent a graph of vertex order n_1, the second index α_1 being used to denote its particular geometrical-topological structure. Graph (ii) is similarly defined. Graphs symbolically drawn in this way may be each thought of as either linked or unlinked. If we wish to focus

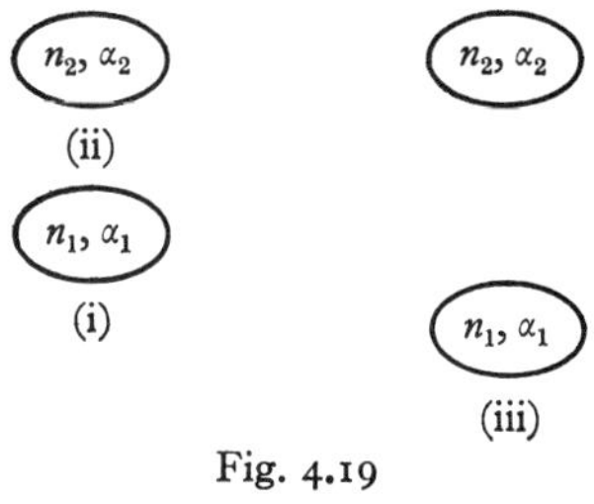

Fig. 4.19

particular attention on two parts of an unlinked diagram then we will draw them separately as indicated in (iii). The parts themselves may be either linked or unlinked. Thus (iii) represents a graph of order $n_1 + n_2$. It is certainly unlinked, falling into at least two parts with vertex orders n_1, n_2 and structures α_1, α_2 respectively. (For actual examples of such graphs as (iii), reference may be made to all terms of (4.16.10), including the sixth.) Then we have

$$\left(\begin{array}{c} \boxed{n_2, \alpha_2} \\ \boxed{n_1, \alpha_1} \end{array} \right) = \frac{n_1! n_2!}{(n_1+n_2)!} \left(\boxed{n_1, \alpha_1} \right) \left(\boxed{n_2, \alpha_2} \right). \quad (4.16.4)$$

This result is easily obtained from (4.15.17) and rule (ii) which follows it. By induction, one readily generalizes (4.16.4) to give

$$\left(\begin{array}{c} \boxed{n_i, \alpha_i} \\ \vdots \\ \boxed{n_1, \alpha_1} \end{array} \right) = \frac{n_1! n_2! \ldots n_i!}{(n_1+n_2+\ldots+n_i)!} \left(\boxed{n_1, \alpha_1} \right) \left(\boxed{n_2, \alpha_2} \right) \ldots \left(\boxed{n_i, \alpha_i} \right). \quad (4.16.5)$$

We now come to the principal result of this section. There are more general forms of this theorem (see especially Hubbard) but the following is sufficient for almost all purposes.

4.16.1. *Statement of theorem*

Let $\langle U(t, -\infty) \rangle_L$ be the series obtained from $\langle U(t, -\infty) \rangle$ by retaining only the linked graphs. Explicitly, if we write

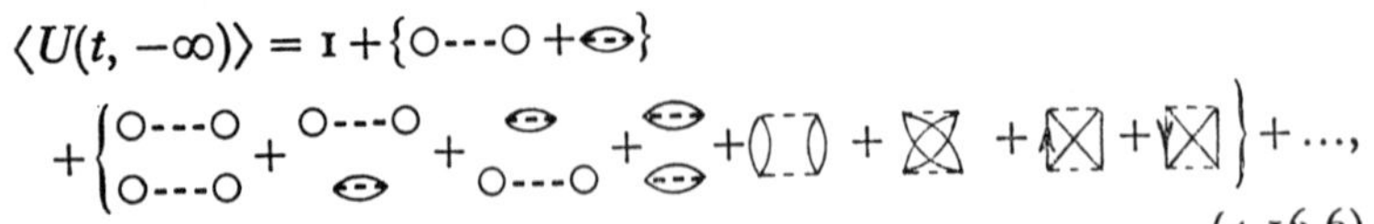

$$\langle U(t, -\infty) \rangle = 1 + \{\circ\text{---}\circ + \Leftrightarrow\}$$
$$+ \left\{ \begin{array}{c} \circ\text{---}\circ \\ \circ\text{---}\circ \end{array} + \begin{array}{c} \circ\text{---}\circ \\ \Leftrightarrow \end{array} + \begin{array}{c} \Leftrightarrow \\ \circ\text{---}\circ \end{array} + \begin{array}{c} \Leftrightarrow \\ \Leftrightarrow \end{array} + \text{Ⓞ} + \boxtimes + \boxtimes + \boxtimes \right\} + \ldots, \quad (4.16.6)$$

then

$$\langle U(t, -\infty) \rangle_L = \{\circ\text{---}\circ + \Leftrightarrow\} + \{\text{Ⓞ} + \boxtimes + \boxtimes + \boxtimes\} + \ldots. \quad (4.16.7)$$

Then, the theorem states that

$$\langle U(t, -\infty) \rangle = \exp\{\langle U(t, -\infty) \rangle_L\}. \quad (4.16.8)$$

An immediate corollary is to use (4.5.6) to obtain

$$\Delta E = \left[i \frac{\partial}{\partial t} \langle U(t, -\infty) \rangle_L \right]_{t=0}. \quad (4.16.9)$$

4.16.2. Proof

To prove (4.16.8), we sum all graphs of (4.16.6) which consist of m unlinked parts and show that the result is $\{\langle U(t, -\infty) \rangle_L\}^m/m!$.

In order to appreciate the general argument, it is desirable to illustrate the method with a variety of low-order examples which bring out the various features involved. The case $m = 0$ is trivial, both sides being unity. The case $m = 1$ also follows immediately, because of our definition (4.16.7) of the linked series. Thus $m = 2$ presents the first non-trivial problem. Our intention, below, is to prove the $m = 2$ case, illustrate (without clinching the proof) the $m = 3$ case, and then with this experience behind us, go on to prove the theorem generally.

Thus we return to the $m = 2$ case, the appropriate terms of (4.16.6) being

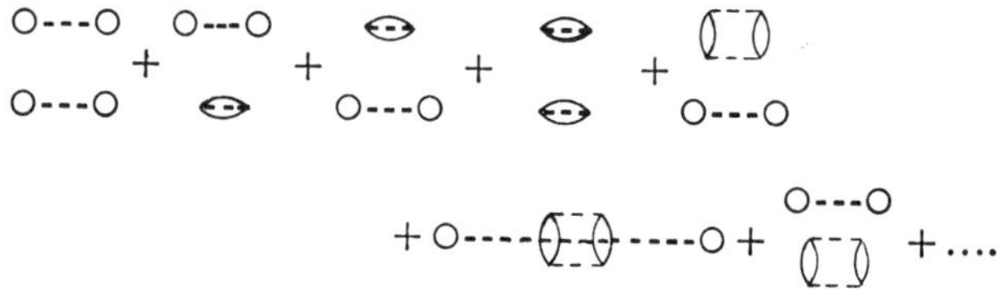

[diagrams] ... (4.16.10)

We have to show that this series is identical with

$$\frac{1}{2!} \{\langle U(t, -\infty) \rangle_L\}^2 = \frac{1}{2!} \left\{ (\circ\text{---}\circ)^2 + (\ominus)^2 + 2(\circ\text{---}\circ)(\ominus) + 2(\circ\text{---}\circ)\left(\begin{array}{c}\text{\small graph}\end{array}\right) + \ldots \right\}. \quad (4.16.11)$$

Let us begin by testing for the terms shown. On using (4.16.4), the first term of (4.16.10) becomes

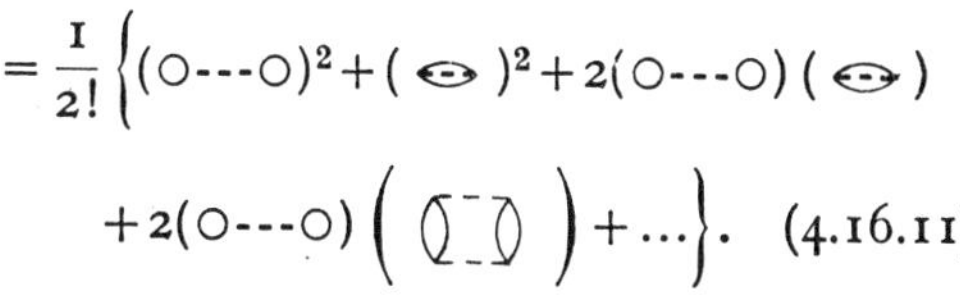

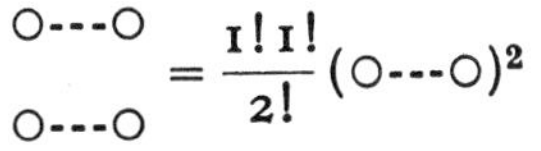

$$\begin{array}{c}\circ\text{---}\circ\\ \circ\text{---}\circ\end{array} = \frac{1!\,1!}{2!}(\circ\text{---}\circ)^2 \quad (4.16.12)$$

which checks with the corresponding term in (4.16.11). The fourth term of (4.16.10) is treated similarly. Next we check the third term of (4.16.11). This arises from the second and third members of (4.16.10), their total contribution being

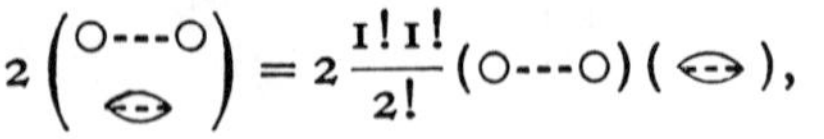

$$2 \begin{pmatrix} \circ\text{-}\text{-}\text{-}\circ \\ \ominus \end{pmatrix} = 2\frac{1!1!}{2!}(\circ\text{-}\text{-}\text{-}\circ)(\ominus), \qquad (4.16.13)$$

as required.

The arguments for graphs with three vertices are essentially identical; we will discuss, for concreteness, the origin of that term shown in fourth position in (4.16.11). This arises from the fifth, sixth and seventh terms of (4.16.10), which, on using (4.16.4) give a total contribution of

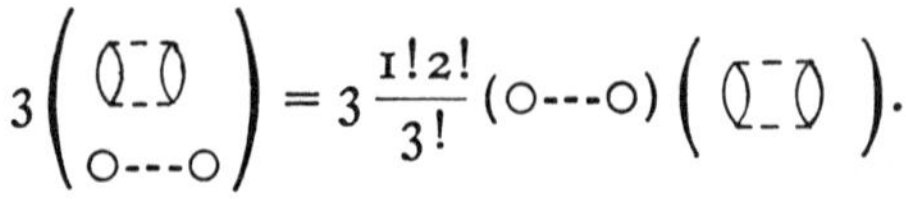

$$3 \begin{pmatrix} \overline{()}\,\overline{)} \\ \circ\text{-}\text{-}\text{-}\circ \end{pmatrix} = 3\frac{1!2!}{3!}(\circ\text{-}\text{-}\text{-}\circ)\left(\overline{()}\,\overline{)} \right). \qquad (4.16.14)$$

Once more, as in (4.16.13), the coefficient is unity, as desired.

In the quadratic form (4.16.11), every off-diagonal coefficient is unity. Thus, to help us appreciate why this is so, it is desirable to consider yet another example, namely that arising from the graphs of Figs. 4.13(iii) and 4.16. The family from (4.16.10) which contribute to this term is shown in Fig. 4.20. One might compute the size of this family in the following suggestive way. Suppose we start with the diagram on the left in Fig. 4.20 and permute the

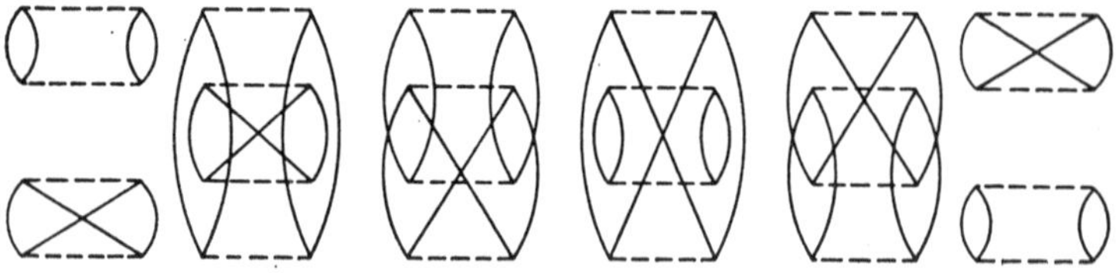

Fig. 4.20

vertices while keeping the contraction lines attached, but always insisting that within each subdiagram the original vertex (time) order is maintained. The total number of unrestricted vertex permutations is 4!. But, in each subgraph, there are 2! forbidden internal permutations. Hence the number of allowed graphs generated is

$4!/(2!)(2!) = 6$. Using (4.16.4), the total contribution from this family is

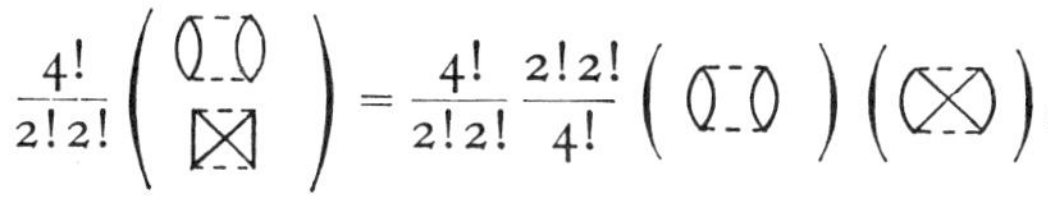

$$\frac{4!}{2!2!}\left(\begin{array}{c}\text{Q-O}\\ \bowtie\end{array}\right) = \frac{4!}{2!2!}\frac{2!2!}{4!}\left(\text{Q-O}\right)\left(\otimes\right), \quad (4.16.15)$$

which has coefficient unity as required.

Every diagonal term in (4.16.11) has coefficient $\frac{1}{2}$. To illustrate how this comes about, suppose the exchange subgraph in Fig. 4.20 were replaced by a direct graph, so that the two subdiagrams are of same form. Then, the six distinct composite graphs of Fig. 4.20 degenerate into three distinct pairs. The topological equivalence of the subdiagrams has cost us a factor of 2 and thus the total contribution from this family of three members is

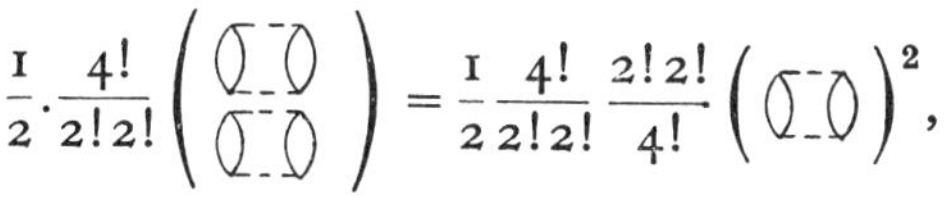

$$\frac{1}{2}\cdot\frac{4!}{2!2!}\left(\begin{array}{c}\text{Q-O}\\ \text{Q-O}\end{array}\right) = \frac{1}{2}\frac{4!}{2!2!}\frac{2!2!}{4!}\left(\text{Q-O}\right)^2, \quad (4.16.16)$$

the right-hand side coefficient being $\frac{1}{2}$ as anticipated.

Having illustrated all relevant points, the proof of the equivalence of (4.16.10) and (4.16.11) is now easy. Let us consider a general product $((n_1,\alpha_1))((n_2,\alpha_2))$ occurring in (4.16.11). The terms of (4.16.10) which contribute to the coefficient of this product are generated by taking the (n_1+n_2)th-order graph shown in Fig. 4.19 (iii) and permuting the vertices so as to obtain all distinct graphs, much as we did in obtaining the graphs of Fig. 4.20. First, keeping the contraction lines held fast to the vertex lines, we permute the latter under the restriction that the internal (time) order of the vertex lines within each subgraph are not to be varied. The number of ways of permuting (n_1+n_2) vertices without restriction is $(n_1+n_2)!$. But in one subgraph there are $n_1!$ forbidden internal permutations, while in the other there are $n_2!$. Thus, we generate $(n_1+n_2)!/n_1!n_2!$ graphs. Unless there are topological equivalences among the subgraphs, these graphs are all distinct, and their total contribution is thus

$$\frac{(n_1+n_2)!}{n_1!n_2!}\left(\begin{array}{c}(n_2,\alpha_2)\\ (n_1,\alpha_1)\end{array}\right) = \frac{(n_1+n_2)!}{n_1!n_2!}\frac{n_1!n_2!}{(n_1+n_2)!}((n_1,\alpha_1))((n_2,\alpha_2)),$$

(4.16.17)

on using (4.16.4). The coefficient is unity, as required. On the other hand, if topologically equivalent subgraphs occur, we must allow, in the above-stated procedure for generating a family of graphs, for those permutations, the net effect of which is to do no more than interchange such subgraphs with each other, thus leaving the composite graph invariant. Since, in the present case, only two subgraphs can be topologically equivalent, when this situation occurs, one must divide by 2! the formula derived above for the total number of graphs in the family. In this way we arrive at the coefficient $\frac{1}{2}$ for the diagonal terms in (4.16.11). Thus, the equivalence of the latter with (4.16.10) is proved.

Fortified with the above experience, we may now tackle the problem in its generality. To begin with, we use the multinomial theorem to raise (4.16.7) to its mth power. One has

$$\frac{\{\langle U(t,-\infty)\rangle_L\}^m}{m!} = \frac{1}{m!}\sum_{x+y+z+\ldots=m} \times \frac{m!}{x!\,y!\,z!\ldots}(\circ\text{---}\circ)^x\,(\;\;)^y\,(\;\;)^z\ldots. \quad (4.16.18)$$

 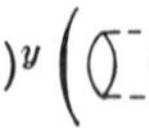

Now let us ask what graphs of (4.16.6) contribute to the coefficient of $(\circ\text{---}\circ)^x\,(\;\;)^y\,(\;\;)^z\ldots$  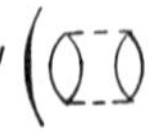 in (4.16.18). Clearly, the family concerned is generated by vertex permutation of the graph schematically drawn in Fig. 4.21. Since the total vertex order (Fig. 4.21) is

$$1.x+1.y+2z+\ldots n_i t+\ldots,$$

the possible number of unrestricted permutations is

$$(1.x+1.y+2z+\ldots+n_i t+\ldots)!.$$

But, as in our low-order examples, we must allow for (i) forbidden internal permutations within each subgraph (this means dividing by $(1!)^x(1!)^y(2!)^z\ldots(n_i!)^t\ldots$), and (ii) permutations which merely rearrange identical subgraphs (this means dividing by $x!\,y!\,z!\ldots t!\ldots$) in order to find the total number of members in the family contributing to the selected term. Thus, there are

$$\frac{(1.x+1.y+2z+\ldots n_i t+\ldots)!}{(1!)^x(1!)^y(2!)^z\ldots(n_i!)^t\ldots x!\,y!\,z!\ldots t!\ldots} \quad (4.16.19)$$

members, each contributing to $\langle U(t, -\infty)\rangle$ a term cf. (4.16.5))

$$\frac{(1!)^x(1!)^y(2!)^z\ldots(n_i!)^t\ldots}{(1.x+1.y+2z+\ldots+n_it+\ldots)!}$$ 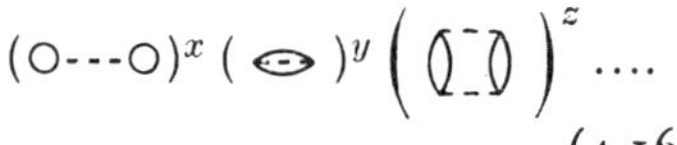 (4.16.20)

The total contribution is obtained by multiplying (4.16.20) by (4.16.19) and in this way, we obtain the coefficient of

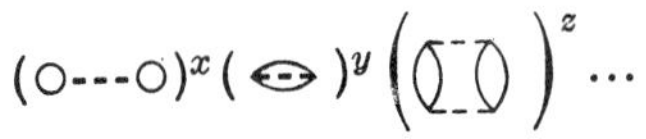

exhibited in (4.16.18). Since this was a quite arbitrary term, we have shown that the sum of all graphs in $\langle U(t, -\infty)\rangle$ consisting of m unlinked subgraphs, is $\{\langle U(t, -\infty)\rangle_L\}^m/m!$. Thus, on summing over m, (4.16.8) follows.

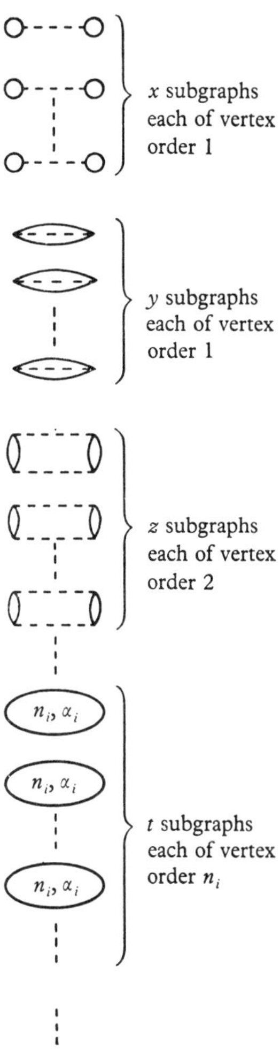

Fig. 4.21

4.16.3. *Graph degeneracy*

Because of the corollary (4.16.9) we now turn our attention to the linked series (4.16.7). It will be shown that certain degeneracies occur in this series which streamline our eventual explicit calculation of the various terms.

(i) *Explicit degeneracies in* (4.16.7). Let us, as an illustration, consider (6, 1) and (6, 2) of Fig. 4.15. The contribution of (6, 1) to (4.15.11) is obtained via (4.15.14); that for (6, 2) is easily written down using our rules. The only difference between the expressions is that the roles of certain dummy suffices are interchanged in such a manner that the contributions to (4.15.11) stay the same. Explicitly, P interchanges with Q and R with S. Topologically, we see that the structures of these graphs are the same (one may think of the (3, 4) vertex rotated through 180° about an axis lying in the plane of the paper). A discussion similar to this is possible for graphs (6, 3) and (6, 4).

The above are examples of an nth-order (vertex) result. Topologically equivalent diagrams (in the above sense) occur in families of size 2^{n-1}, each member giving the same contribution to

$$\langle U(t, -\infty)\rangle_L.$$

A third-order example is shown in Fig. 4.22.

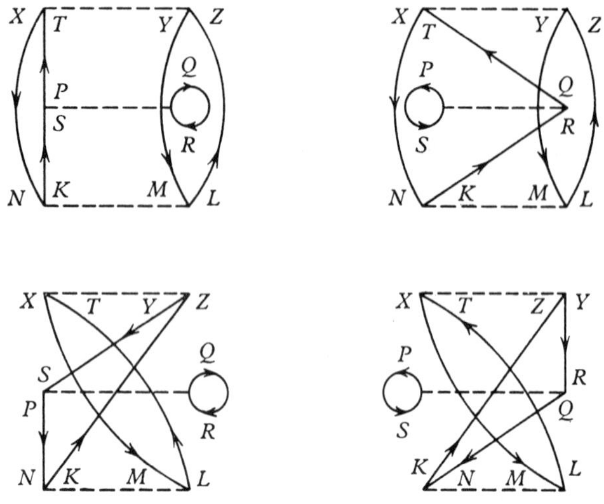

Fig. 4.22. Third-order degenerate graphs (explicit degeneracy).

To prove the general statement, we observe that one topologically invariant vertex switch of the kind described above (let us, for concreteness, think of graphs (1, 1) and (1, 2) of Fig. 4.22) corresponds to changing the delta and Green function parts of the integrands only by interchanging the labels of the outgoing and of the incoming lines at the vertex of interest ($P \leftrightarrow Q$, $R \leftrightarrow S$ in the example). Such an interchange of labels leaves all matrix elements invariant, and thus all we have done is interchanged dummy suffices and left the mathematical contribution to $\langle U_L(t, -\infty)\rangle_L$ the same. For fixed vertex (1, 2), say, there are $(n-1)$ possible interchanges and thus the size of an nth-order family of the above type is 2^{n-1}.

(ii) *Implicit degeneracies in* (4.16.7). As has been stated, every nth-order graph drawn as in Fig. 4.17 represents $n!$ diagrams with all possible time orderings of the vertices. Thus, graph (6, 1) of Fig. 4.15 represents the two situations shown in Fig. 4.23, while an illustration of a third-order set is given in Fig. 4.24, the first of these being the usual representative of the whole family.

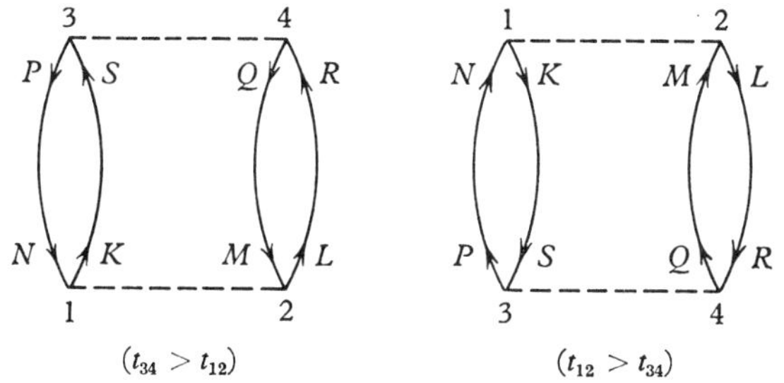

Fig. 4.23. Implicit degeneracy in second-order graphs.

Now let us select any nth-order linked graph appropriately arrowed, but devoid, for the moment, at least, of lettering and numbering. In other words, we are focussing attention on a rather basic entity, namely a graphical structure representing a physical process. Let us allocate the usual labels (1, 2), (3, 4), ..., $(2n-1, 2n)$ in all possible time sequences to obtain $n!$ graphs. The letters may be supplied to each using the usual convention (see rule (i)). This process sometimes generates the same family as was obtained above by permuting the vertices, a simple example of such a case being the second-order process shown in Fig. 4.23. But this is not necessarily true. For example, the graphs of Fig. 4.25 have been obtained from a given structure and of these only half overlap with the family of Fig. 4.24 obtained by vertex permutation.

The essential observations we have to make here concerning graph generation by vertex permutation and by structure labelling are: (i) there is a one–one correspondence between the totality of individual graphs generated by either method (even though, as we have seen, this is not true of complete families); and (ii) defining the contribution of any such explicitly time-ordered graph in the obvious way, namely, as in rule (ii) except that the time integration

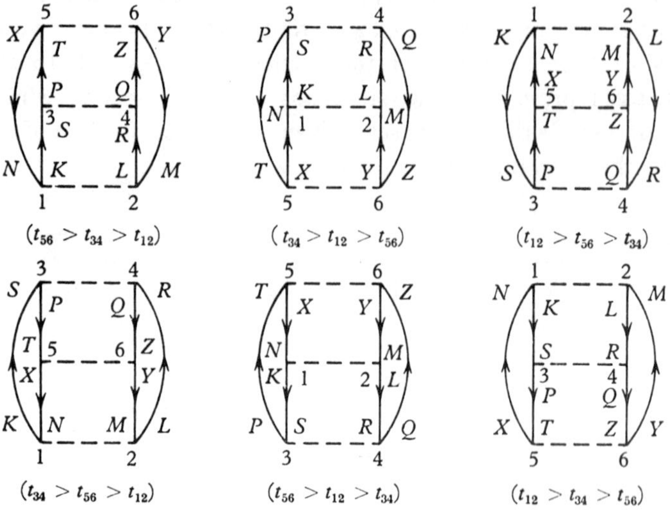

Fig. 4.24. Third-order graphs. Implicit degeneracy.

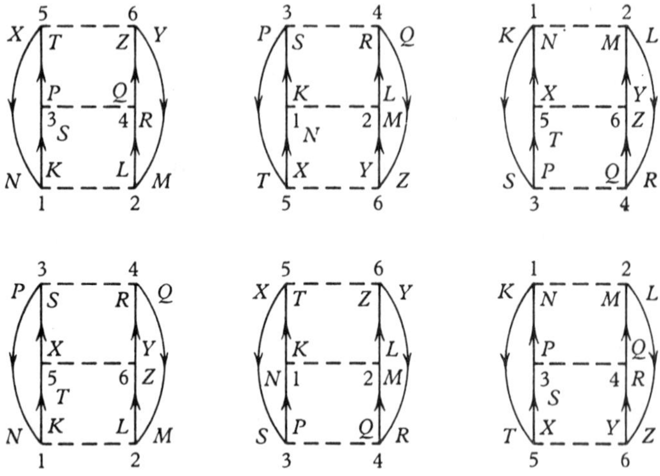

Fig. 4.25

takes place only over the specified time sequence, then, the contributions of all members of the same family generated by structure labelling are the same.

Thus, in (4.15.17), one may write

$$\frac{1}{n!}\int_{-\infty}^{t} dt_{2n-1\,2n}\cdots\int_{-\infty}^{t} dt_{34}\int_{-\infty}^{t} dt_{12} = \int_{-\infty}^{t} dt_{2n-1\,2n}\cdots \times\int_{-\infty}^{t_{56}} dt_{34}\int_{-\infty}^{t_{34}} dt_{12}. \tag{4.16.21}$$

That a result of this kind should hold is reasonable because of the step taken a long time ago in going from (4.4.7) to (4.4.8). (See the final paragraph of section 4.11.)

Let us end this section with a brief summary of the position we have reached so far. Our intention is to calculate ΔE as given by (4.5.6). Thus, we began with an examination of the unperturbed ground-state matrix element of a perturbation series development of the U matrix defined through (4.7.4) and (4.16.6). On classifying the terms as either linked or unlinked, one was able (see (4.16.8)) to rewrite the original U-matrix average in terms of a more useful series containing only the linked terms. This new series, (4.16.7), and its relationship to ΔE through (4.16.9), become the objects of our exclusive attention in the remainder of this chapter. We shall see that by removing the unlinked terms we have entirely eliminated the non-physical terms of section 4.3 from the theory.

Above, we showed that certain degeneracies occur in the linked series which will aid our computations, and, especially, the use of (4.16.21) in (4.15.17) will be important in the following section where we proceed to the final stage of our calculation by performing the time integrations.

4.17. Time integrations

Our aim, now, is to use (4.16.21) to perform the time integrations shown in (4.15.17). Rule (ii) tells us that the time-dependence of this average is contained in a product of unperturbed propagators, each of the kind defined by (4.13.17). On collecting together the various time exponents, we always find we are faced with evaluating the integral over $t_{2n-1\,2n} > \ldots > t_{34} > t_{12}$ of a product of exponential functions. At this stage, we need to use the adiabatic theorem (see (4.5.2) and the remarks preceding it). For example,

$$\begin{aligned}\int_{-\infty}^{t} e^{i\beta t_{12}}\,dt_{12} &\equiv \lim_{\alpha\to+0}\int_{-\infty}^{t} e^{(\alpha+i\beta)t_{12}} \\ &= \lim_{\alpha\to 0}\frac{e^{(\alpha+i\beta)t}}{(\alpha+i\beta)} \\ &= \frac{e^{i\beta t}}{i\beta},\end{aligned} \tag{4.17.1}$$

whereas a more naïve argument would have produced an ambiguous answer. This result is easily generalized. One has

$$\int_{-\infty}^{t} dt_m \ldots \int_{-\infty}^{t_3} dt_2 \int_{-\infty}^{t_2} dt_1\, e^{i\beta_m t_m}\ldots e^{i\beta_2 t_2} e^{i\beta_1 t_1}$$

$$\equiv \frac{e^{i(\beta_m+\ldots+\beta_2+\beta_1)t}}{i^m(\beta_m+\ldots+\beta_2+\beta_1)\ldots(\beta_3+\beta_2+\beta_1)(\beta_2+\beta_1)\beta_1}. \tag{4.17.2}$$

With these preliminaries, let us begin a systematic discussion of the problem. As usual, we begin with the low-order terms before tackling the general problem.

First order. Here, we are concerned with (4.15.7) and the direct and exchange contributions to it, defined respectively by (4.15.9) and (4.15.10). We have

$$\begin{aligned}\text{O---O} = -\frac{i}{2}\int_{-\infty}^{t} dt_{12} \sum_{KLMN} \langle KL|v|NM\rangle(-i)^2 \\ \times \delta_{KN}\delta_{LM} G_0(K,-0)\,G_0(L,-0).\end{aligned} \tag{4.17.3}$$

Now let us substitute for the G_0's using (4.13.17), taking care to insert the implications of our time arguments being -0. One obtains

$$\text{O---O} = -\frac{i}{2}\int_{-\infty}^{t} dt_{12} \sum_{k,l<k_f} \langle KL|v|KL\rangle(-i)^2(-i)^2 \tag{4.17.4}$$

and thus, on introducing a notation suggested by (4.16.9), we find

$$(\Delta E)_{1d} \equiv \left[i\frac{\partial}{\partial t}\,\text{O---O}\right]_{t=0} = \frac{1}{2}\sum_{k,\,l<k_f}\langle KL|v|KL\rangle. \tag{4.17.5}$$

In a similar manner, we have

$$(\Delta E)_{1e} \equiv \left[i \frac{\partial}{\partial t} \ominus \right]_{t=0} = -\frac{1}{2} \sum_{k,l<k_f} \langle KL|v|LK \rangle. \quad (4.17.6)$$

The differences between (4.15.9) and (4.15.10) are reflected in (4.17.5) and (4.17.6). It is of interest, here, to recall the remarks made at the beginning of section 4.16 on the need for a linked cluster theorem. If we had employed (4.5.5) to calculate ΔE, we would have used (4.17.4), which is divergent. On the other hand, in the form (4.16.9), the time derivatives of (4.17.4), etc., appear and the two contributions (4.17.5) and (4.17.6) evaluated so far turn out to be $\propto N$.

Second order. Now, we consider (4.15.11), and begin with the term represented by graph (iii) of Fig. 4.13. Using (4.15.14), we obtain

$$= \frac{(-i)^2}{2^2} \int_{-\infty}^{t} dt_{34} \int_{-\infty}^{t_{34}} dt_{12} \sum_{\substack{KLMN\\PQRS}} \langle KL|v|NM \rangle \langle PQ|v|SR \rangle$$
$$\times \delta_{KS} \delta_{LR} \delta_{MQ} \delta_{NP} G_0(K, t_{34}-t_{12}) G_0(L, t_{34}-t_{12})$$
$$\times G_0(M, t_{12}-t_{34}) G_0(N, t_{12}-t_{34}), \quad (4.17.7)$$

where (4.16.21) has been used to omit the 2! and restrict the region of integration to $t_{34} > t_{12}$. On using (4.13.17) to replace the G_0's in the above expression, we have, after a few elementary simplifications,

$$= \frac{(-i)^2}{2^2} \int_{-\infty}^{t} dt_{34} \int_{-\infty}^{t_{34}} dt_{12} \sum_{\substack{k,l>k_f\\m,n<k_f}} \langle KL|v|NM \rangle \langle NM|v|KL \rangle$$
$$\times e^{-i\epsilon_{KLMN} t_{12}} e^{i\epsilon_{KLMN} t_{34}}, \quad (4.17.8)$$

where $$\epsilon_{KLMN} = \epsilon_K + \epsilon_L - \epsilon_M - \epsilon_N = -\epsilon_{MNKL}. \quad (4.17.9)$$

Now let us do the t_{12}-integration using (4.17.1). We find

$$= \frac{(-i)^2}{2^2} \int_{-\infty}^{t} dt_{34} \sum_{\substack{k,l>k_f\\m,n<k_f}} |\langle KL|v|NM \rangle|^2 \frac{1}{i\epsilon_{KLMN}}. \quad (4.17.10)$$

Because of the type 1 degeneracy discussed in section 4.16, the graph , which is topologically equivalent to the above and descriptive of the same physical process, contributes the same amount to $\langle U(t, -\infty)\rangle_L$. We will call the sum of these two equal terms the direct second-order term and designate the contribution to ΔE by $(\Delta E)_{2d}$. Then,

$$(\Delta E)_{2d} = \left[i \frac{\partial}{\partial t} \left(\right. \right.$$
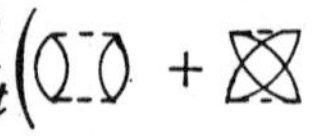
$$\left. \left. \right) \right]_{t=0} = \frac{1}{2} \sum_{\substack{k,l>k_f \\ m,n<k_f}} \frac{|\langle KL|v|NM\rangle|^2}{\epsilon_{MNKL}}. \tag{4.17.11}$$

Similarly, the other two (topologically equivalent) second-order (exchange) graphs of (4.16.7) contribute equally. Using Fig. 4.16 and (4.15.16), we find their total contribution to (4.16.9) to be

$$(\Delta E)_{2e} = \left[i \frac{\partial}{\partial t} \left(\right. \right.$$
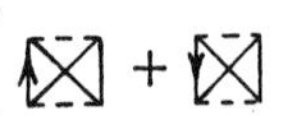
$$\left. \left. \right) \right]_{t=0} = -\frac{1}{2} \sum_{\substack{k,l>k_f \\ m,n<k_f}} \times \frac{\langle KL|v|NM\rangle\langle MN|v|KL\rangle}{\epsilon_{MNKL}}. \tag{4.17.12}$$

nth order. Selecting any nth-order graph from the linked series, necessarily, for a non-zero contribution, there can be no equal time contractions at the extreme vertices. Thus, we can begin a tentative sketch of our graph as in Fig. 4.26, the purpose of the arrows bearing Greek labels being explained later. The intermediate structure cannot be drawn in without being more specific. Suppose, however, there are h hole lines (travelling backwards in time), then there will be $2n-h$ particle lines (travelling forward in time). On account of their -0 time arguments, equal-time contractions are to be counted as hole lines.

Using (4.15.17), (4.16.21) and rule (ii), and an abbreviated symbol for Fig. 4.26, we have

$$\boxed{n,\alpha} = \frac{(-i)^n}{2^n} \int_{-\infty}^{t} dt_{2n-1\,2n} \cdots \int_{-\infty}^{t_{56}} dt_{34} \int_{-\infty}^{t_{34}} dt_{12} \sum_{\substack{KLMN \\ \cdots\cdots \\ XYZT}}$$
$$\times \langle KL|v|NM\rangle \ldots \langle XY|v|TZ\rangle$$
$$\times (-1)^l \prod_{\substack{2n-h \text{ particle lines} \\ (\text{from } t_{\lambda\mu} \text{ to } t_{\nu\sigma})}} \{-iG_0(A, t_{\nu\sigma} - t_{\lambda\mu})\,\delta_{AB}\}$$
$$\times \prod_{\substack{h \text{ hole lines} \\ (\text{from } t_{\lambda'\mu'} \text{ to } t_{\nu'\sigma'})}} \{-iG_0(A', t_{\nu'\sigma'} - t_{\lambda'\mu'})\,\delta_{A'B'}\}, \tag{4.17.13}$$

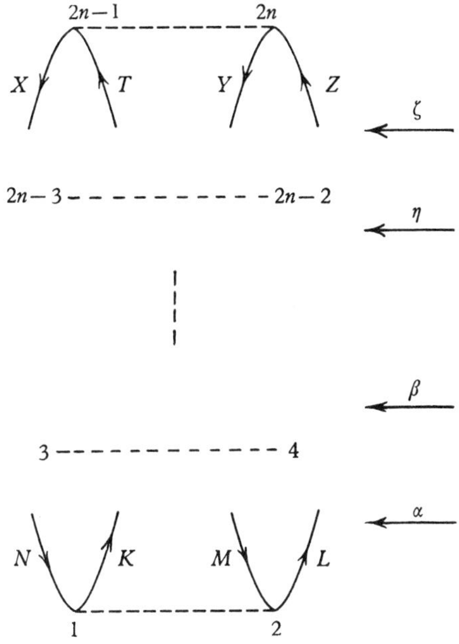

Fig. 4.26

the A's, B's, A''s and B''s simply representing selections from the set $K, L, \ldots, X, Y, Z, T$, and similarly the $t_{\nu\sigma}$'s, $t_{\lambda\mu}$'s, $t_{\nu'\sigma'}$'s and $t_{\lambda'\mu'}$'s being taken from the set $t_{12}, t_{34}, \ldots, t_{2n-1\,2n}$.

Now, by (4.13.17), provided we restrict the particle (i.e. $A, B, \ldots$) sums to states above the Fermi surface, each $-i\,G_0(A, t_{\nu\sigma}-t_{\lambda\mu})$ can be replaced by $\exp\{-i\epsilon_A(t_{\nu\sigma}-t_{\lambda\mu})\}$. Similarly, provided we restrict the hole (i.e. $A', B', \ldots$) sums to states below the Fermi surface, each $-i\,G_0(A', t_{\nu'\sigma'}-t_{\lambda'\mu'})$ can be replaced by

$$-\exp\{-i\epsilon_{A'}(t_{\nu'\sigma'}-t_{\lambda'\mu'})\}.$$

Restricting the momenta sums in the above way, and allowing for sign, the particle and hole G_0's contribute the same kinds of factors and the two products shown in (4.17.13) can be merged thus:

$$\boxed{n, \alpha} = \frac{(-i)^n}{2^n}\int_{-\infty}^{t} dt_{2n-1\,2n} \cdots \int_{-\infty}^{t_{56}} dt_{34} \int_{-\infty}^{t_{34}} dt_{12} \sum_{\substack{\text{particles above F.S.}\\ \text{holes below F.S.}}}$$
$$\times \langle KL|v|NM\rangle \ldots \langle XY|v|TZ\rangle \prod_{\text{all lines}} \delta_{AB}(-1)^{l+h} \prod_{\substack{\text{all lines}\\ (\text{from } t_{\lambda\mu} \text{ to } t_{\nu\sigma})}}$$
$$\times \exp\{-i\epsilon_A(t_{\nu\sigma}-t_{\lambda\mu})\}. \qquad (4.17.14)$$

The next step is to observe that because of our convention of labelling the outward going lines with $K, L, P, Q, \ldots, X, Y$ and the incoming ones with $M, N, R, S, \ldots, Z, T$, we can collect terms in the final product in (4.17.14) and rewrite it (using the notation of (4.17.9)) as

$$e^{i\epsilon_{XYZT} t_{2n-1\,2n}} \ldots e^{i\epsilon_{PQRS} t_{34}} e^{i\epsilon_{KLMN} t_{12}}. \tag{4.17.15}$$

Then, using (4.17.2), the time integrations are straightforward. On doing $(n-1)$ integrations, we find

$$\text{\textcircled{n,α}} = \frac{(-i)^{n+l+h}}{2^n} \int_{-\infty}^{t} dt_{2n-1\,2n} \sum_{\substack{\text{particles above F.S.}\\ \text{holes below F.S.}}} \times \frac{\langle KL|v|NM\rangle \ldots \langle XY|v|TZ\rangle \prod\limits_{\text{all lines}} \delta_{AB}}{i^{n-1}\cancel{(\epsilon_{XYZT}+\ldots+\epsilon_{PQRS}+\epsilon_{KLMN})}\ldots(\epsilon_{PQRS}+\epsilon_{KLMN})\,\epsilon_{KLMN}}, \tag{4.17.16}$$

the deletion being merely a means of indicating at what point (reading from right to left) the denominator product terminates. Actually,

$$\epsilon_{XYZT} + \ldots + \epsilon_{PQRS} + \epsilon_{KLMN} = 0 \tag{4.17.17}$$

since, on using (4.17.9), we find each singly indexed ϵ occurring once with a plus sign and once with a minus sign, if we remember the δ_{AB} product which is present in (4.17.16). Equations (4.17.17) and (4.17.2) together explain why there is no residual time-dependence in the integrand of (4.17.16).

It is now an easy matter to write down the contribution of our graph to ΔE. Once more, using the degeneracy properties of section 4.16, we have

$$(\Delta E)_{n\alpha} = \left[i \frac{\partial}{\partial t} \left(\text{\textcircled{n,α}} + \text{all type 1 degeneracies} \right) \right]_{t=0}$$

$$= \frac{(-1)^{l+h}}{2} \sum_{\substack{\text{particles above F.S.}\\ \text{holes below F.S.}}} \times \frac{\langle KL|v|NM\rangle \langle PQ|v|SR\rangle \ldots \langle XY|v|TZ\rangle \prod\limits_{\text{all lines}} \delta_{AB}}{\epsilon_{MNKL}(\epsilon_{MNKL}+\epsilon_{RSPQ}) \ldots \cancel{(\epsilon_{MNKL}+\ldots+\epsilon_{ZTXY})}}. \tag{4.17.18}$$

The reversal of indices on the ϵ's should be noted. This is an essentially trivial matter which uses the symmetry (cf. (4.17.9)) of the $(n-1)$ terms in the denominator product to absorb a factor of $(-1)^{n-1}$.

Equation (4.17.18), telling us how to calculate the contribution of a general graph to ΔE, is the fundamental result of this chapter, the sum of all such terms being symbolically written as in (4.3.5). In applications one soon learns to apply it with a minimum of formality. Thus, it is usual to omit the $\Pi\delta_{AB}$ term and duplicate the momentum-spin indices appropriately. Furthermore, from a given graph, the energy factors can be rather easily read off. For, let us consider a time α (see Fig. 4.26) intermediate between t_{12} and t_{34}. The sum of the hole energies minus the sum of the particle energies of lines crossing the time α axis is ϵ_{MNKL}, the first term of the denominator in (4.17.17). The next energy denominator in (4.17.17) is $\epsilon_{MNKL}+\epsilon_{RSPQ}$. We assert that to obtain this quantity, we add the hole energies and subtract off the particle energies crossing the equal time line β. To see this let us examine the possibilities. If there are no contractions, all lines from 3 and 4 are located to the north of this vertex and quite independent to the K, L, M, N lines which proceed themselves up the diagram. The sum of the hole energies at β is $\epsilon_M+\epsilon_N+\epsilon_R+\epsilon_S$ while the sum of the particle energies is $\epsilon_K+\epsilon_L+\epsilon_P+\epsilon_Q$ and subtraction gives $\epsilon_{MNKL}+\epsilon_{RSPQ}$. On the other hand, there might be contractions. Every contraction results in a cancellation of one of the particle energies exactly with one of the hole energies and the non-appearance of the corresponding lines at the time β. Taken in conjunction, these observations prove our assertion. We may now go on and establish the result, at any intermediate position, that the appropriate energy denominator is got by adding all the hole energies and subtracting off the particle energies associated with lines crossing some intermediate equal time marker.

We should note, also, when applying (4.17.17), that a direct equal time contraction counts as one loop (on account of rule (ii)) and one hole (because of its time argument -0), and thus contributes a total signature of $(-1)^2 = 1$.

As an example of the use of (4.17.18), let us consider the third-order family of 2^2 topologically equivalent graphs shown in Fig.

4.27 (i). The total contribution is four times that of the first, the lettered diagram for which is given in Fig. 4.18. Thus, the total contribution to ΔE is $(\Delta E)_{3r(\mathrm{i})}$, say, where

$$(\Delta E)_{3r(\mathrm{i})} = 4\frac{(-1)^{3+3}}{2} \sum_{\substack{klqstz>k_f \\ mnprxy<k_f}} \langle KL|v|NM\rangle\langle PQ|v|SR\rangle\langle XY|v|TZ\rangle \times \frac{\delta_{KT}\,\delta_{LS}\,\delta_{MP}\,\delta_{NX}\,\delta_{QZ}\,\delta_{RY}}{\epsilon_{MNKL}(\epsilon_{MNKL}+\epsilon_{RSPQ})}$$

$$= \frac{4(-1)^{3+3}}{2} \sum_{\substack{klq>k_f \\ mnr<k_f}} \frac{\langle KL|v|NM\rangle\langle MQ|v|LR\rangle\langle NR|v|KQ\rangle}{(\epsilon_M+\epsilon_N-\epsilon_K-\epsilon_L)(\epsilon_N+\epsilon_R-\epsilon_K-\epsilon_Q)}. \tag{4.17.19}$$

Because of our discussion above, it is possible to write down the final line of (4.17.19) without the intermediate expression. The

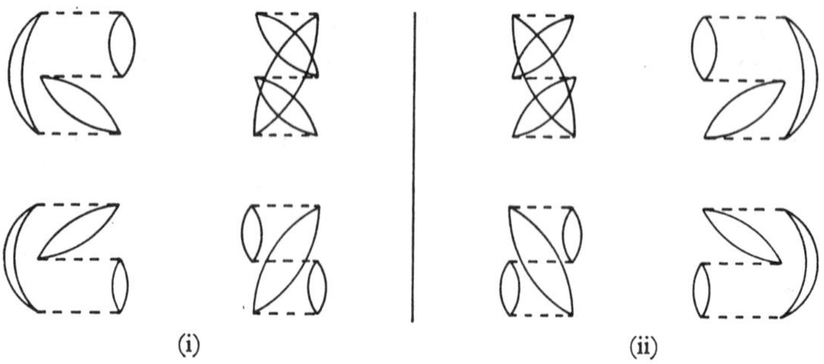

Fig. 4.27

family of Fig. 4.27 (ii) is just the mirror image of family (i) and gives an equal contribution to ΔE. The two families together yield the total so-called ring contribution which is therefore twice (4.17.19). The ring graphs of all orders are important in the theory of the high density electron gas, as we shall see in Chapter 5.

Problems

P.4 (i). With A_k and $A_k^\dagger$ as defined in P.3 (i), draw all diagrams appropriate to a perturbation expansion when the perturbing term is

$$H' = \lambda_q\left(\sum_{\mathbf{k}}\left\{A_{\mathbf{k}}^\dagger + A_{\mathbf{k}}\right\}\right)^2.$$

P.4 (ii). Show explicitly that

$$\int_{t_0}^{t} dt_n \int_{t_0}^{t_n} dt_{n-1} \dots \int_{t_0}^{t_2} dt_1 \, V(t_1) \dots V(t_n)$$
$$= \frac{1}{n!} \int_{t_0}^{t} dt_n \int_{t_0}^{t} dt_{n-1} \dots \int_{t_0}^{t} dt_1 \, P(V(t_1) \dots V(t_n))$$

for the case $n = 2$.

Establish the general result by induction.

P.4 (iii). If a_k, $a_k^\dagger$ are the annihilation and creation operators associated with single-particle level $|\mathbf{k}\rangle$, the ground state of an N particle system is

$$|G\rangle = \prod_{k=1}^{N} a_k^\dagger |0\rangle,$$

where $|0\rangle$ is the vacuum state and $k = 1 \dots N$ are the N single-particle levels with lowest energy.

Show that any N-particle state not orthogonal to $|G\rangle$ can be specified in either of the forms

$$|S\rangle = \left\{ \prod_{h=1}^{N} \left(1 + \sum_{p=N+1}^{\infty} C_{ph} \, a_p^\dagger \, a_h \right) a_h^\dagger \right\} |0\rangle$$

or

$$|S\rangle = \left\{ \exp \left(\sum_{h=1}^{N} \sum_{p=N+1}^{\infty} C_{ph} \, a_p^\dagger \, a_h \right) \right\} |G\rangle.$$

Evaluate $\langle G|S\rangle$ and $\langle S|S\rangle$ and give a meaning to C_{ph}.

P.4 (iv). Prove that the contribution to the ground-state energy of a connected diagram containing n independent hole lines is proportional to $\rho^n \Omega$, where ρ is the density and Ω the volume occupied by the system.

CHAPTER 5

FERMI FLUIDS

5.1. Introduction

In this chapter, we shall be concerned with essentially homogeneous Fermion systems. Clearly to give a basic microscopic theory, we must immediately specify the detailed interactions. On the other hand, a more general, though phenomenological, description is also possible without specifying the forces right at the outset. This is the Landau theory of Fermi liquids.

The systems of prime physical interest are the uniform electron gas and liquid helium three. In the former case, of immediate relevance to the problem of the electronic structure of metals, the forces are purely Coulombic, and, as we shall see, are amenable to discussion via the many-body perturbation theory of Chapter 4. Therefore, our main attention will be devoted to this problem. However, the second approach is briefly discussed at the end of the chapter, with particular reference to liquid helium three.

5.2. Physical description of screening in uniform electron gas

We turn immediately then to the microscopic theory of the uniform electron gas. This is the model of a metal due originally to Sommerfeld, which views the metal, in first approximation, as a gas of electrons moving in a uniform background of positive charge, chosen to maintain electrical neutrality. The 'smearing out' of the lattice ions into a uniform background leads to enormous simplifications, though at the expense of Brillouin zone structure, so basic to many properties of metals. Nevertheless, it gives us a simple framework within which to assess the main features of the electron-electron interactions, and on this aspect of the problem of metals we focus our whole attention here.

We shall see below that the effective forces operating in the gas are of much shorter range than the bare Coulomb interaction. Indeed, we can immediately use the density matrix perturbation theory of Chapter 1 to discuss the screening of a static charge in a high-density Fermi gas. In practice, of course, the shielding we wish to describe is that around one of the electrons in the gas, as it ploughs through the metal. This problem requires a time-dependent treatment, and indeed, as Ehrenreich & Cohen (1959) first showed, can again be handled precisely in a Hartree–Fock time-dependent framework (see Appendix 5 A. 2).

We then go on to consider the use of the many-body perturbation theory of Chapter 4. This leads us to the theory of Gell-Mann & Brueckner (1957; see also the earlier calculation by Macke (1950)), based on summing selected terms in the perturbation expansion. Following this, we consider at some length the model high-density Hamiltonian, due to Sawada (1957), as well as the dielectric function formulation of the high-density gas. We shall see that this leads to the collective modes of the gas, as well as the independent particle aspects. Indeed, some of the main results can be obtained by elementary classical methods. One such approach will be encountered in Chapter 8 where we have phonons specifically in mind, but the theory is the same, except that we must insert Coulombic interactions appropriately. The methods of this chapter give the classical plasma frequency as a first approximation and point the way to a systematic treatment of corrections.

To obtain a preliminary orientation on the nature of the effective interaction between electrons in a Fermi gas, we replace the electron we have singled out by a static charge without spin properties. We can then readily perform a Hartree calculation for the potential $V(\mathbf{r})$ due to that 'electron' at position $\mathbf{r}$ relative to it, using the Dirac density matrix theory discussed in Chapter 1.

There, we obtained the density $\rho(\mathbf{r})$ of electrons surrounding the 'electron' on which we are sitting, in terms of the unperturbed density ρ_0, as (cf. (1.7.10))

$$\rho(\mathbf{r}) = \rho_0 - \frac{k_f^2}{2\pi^3} \int V(\mathbf{r}') \frac{j_1(2k_f|\mathbf{r}-\mathbf{r}'|)}{|\mathbf{r}-\mathbf{r}'|^2} d\mathbf{r}'. \tag{5.2.1}$$

We have worked here only in a linear theory, and we shall now make the additional assumption, also discussed in Chapter 1, section 1.9, that $V(\mathbf{r})$ varies slowly in space (Thomas–Fermi approximation). Then we find

$$\rho(\mathbf{r}) = \rho_0 - \frac{q_0^2}{4\pi} V(\mathbf{r}), \tag{5.2.2}$$

where $q_0^2 = 4k_f/\pi$. Combining this with Poisson's equation

$$\nabla^2 V = 4\pi(\rho_0 - \rho) - 4\pi\delta(\mathbf{r}), \tag{5.2.3}$$

we have the simple result

$$V = \frac{1}{r}\exp\{-q_0 r\}, \tag{5.2.4}$$

and hence we conclude that the presence of the electron at the origin will hardly be felt at distances beyond the characteristic screening length q_0^{-1}. Since, in good metals, $q_0^{-1} \sim 1$ Å, the effect of the Coulomb interactions is seen, in practice, to be screened out very rapidly.

The physical reason for this behaviour is very simple. The chosen particle electrostatically repels other electrons from its immediate vicinity. The net effect of an electronic charge together with a diminution of negative charge around it is a much reduced force of screened Coulomb type. One can expect corrections to this result for (*a*) moving and spinning electrons and (*b*) for the deficiencies of the Thomas–Fermi method itself, but the qualitative picture remains.

5.2.1. *Fourier components of screened interaction*

The Fourier transforms $V(k)$ of the pure and screened Coulomb potentials are respectively $4\pi/k^2$ and $4\pi/(k^2+q_0^2)$. Let us now correct the latter result for the deficiencies of the Thomas–Fermi approximation. To do so, we simply have to use (5.2.1) together with Poisson's equation, rather than (5.2.2). Unfortunately the problem cannot now be solved in closed form in $\mathbf{r}$ space, but the Fourier components $V(k)$ can be obtained. Thus we notice that (5.2.1) and (5.2.3) take the form, in terms of $V(k)$,

$$\int k^2 V(k)\, e^{i\mathbf{k}.\mathbf{r}}\, d\mathbf{k} + \frac{2k_f^2}{\pi^2}\int V(k)\, e^{i\mathbf{k}.\mathbf{r}'} \frac{j_1(2k_f|\mathbf{r}-\mathbf{r}'|)}{|\mathbf{r}-\mathbf{r}'|^2}\, d\mathbf{r}'\, d\mathbf{k}$$
$$= 4\pi \int e^{i\mathbf{k}.\mathbf{r}}\, d\mathbf{k}. \tag{5.2.5}$$

But the definite integral over $\mathbf{r}'$ required in the second term on the left-hand side of (5.2.5) is given by

$$J(k_f, k) = \frac{2k_f^2}{\pi}\int e^{i\mathbf{k}.\mathbf{r}} \frac{j_1(2k_f r)}{r^2}\, d\mathbf{r}$$
$$= \left[\frac{2k_f}{\pi} + \left(\frac{2k_f^2}{\pi k} - \frac{k}{2\pi}\right)\ln\left|\frac{k+2k_f}{k-2k_f}\right|\right], \tag{5.2.6}$$

and hence we find from (5.2.5) the result

$$V(k) = \frac{4\pi}{k^2 + J(k_f, k)}. \tag{5.2.7}$$

The connection with the Thomas–Fermi screening is obtained when we approximate $J(k_f, k)$ by its value at $k = 0$. From (5.2.6) it then follows that the second term on the right-hand side contributes $2k_f/\pi$, as does the first term, yielding immediately the Fourier components $4\pi/(k^2 + q_0^2)$ of the screened Coulomb potential. The main difference between this Thomas–Fermi result and (5.2.7) occurs at $k = 2k_f$, where, according to (5.2.7), $V(k)$ has a discontinuous first derivative. But we know (see, for example, Lighthill, 1958) that such behaviour influences the asymptotic form of the Fourier transform very profoundly. In fact, Lighthill's results apply almost immediately to (5.2.7) and yield the asymptotic form for the interaction

$$V(r) \sim \frac{\pi}{k_f(1 + 2\pi k_f)^2} \frac{\cos 2k_f r}{r^3}, \tag{5.2.8}$$

a result that we see is basically different from a screened Coulomb form. The presence of such long-range oscillations in the displaced charge round a perturbation in an electron gas was first noticed by Blandin, Daniel & Friedel (1959), and the self-consistent oscillatory screening having the character of (5.2.8) was given by several authors independently (Langer & Vosko, 1959; March & Murray, 1960, 1961). The basic result (5.2.7) in **k**-space was already known to Bardeen (1937) and to Lindhard (1954).

We shall come back to the form (5.2.7) again, in sections 5.5 and 5.7, but, to summarize, we see that the static screening, which according to (5.2.4) is rather localized near the charge, with a screening radius q_0^{-1}, must be supplemented by the long-range oscillatory behaviour of (5.2.8). This last effect is fundamentally a diffraction phenomenon, the electron wave nature being essentially disregarded in the Thomas–Fermi calculation (roughly equivalent to the approximation of geometrical optics). In general, as we shall see, the static Fourier components $V(k)$ of (5.2.7) must be generalized to include time (or frequency) dependence, before the principal results of the electron gas theory can be obtained.

Having established that electron-electron interactions are heavily screened in a Fermi gas, we turn next to the calculation of the ground-state energy, using the many-body perturbation theory of Chapter 4. We shall see that the long range of the Coulomb interactions precludes immediate application of these results. However, the physical picture which emerged above gives us the clue to the modifications which have to be made.

5.3. Gell-Mann & Brueckner calculation of ground-state energy

We consider N electrons in a cube of side L. It is usual to define a mean interparticle separation r_s by the equation

$$\tfrac{4}{3}\pi r_s^3 = \frac{L^3}{N}. \tag{5.3.1}$$

The Hamiltonian will be taken to be that giving energies in atomic units or double Rydbergs:

$$H = -\frac{1}{2}\sum_1^N \nabla_{\mathbf{r}}^2 + \sum_{i<j} \frac{1}{|\mathbf{r}_i - \mathbf{r}_j|}. \tag{5.3.2}$$

5.3.1. *Coupling constant*

It is often valuable to refer to coupling strength rather than particle density. To see what is meant by this, we make the simple linear change of variable

$$\boldsymbol{\xi} = \left(\frac{3}{4\pi}\right)^{\frac{1}{3}} \frac{\mathbf{r}}{r_s}. \tag{5.3.3}$$

Then, using (5.3.2), one can write down a related Hamiltonian

$$H' = -\frac{1}{2}\sum_1^N \nabla_{\xi_i}^2 + \lambda \sum_{i<j} \frac{1}{|\boldsymbol{\xi}_i - \boldsymbol{\xi}_j|}, \tag{5.3.4}$$

where

$$H = \frac{1}{\lambda^2} H', \quad \lambda = \left(\frac{4\pi}{3}\right)^{\frac{1}{3}} r_s. \tag{5.3.5}$$

The original cube transforms into one of volume N in $\boldsymbol{\xi}$ space, the average particle density in the latter thus being unity.

Our prime interest will focus on the Hamiltonian (5.3.4) in which, clearly, λ (or r_s) plays the role of a coupling constant. It is then an easy matter to relate results concerning H' to those for H. For example, to convert energies, one uses a factor λ^2 according to (5.3.5). We see immediately that high densities (small r_s) in the original electron gas corresponds to weak coupling in (5.3.4) and thus we are in a position to consider the use of perturbation theory.

5.3.2. *Second quantized formalism*

The formalism of second quantization which we discussed in Chapter 3 turns out to be extremely convenient, so we now rewrite (5.3.4) as

$$H' = \sum_K \epsilon_K a_K^\dagger a_K + \frac{\lambda}{2} \sum_{K_1 K_2 K_3 K_4} \langle K_1 K_2 | v | K_3 K_4 \rangle a_{K_1}^\dagger a_{K_2}^\dagger a_{K_4} a_{K_3}, \tag{5.3.6}$$

where the a's and $a^\dagger$'s are the usual Fermion annihilation and creation operators, $K = (k, \sigma)$, $\epsilon_K = \epsilon_k = \frac{1}{2}k^2$ and $v(\xi) = 1/\xi$. The matrix element is defined as

$$\langle K_1 K_2 | v | K_3 K_4 \rangle = \iint d\mathbf{x}_1 \, d\mathbf{x}_2 \, \phi_{K_1}^*(x_1) \, \phi_{K_2}^*(x_2) \times \frac{1}{|\boldsymbol{\xi}_1 - \boldsymbol{\xi}_2|} \phi_{K_3}(x_1) \phi_{K_4}(x_2), \tag{5.3.7}$$

where $$\phi_K(\mathbf{x}) = \frac{1}{\sqrt{N}} e^{i\mathbf{k}.\boldsymbol{\xi}} \chi_\sigma(s), \quad d\mathbf{x} = d\boldsymbol{\xi} \, ds. \tag{5.3.8}$$

The evaluation of (5.3.7) using (1.3.6) yields

$$\langle K_1 K_2 | V | K_3 K_4 \rangle = \frac{1}{N} \delta_{\mathbf{k}_1+\mathbf{k}_2, \mathbf{k}_3+\mathbf{k}_4} \delta_{\sigma_1 \sigma_3} \delta_{\sigma_2 \sigma_4} v(\mathbf{k}_1 - \mathbf{k}_3), \tag{5.3.9}$$

where $$v(\mathbf{k}) = \int \frac{d\mathbf{r}}{r} e^{-i\mathbf{k}.\mathbf{r}} = \frac{4\pi}{k^2}. \tag{5.3.10}$$

(Strictly, the Fourier integral in (5.3.10) does not exist. The difficulty is resolved by supposing that initially our particles interact via screened Coulomb interactions. All computations are done and then finally the screening factor is put equal to zero in our results. The effect is to take throughout $v(k)$ as given by (5.3.10).)

Bearing in mind the above conservation conditions, we may rewrite (5.3.6) as

$$H' = \sum_{\mathbf{k}, \sigma} \epsilon_{\mathbf{k}} a_{\mathbf{k}\sigma}^\dagger a_{\mathbf{k}\sigma} + \frac{\lambda}{2N} \sum_{\mathbf{k}_1, \mathbf{k}_2, \mathbf{q}, \sigma_1, \sigma_2} v(\mathbf{q}) \, a_{\mathbf{k}_1+\mathbf{q}\sigma_1}^\dagger a_{\mathbf{k}_2-\mathbf{q}\sigma_2}^\dagger a_{\mathbf{k}_2\sigma_2} a_{\mathbf{k}_1\sigma_1}. \tag{5.3.11}$$

The potential energy is a weighted sum over operators representing the scattering of pairs of particles $(\mathbf{k}_1, \sigma_1)$, $(\mathbf{k}_2, \sigma_2)$ to $(\mathbf{k}_1 + \mathbf{q}, \sigma_1)$ and $(\mathbf{k}_2 - \mathbf{q}, \sigma_2)$ respectively with momentum transfer $\mathbf{q}$.

In the Hamiltonian (5.3.11), which will be written

$$H' = H_0 + V, \tag{5.3.12}$$

we see that V is a small perturbation when the coupling constant λ is small, the unperturbed Schrödinger equation being just

$$H_0|\Phi_0\rangle = E_0|\Phi_0\rangle, \tag{5.3.13}$$

where $|\Phi_0\rangle$ is the usual Slater determinant of plane waves, with orbitals occupied up to the Fermi level $k_f = (3\pi^2)^{\frac{1}{3}}$. The energy of the perturbed system is $E_0 + \Delta E$, where ΔE is specified by the Goldstone formula which we proved in Chapter 4. Specifically, this energy shift is obtained by summing all terms of the type (4.17.18). Symbolically, we write

$$\Delta E = \sum_n \langle \Phi_0 | V \left(\frac{1}{E_0 - H_0} V \right)^n |\Phi_0\rangle_L, \tag{5.3.14}$$

where L indicates that only linked graphs are to be taken.

We now proceed to investigate the various terms of (5.3.14). We shall find that first- and second-order contributions require special consideration, but that, at the high densities we are considering, the remaining orders can be treated systematically.

5.3.3. *First-order calculation*

The first-order terms are straightforward. The direct term (see Fig. 5.1) is a constant (albeit infinite), but the presence of the neutralizing positive background charge removes it completely. The exchange term is given by (cf. (4.17.6))

$$\frac{2\mathcal{E}^{(1)}_{\text{ex.}}}{\lambda^2 N} = -\frac{1}{\lambda N^2} \sum_{\substack{k_1 < k_f \\ k_2 < k_f \\ \sigma_1, \sigma_2}} \delta_{\sigma_1 \sigma_2} \frac{4\pi}{|\mathbf{k}_1 - \mathbf{k}_2|^2}$$

$$= -\frac{0.916}{r_s}, \tag{5.3.15}$$

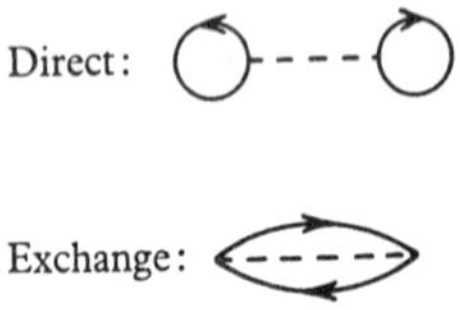

Fig. 5.1

the results following after some manipulation (cf. problem P.2(iv) of Chapter 2).

The expression (5.3.15) is just the potential energy obtained by using a plane-wave determinant variationally in this problem. Such a wave function characterizes independent particles and thus, while it contains the statistical property that electrons of parallel spin avoid each other to some extent, it does not reflect the property that Coulombic repulsion should also keep electrons apart (cf. Chapter 2). The inclusion of the latter correlations is our aim in the following and, in particular, we wish to evaluate the correlation energy, which is the energy to be added to the sum of the Fermi energy and the exchange energy (5.3.15) to give the exact energy. Thus defined, the correlation energy is negative.

5.3.4. *Second-order terms*

Two processes contribute to

$$\langle\Phi_0|V\frac{1}{E_0-H_0}V|\Phi_0\rangle,$$

namely those shown in Fig. 5.2.

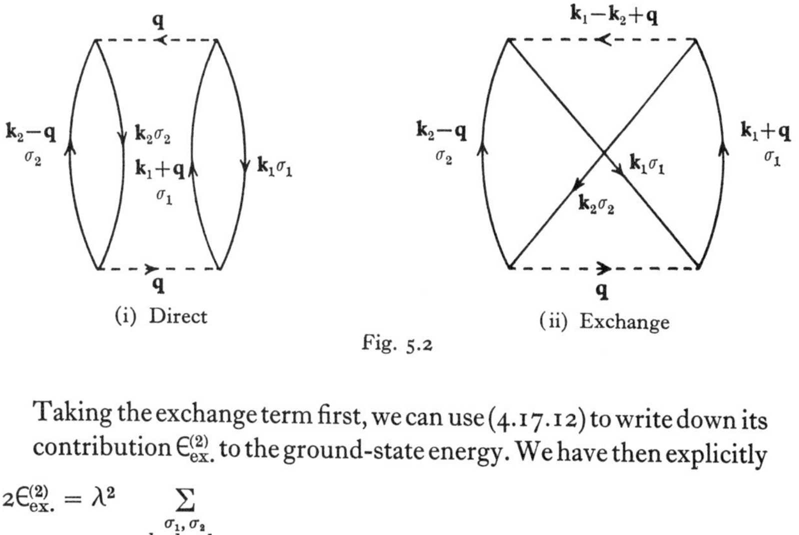

Fig. 5.2

Taking the exchange term first, we can use (4.17.12) to write down its contribution $\epsilon^{(2)}_{\text{ex.}}$ to the ground-state energy. We have then explicitly

$$2\epsilon^{(2)}_{\text{ex.}} = \lambda^2 \sum_{\substack{\sigma_1,\sigma_2\\ k_1,k_2<k_f\\ |\mathbf{k}_1+\mathbf{q}|,|\mathbf{k}_2-\mathbf{q}|>k_f}} \times\frac{(\mathbf{k}_2\,\sigma_2\,\mathbf{k}_1\,\sigma_1|v|\mathbf{k}_1+\mathbf{q}\sigma_1\,\mathbf{k}_2-\mathbf{q}\sigma_2)(\mathbf{k}_1+\mathbf{q}\sigma_1\,\mathbf{k}_2-\mathbf{q}\sigma_2|v|\mathbf{k}_1\,\sigma_1\,\mathbf{k}_2\sigma_2)}{\epsilon_{\mathbf{k}_1+\mathbf{q}}+\epsilon_{\mathbf{k}_2-\mathbf{q}}-\epsilon_{\mathbf{k}_1}-\epsilon_{\mathbf{k}_2}}$$

$$= \frac{\lambda^2}{N^2}\sum_{\substack{k_1,k_2<k_f\\ |\mathbf{k}_1+\mathbf{q}|,|\mathbf{k}_2-\mathbf{q}|>k_f}} 2\cdot\frac{4\pi}{(\mathbf{k}_1-\mathbf{k}_2+\mathbf{q})^2}\cdot\frac{(4\pi/q^2)}{\frac{(\mathbf{k}_1+\mathbf{q})^2}{2}+\frac{(\mathbf{k}_2-\mathbf{q})^2}{2}-\frac{k_1^2}{2}-\frac{k_2^2}{2}},$$

the first factor 2 after the summation sign coming from a summation over spins. Thus, substituting $-\mathbf{k}_2$ for $\mathbf{k}_2$,

$$\frac{\epsilon^{(2)}_{\text{ex.}}}{N} = -\frac{16\pi^2\lambda^2}{(2\pi)^9}\iiint_{\substack{k_1,\,k_2<k_f\\|\mathbf{k}_1+\mathbf{q}|,\,|\mathbf{k}_2+\mathbf{q}|>k_f}} d\mathbf{k}_1\,d\mathbf{k}_2\,d\mathbf{q}\frac{1}{(\mathbf{k}_1+\mathbf{k}_2+\mathbf{q})^2}\cdot\frac{1}{q^2+\mathbf{q}.(\mathbf{k}_1+\mathbf{k}_2)}\cdot\frac{1}{q^2} \quad (5.3.16)$$

where we have made the usual replacement

$$\frac{1}{N}\Sigma \to \frac{1}{(2\pi)^3}\int.$$

The details of the evaluation of this integral are given by Brueckner (1959). The essential feature, from our point of view, is that it is convergent and leads to

$$\frac{2\epsilon^{(2)}_{\text{ex.}}}{\lambda^2 N} = -0{\cdot}046. \quad (5.3.17)$$

Now let us consider the direct term in Fig. 5.2. This is (cf. (4.17.11))

$$2\epsilon^{(2)}_{\text{dir.}} = \lambda^2 \sum_{\substack{\sigma_1,\,\sigma_2\\k_1,\,k_2<k_f\\|\mathbf{k}_1+\mathbf{q}|,\,|\mathbf{k}_2-\mathbf{q}|>k_f}}$$

$$\times\frac{(\mathbf{k}_1\,\sigma_1\,\mathbf{k}_2\,\sigma_2|v|\mathbf{k}_1+\mathbf{q}\sigma_1\,\mathbf{k}_2-\mathbf{q}\sigma_2)(\mathbf{k}_1+\mathbf{q}\sigma_1\,\mathbf{k}_2-\mathbf{q}\sigma_2|v|\mathbf{k}_1\,\sigma_1\,\mathbf{k}_2\,\sigma_2)}{\epsilon_{\mathbf{k}_1+\mathbf{q}}+\epsilon_{\mathbf{k}_2-\mathbf{q}}-\epsilon_{\mathbf{k}_1}-\epsilon_{\mathbf{k}_2}}$$

$$= \frac{\lambda^2}{N^2}\sum_{\substack{k_1,\,k_2<k_f\\|\mathbf{k}_1+\mathbf{q}|,\,|\mathbf{k}_2-\mathbf{q}|>k_f}} 4\cdot\frac{4\pi}{q^2}\cdot\frac{1}{\dfrac{(\mathbf{k}_1+\mathbf{q})^2}{2}+\dfrac{(\mathbf{k}_2-\mathbf{q})^2}{2}-\dfrac{k_1^2}{2}-\dfrac{k_2^2}{2}}\cdot\frac{4\pi}{q^2}.$$

Hence, much as before

$$\frac{\epsilon^{(2)}_{\text{dir.}}}{\lambda^2 N} = \frac{32\pi^2}{(2\pi)^9}\iiint_{\substack{k_1,\,k_2<k_f\\|\mathbf{k}_1+\mathbf{q}|,\,|\mathbf{k}_2+\mathbf{q}|>k_f}} d\mathbf{k}_1\,d\mathbf{k}_2\,d\mathbf{q}\cdot\frac{1}{q^4}\cdot\frac{1}{q^2+\mathbf{q}.(\mathbf{k}_1+\mathbf{k}_2)}. \quad (5.3.18)$$

We will now show that (5.3.18) is, in fact, divergent. The nature of this divergence will turn out to be of great significance, so this matter will be considered in some detail. The following discussion parallels that of Brueckner (1959). We see that the potential danger of divergence arises from small q values in the integrand. Let us then consider

$$I(q) = \iint_{\substack{k_1,\,k_2<k_f\\|\mathbf{k}_1+\mathbf{q}|,\,|\mathbf{k}_2+\mathbf{q}|>k_f}} \frac{d\mathbf{k}_1\,d\mathbf{k}_2}{q^2+\mathbf{q}.(\mathbf{k}_1+\mathbf{k}_2)} \quad (q\text{ small}). \quad (5.3.19)$$

If q is small, because of the restrictions on the domains of the variables, $k_i \sim k_f$. Thus we may write $k_i = k_f - \delta$ ($\delta > 0$). No suffix is required on δ because of the symmetric roles of k_1 and k_2. The condition $|\mathbf{k}_i + \mathbf{q}| > k_f$ thus gives, on expanding and neglecting second-order terms, $\delta < q\mu_i$, where $\mu_i = \mathbf{k}_i . \hat{\mathbf{q}}$. Thus, we have

$$k_f - q\mu_i < k_i < k_f \quad (q \text{ small})$$

and so, to leading order in q,

$$I(q) = \int_0^1 2\pi\, d\mu_1 \int_{k_f - q\mu_1}^{k_f} k_1^2\, dk_1 \int_0^1 2\pi\, d\mu_2 \int_{k_f - q\mu_2}^{k_f} k_2^2\, dk_2 \times \frac{1}{q^2 + qk_1\mu_1 + qk_2\mu_2}.$$

Now the k_i integrations take place over a narrow rim. Thus, to leading order, we can replace the k_i in the integrand by k_f. To this order, then, dropping the q^2 term in the denominator,

$$I(q) = (2\pi k_f^2)^2 \frac{q}{k_f} \int_0^1 d\mu_1 \int_0^1 d\mu_2 \frac{\mu_1 \mu_2}{\mu_1 + \mu_2} = \tfrac{8}{3}\pi^2 k_f^3 (1 - \ln 2)\, q. \quad (5.3.20)$$

From (5.3.19) and (5.3.20), we see that the integral (5.3.18) is logarithmically divergent. Formally, we may write

$$\frac{\epsilon_{\text{dir.}}^{(2)}}{\lambda^2 N} = \frac{32\pi^2}{(2\pi)^9} \cdot \tfrac{8}{3}\pi^2 k_f^3 (1 - \ln 2) \int 4\pi q^2\, dq \cdot \frac{1}{q^4} \cdot q = \frac{2}{\pi^2}(1 - \ln 2) \int \frac{dq}{q}, \quad (5.3.21)$$

where the important behaviour is at the lower q limit.

The discussion of screening given in section 5.2 suggests a way out of our difficulties. The divergency arose because of the long-range nature of the Coulomb potential with its large Fourier components for small q. However, as we have seen, in an electron gas, the Coulomb potential is hardly experienced beyond a characteristic screening length q_0^{-1} and this is reflected by a screening out of the large Fourier components for $q < q_0$. This situation is not qualitatively changed on considering the modified system of 5.3.1. On taking an effective electronic charge of $\lambda^{\frac{1}{2}}$ instead of unity, one can easily generalize (5.2.4). A revised screening constant $q_0' = q_0 \lambda^{\frac{1}{2}}$ is then found, where $q_0^2 = 4k_f/\pi$, as before, although now $k_f = (3\pi^2)^{\frac{1}{3}}$, appropriate to unit particle density.

If we could modify our potential to one of appropriately screened type, (5.3.21) would contain a term $\int_{q_0'} \frac{dq}{q}$, which is no longer divergent. If we note that the lower limit is proportional to $r_s^{\frac{1}{2}}$, then the integral has the form $\ln r_s +$ constant, a convergent result which, however, tends to infinity as $r_s \to 0$. If this is the leading correction to the Hartree–Fock result at high densities, then no wonder our power series development fails. We have made a wrong expansion. It follows, therefore, that to get meaningful results we must re-sum in a different way. The method of summing all terms of primary significance at high densities will become apparent below. In fact, one gets the vital clue when studying third-order terms.

5.3.5. *Third-order contributions*

The possible processes are described graphically in Fig. 5.3. We will now consider their contributions to

$$\langle \Phi_0 | V \frac{1}{E_0 - H_0} V \frac{1}{E_0 - H_0} V | \Phi_0 \rangle.$$

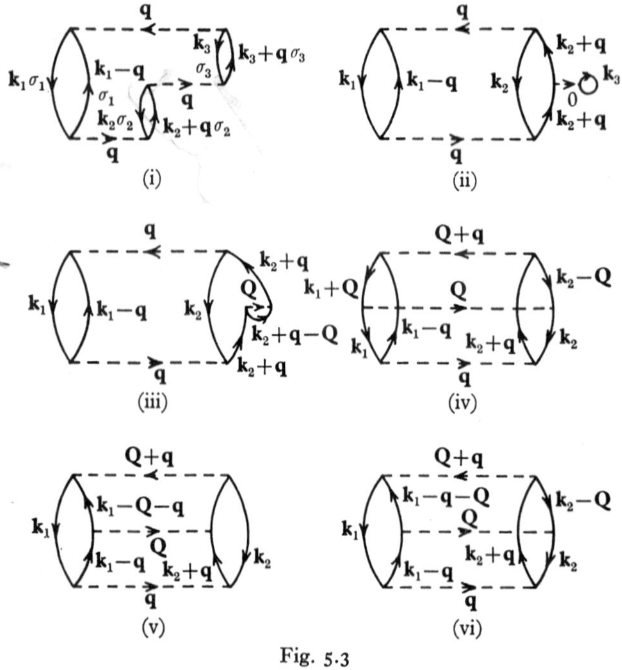

Fig. 5.3

Let us divide the graphs into two classes, ring graphs in which the momentum transfer along every interaction line is the same, and non-ring graphs. Thus the second-order direct diagram (Fig. 5.2 (i)) and Fig. 5.3 (i) above are both ring graphs. All other second- and third-order graphs are of non-ring type. Our intention is to show that for third order and above, ring diagrams contribute to the energy per particle a term of order unity, while contributions from non-ring graphs tend to zero with r_s.

To make this clear, consider the ring diagram (Fig. 5.3 (i)). Its contribution to the correlation energy is [recall (4.17.19) and remarks pertaining thereto] given by

$$\frac{\epsilon^{(3)}_{\text{ring}}}{\lambda^2 N} = \frac{4\lambda}{N^4} \sum_{\substack{\sigma_1\sigma_2\sigma_3 \\ k_1,\,k_2,\,k_3<k_f \\ |\mathbf{k}_1-\mathbf{q}|,\,|\mathbf{k}_2+\mathbf{q}|,\,|\mathbf{k}_3+\mathbf{q}|>k_f}}$$

$$\times \frac{\begin{array}{l}\langle \mathbf{k}_1\sigma_1\mathbf{k}_3\sigma_3|v|\mathbf{k}_1-\mathbf{q}\sigma_1\mathbf{k}_3+\mathbf{q}\sigma_2\rangle \\ \quad\times\langle \mathbf{k}_2\sigma_2\mathbf{k}_3+\mathbf{q}\sigma_3|v|\mathbf{k}_2+\mathbf{q}\sigma_2\mathbf{k}_3\sigma_3\rangle \\ \qquad\times\langle \mathbf{k}_2+\mathbf{q}\sigma_2\mathbf{k}_1-\mathbf{q}\sigma_1|v|\mathbf{k}_2\sigma_2\mathbf{k}_1\sigma_1\rangle\end{array}}{(\epsilon_{\mathbf{k}_3+\mathbf{q}}+\epsilon_{\mathbf{k}_1-\mathbf{q}}-\epsilon_{\mathbf{k}_1}-\epsilon_{\mathbf{k}_3})(\epsilon_{\mathbf{k}_2+\mathbf{q}}+\epsilon_{\mathbf{k}_1-\mathbf{q}}-\epsilon_{\mathbf{k}_1}-\epsilon_{\mathbf{k}_2})}$$

$$\sim r_s \iiiint_{\substack{k_1,\,k_2,\,k_3<k_f \\ |\mathbf{k}_1+\mathbf{q}|,\,|\mathbf{k}_2+\mathbf{q}|,\,|\mathbf{k}_3+\mathbf{q}|>k}}$$

$$\times\, d\mathbf{k}_1\, d\mathbf{k}_2\, d\mathbf{k}_3\, d\mathbf{q}\,.\frac{1}{q^2}\cdot\frac{1}{q^2+\mathbf{q}.(\mathbf{k}_3+\mathbf{k}_1)}\cdot\frac{1}{q^2}\cdot\frac{1}{q^2+\mathbf{q}.(\mathbf{k}_2+\mathbf{k}_1)}\cdot\frac{1}{q^2}$$

$$\sim r_s\int_{q_0'}\frac{d\mathbf{q}}{q^6}\cdot q \quad (q_0' = q_0\lambda^{\frac{1}{2}};\ \text{see end of section 5.3.4})$$

$$\sim \frac{r_s}{q_0'^2} \sim \text{const.},\quad \text{since } q_0' \sim r_s^{\frac{1}{2}}.$$

We see that the divergency is caused by a piling up of $1/q^2$ factors, one for each interaction line. Thus the contribution from any non-ring diagram will be weaker (since the number of $1/q^2$ factors will be fewer) and thus must vanish as $r_s \to 0$.

This argument turns out to be general and thus the leading terms $A \ln r_s + B$ in the correlation energy are evaluated exactly by calculating the contributions from all ring graphs and adding $\epsilon^{(2)}_{\text{ex.}}$ to it. Actually the series for the ring graphs may be summed for given q directly, as shown in Appendix 5A. 1, to obtain an expression which is then integrable over all q. The result for the correlation energy is

$$\frac{2\epsilon_{\text{corr.}}}{\lambda^2 N} = 0{\cdot}0622 \ln r_s - 0{\cdot}096 + O(r_s), \tag{5.3.22}$$

this final expression thus representing the correlation energy in Rydbergs per particle. This is, of course, the central result of the theory.

5.4. Sawada Hamiltonian

The calculation of Gell-Mann & Brueckner, described above, focussed attention on a single-particle determinant $|\Phi_0\rangle$, and corrections to the energy constructed from it.

But much of the progress in the electron gas problem stemmed from the fundamental observation of Bohm & Pines (1952) that the long-range Coulomb forces, while affecting the single-particle behaviour noticeably, also led to organized or collective oscillations of the electron gas. In general, these collective modes are not excited for this would require energy substantially greater than the Fermi energy, but they nevertheless play an important role as we shall see in detail below. In the simple treatments of plasma oscillations (cf. Raimes, 1961), a fundamental plasma frequency ω_p comes in immediately, related to the gas density ρ by

$$\omega_p = \left(\frac{4\pi\rho\, e^2}{m}\right)^{\frac{1}{2}}. \tag{5.4.1}$$

(This will essentially be derived by elementary arguments in chapter 8, cf. equation 8.2.74.)

One of the achievements of Sawada's work, which we consider below, is that the part played by the co-operative effects due to the Coulomb forces is clearly exhibited. This is in contrast to the perturbative approach of the previous section, where no clear separation between plasma modes and single-particle states emerged.

5.4.1. *Effective potential energy at high densities*

The starting-point for the simplification of the Hamiltonian is the recognition that many terms in (5.3.11) contribute only to non-ring diagrams, and can therefore be ignored in the high density theory. If we examine any ring graph, we see each vertex is characterized by the same momentum transfer $\mathbf{q}$ while creation and annihilation take place in particle hole pairs. No other kind of scattering occurs. Thus, following Sawada, we reject from the general form of the potential energy all scattering processes of other kinds.

To see how to do this let us rewrite the potential energy term of (5.3.11) in the alternative form

$$V = \frac{\lambda}{2N} \sum_{\substack{\mathbf{k}_1'-\mathbf{k}_1 \\ =\mathbf{k}_2-\mathbf{k}_2'=\mathbf{q}}} v(q)\, a^\dagger_{\mathbf{k}_1'\sigma_1} a^\dagger_{\mathbf{k}_2'\sigma_2} a_{\mathbf{k}_2\sigma_2} a_{\mathbf{k}_1\sigma_1}, \tag{5.4.2}$$

where the summations over the momenta are restricted only by the momentum conservation condition indicated. We now drop a large class of irrelevant terms by explicitly retaining only scattering processes of the type shown in the ring graph Fig. 5.3 (i), where it will be seen that all momentum transfers are equal. Then V simplifies to (subject to momentum being conserved as in (5.4.2))

$$V_s = \frac{\lambda}{2N}\left\{\sum_{\substack{k_1',\, k_2'<k_f \\ k_1,\, k_2>k_f}} + \sum_{\substack{k_1',\, k_2<k_f \\ k_2',\, k_1>k_f}} + \sum_{\substack{k_1,\, k_2'<k_f \\ k_2,\, k_1'>k_f}} + \sum_{\substack{k_1,\, k_2<k_f \\ k_1',\, k_2'>k_f}}\right\}$$

$$\times\, v(q)\, a^\dagger_{\mathbf{k}_1'\sigma_1} a^\dagger_{\mathbf{k}_2'\sigma_2} a_{\mathbf{k}_2\sigma_2} a_{\mathbf{k}_1\sigma_1} \tag{5.4.3}$$

$$= \frac{\lambda}{2N} \sum_{(\text{Excl.})}' v(q)\, \{a^\dagger_{\mathbf{k}_1'\sigma_1} a^\dagger_{\mathbf{k}_2'\sigma_2} a_{\mathbf{k}_2'+\mathbf{q}\sigma_2} a_{\mathbf{k}_1'-\mathbf{q}\sigma_1}$$

$$+\, a^\dagger_{\mathbf{k}_1'\sigma_1} a^\dagger_{\mathbf{k}_2-\mathbf{q}\sigma_2} a_{\mathbf{k}_2\sigma_2} a_{\mathbf{k}_1'-\mathbf{q}\sigma_1}$$

$$+\, a^\dagger_{\mathbf{k}_1+\mathbf{q}\sigma_1} a^\dagger_{\mathbf{k}_2'\sigma_2} a_{\mathbf{k}_2'+\mathbf{q}\sigma_2} a_{\mathbf{k}_1\sigma_1} + a^\dagger_{\mathbf{k}_1+\mathbf{q}\sigma_1} a^\dagger_{\mathbf{k}_2-\mathbf{q}\sigma_2} a_{\mathbf{k}_2\sigma_2} a_{\mathbf{k}_1\sigma_1}\}, \tag{5.4.4}$$

where Σ' denotes that momenta indices written as a sum refer to particles and single momentum indices refer to holes. (Excl.) means that the Exclusion Principle must be satisfied. Thus, for example, in the first term of (5.4.4), $\mathbf{k}_1' \neq \mathbf{k}_2'$ if $\sigma_1 = \sigma_2$. The rule for writing down the terms of expression (5.4.4) is to keep the same description of the holes in (5.4.3) and to re-define the particle indices in terms of the momentum conservation law displayed in (5.4.2). Equation (5.4.4) can be readily rewritten as follows:

$$V_s = \frac{\lambda}{2N} \sum_{\mathbf{q}} v(\mathbf{q}) \sum_{\substack{\mathbf{k},\, \mathbf{k}' \\ \sigma\sigma'}} \{a^\dagger_{\mathbf{k}\sigma} a^\dagger_{-\mathbf{k}'\sigma'} a_{-\mathbf{k}'-\mathbf{q}\sigma'} a_{\mathbf{k}+\mathbf{q}\sigma} + a^\dagger_{\mathbf{k}\sigma} a^\dagger_{\mathbf{k}'+\mathbf{q}\sigma'} a_{\mathbf{k}'\sigma'} a_{\mathbf{k}+\mathbf{q}\sigma}$$

$$+\, a^\dagger_{-\mathbf{k}-\mathbf{q}\sigma} a^\dagger_{-\mathbf{k}'\sigma'} a_{-\mathbf{k}'-\mathbf{q}\sigma'} a_{-\mathbf{k}\sigma} + a^\dagger_{-\mathbf{k}-\mathbf{q}\sigma} a^\dagger_{\mathbf{k}'+\mathbf{q}\sigma'} a_{\mathbf{k}'\sigma'} a_{-\mathbf{k}\sigma}\}$$

$$= \frac{\lambda}{2N} \sum_{\mathbf{q}} v(\mathbf{q}) \sum_{\substack{\mathbf{k},\, \mathbf{k}' \\ \sigma\sigma' \\ (\text{Excl.})}} \{d_{\mathbf{q}}(\mathbf{k}\sigma) + d^\dagger_{-\mathbf{q}}(-\mathbf{k}\sigma)\}\{d_{-\mathbf{q}}(-\mathbf{k}'\sigma') + d^\dagger_{\mathbf{q}}(\mathbf{k}'\sigma')\}, \tag{5.4.5}$$

where
$$d_{\mathbf{q}}(\mathbf{k}\sigma) = a^{\dagger}_{\mathbf{k}\sigma}\, a_{\mathbf{k}+\mathbf{q}\sigma}. \tag{5.4.6}$$
The operator d defined by (5.4.6) can evidently be regarded as a destruction operator for an electron-hole pair, while its Hermitian conjugate $d^{\dagger}$ corresponds to the creation of such a pair. We shall find that these operators afford powerful tools for discussing the elementary excitations of the gas.

One further point should be made concerning the form we have taken for V_s. We have ensured that all vertices are of the type associated with ring diagrams, that is of the forms shown in Fig. 5.4. However, not only ring diagrams consist of vertices entirely of this form. To construct an example, one need only exchange two particle lines in a ring graph in the manner shown in Fig. 5.5. The structure

Fig. 5.4. Vertices of ring diagrams.

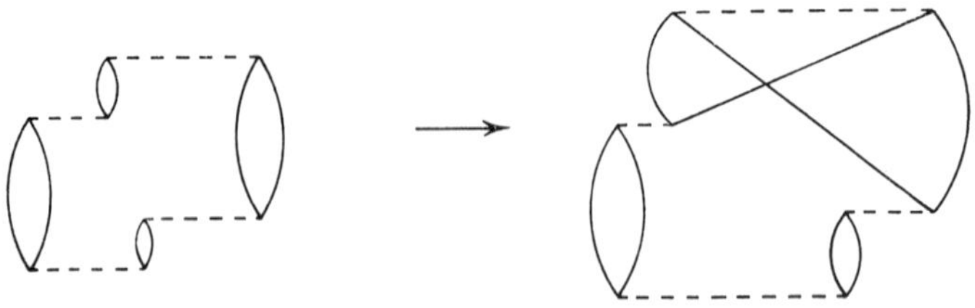

Fig. 5.5

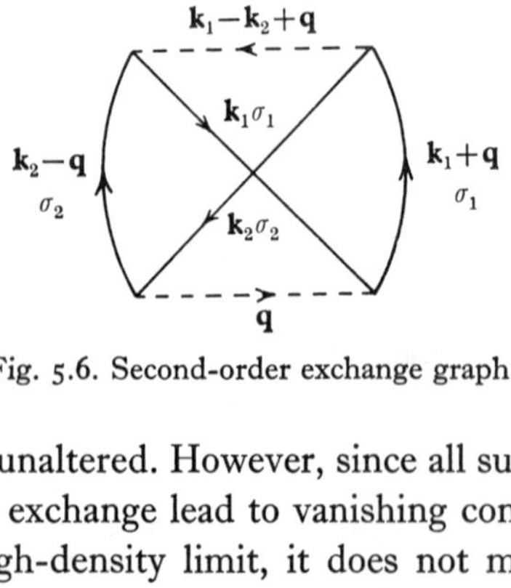

Fig. 5.6. Second-order exchange graph.

of the vertices is unaltered. However, since all such graphs except the second-order exchange lead to vanishing contributions to the energy in the high-density limit, it does not matter if they are included or not. The only matter to be settled is whether the second-order exchange graph contributes. The answer is no, since the two hole lines drawn mean $\mathbf{k}_1 \neq \mathbf{k}_2$ (to satisfy the exclusion principle) and this in turn means

$$\mathbf{k}_1 - \mathbf{k}_2 + \mathbf{q} \neq \mathbf{q},$$

i.e. the momentum transfers are not the same. In (5.4.4), however, they manifestly are the same.

Our conclusion is then that

$$H_s = T + V_s, \tag{5.4.7}$$

giving, as it does, the leading logarithmic term in the correlation energy, describes an electron gas at high density. To get the correlation energy exactly, as far as the constant term in the series, one adds $\epsilon^{(2)}_{\text{ex.}}$ as given by (5.3.16) to the exact ground-state energy of H_s.

It is clear that our next task is to investigate the problem posed by H_s. This turns out to be amenable to analytical treatment.

5.4.2. *Elementary excitations*

We shall draw attention here to the main steps in the argument, the detail being presented in Appendix 5A. 3. Suppose the ground state of a system is given by

$$H\Psi = E\Psi. \tag{5.4.8}$$

We wish now to construct the excited states. To do so, consider an operator Ω_K (a normal mode) such that

$$[H, \Omega_K^\dagger]\Psi = \omega_K \Omega_K^\dagger \Psi. \tag{5.4.9}$$

Combining (5.4.8) and (5.4.9) it follows immediately that

$$H\Omega_K^\dagger \Psi = (E + \omega_K)\Omega_K^\dagger \Psi, \tag{5.4.10}$$

and hence $\Omega_K^\dagger \Psi$ is an excited state of the system with excitation energy ω_K. Further, taking the conjugate of (5.4.9) and using (5.4.8) leads to the result

$$H\Omega_K \Psi = (E - \omega_K)\Omega_K \Psi. \tag{5.4.11}$$

Since (5.4.8) defines the ground state, and not (5.4.11), we must have

$$\Omega_K \Psi = 0. \tag{5.4.12}$$

If we now argue that a suitable form can be found for the ground-state wave function Ψ, then, in principle at least, (5.4.10) will allow us to construct the excited states and their corresponding excitation energies ω_K.

First, we can readily see how the above method works for independent Fermions, in which case, of course, we know the answer. The simplest (particle conserving) excitations are particle-hole pairs of the type described by (5.4.6). We can test this easily by checking that

$$[T, d_{\mathbf{q}}^{\dagger}(\mathbf{k}\sigma)] = (\epsilon_{\mathbf{k}+\mathbf{q}\sigma} - \epsilon_{\mathbf{k}\sigma})\, d_{\mathbf{q}}^{\dagger}(\mathbf{k}\sigma)$$
$$\equiv \omega_{\mathbf{q}}(\mathbf{k}\sigma)\, d_{\mathbf{q}}^{\dagger}(\mathbf{k}\sigma) \quad (k < k_f, |\mathbf{k}+\mathbf{q}| > k_f), \quad (5.4.13)$$

and, as we know, $(\epsilon_{\mathbf{k}+\mathbf{q}\sigma} - \epsilon_{\mathbf{k}\sigma})$ is the excitation energy associated with that state defined by operating with $d_{\mathbf{q}}^{\dagger}(\mathbf{k}\sigma)$ on the unperturbed Fermi sea.

Now let us apply the method to the Sawada Hamiltonian (5.4.7). The first thing to decide is what to take for Ω_K. We have a clue here, since experiment suggests that often the low-lying single-particle excitations are very like those of a system of non-interacting Fermions, and it then seems natural to inquire how well $d_{\mathbf{q}}(\mathbf{k}\sigma)$ will serve our purposes. We already have $[T, d_{\mathbf{q}}^{\dagger}(\mathbf{k}\sigma)]$ from (5.4.13) and it remains to examine $[V_s, d_{\mathbf{q}}^{\dagger}(\mathbf{k}\sigma)]$. For details, the reader may consult Appendix 5A. 2, where it is shown that

$$[V_s, d_{\mathbf{q}}^{\dagger}(\mathbf{k}\sigma)] = \frac{\lambda}{N} v(q) \sum_{\mathbf{k}'\sigma'} \{d_{\mathbf{q}}(\mathbf{k}'\sigma') + d_{-\mathbf{q}}^{\dagger}(-\mathbf{k}'\sigma')\}. \quad (5.4.14)$$

Although we have not yet found equations of the form of (5.4.9) we only have to notice now that while the d's and the $d^{\dagger}$'s are not directly suitable choices for Ω_K, certain linear combinations of them are. For, writing

$$\eta_q(\omega) = \sum_{\substack{\sigma,\, k<k_f \\ |\mathbf{k}+\mathbf{q}|>k_f}} \{C_{\mathbf{q}}(\mathbf{k}\sigma, \omega)\, d_{\mathbf{q}}(\mathbf{k}\sigma) + D_{\mathbf{q}}(\mathbf{k}\sigma, \omega)\, d_{-\mathbf{q}}^{\dagger}(-\mathbf{k}\sigma)\} \quad (5.4.15)$$

and requiring that $$[H_s, \eta_q^{\dagger}(\omega)] = \omega \eta_q^{\dagger}(\omega) \quad (5.4.16)$$

we find, on collecting terms

$$\sum_{\substack{\sigma,\, k<k_f \\ |\mathbf{k}+\mathbf{q}|>k_f}} \{\omega - \omega_{\mathbf{q}}(\mathbf{k}\sigma)\}\, C_{\mathbf{q}}^{*}(\mathbf{k}\sigma, \omega)\, d_{\mathbf{q}}^{\dagger}(\mathbf{k}\sigma)$$
$$+ \{\omega + \omega_{\mathbf{q}}(\mathbf{k}\sigma)\}\, D_{\mathbf{q}}^{*}(\mathbf{k}\sigma, \omega)\, d_{-\mathbf{q}}(-\mathbf{k}\sigma) - \frac{\lambda}{N} v(q) \{C_{\mathbf{q}}^{*}(\mathbf{k}\sigma, \omega)$$
$$- D_{\mathbf{q}}^{*}(\mathbf{k}\sigma, \omega)\} \sum_{\sigma',\, k'<k_f} \{d_{\mathbf{q}}(\mathbf{k}'\sigma') + d_{-\mathbf{q}}(-\mathbf{k}'\sigma')\} = 0. \quad (5.4.17)$$

This is just an eigenvalue equation for the excitation energies ω. It can be verified readily that the choice

$$C_{\mathbf{q}}^{*}(\mathbf{k}\sigma,\omega)=\frac{\text{constant}}{\omega-\omega_{\mathbf{q}}(\mathbf{k}\sigma)},\quad D_{\mathbf{q}}^{*}(\mathbf{k}\sigma,\omega)=\frac{\text{constant}}{\omega+\omega_{\mathbf{q}}(\mathbf{k}\sigma)} \tag{5.4.18}$$

satisfies (5.4.17) provided

$$\epsilon_{\mathbf{q}}(\omega)\equiv 1-\frac{\lambda}{N}v(q)\sum_{\substack{\sigma,\,k<k_f\\ |\mathbf{k}+\mathbf{q}|>k_f}}\left(\frac{1}{\omega-\omega_{\mathbf{q}}(\mathbf{k}\sigma)}-\frac{1}{\omega+\omega_{\mathbf{q}}(\mathbf{k}\sigma)}\right)=0. \tag{5.4.19}$$

Equation (5.4.19) is the Bohm–Pines dispersion relation, which we discuss in detail below.

5.4.3. *Discussion of plasma mode*

Using our (spin independent) definition of $\omega_{\mathbf{q}}(\mathbf{k}\sigma)$, given by (5.4.13), we may write (5.4.19) alternatively as

$$\alpha_{\mathbf{q}}(\omega)\equiv\frac{2\lambda}{N}v(q)\sum_{\substack{k<k_f\\ |\mathbf{k}+\mathbf{q}|>k_f}}\left(\frac{1}{\omega-[\frac{1}{2}q^2+\mathbf{q}\,.\,\mathbf{k}]}-\frac{1}{\omega+[\frac{1}{2}q^2+\mathbf{q}\,.\,\mathbf{k}]}\right)=1, \tag{5.4.20}$$

where (5.4.20) evidently defines α_q.

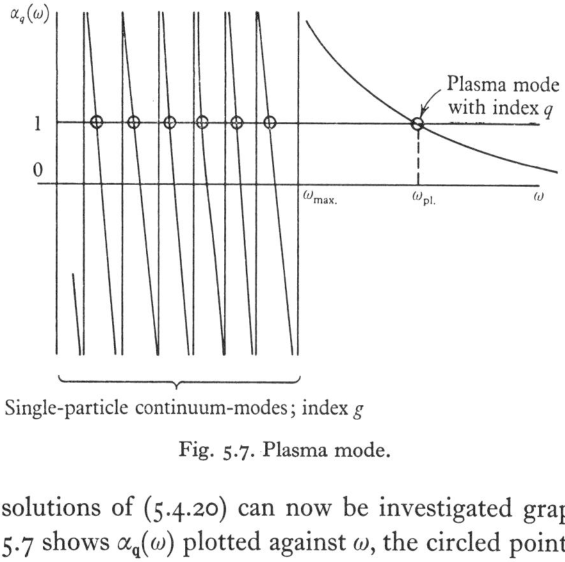

Fig. 5.7. Plasma mode.

The solutions of (5.4.20) can now be investigated graphically. Figure 5.7 shows $\alpha_{\mathbf{q}}(\omega)$ plotted against ω, the circled points corresponding to the intersections with the ordinate value of unity. The main feature of the solutions is that all except one are only slightly displaced from the independent-particle values $\omega_{\mathbf{q}}(\mathbf{k}\sigma)$ and like the latter, form a quasi-continuum from 0 to $\omega_{\max.}$ determined by the Fermi momentum. The exceptional point is on the far right where $\omega = \omega_{\mathbf{p}}$ is given by that branch of $\alpha_{\mathbf{q}}(\omega)$ which separates off from the continuum and takes on the value unity well above $\omega_{\max.}$.

The former kind of excited state is called a single-particle excitation because, near $\omega = \omega_{\mathbf{q}}(\mathbf{k}\sigma)$, we see from (5.4.17) that $d_{\mathbf{q}}(\omega)$ enters into $\eta_q(\omega)$ as given by (5.4.15) with overwhelming weight. Essentially then $\eta_q(\omega) \propto d_{\mathbf{q}}(\omega)$ and as has already been remarked, the energy is of independent particle character. However, the exceptional solution, being removed from the continuum, does not have this characteristic and the particles more or less contribute equally to the coefficients in (5.4.15). For this reason, this state is basically descriptive of collective behaviour, and is the plasma mode referred to earlier. (The time-dependence follows at once by relating the commutator (5.4.16) to the time derivative of $\eta_q(\omega)$.) It corresponds to a coherent movement of the electrons due to a change in all wave vectors by about the same amount $\mathbf{q}$.

The long wavelength (small q) limit of the plasma modes may be investigated in the following way. Since $\omega_p > \omega_{\mathbf{q}}(\mathbf{k}\sigma)$, we can expand (5.4.20) in powers of $\omega_{\mathbf{q}}(k)/\omega_p$:

$$\frac{2\lambda}{N} v(q) \sum_{\substack{k<k \\ |\mathbf{k}+\mathbf{q}|>k_f}} \left\{2\frac{\omega_{\mathbf{q}}(k)}{\omega_p^2} + 2\frac{\omega_{\mathbf{q}}^3(k)}{\omega_p^4} + \ldots\right\} = 1. \qquad (5.4.21)$$

Then, writing in explicitly that $\omega_{\mathbf{q}}(k) = \frac{1}{2}q^2 + \mathbf{q}.\mathbf{k}$ and $v(q) = 4\pi/q^2$, and noting that $\Sigma\mathbf{q}.\mathbf{k}$ vanishes because the summand is odd in $\mathbf{k}$, we find

$$\frac{2\lambda}{N}\cdot\frac{4\pi}{q^2} \sum_{\substack{k<k_f \\ |\mathbf{k}+\mathbf{q}|>k_f}} \left\{\frac{q^2}{\omega_p^2} + \frac{(\mathbf{q}.\mathbf{k})^3}{\omega_p^4} + \ldots\right\} = 1, \qquad (5.4.22)$$

or

$$\left\{\frac{2\lambda}{N}\cdot 4\pi \sum_{k<k_f} \frac{1}{\omega_p^2}\right\} + \text{higher terms in } q = 1. \qquad (5.4.23)$$

For very long waves, then, we have

$$\frac{2\lambda}{N}\cdot 4\pi \cdot \frac{1}{\omega_p^2}\cdot\frac{N}{2} = 1, \qquad (5.4.24)$$

or,

$$\omega_p = (4\pi\lambda)^{\frac{1}{2}}, \qquad (5.4.25)$$

which corresponds to the classical plasma oscillation frequency for a unit density gas with electronic charge $\lambda^{\frac{1}{2}}$. To convert to the original problem we now recall the rule after equation (5.3.5) and divide (5.4.25) by λ^2. This then reproduces the familiar result (5.4.1), with which we introduced the discussion of collective excitations.

Actually (5.4.20) can be integrated analytically (cf. Hubbard, 1958; Sawada, Brueckner, Fukuda & Brout, 1957). One finds that as q increases, $\omega_p(q)$ varies little from the classical value until it becomes highly unstable and effectively merges into the continuum at about $q = k_f$. For further details, we refer the reader to Pines (1961), Nozières & Pines (1958) and the other authors cited above.

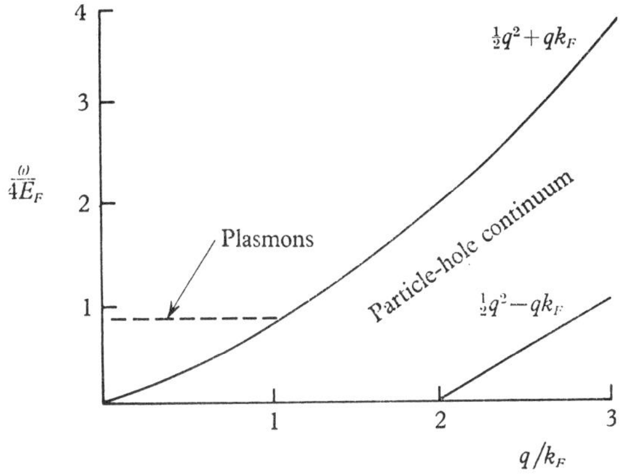

Fig. 5.8. Excitation spectrum.

We are thus in a position to understand the features of the excitation spectrum shown in Fig. 5.8. It should be stressed again that, for $q \ll k_f$, ω_p is several multiples of the Fermi energy, and thus plasma oscillations are somewhat difficult to excite. To this extent the system behaves much like a set of independent Fermions. Of course, the theory described above is a weak coupling approximation. But in practice it is found to contain many of the features discovered experimentally in metals.

5.5. Dielectric function of high-density Fermi gas

We turn now to an alternative way of studying the properties of an electron gas by finding the response of the system to a weak external perturbation. It will be convenient to transform the Hamiltonian of the electron gas as

$$H_0 = \Sigma \frac{p_i^2}{2m} + \mathscr{H}_{\text{coul.}}$$

$$\equiv \Sigma \frac{p_i^2}{2m} + \sum_k \frac{2\pi e^2}{k^2}(\rho_{\mathbf{k}}\rho_{-\mathbf{k}} - N), \tag{5.5.1}$$

where
$$\rho_{\mathbf{k}} = \sum_i e^{-i\mathbf{k}.\mathbf{r}_i} \tag{5.5.2}$$

are the Fourier components of the electron density $\rho(\mathbf{r})$, the so-called density fluctuations. The applied perturbation is taken to be a (non-localized) oscillatory test charge density

$$\rho_1 \equiv \{A_{\mathbf{k}} e^{-i(\mathbf{k}.\mathbf{r}+\omega t)} + \text{complex conjugate}\}. \tag{5.5.3}$$

Its interaction with the electron gas is given by

$$H_1 = e^2 \int \frac{\rho_1(\mathbf{r})\rho(\mathbf{r}')}{|\mathbf{r}'-\mathbf{r}|} d\mathbf{r}\, d\mathbf{r}' = e^2 e^{-i\omega t} A_{\mathbf{k}} \int d\mathbf{r}' \rho(\mathbf{r}') \int \frac{e^{-i\mathbf{k}.\mathbf{r}}}{|\mathbf{r}'-\mathbf{r}|} d\mathbf{r}$$
$$+ \text{complex conjugate}$$
$$= e^2 e^{-i\omega t} A_{\mathbf{k}} \int d\mathbf{r}' \rho(\mathbf{r}') e^{i\mathbf{k}'.\mathbf{r}'} \frac{4\pi}{k^2} + \text{complex conjugate}$$
$$= \frac{4\pi e^2}{k^2} A_{\mathbf{k}} \rho_{-\mathbf{k}} e^{i\omega t} + \text{complex conjugate}. \tag{5.5.4}$$

As we remarked in Chapter 4, we can view the result (5.5.4) in terms of switching on the interactions slowly from $t = -\infty$ until the interacting stationary state is reached at $t = 0$. Then, factor $e^{\eta t}$ can be thought of as multiplying (5.5.4), with the limit $\eta \to 0$ yielding the desired result. This device will be necessary in dealing with the time-dependent response of the system below, and this we now consider.

We suppose that the weakness of our probe means we can use low-order perturbation theory in the following way. Assuming that

$$H_0 \Psi_n = E_n \Psi_n \tag{5.5.5}$$

describes the interacting electron gas without the perturbation (the system, in fact, in which we are primarily interested), we may write the perturbed ground-state wave function as

$$\Psi(t) = \Sigma \Psi_n e^{-iE_n t/\hbar} a_n(t). \tag{5.5.6}$$

Solving this by first-order time-dependent perturbation theory under the conditions $a_0(-\infty) = 1$, $a_n(-\infty) = 0$ $(n \neq 0)$ we find in the usual way

$$a_n(t) = \frac{-4\pi e^2}{\hbar k^2}\left[\frac{A_{\mathbf{k}}\langle n|\rho_{-\mathbf{k}}|0\rangle\, e^{i(-\omega+\omega_{n0})t+\eta t}}{-\omega+\omega_{n0}-i\eta} + \frac{A_{\mathbf{k}}\langle n|\rho_{\mathbf{k}}|0\rangle\, e^{i(\omega_{n0}+\omega)t+\eta t}}{\omega_{n0}+\omega-i\eta}\right],$$

$$\text{where} \quad \omega_{n0} = \frac{(E_n - E_0)}{\hbar}, \tag{5.5.7}$$

are the exact frequencies of the density fluctuations $\rho_{\mathbf{k}}$.

Calculating the expectation value of $\rho_{\mathbf{k}}$ with respect to the perturbed ground-state wave function calculated to first order in $A_{\mathbf{k}}$ (translational invariance means $\langle n|\rho_{-\mathbf{k}}|0\rangle = 0$ if $\langle n|\rho_{\mathbf{k}}|0\rangle \neq 0$) we find

$$\langle \rho_{\mathbf{k}} \rangle = \langle \Psi(t)\,|\rho_{\mathbf{k}}|\Psi(t)\rangle,$$

which from (5.5.6) and (5.5.7) may be written

$$\langle \rho_{\mathbf{k}} \rangle = \frac{-4\pi e^2}{\hbar k^2} A_{\mathbf{k}} e^{-i\omega t+\eta t}\left[\sum_n |\langle n|\rho_{\mathbf{k}}|0\rangle|^2 \left\{\frac{1}{-\omega+\omega_{n0}-i\eta} + \frac{1}{\omega_{n0}+\omega+i\eta}\right\}\right]. \tag{5.5.8}$$

This analysis focusses attention on the quantity multiplying the amplitude $A_{\mathbf{k}}$ on the right-hand side of (5.5.8), and in particular

$$\frac{4\pi e^2}{\hbar k^2}\sum_n |\langle n|\rho_{\mathbf{k}}|0\rangle|^2 \left\{\frac{1}{\omega-\omega_{n0}+i\eta} - \frac{1}{\omega+\omega_{n0}+i\eta}\right\} \equiv \frac{1}{\epsilon(\mathbf{k},\omega)} - 1, \tag{5.5.9}$$

which serves as a definition of the dielectric function $\epsilon(\mathbf{k}, \omega)$, the $\epsilon_{\mathbf{k}}(\omega)$ of (5.4.19) being the lowest-order approximation to this expression.

5.5.1. *Properties of dielectric function*

Dispersion relation. From (5.5.8) and (5.5.9) we have

$$\frac{1}{\epsilon(\mathbf{k},\omega)} - 1 = \frac{\langle \rho_{\mathbf{k}} \rangle}{A_{\mathbf{k}} e^{-i\omega t}}. \tag{5.5.10}$$

This can be thought of as the link with the macroscopic case (compare the discussion of the dielectric constant in Appendix 5A. 3). For the moment, however, it is more important to note that, in general, we see from (5.5.8) that as $A_{\mathbf{k}} \to 0$, then $\langle \rho_{\mathbf{k}} \rangle \to 0$. However, this argument fails if the term $\langle \rho_{\mathbf{k}} \rangle / (A_{\mathbf{k}} e^{i\omega t})$ should be infinite. In view of (5.5.10), this is true if

$$\epsilon(\mathbf{k}, \omega) = 0. \tag{5.5.11}$$

Thus, the system continues to oscillate even when the perturbation is removed, provided k and ω satisfy (5.5.11). This equation then is the exact dispersion relation for an electron gas.

5.5.2. *Electron-electron interaction energy and sum rules*

Using the mathematical identities

$$\frac{1}{x-i\eta} = \frac{P}{x} + i\pi\delta(x); \quad \frac{1}{x+i\eta} = \frac{P}{x} - i\pi\delta(x), \tag{5.5.12}$$

where P denotes the principal value, and taking imaginary parts in (5.5.10) we obtain

$$\operatorname{Im} \frac{1}{\epsilon(\mathbf{k}, \omega)} = \frac{-4\pi^2 e^2}{\hbar k^2} \sum_n |\langle n|\rho_{\mathbf{k}}|0\rangle|^2 \{\delta(\omega - \omega_{n0}) + \delta(\omega + \omega_{n0})\}, \tag{5.5.13}$$

and thus

$$-\int \operatorname{Im} \frac{1}{\epsilon(\mathbf{k}, \omega)} d\omega = \frac{4\pi^2 e^2}{\hbar k^2} \Sigma |\langle n|\rho_{\mathbf{k}}|0\rangle|^2 = \frac{4\pi^2 e^2}{\hbar k^2} \langle 0|\rho_{\mathbf{k}} \rho_{-\mathbf{k}}|0\rangle, \tag{5.5.14}$$

the latter step following from the completeness of the set Ψ_n and the definition of $\rho_{\mathbf{k}}$. The electron-electron interaction energy in the ground state may be found in terms of the dielectric constant, and explicitly we have

$$E_{\text{int.}} = \langle 0|\mathscr{H}_{\text{coul.}}|0\rangle = -\sum_{\mathbf{k}} \left[\frac{\hbar}{2\pi} \int_0^\infty \operatorname{Im} \frac{1}{\epsilon(\mathbf{k}, \omega)} d\omega + \frac{2\pi N e^2}{k^2} \right]. \tag{5.5.15}$$

In principle, the ground-state energy can be obtained from complete knowledge of $E_{\text{int.}}$ as a function of r_s, or of the effective coupling constant e^2. A further result, which we use below, can be obtained by multiplying (5.5.13) by ω and integrating over ω. We then find

$$\int_0^\infty \omega \operatorname{Im} \frac{1}{\epsilon(\mathbf{k}, \omega)} d\omega = \frac{-4\pi^2 e^2}{\hbar k^2} \sum_n |\langle n|\rho_{\mathbf{k}}|0\rangle|^2 \omega_{n0}. \quad (5.5.16)$$

A well-known result, the f-sum rule, now tells us that (cf. Pines, 1961)

$$\sum_n \frac{2m}{\hbar k^2} \omega_{n0} |\langle n|\rho_{\mathbf{k}}|0\rangle|^2 = N \quad (5.5.17)$$

and hence

$$\int_0^\infty \omega \operatorname{Im} \frac{1}{\epsilon(\mathbf{k}, \omega)} d\omega = -\frac{\pi}{2}\left(\frac{4\pi N e^2}{m}\right). \quad (5.5.18)$$

From our general discussion, it is possible to derive an exact form for $\epsilon(\mathbf{k}, \omega)$ as $\omega \to \infty$. The result, however, is the same as given by the random phase approximation which we consider below, and we shall not give the proof.

5.6. Van Hove correlation function

The above discussion of the dielectric function $\epsilon(\mathbf{k}, \omega)$ can be used as a convenient way of introducing the important correlation function $S(\mathbf{k}, \omega)$ due to Van Hove. We shall meet this subsequently on numerous occasions and we shall not discuss it in its most general terms at this point. Suffice it to say that equation (5.5.14) focusses attention on the quantity

$$\langle 0|\rho_{\mathbf{k}}\rho_{-\mathbf{k}}|0\rangle,$$

and as we shall see later in detail, this is N times the structure factor $S(\mathbf{k})$. The Van Hove correlation function effects a generalization of this such that the frequency-dependent Van Hove function is related to $S(\mathbf{k})$ through

$$S(\mathbf{k}) = \int_0^\infty S(\mathbf{k}, \omega)\, d\omega. \quad (5.6.1)$$

Thus, from (5.6.1) and (5.5.14), it is natural to write

$$\frac{-4\pi^2 e^2}{\hbar k^2} S(\mathbf{k}, \omega) = \operatorname{Im} \frac{1}{\epsilon(\mathbf{k}, \omega)} \quad (\omega > 0). \quad (5.6.2)$$

The result (5.5.18) above clearly gives us for the first moment of ω:

$$\int_0^\infty \omega\, S(\mathbf{k}, \omega)\, d\omega = \frac{N\hbar k^2}{2m}. \quad (5.6.3)$$

The descriptions of the electron gas in terms of $\epsilon(\mathbf{k}, \omega)$ or $S(\mathbf{k}, \omega)$ are seen then to be essentially equivalent, but, as remarked, the Van Hove function is of wide applicability (see Chapter 8).

5.7. Relation to high-density theory

The dielectric formalism developed above has the merit of generality but, nevertheless, to use an equation such as (5.5.13) in practice we must approximate the exact wave functions Ψ_n. In the high-density limit, the approximation involves putting ω_{n0} equal to the free-electron values, and while (5.5.10) can be evaluated by this procedure, we have, in fact anticipated the result in the treatment via the Sawada Hamiltonian in section 5.4. Equation (5.4.19) gives in reality $\epsilon(\mathbf{k}, \omega)$ in the high-density limit, and we shall now show that various limiting cases follow immediately:

(*a*) The Hartree treatment of shielding of a static charge given in section 5.2 is recovered when $\omega = 0$. Each component $V(\mathbf{k})$ of that treatment can be thought of as screened appropriately to become $V(\mathbf{k})/\epsilon(\mathbf{k}, 0)$. The static dielectric constant then follows from either (5.4.20) or (5.2.7) as

$$\epsilon(\mathbf{k}, 0) = 1 + \frac{q_0^2}{k^2}\left[\frac{1}{2} + \frac{k_f}{2k}\left(1 - \frac{k^2}{4k_f^2}\right)\ln\left|\frac{k+2k_f}{k-2k_f}\right|\right]. \tag{5.7.1}$$

(*b*) For very high frequencies, it is again straightforward to evaluate (5.4.20) and we find

$$\epsilon(\mathbf{k}, \omega) \approx 1 - \frac{\omega_p^2}{\omega^2}, \tag{5.7.2}$$

a result which, as we emphasized above, is generally valid in this limit.

(*c*) For small k, again from (5.4.20), we find

$$\epsilon(\mathbf{k}, \omega) \approx 1 + \frac{q_0^2}{k^2}. \tag{5.7.3}$$

5.8. Low-density electron gas

The previous discussion has assumed that the coupling between electrons is weak, and that, in some sense, perturbation theory can be applied. However, real metal densities lie in a range of r_s given by

$2 < r_s < 5{\cdot}5$, using atomic units, whereas the high-density theory is valid for $r_s \ll 1$.

There is now a variety of evidence, experimental and theoretical, that the high-density properties discussed above, with the important exception of (5.3.22), are in general reflected in the actual behaviour of conduction electrons in metals. Nevertheless, it is of some interest to consider the strong coupling or low-density limit.

The arguments on which our discussion will be based are intuitive. Let us briefly recall the conclusions of Chapter 2. When we describe a molecule, H_2, we could use, as a starting approximation, either molecular orbitals, in which each electron belonged to the whole system and had an orbital extending over the molecule, or we could use a localized orbital description. Clearly, in the electron gas, the molecular orbitals of the H_2 molecule become plane waves $e^{i\mathbf{k}.\mathbf{r}}$, on which all the high-density arguments are based. Now it is true that a determinant of plane waves can always be rewritten in terms of localized orbitals, the Wannier functions. But these functions when derived from plane waves, are very poor approximations, as we shall see below, in a low-density electron gas.

This all suggests, therefore, that we should turn our attention away from molecular orbitals, and, as we did in the Heitler–London treatment of the H_2 molecule, think physically about the role of electron interactions. In that case, we argued that, when the atoms were far apart, Coulomb repulsions between the two electrons would keep these on their own atoms, and suppress any tendency for them to occupy molecular orbitals. Naturally, the physical picture is rather different in the electron gas, but nevertheless, just as the effective way of minimizing the Coulomb energy was to keep the electrons in H_2 on their own atoms, so, in the strong coupling limit of the electron gas, we can argue that we must confine each electron to its 'own region' of space.

Then, as Wigner (1934, 1938) pointed out, when the electron interactions become sufficiently strong, the most effective way for the electrons to avoid one another is to go on to the sites of a lattice. In the extreme limit of strong coupling, the energy of such an electron lattice is purely electrostatic, and the lowest energy found, for the structures so far investigated, is for a body-centred cubic lattice (see, for example, Fuchs, 1935). The energy/particle in Rydbergs may then be written

$$E/N = -\frac{1{\cdot}792}{r_s}, \tag{5.8.1}$$

which is substantially lower than the exchange energy given by $-0{\cdot}916 r_s^{-1}$ Rydbergs. As we now relax the coupling a little, clearly the electrons will vibrate about these lattice sites, and for a careful wave mechanical discussion we refer the reader to Carr (1961). For the present purposes, we shall illustrate the main physical points by a simpler discussion, based on essentially an Einstein model of such oscillations.

5.8.1. *Wigner orbitals and total energy*

We shall suppose that the coupling is so strong that the electrons vibrate but little about their respective lattice sites. We can then construct the field in which they vibrate by looking at a Wigner–Seitz cell, arguing that, due to its high symmetry, it may be replaced by a sphere, and that, within the sphere, the only significant potential acting on an electron is that due to the positive background charge within the sphere. The other cells, due to their high symmetry, make only multipole contributions to the potential within the central cell we are considering. Thus the potential energy in which the electron moves is simply

$$-\frac{e^2 r^2}{2 r_s^2} + \text{const.} \tag{5.8.2}$$

and the ground-state wave function is that of a three-dimensional isotropic harmonic oscillator, given by

$$\psi = \left(\frac{\alpha}{\pi}\right)^{\frac{3}{4}} e^{-\frac{1}{2}(\alpha r^2)}, \quad \alpha = r_s^{-\frac{3}{2}}. \tag{5.8.3}$$

In this approximation, the kinetic energy/particle, $T/N = 3/(2r_s^{\frac{3}{2}})$ Rys, and hence the total energy is given by (energy of harmonic oscillator is half potential and half kinetic)

$$\frac{E}{N} = -\frac{1{\cdot}792}{r_s} + \frac{3}{r_s^{\frac{3}{2}}} + \dots. \tag{5.8.4}$$

A more careful calculation, based on a correct lattice dynamical analysis, in contrast with the above use of the Einstein model, gives 2·66 for the coefficient of the term in $r_s^{-\frac{3}{2}}$ (Carr, 1961; Coldwell-Horsfall & Maradudin, 1963).

Having discussed the energy per particle in the limits of weak and strong coupling, we turn to a brief discussion of the régime of intermediate density.

5.9. Approaches for intermediate densities

So far, such calculations as exist remain something in the nature of interpolation schemes. The early work of Wigner, referred to above, led to correlation energies which have been generally substantiated by subsequent workers. Wigner's rough interpolation used

$$\epsilon_{\text{correlation}} = \frac{-0{\cdot}88}{r_s+7{\cdot}8}\,\text{Ry}, \tag{5.9.1}$$

but this was before the high-density theory was developed, and (5.9.1) is too crude as $r_s \to 0$.

The important early papers of Hubbard remain still among the most careful estimates (see also very recent work by Rice (1965) in which exchange terms along the lines of Hubbard's discussion are calculated), but interpolation procedures have been fully investigated by Nozières & Pines (1958) and Carr *et al.* (1961). Alternative approaches, based on a many-body wave function of the form of a determinant multiplied by products of two-particle correlation functions, have been studied, using collective co-ordinates, particularly by Gaskell (1961), who again finds correlation energies in good agreement with other workers.

To summarize the results of these approaches, we show in Fig. 5.9 a selection of the results so far obtained. It is worth pointing out that, once the total energy/particle E is known as a function of r_s, the kinetic and potential energies can be obtained from the virial theorem (March, 1958), which takes the form

$$2T+V = -r_s\frac{dE}{dr_s}. \tag{5.9.2}$$

Equation (5.9.2) is almost obvious when we note that the left-hand side equal to zero is the customary form of the virial theorem for a system in equilibrium under the influence of purely Coulomb forces. The right-hand side can be regarded as the virial $3p\Omega$ of the

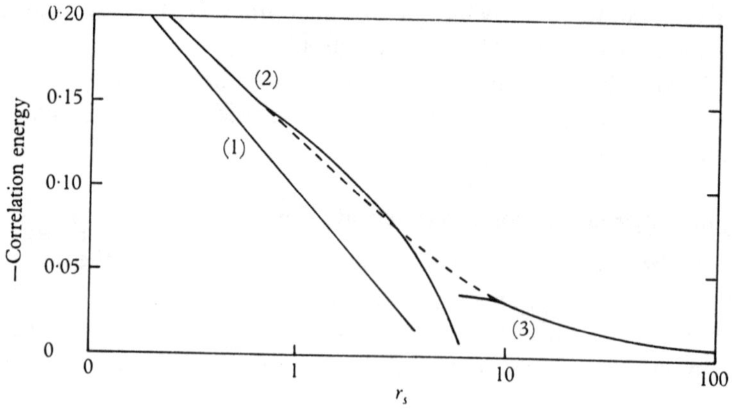

Fig. 5.9. Correlation energy as function of gas density. Units are Rydbergs with r_s in atomic (Bohr) units. (1) Gell-Mann & Brueckner result: $0{\cdot}0622 \ln r_s - 0{\cdot}096$. (2) Best available high density results (Carr & Maradudin, 1964). (3) Low density theory (Carr *et al.* 1961). Dashed curve is an interpolation.

pressure p arising from the bombardment of the walls of the 'container' by the electron gas. At absolute zero, $p = -dE/d\Omega$, from thermodynamics, and hence (5.9.2) follows.

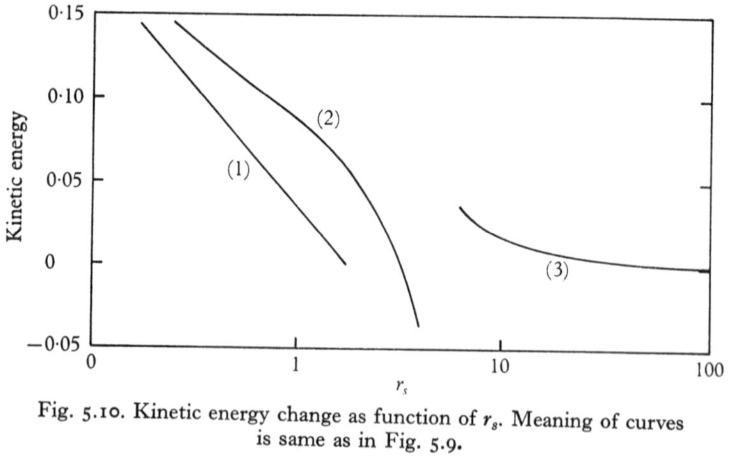

Fig. 5.10. Kinetic energy change as function of r_s. Meaning of curves is same as in Fig. 5.9.

Figures (5.10) and (5.11) show the total correlation energy separated into kinetic and potential energies respectively. In particular, the limiting forms of T and V as $r_s \to 0$ can be found from the Gell-Mann & Brueckner energy and are

$$T = \bar{E}_f - A \ln r_s - (A+C) + \dots \qquad (5{\cdot}9{\cdot}3)$$

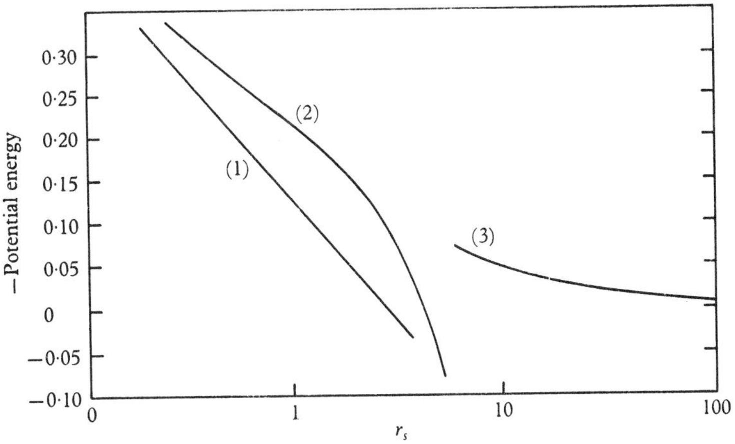

Fig. 5.11. Potential energy change as function of r_s. Meaning of curves is same as in Fig. 5.9.

and $$V = V_{ex} + 2A \ln r_s + (A + 2C), \qquad (5.9.4)$$
where $A = (2/\pi^2)(1 - \ln 2)$, $C = -0{\cdot}096$, $\bar{E}_f = 2{\cdot}21 r_s^{-2}$
and $$V_{ex} = -0{\cdot}916 r_s^{-1}.$$

As we have stressed, correlations must always promote particles into states outside the Fermi sphere, and hence increase the kinetic energy, and this is clearly seen in Fig. 5.10. A physical understanding of the change in T from the Fermi energy $\propto r_s^{-2}$ for high density to the vibrational form $r_s^{-\frac{3}{2}}$ at low density is found in the argument that because of the localization implied by the wave function (5.6.3), the electrons only occupy a fraction $r_s^{-\frac{3}{4}}$ of the total volume, and since the Fermi energy is $\propto (N/\Omega)^{\frac{2}{3}}$, we find that the kinetic energy $\propto (1/r_s^2)(r_s^{\frac{3}{4}})^{\frac{2}{3}}$ or $\propto r_s^{-\frac{3}{2}}$ as required (cf. Carr, 1961).

The way in which such localization is achieved, as we pass through the intermediate density range, must be a very complex process indeed. For the present, it is our opinion that the best method of tackling the intermediate density range resides in the application of the variational theorem (cf. Gaskell's work; though here, for practical evaluation, the random phase approximation is again invoked). Increasing attention is to be expected in the direction of variational methods based on Green functions or density matrices. Many unsolved problems remain, connected with the formulation of necessary and sufficient conditions that a second-order Green function or density matrix shall be variationally valid (see, for example, Mayer, 1955; Tredgold, 1957; Young & March, 1960).

5.10. Dependence of pair function and momentum distribution on gas density

Having discussed the high-density gas fully, and the strong coupling limit in a semi-quantitative way, we turn finally to the forms of the pair function and momentum distribution as the interparticle separation r_s is varied.

5.10.1. *Momentum distribution*

We consider first the probability $P(\mathbf{k})$ of occupation of a state $\mathbf{k}$. Clearly, when the density becomes very high, $P(k)$ tends to the usual Fermi distribution

$$\left.\begin{aligned} P(k) &= 1 \quad (k < k_f), \\ &= 0 \quad (k > k_f). \end{aligned}\right\} \qquad (5.10.1)$$

When we switch on the interactions, the development of (5.10.1) is difficult to deal with rigorously. In any perturbative treatment it is evident that the discontinuity will persist at k_f, though it will be reduced, since interactions inevitably excite some particles outside the Fermi sphere.

Daniel & Vosko (1960) have made a calculation of $P(k)$ by means of high-density perturbation theory. We shall not report on it in detail here but simply note that they find a reduction in the discontinuity as the density is increased. We show their results in curve 1 of Fig. 5.12, for $r_s = 2$.

We next turn to the strong coupling limit, discussed in a semi-quantitative way using the Wigner orbitals obtained in section 5.6.1 by March & Sampanthar (1962). Here, the momentum distribution turns out to be simply the Fourier transform of the Wigner orbital, given by

$$P(K) = \frac{3\pi^{\frac{1}{2}}}{r_s^{\frac{3}{4}}} \exp\left\{-\left(\frac{9\pi}{4}\right)^{\frac{2}{3}} r_s^{-\frac{1}{2}} K^2\right\}, \qquad (5.10.2)$$

where $k/k_f = K$. It is worth noting that the corresponding spinless first-order density matrix is given by

$$\gamma(\mathbf{r}'\mathbf{r}) = \frac{k_f^3}{3\pi^2} \exp\left\{-\frac{|\mathbf{r}' - \mathbf{r}|^2}{4r_s^{\frac{3}{2}}}\right\}, \qquad (5.10.3)$$

giving a uniform particle density $k_f^3/3\pi^2$ as required.

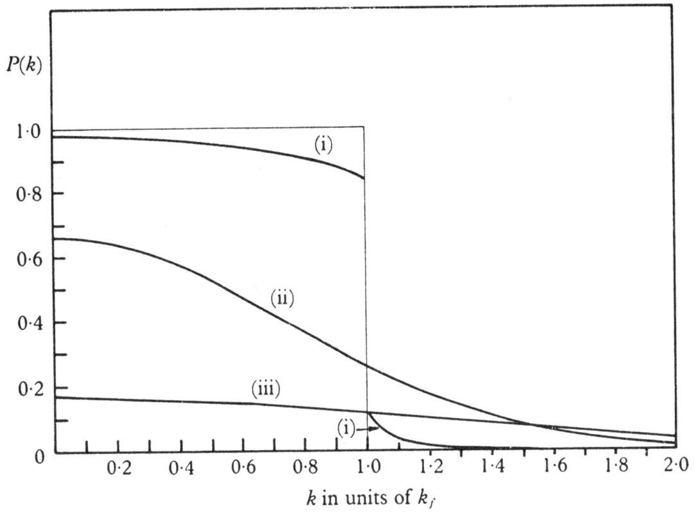

Fig. 5.12. Momentum distribution. (i) High-density theory $r_s = 2$. (ii) Low-density result (5.10.2) for $r_s = 4$. (iii) Low-density result (5.10.2) for $r_s = 100$.

The range of validity of (5.10.2) and (5.10.3) is restricted, because the Wigner orbitals have been assumed orthogonal. A necessary, though not sufficient condition for the applicability of (5.10.2), is that $0 < P(K) < 1$, and thus we must have

$$\frac{3\pi^{\frac{1}{2}}}{r_s^{\frac{3}{4}}} < 1, \tag{5.10.4}$$

or $r_s > 9.3$.

The point we wish to stress is that in the strong coupling limit there is no longer any sign of the discontinuity characteristic of weakly interacting Fermions. For sufficiently low density, this conclusion may be shown to be unchanged when the Wigner orbitals are made orthogonal to one another.

5.10.2. *Meaning of a Fermi surface*

The evidence discussed above supports the view that, as we follow the momentum distribution as it develops from the limiting case of small r_s, there must come a critical coupling strength, or a critical density, at which the discontinuity in the momentum distribution is reduced to zero. For lower densities, it then appears that the concept of a Fermi surface will no longer be useful. No quantitative evaluation of the critical density has so far proved possible. Questions also remain as to the nature and order of the 'transition' occurring at the critical density. All the evidence in metals points to the fact that the transition occurs for $r_s \gg 5.5$, this value corresponding to the lowest electron density metal, caesium.

5.10.3. *Pair function*

For non-interacting Fermions, the pair function is readily obtained from the second-order spinless density matrix, and, normalized to unity at large distances, it is given by

$$g(\mathbf{r}'\mathbf{r}) = 1 - \frac{9}{2}\left\{\frac{j_1(k_f|\mathbf{r}'-\mathbf{r}|)}{k_f|\mathbf{r}'-\mathbf{r}|}\right\}^2, \tag{5.10.5}$$

where $$j_1(\rho) = \frac{\sin\rho - \rho\cos\rho}{\rho^2}.$$

A high-density perturbative calculation to include the effect of interactions in the result (5.10.5) has been made by several workers (Ueda, 1961; March & Sampanthar, 1962), Ueda plotting some modified forms of (5.10.5) for various r_s.

However, some analytical progress concerning the effect of interactions can be made if we follow the collective co-ordinate approach adopted by Gaskell (1961, 1962). From this work, it may be shown that, whereas the unperturbed form (5.10.5) falls off at large separation r like $[\cos^2(k_f r)]/r^4$, or $[1+\cos 2k_f r]/r^4$, the effect of the interactions is to remove the r^{-4} term, and leave the form $r^{-4}\cos 2k_f r$.‡ This is in interesting contrast to the distribution of electrons round a static, spinless charge, which, as we saw in section 5.2.1, falls off as $r^{-3}\cos 2k_f r$.

These calculations all have features which restrict them to the high-density régime. In the low-density electron gas, the pair function can be constructed immediately from the Wigner orbitals, as the spherical average of the sum of the squares of these orbitals, centred on every lattice site of the body-centred cubic lattice other than the origin. The results are shown in curves 1 and 2 of Fig. 5.13, while the non-interacting Fermi hole is shown in curve 3. The oscillations in the Fermi hole function, which incidentally always lies

‡ Gaskell, Jones & March (to be published).

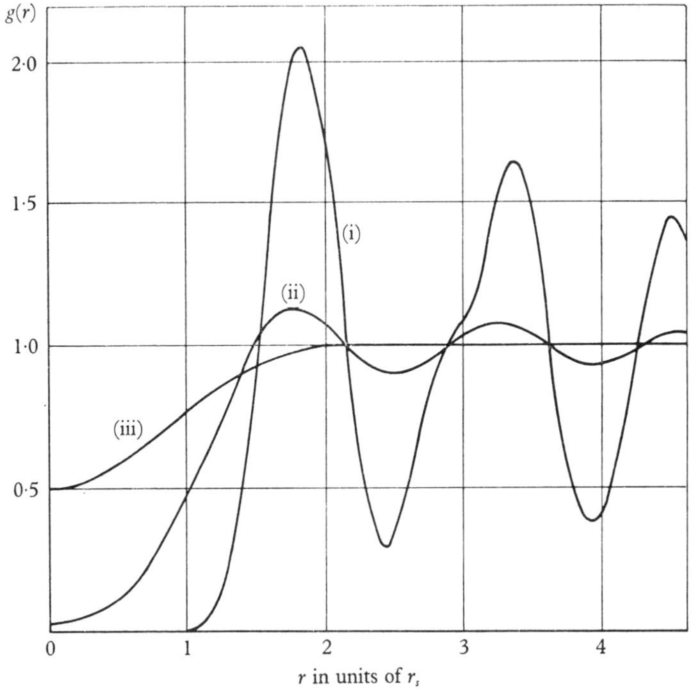

Fig. 5.13. Pair function. (i) Low-density result for $r_s = 100$. (ii) Low-density result for $r_s = 4$. (iii) Fermi hole.

below unity, are not distinguishable to graphical accuracy, but it is clearly seen that as the density is lowered, these oscillations develop and g can greatly exceed unity. Curve 2, constructed from the low-density theory, has been extrapolated to $r_s = 4$, and it should be remembered that this is beyond the limits of validity of the low-density theory. It probably illustrates the general behaviour of the pair function for real metal densities (cf. Gaskell, 1961, 1962), though it almost certainly exaggerates the amplitude of the oscillations considerably.

5.11. Thermodynamic properties of electron gas

We shall now consider various properties of an interacting electron gas. The most accurate calculations, as is to be expected from the earlier discussion, are valid only at high densities. Where possible, however, we shall consider the low-density results also.

5.11.1. *Specific heat*

It is well known from elementary band theory (see, for example, Ziman, 1964, p. 124) that the electronic specific heat C_v is given in terms of the density of states at the Fermi surface $N(\epsilon_f)$ by

$$C_v = (\tfrac{1}{3}\pi^2)\, k^2 T N(\epsilon_f). \qquad (5.11.1)$$

An elementary calculation then allows us to rewrite this in terms of the Fermi momentum k_f as

$$\frac{C_v}{N} = \frac{\pi^2 k^2 T}{k_f (d\epsilon/dk)_{k_f}}. \qquad (5.11.2)$$

Here $\epsilon(k)$ is the excitation energy of the quasi-particles in general. The calculation of this quantity has been discussed in an approximate way on the basis of the plasma oscillations treatment of Bohm & Pines. It is shown in Appendix 5A. 1 that, actually at the Fermi momentum, $(d\epsilon|dk)_{k_f}$ can be calculated precisely from the high-density theory, as was first done by Gell-Mann (1957). The basic result is

$$\left(\frac{d\epsilon}{dk}\right)_{k_f} = \left(\frac{9\pi}{4}\right)^{\frac{1}{3}} \frac{2}{\pi r_s} \int_{-1}^{+1} \frac{x\,dx}{2\left[1+4\left(\frac{4}{9\pi}\right)^{\frac{1}{3}} \frac{r_s}{\pi} \frac{1}{2(1-x)}\right]}, \qquad (5.11.3)$$

and if we evaluate the leading terms of this, and use the result in (5.11.2), we readily obtain

$$C_v = C_v^0[1 + 0{\cdot}083 r_s(-\ln r_s - 0{\cdot}203) + \ldots]^{-1}. \qquad (5.11.4)$$

Here C_v^0 is the result for a non-interacting Fermi gas. Equation (5.11.4), unfortunately, cannot be usefully compared with experiment, because (*a*) it is restricted to small r_s, and (*b*) it neglects Bloch wave effects or Brillouin zone boundaries.

In the low-density limit, the electrons vibrate, as we have seen, about regular lattice sites, and it is therefore clear that, by a correct collective treatment of the normal modes, they make a contribution $\propto T^3$ to C_v. A more detailed calculation by Carr (1961) shows that the 'low'-temperature specific heat per particle is

$$C_v = 62 k r_s^{\frac{9}{2}} [(kT)_{\text{Rydbergs}}]^3, \qquad (5.11.5)$$

though a small modification of the coefficient (from 62 to 56) appears in a slightly different approach by Coldwell-Horsfall & Maradudin (1960).

The smallness of C_v in (5.11.5) comes from the fact that the sound velocity is large, and the unusual feature of the lattice dynamical calculation is that only the transverse modes contribute to C_v, since the longitudinal mode behaves like an optical branch. The 'Debye' temperature associated with (5.11.5) is readily obtained as

$$2{\cdot}45 \times 10^5 r_s^{-\frac{3}{2}}\,^\circ\mathrm{K}.$$

In spite of the theoretical interest in (5.11.5) as the limit of strong coupling, it appears to have little relevance to experiment at the present time, and we shall not pursue the discussion further.

5.11.2. *Magnetic properties*

(*a*) *Paramagnetic susceptibility.* In the Gell-Mann & Brueckner calculation of the energy (see Appendix 5A. 1: equation 5A. 1.13), we introduced a function $Q_q(u)$. When we switch on a magnetic field, as in the elementary treatments of Pauli spin susceptibility, the electrons of ↑ spin occupy a Fermi sphere of radius k_f^+ and those of spin ↓ one of radius k_f^-. In the expression obtained for the correlation energy per particle, in the approximation in which all ring diagrams are summed, we simply have to replace $Q_q(u)$ by

$$\tfrac{1}{2}(Q_q^+(u)+Q_q^-(u)),$$

where the + and − refer to the different Fermi wave numbers k_f^+ and k_f^-.

The correlation energy/particle in Rydbergs is then obtained immediately, in terms of the relative magnetization $\zeta = (N_+ - N_-)/N$, in an obvious notation, as

$$\epsilon_{\mathrm{corr.}}(\zeta) = -\frac{3}{4\pi^5}\int_0^\infty \frac{dq}{q}\int_{-\infty}^{\infty} du \sum_{n=2}^{\infty} \frac{(-1)^n}{n} \times \left[\frac{Q_q^+(u)+Q_q^-(u)}{2}\right]^2 \left(\frac{\alpha r_s}{\pi q^2}\right)^{n-2}. \quad (5.11.6)$$

To approximate this we can replace Q_q by Q_0, and the upper limit of the q integration by unity, provided we then add a correction

$\Delta(\zeta)$ to include the contribution from the second-order term arising from the range of q greater than unity. We then find

$$\epsilon_{\text{corr.}}(\zeta) = -\frac{3}{4\pi^5}\int_0^1 \frac{dq}{q}\int_{-\infty}^{\infty} du \sum_{n=2}^{\infty} \frac{(-1)^n}{n}\left[\frac{Q_0^+(u)+Q_0^-(u)}{2}\right]^n \times \left(\frac{\alpha r_s}{\pi q^2}\right)^{n-2} + \Delta(\zeta), \quad (5.11.7)$$

where

$$\Delta(\zeta) = \frac{3}{8\pi^5}\int_0^1 \frac{dq}{q}\int_{-\infty}^{\infty} du \left[\frac{Q_0^+(u)+Q_0^-(u)}{2}\right]^2 - \frac{3}{32\pi^6}\int \frac{d\mathbf{q}}{q^3}\int_{-\infty}^{\infty} du \left[\frac{Q_q^+(u)+Q_q^-(u)}{2}\right]^2. \quad (5.11.8)$$

Evaluating these expressions to leading order in ζ, we find

$$\epsilon_{\text{corr.}}(\zeta) = \epsilon_{\text{corr.}}(0) - \frac{\zeta^2}{6\pi^2}\left\{\ln\frac{4\alpha r_s}{\pi} + \langle \ln R\rangle_{\text{av.}}\right\} + \frac{\zeta^2}{3\pi^3}\{\ln 2 + \tfrac{1}{2}\}, \quad (5.11.9)$$

where

$$\langle \ln R\rangle_{\text{av.}} = \frac{\int_{-\infty}^{\infty} du \frac{R(u)}{(1+u^2)^2}\ln R}{\int_{-\infty}^{\infty} du \frac{R(u)}{(1+u^2)^2}}. \quad (5.11.10)$$

R was introduced in Appendix 5A. 1 as simply

$$R(u) = 1 - u\tan^{-1}\left(\frac{1}{u}\right). \quad (5.11.11)$$

Adding the contribution $\epsilon_{\text{corr.}}(\zeta)-\epsilon_{\text{corr.}}(0)$ to the changes in kinetic and exchange energies to order ζ^2, the magnetic susceptibility can be calculated in the usual way.

It is interesting to note that the ratio of the paramagnetic spin susceptibility χ to the non-interacting Fermi gas value χ_0, when evaluated in this way, is not the same as for the specific heat ratio C_v/C_v^0 calculated in the previous section. One might have been tempted to argue that since C_v and χ were proportional to the density of states at the Fermi surface, this should have been the case. As Brout & Carruthers (1964) stress, in addition to this effect, the gyromagnetic ratio, originally γ_0, changes to a value $\gamma(k_f)$ which they evaluate as

$$\frac{\gamma(k_f)}{\gamma_0} = \left\{ 1 - \left(\frac{9\pi}{4}\right)^{\frac{1}{3}} \frac{2}{\pi r_s} \int_{-1}^{+1} \frac{dx}{2(1-x)\left[1+4\left(\frac{4}{9\pi}\right)^{\frac{1}{3}} \frac{r_s}{\pi} \frac{1}{2(1-x)}\right]} \right\}^{-1} \tag{5.11.12}$$

If we multiply the ratio of specific heats by (5.11.12) we obtain the correct high density form for the Pauli spin susceptibility, as calculated above.

(*b*) *Diamagnetism*. A more difficult problem is the influence of electron interactions on the Landau diamagnetism. We shall not go into details here, but simply record the fact that Kanazawa *et al.* (1960) have made a careful study of the problem using the Green function techniques described in Chapter 10. Their result is that the diamagnetism is modified in the high-density limit, a term in $r_s \ln r_s$ modifying the Landau result in leading order. The coefficient of this term was obtained earlier by March & Donovan (1954) from the Bohm–Pines theory. The treatment of Kanazawa is also applicable however at higher order. The reader is referred to the original papers for further details.

(*c*) *Low density results*. As shown in detail by Carr (1961), and as was used earlier by March & Young (1959) in constructing a model of a low density gas, it seems that for all practical purposes the low density gas is antiferromagnetic. In other words, we can think of the body-centred cubic lattice as made up of two interpenetrating simple cubic lattices, with ↑ spins on one lattice, and ↓ spins on the other.

In fact, unless the Wigner orbitals overlap, the ferromagnetic and the antiferromagnetic states have the same energy. It is a straightforward if lengthy matter to study the exchange integral with the Wigner oscillator orbitals, and in this way Carr gives as a rough estimate for the Neel temperature

$$T_N \sim 1{\cdot}6 \times 10^5 [13 r_s^{-1} - 3{\cdot}2 r_s^{-\frac{3}{4}}] e^{-1{\cdot}55 r_s^{\frac{1}{2}}} \,{}^\circ\mathrm{K},$$

but we shall not give the details here.

The calculations indicate ferromagnetism in the limit $r_s \to \infty$, but it occurs only for r_s values so large that the Curie temperature is infinitesimal.

Finally, we should stress that the stability of the electron gas against magnetic transitions is a very difficult problem to study rigorously. Even in the Hartree–Fock approximation, there is no universal agreement, though Overhauser (1962) has demonstrated conclusively that the usual paramagnetic state is unstable with respect to spin density waves, even in the limit $r_s \to 0$. Of course, Bloch's earlier work had shown that, with neglect of correlations again, the exchange energy was sufficiently favoured by parallel spin alignment to cause a transition to a ferromagnetic state at a gas density corresponding to $r_s \sim 6$. However, correlation effects are vital in considering the correct magnetic properties of the ground states, and, with the small reservation as $r_s \to \infty$ made above, ferromagnetism is never favoured in a Sommerfeld electron gas (see, however, Misawa, 1965).

5.12. Energy losses of fast electrons

We show here that a fast electron tranversing an electron gas will suffer characteristic energy losses which are directly related to the plasma oscillations. The electron must be fast to validate the Born approximation and then we can show, rather easily, that the rate of loss of energy to the metal is related directly to the frequency and wave number-dependent dielectric constant.

We can say first that the interaction of the fast electron at $\mathbf{R}$ with the metal is given by

$$\begin{aligned} H_{\text{interaction}} &= \sum_i \frac{e^2}{|\mathbf{r}_i - \mathbf{R}|} \\ &= \sum_{j\mathbf{q}}{}' \frac{4\pi e^2}{\Omega q^2} e^{i\mathbf{q}.(\mathbf{r}_j - \mathbf{R})} \\ &= \sum_{\mathbf{q}}{}' \frac{4\pi e^2}{q^2 \Omega} \rho_{-\mathbf{q}} e^{-i\mathbf{q}.\mathbf{R}}. \end{aligned} \tag{5.12.1}$$

If the electron is only scattered through a small angle, we can write $\mathbf{R} = \mathbf{v}t$, so that the effective energy transfer is $\omega = \mathbf{q}.\mathbf{v}$. Alternatively, the energy loss ω is approximately

$$\omega = \frac{k^2}{2m} - \frac{(\mathbf{k}-\mathbf{q})^2}{2m} = \frac{\mathbf{k}.\mathbf{q}}{m} - \frac{q^2}{2m} \doteqdot \mathbf{v}.\mathbf{q}, \tag{5.12.2}$$

since **k** is much larger than **q**. Then we have from standard time-dependent perturbation theory, that the probability that the electron transfers momentum **q** and energy $\omega = \mathbf{q}.\mathbf{v}$ to the system is

$$\frac{dP(\mathbf{q},\omega)}{dt} = 2\pi\left(\frac{4\pi e^2}{q^2\Omega}\right)^2 \sum_n (\rho_\mathbf{q})^2_{n0}\,\delta(\omega_{n0}-\omega), \qquad (5.12.3)$$

where the summation is over all accessible states n. The energy loss is just

$$\frac{d}{dt}W(\mathbf{q},\omega) = \omega\frac{dP(\mathbf{q},\omega)}{dt}. \qquad (5.12.4)$$

It then follows that

$$\frac{d}{dt}W(\mathbf{q},\omega) = -\frac{8\pi e^2}{q^2\Omega}\,\omega\,\mathrm{Im}\left(\frac{1}{\epsilon(\mathbf{q},\omega)}\right), \qquad (5.12.5)$$

which is the desired result. Using the relation between the dielectric function and $S(\mathbf{q},\omega)$, (5.12.5) shows that the energy loss is fundamentally given by the Van Hove function.

Equation (5.12.5) can be used as a basis for the experimental determination of plasma frequencies, since the latter are determined by the resonances in this formula (Ritchie, 1957; recall also (5.5.11)).

This concludes our discussion of the microscopic theory. The phenomenological theory given in section 5.13 below is, in principle, applicable to the electron gas, but in view of the extensive discussion given above, we shall restrict ourselves to a description of the general theory, with a brief account of its application to liquid ^{3}He. This is not inappropriate because the theory of Fermi liquids was originally invented by Landau to explain the low-temperature properties of this quantal fluid.

5.13. Fermi liquid theory and ^{3}He

In this theory, the use of the term 'liquid' is meant to imply that the simplest view of an excited state, as a set of independent quasiparticles, no longer suffices and that it is necessary to include quasiparticle interactions. Landau's original formulation (Landau, 1957, 1959) was intuitive and semi-phenomenological. He assumed (see (5.13.2) and (5.13.3) below) single quasiparticle energies ϵ^0_K and quasiparticle interactions $f(K, K')$, and correlated various physically observable quantities in terms of them. The theory has been successful in predicting a number of correct functional dependences particularly with respect to temperature) and might even be said to be semi-quantitative, allowing for the crudity of our knowledge of $f(K, K')$, the K's referring to momentum and spin.

We must assume at the outset that the interacting system has a single-particle spectrum of Fermi type. (This is not necessarily true, even though the constituent particles have spin one half. To quote Landau's example, liquid deuterium has a Bose spectrum (cf. Chapter 8) because atoms of deuterium combine to form molecules.) Thus we may suppose, as discussed in chapter 1, section 1.11, that we have reduced the original many-body Hamiltonian, for the purpose of computing the one-particle spectrum, to that of a set of N quasi-Fermions carrying the same space-spin indices K as their non-interacting counterparts, and with quasi-particle Fermi level k_f as for independent particles.

Now let us consider a state of the system near to that of the ground state and such that its quasi-particle distribution is

$$n_K = n_K^0 + \delta n_K, \tag{5.13.1}$$

n_K^0 being the usual Fermi step-function. Then, a starting-point of the Landau theory is to specify the energy of the perturbed state by

$$\delta E = E - E_0 = \sum_K \epsilon_K^0 \,\delta n_K + \frac{1}{2} \sum_{K \neq K'} f(K, K')\, \delta n_K \,\delta n_{K'}, \tag{5.13.2}$$

and the single-particle spectrum by

$$\delta \epsilon_K = \epsilon_K - \epsilon_K^0 = \sum_{K'(\neq K)} f(K, K')\, \delta n_{K'}, \tag{5.13.3}$$

where E_0 is the energy of the (interacting) ground state, ϵ_K^0 is the single-particle excitation energy when no other excitations are present and $f(K, K')$ is the interaction energy between two distinct excitations K and K'. The similarity of these equations with the Hartree–Fock results of Chapter 3, will be noted.

The explicit connection with the thermodynamics is made via the relationship

$$n_K = \{1 + e^{(\epsilon_K - \epsilon_F)/k_B T}\}^{-1}, \tag{5.13.4}$$

which follows by the usual arguments of statistical mechanics on assuming a Fermi-type spectrum. Equation (5.13.4) is, in general, because of (5.13.3) an implicit equation for n_K. However, it simplifies at sufficiently low temperatures, when ϵ_K can be replaced by ϵ_K^0.

It should be noted that we do not need complete knowledge of ϵ_K^0 and $f(K, K')$ because in perturbations of physical interest, the momentum vectors are restricted to lie on the Fermi surface. Also, for isotropic systems, the only angular variable χ, is that between $\mathbf{k}$ and $\mathbf{k}'$. Thus, suppressing the k_f-dependence for convenience, we write down the two physically useful functions

$$[f(K, K')]_{k=k'=k_f} = f(\chi, \sigma, \sigma') \tag{5.13.5}$$

and

$$[\partial \epsilon_K^0/\partial k]_{k=k_f} = v_f \equiv k_f/m^*, \tag{5.13.6}$$

where m^* is the effective mass, to be discussed more fully below.

We now show how a number of physical phenomena may be correlated in terms of these two functions, our main concern here being with spin-independent properties when (see (5.13.8), (5.13.20) and (5.13.28) below), only the spin trace over (5.13.5) is relevant.

5.13.1. *Effective mass*

The first point to understand is that ϵ^0 and f are not completely independent for reasons connected with Galilean invariance. The following form of the argument is due to Thouless (1961). Suppose each particle in the system is given a small momentum $\delta\mathbf{q}$. The potential energy is unaltered, and so the total energy increase, all kinetic, is $N(\delta q)^2/2m$. But what we have done is to displace the

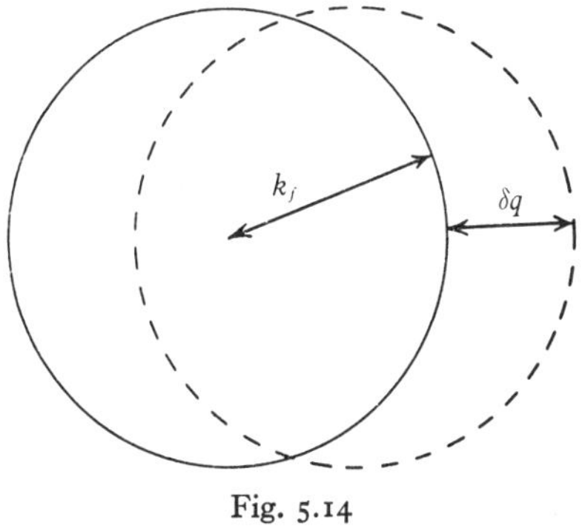

Fig. 5.14

quasiparticle sphere rigidly as indicated in Fig. 5.14, thus creating quasiparticles on the right and destroying them on the left. Then, an alternative way to calculate the energy increase is to insert

$$\delta n_K = \begin{cases} 1 \text{ for } k_f < k < k_f + \delta q \cos\theta, & \theta \text{ in quadrants } 1, 4, \\ -1 \text{ for } k_f + \delta q \cos\theta < k < k_f, & \theta \text{ in quadrants } 2, 3, \\ 0 \quad \text{otherwise,} & \end{cases} \tag{5.13.7}$$

into (5.13.2). One finds (on cancelling $(\delta q)^2$ throughout and taking the limit) that

$$\frac{1}{m} = \frac{1}{m^*} + \frac{k_f}{2(2\pi)^3} \int_0^\pi \sum_{\sigma, \sigma'} f(\chi, \sigma, \sigma') \cos \chi \, 2\pi \sin \chi \, d\chi. \quad (5.13.8)$$

The first term on the right comes from the independent quasi-particle contribution, m^* arising because (cf. (5.13.6), to leading order, $\epsilon^0_K = \epsilon^0_f + (k_f/m^*)(k-k_f)$. The final term results from the quasi-particle interactions. The angle χ appearing in (5.13.8) is given by

$$\cos \chi = \cos \theta \cos \theta' + \sin \theta \sin \theta' \cos(\phi - \phi'), \quad (5.13.9)$$

where θ and ϕ are the angles defining **k** in polar co-ordinates.

5.13.2. *Specific heat*

Because of (5.13.4) and the simplification noted thereafter, the low-temperature specific heat may be derived as for independent Fermions, except, of course, that m^* instead of m appears in the density of states function. Thus

$$C_v = \tfrac{1}{3} m^* k_f k_B^2 T, \quad (5.13.10)$$

and m^* may be determined experimentally.

5.13.3. *Compressibility and first sound*

The standard formula for the sound velocity c is

$$c^2 = \frac{1}{m} \frac{dp}{d\rho}, \quad (5.13.11)$$

where p is the pressure and ρ the density. But because of the present formulation (in particular (5.13.3)), it is desirable to relate this result to the chemical potential $\mu = \epsilon_f$, and its variation with respect to quasi-particle number. We have

$$\left(\frac{\partial \mu(\rho)}{\partial N}\right)_\Omega = \frac{1}{\Omega} \frac{\partial \mu}{\partial \rho} = \frac{1}{N\Omega} \left(\frac{\partial (\mu N)}{\partial \rho}\right)_N$$
$$= \frac{1}{N\Omega} \left(\frac{\partial (P\Omega)}{\partial \rho}\right)_N = \frac{1}{N} \left(\frac{\partial P}{\partial \rho}\right)_N = \frac{m}{N} c^2. \quad (5.13.12)$$

Thus, let us keep the volume fixed and increase the particle number from N to $N + \delta N$. Then μ changes for two reasons. First, there is a simple change of Fermi level from k_f to $k_f + \delta k_f$ governed by

$$k_f = (3\pi^2 n)^{\frac{1}{3}} \quad (n = N/\Omega), \quad (5.13.13)$$

and, secondly, there is created a thin spherical shell of additional quasi-particle states with which a given particle at the Fermi surface interacts. In first order, the two effects are additive. More formally,

$$\frac{\partial \mu}{\partial N} = \frac{\partial \epsilon_f}{\partial N} = \frac{\partial \epsilon_f}{\partial k_f}\frac{\partial k_f}{\partial N} + \frac{\partial \epsilon_f}{\partial n}\frac{\partial n}{\partial N}. \tag{5.13.14}$$

On using (5.13.13), $\partial k_f/\partial N$ and $\partial n/\partial N$ become $k_f/3N$ and $1/\Omega$ respectively. Also, noting that in (5.13.14) we must calculate $\partial \epsilon_f/\partial k_f$ at constant quasi-particle number, writing $\delta n_K = 0$ in (5.13.3) and using (5.13.6) gives

$$\partial \epsilon_f/\partial k_f = \partial \epsilon_f^0/\partial k_f = k_f/m^*. \tag{5.13.15}$$

Finally, let us examine $\partial \epsilon_f/\partial n$. In (5.13.3) only the quasi-particle term contributes and the small change in ϵ_f from the thin shell is

$$\delta \epsilon_f = \sum_{K'(\neq K)} f(\chi, \sigma, \sigma')\, \delta n_{K'}, \tag{5.13.16}$$

where

$$\delta n_K = \begin{cases} 1 & \text{if } k_f < k < k_f + \delta k_f, \\ 0 & \text{otherwise.} \end{cases} \tag{5.13.17}$$

Thus, (5.13.16) becomes

$$\delta \epsilon_f = \frac{1}{(2\pi)^3} k_f^2\, \delta k_f \int_0^\pi \sum_{\sigma'} f(\chi, \sigma, \sigma')\, 2\pi \sin\chi\, d\chi. \tag{5.13.18}$$

Now using (5.13.13) in the form $k_f^2\, \delta k_f = \pi^2 \delta n$ and, for reasons of symmetry, since $\delta \epsilon_f$ does not depend on spin, replacing $\sum_{\sigma'}$ by $\frac{1}{2}\sum_{\sigma,\sigma'}$, we find

$$\frac{\partial \epsilon_f}{\partial n} = \frac{\pi^2}{2(2\pi)^3} \int_0^\pi \sum_{\sigma,\sigma'} f(\chi, \sigma, \sigma')\, 2\pi \sin\chi\, d\chi. \tag{5.13.19}$$

Thus, collecting the various terms in (5.13.14), we find that the velocity of sound c, as defined through (5.13.12), is given by

$$c^2 = \frac{k_f^2}{3mm^*} + \frac{k_f^3}{6m}\frac{1}{(2\pi)^3} \int_0^\pi \sum_{\sigma,\sigma'} f(\chi, \sigma, \sigma')\, 2\pi \sin\chi\, d\chi. \tag{5.13.20}$$

5.13.4. *Zero sound*

Landau introduced the terminology 'zero' sound when predicting collective excitations in liquid ^{3}He analogous to plasma oscillations in an electron gas. An elementary example of the phenomenon is discussed in Chapter 8, section 8.3.7 (see also the end of section

10.14.1). There, however, the interparticle potential had convergent Fourier components in the long wavelength limit. Thus, while the model is clearly related to actual physical systems (for example, as the concluding part of a Brueckner calculation of the type discussed in chapter 6), it is not immediately amenable to quantitative application. On the other hand, the Landau theory of collective oscillations does not suffer from this disadvantage.

It is important to distinguish clearly between zero sound and ordinary (first) sound of the kind governed by equation (5.13.12). In the latter case, we know that collisions are necessary for the propagation of the wave, such formulae as (5.13.11) being derived on the basis of this mechanism. The higher the collision rate, the better the sound propagates. If τ is the mean time between collisions, then we require $\omega\tau \ll 1$, where ω is the frequency of propagation. Now as we shall see in detail in Chapter 10, a quasi-particle with momentum index $\mathbf{k}$, has a lifetime proportional to $(k-k_f)^{-2}$, and this means $\tau \propto T^{-2}$. Thus, as the temperature falls, collisions become rarer and eventually the sound waves cannot propagate.

On the other hand, at zero temperature, the density fluctuations are uncoupled to a good approximation, just as are plasmons in metals and phonons in solids. As the temperature is raised, more collective modes are excited and these interact, tending to destroy the well-defined excited states. Thus, the effect of the increased collision rate is disruptive, and the criterion for propagation in this case is $\omega\tau \gg 1$. Note that the conventional theory of sound in solids (namely, the phonon theory mentioned three sentences ago) is more analogous to the unfamiliar zero sound than the familiar first sound of classical kinetic theory.

To summarize, one might expect that as a Fermi liquid is cooled, first sound, initially easily observed, becomes unable to propagate. Then, at a lower temperature, and with a different velocity, in general, zero sound should be observed. Very recently (Abel, Anderson & Wheatley, 1966), this effect has been found in liquid ^{3}He and is in agreement with the Landau theory.

To make the above discussion of zero sound quantitative, it is supposed that δn_K, as written in (5.13.1), may now be spatially varying on a macroscopic scale. Thus, within any small element of space this variation is negligible and the previous Landau philosophy applies. This is clearly reasonable for short-range interactions such as in liquid helium, but one must be cautious in the case of an electron gas (a problem which will not be pursued here in view of the very detailed microscopic theory given earlier). In what follows, those quantities which spatially vary are written explicitly as functions of $\mathbf{r}$; functions not carrying such an argument are independent of position. Thus, we rewrite (5.13.1) as

$$n_K(\mathbf{r}) = n_K^0 + \delta n_K(\mathbf{r}) \tag{5.13.21}$$

and (5.13.3) as

$$\epsilon_K(\mathbf{r}) = \epsilon_K^0 + \sum_{K'(\neq K)} f(K, K')\, \delta n_{K'}(\mathbf{r}). \tag{5.13.22}$$

The next step is to substitute these two expressions in the standard transport equation

$$\frac{\partial n_K(\mathbf{r})}{\partial t} + \frac{\partial n_K(\mathbf{r})}{\partial \mathbf{r}} \cdot \frac{\partial \epsilon_K(\mathbf{r})}{\partial \mathbf{k}} - \frac{\partial n_K(\mathbf{r})}{\partial \mathbf{k}} \cdot \frac{\partial \epsilon_K(\mathbf{r})}{\partial \mathbf{r}} = I(n_K(\mathbf{r})), \tag{5.13.23}$$

where I is the usual collision term. If we now linearize, we obtain

$$\frac{\partial}{\partial t}\delta n_K(\mathbf{r}) + \frac{\partial}{\partial \mathbf{r}}\delta n_K(\mathbf{r}) . \frac{\partial}{\partial \mathbf{k}}\epsilon_K^0 - \frac{\partial}{\partial \mathbf{k}} n_K^0 . \sum_{K'} f(K, K') \frac{\partial}{\partial \mathbf{r}} \delta n_{K'}(\mathbf{r})$$
$$= \text{const.}\, \frac{\delta n_K(\mathbf{r})}{\tau}, \tag{5.13.24}$$

where we have assumed the existence of a collision time τ. For present purposes we need not worry about defining the collision term in detail, because in the limit $\omega\tau \gg 1$ it becomes negligible. By directly testing we find

$$\delta n_K(\mathbf{r}) = \nu(\theta, \phi, \sigma)\, \partial n_K^0 / \partial \epsilon_K\, e^{i(\mathbf{q}.\mathbf{r} - \omega t)} \tag{5.13.25}$$

is a solution of (5.13.23) for $\omega\tau \gg 1$, provided

$$\left(\frac{\partial \epsilon_K^0}{\partial \mathbf{k}} . \mathbf{q} - \omega\right) \nu(\theta, \phi, \sigma)$$
$$= \left(\frac{\partial \epsilon_K^0}{\partial \mathbf{k}} . \mathbf{q}\right) \sum_{K'} f(K, K') \frac{\partial n_{K'}}{\partial \epsilon_{K'}} \nu(\theta', \phi', \sigma'). \tag{5.13.26}$$

On setting $k = k_f$, and using the low-temperature approximation $\partial n_K / \partial \epsilon_K \approx -\delta(\epsilon_K - \epsilon_f)$, equation (5.13.26) reduces to

$$(s - \cos\theta)\, \nu(\theta, \phi, \sigma) = (2\pi)^{-3} m^* k_f \cos\theta \sum_{\sigma'} \int f(\chi, \sigma, \sigma')$$
$$\times \nu(\theta', \phi', \sigma') \sin\theta'\, d\theta'\, d\phi', \tag{5.13.27}$$

where $s = \omega/qv_f$ (the ratio of zero sound velocity to Fermi velocity), θ and ϕ define the direction of $\mathbf{v}_f$ relative to $\mathbf{q}$, the zero sound-wave index and χ is given by (5.13.9).

Equation (5.13.27) is the fundamental eigenequation governing the collective oscillations. Its eigenvalues s define the allowed propagation frequencies ω, and its eigenfunctions ν, through (5.13.25), give information about details of the propagation. The latter are decidedly peculiar but we will not pursue this matter here.

To illustrate the way the Landau theory is used in practice, let us expand the spin trace of f, in Legendre polynomials, as follows:

$$(m^*k_f/4\pi^2) \sum_{\sigma,\sigma'} f(\chi, \sigma, \sigma') = \sum_l F_l P_l(\cos\chi). \qquad (5.13.28)$$

Then, equation (5.13.20) gives

$$F_0 = (3mm^*c^2/k_f^2) - 1 \qquad (5.13.29)$$

and (5.13.8) gives

$$F_1 = 3\{(m^*/m) - 1\}. \qquad (5.13.30)$$

The quantities k_f, c and m^* are experimentally available (the latter from the specific heat, using (5.13.10)) and thus F_0 and F_1 may be numerically calculated. Now assuming it is possible to approximate the right-hand side of (5.13.29) by its first two terms (there is little evidence in favour of this!), (5.13.27) is rather easily solved (see Abrikosov & Khalatnikov (1959) for details) for a ϕ- and σ-independent ν and a zero sound velocity of the order of 10 % higher than that of first sound. A second (this time, ϕ-dependent) solution of (5.13.27) also exists for $F_1 > 6$. Since the latest experimental evidence points to an F_1 approaching this value, the question is open whether such a mode can exist.

Using the above approximation and by plausibly approximating the collision integral in (5.13.23). Abrikosov & Khalatnikov have investigated a number of physical properties under wider ranges of validity than those considered above. The Landau theory appears to give semi-quantitative answers and, in particular, generally predicts correct temperature dependences (for example, of viscosity and thermal conductivity).

To conclude the summary of this empirical approach, it should be mentioned that the spin-dependent aspect of f may be studied by techniques similar to those used above. The application of a magnetic field, for example, produces an imbalance of the two spin populations. In the language of the Landau theory, we create some quasi-particles of one spin and destroy an equal number of the other. Thus (cf. equation (5.13.3)) the effect of this quasi-particle population change on any one particle must be taken into account as well as the free field effect. Also, σ-dependent solutions of (5.13.27) may be found. (By writing $\nu \propto \sigma$, say, we essentially recover the spin-independent formalism, as Landau has pointed out.) As with the ϕ-dependent zero sound mode, however, it is not clear whether such spin-waves can be physically realized.

The microscopic derivation of the Landau theory has, of course, not been considered here. As will become clear when we discuss Green functions in Chapter 10, the basic understanding of the level spectrum ϵ_K^0 is best discussed via the theory of the single-particle Green functions (cf. the propagators of Chapter 4). Similarly, the correlation function $f(K, K')$ can be investigated using two-particle Green functions (Landau, 1959). These matters have been dealt with at great length in the book by Nozières (1963), to which account we refer the reader.

Problems

P.5 (i). Use the virial theorem (5.9.2) to obtain $E(r_s)$ explicitly as an integral over the potential energy $V(r_s)$. This is the same result as would be obtained by thinking of the charge on the electrons being 'switched on'.

P.5 (ii). Show that the pair function $g(r)$ for a non-interacting Fermi gas, given by

$$g(r) = 1 - \frac{9}{2}\left[\frac{j_1(k_f r)}{k_f r}\right]^2$$

is such that the Fourier transform of $(g(r)-1)$, say $S(k)-1$, is given by

$$S(k) = \frac{3k}{4k_f} - \frac{1}{16}\left(\frac{k}{k_f}\right)^3 \quad (k < 2k_f),$$

$$= 1 \quad (k > 2k_f).$$

From this form of $S(k)$, show that the asymptotic behaviour of the exact $g(r)$ comes from two regions:

(*a*) A contribution from small $k \propto r^{-4}$.

(*b*) A contribution from the 'anomaly' at $k = 2k_f$,

$$\propto \frac{\cos 2k_f r}{r^4}.$$

Note how (*a*) and (*b*) combine to yield the correct asymptotic form

$$-\frac{9}{2}\left(\frac{\cos k_f r}{\{k_f r\}^2}\right)^2.$$

P.5 (iii). Discuss in general terms how the behaviour of the pair function will be modified when the interactions are switched on.

P.5 (iv). Devise an experiment which could detect plasma losses in a solid.

Explain the origin of the transparency of the alkali metals in the ultraviolet (Zener, 1933).

P.5 (v).* Describe qualitatively how the dielectric constant of a uniform electron gas in *RPA* would be modified in the presence of a magnetic field. In particular, discuss how the integrated value of the screened potential $V(r)$ round a point charge, i.e. $\int V(\mathbf{r})\,d\mathbf{r}$, is altered from its zero field value.

P.5 (vi).* Discuss in *RPA* the interaction energy of two test charges e_1 and e_2 in a Fermi gas, as a function of their separation. Show that a 'shocking' violation of classical electrostatics can occur.

What physical applications does this result have?

P.5 (vii). Estimate the change due to electron interactions in the Pauli spin susceptibility from the free electron value in metallic Li. Assume that the conduction electrons may be treated as a uniform electron gas with $r_s = 3{\cdot}6a_0$. [N.B. The experimental result is $2{\cdot}5 \times \chi_{p\ \text{free electrons}}$.]

P.5 (viii). Suppose $f = f_0$, a constant, in (5.13.27). Prove that $\nu \propto \cos\theta/(s-\cos\theta)$ is a solution, provided

$$\tfrac{1}{2}s\ln\left|\frac{s+1}{s-1}\right| - 1 = \frac{\pi^2}{m^* k_f f_0}.$$

Use this result and (5.13.20) to show that in the weak coupling limit ($f \sim 0$, $m \sim m^*$), zero sound propagates with the Fermi velocity and that this is $\sqrt{3}$ times the velocity of the first sound.

Then go back to (5.13.27) and ascertain that this weak coupling result is really independent of the assumption that f is constant. (The details are given by Landau (1957).)

CHAPTER 6

NUCLEAR MATTER

6.1. Introduction

In Chapter 4, we deduced a formal expansion for the ground-state energy of a system of interacting Fermions, and we saw in Chapter 5 how it could be applied to an electron gas. There a major success of the theory was in yielding an exact high-density expansion for the energy. The methods of the present chapter are more suitable for systems of low density. We shall see, in fact, that the terms of the perturbation series having the same density dependence can be summed, to give useful results for a low-density system.

6.2. Definition of nuclear matter

To be specific we shall discuss the application of the perturbation method to 'nuclear matter'. To introduce this concept, we remark first that, for light nuclei, the numbers of protons and neutrons tend to be equal, for the most stable configurations. As the mass of the nucleus increases, the number of neutrons exceeds the number of protons, due, of course, to the Coulomb repulsion between the protons, which prevents a large number of them being brought within a small volume. It is this force which prevents nuclei of arbitrarily large mass number being stable.

Despite Coulomb repulsion, the fact that (neutrons and) protons can be brought together to form stable nuclei shows that the specifically nuclear forces are very much stronger than the Coulomb forces. To study the effects of these nuclear forces it is therefore convenient to think of the Coulomb force switched off. Later, we can modify these results by incorporating the (usually small) effects of the Coulomb interaction. Thus, we are led to the concept of nuclear matter, consisting of large but equal numbers, $\frac{1}{2}N$, of protons and neutrons, held together only by the specifically nuclear forces in a large volume Ω. As in Chapter 5, we then ignore surface effects by passing to the limit $N \to \infty$, $\Omega \to \infty$, such that N/Ω is finite. It is expected that the cores of large finite nuclei that exist in nature will bear some resemblance to 'chunks' of this idealized nuclear matter.

It is this nuclear matter, then, to which we shall apply the perturbation method of Chapter 4. Two questions come up immediately:

(i) Is perturbation theory valid?

(ii) Assuming it is, does the Goldstone formula give the correct result?

If we anticipate section 6.12 by answering (i) in the affirmative, we still should look at (ii) before proceeding further, for we proved the linked cluster theorem in Chapter 4 for a simple form of the two-body potential and the inter-nucleon potential (see section 6.3) is a complicated interaction. While an affirmative answer to (ii) cannot be given, as yet, with complete certainty, the work of Kohn & Luttinger (1960), who showed that, for nucleons interacting with tensor forces, the Goldstone formula was correct to second order, seems to indicate that the result is true to higher orders as well. This is therefore assumed in what follows.

6.3. Nuclear forces

By extrapolating the data available for heavy nuclei, we can estimate the value for the binding energy per nucleon of the idealized nuclear matter as $\sim 16\,\text{MeV}$. Furthermore, if r_0, the radius per nucleon, is defined by $\Omega = N(\frac{4}{3}\pi r_0^3)$, then the 'experimental' value of r_0 comes out to be $\sim 10^{-13}\,\text{cm}$. The aim of the microscopic theory, at least in the first instance, will be to understand these results, starting from the information available from scattering experiments on the inter-nucleon force.

The ultimate aim of the theory, of course, is the calculation of all properties of nuclei starting from the two-body forces. But no theory which fails to predict the binding energy per particle and the equilibrium density can be relied upon to make satisfactory predictions of other static properties. We have therefore not thought it appropriate in this book to discuss other parameters of nuclear matter.

6.3.1. *Units*

We shall express energies in MeV in this chapter, and choose

$$1 \text{ Fermi} \equiv 1 \times 10^{-13}\,\text{cm}$$

as the unit of length. It will be convenient to put $\hbar = 1$, and this then completes the system of units. If we choose to ignore the slight difference in mass between protons and neutrons and denote by m the mass of the nucleon, then, in these units

$$1/m = 41{\cdot}5.$$

6.3.2. *Two-nucleon potential*

While a fundamental theory must aim at predicting the forces that act between nucleons, a phenomenological approach aims initially at explaining all experimental data in terms of some suitable two-nucleon potential. Only if ingenuity in choosing such a potential eventually fails will one want to consider (*a*) the possibility of two-body interactions which cannot be represented by a potential, and/or (*b*) many-body forces.

Our starting-point will then be an inter-nucleon potential chosen to fit as much data as is available. We shall not go into details of how this is done, but refer the reader, for example, to an excellent review given in Moravesik (1963). A satisfactory fit of the experimental data is obtained by assuming that:

(1) The nuclear force is charge independent. In other words, the force between two neutrons equals that between two protons, or that between one proton and one neutron (the other conditions remaining the same).

(2) The internucleon potential is strong and short ranged with an attractive tail and a very much stronger repulsive core.

(3) The potential depends on the spin state of the two-nucleon system.

(4) The potential depends on the relative angular momentum of the two-nucleon system.

(5) The potential has a tensor and possibly a spin-orbit term.

The charge independence of the inter-nucleon force allows us to treat the proton and neutron as two states of the nucleon (by analogy with spin). One type of force satisfying the above requirements is that in which the inter-nucleon potential becomes infinitely repulsive at an internuclear separation $r = c$, the hard core radius (see, for example, Gammel & Thaler (1960)). For $r > c$, the shape of the potential is traditionally taken to be of Yukawa form: $V e^{-\mu r}/\mu r$. The strengths and ranges for the potential outside the hard-core radius are in general different for the different inter-nucleon states (triplet-even, triplet-odd, singlet-even, singlet-odd) as well as for the different components (central, tensor and spin-orbit part) of it.

Since such forces contain a large number of parameters, several sets of potentials which fit the scattering data more or less equally well, can be obtained. Two-body scattering data alone seems at present unable to choose between these competing potentials.

Assuming a hard core for the inter-nucleon potential is not the only way in which to obtain agreement with experimental data. If one assumes a velocity dependent potential which becomes repulsive with increasing energy, one can explain the high energy scattering data (which clearly indicates a repulsive part in the inter-nucleon potential). However potentials of this variety (soft potentials) have not yet been fully analysed.

The method we give below, based on partial summation of the perturbation series, is applicable to both soft- and hard-core potentials. This summation is crucial in the treatment of a hard-core potential, but for a soft-core potential, depending on the strength of the coupling, ordinary perturbation theory may suffice.

6.4. Non-interacting nucleons

Since we consider the neutron and proton as two states of the same Fermion, the nucleon, N non-interacting nucleons in volume Ω will of course fill a sphere in momentum space ($\mathbf{k}$ space) of radius k_f where the density ρ_0 is related to k_f by

$$\rho_0 = \frac{2k_f^3}{3\pi^2},$$

since four nucleons, two protons with spins ↑ and ↓ and two neutrons with spins ↑ and ↓, can occupy the same momentum state $\mathbf{k}$.

The unperturbed ground-state energy per particle is

$$E_0/N = \tfrac{3}{10}k_f^2/m$$
$$\sim 25 \text{ MeV for } r_0 \sim 1 \text{ F}.$$

A normalized single-particle level of momentum $\mathbf{k}$ will be denoted by vector $|k\rangle$. The wave function corresponding to it will be $(1/\Omega^{\frac{1}{2}})\, e^{i\mathbf{k}.\mathbf{r}}$. We shall denote by α, β, γ and δ the four states of the nucleon, proton $\uparrow$, proton $\downarrow$, neutron $\uparrow$, neutron $\downarrow$, in that order. $a_{\mathbf{k}\alpha}$ will be the destruction operator corresponding to a proton $\uparrow$ in the level $|k\rangle$ and so on. When we do not wish to specify a particular nucleon state we shall find it convenient to use scalar suffices. Thus a_k will be the destruction operator corresponding to a nucleon in level $|k\rangle$.

Thus the unperturbed Fermi gas state for the N non-interacting nucleus will be

$$|g\rangle = \prod_{\substack{|\mathbf{k}|<|\mathbf{k}_f| \\ \omega=\alpha,\,\beta,\,\gamma,\,\delta}} a^\dagger_{\mathbf{k}\omega}|0\rangle \tag{6.4.1}$$

where $|0\rangle$ denotes the vacuum state. In future, whenever summation over the four nucleon states is involved, we shall not explicitly write that out. Instead we shall indicate it by writing scalar k's instead of vector $\mathbf{k}$'s. Thus we write

$$|g\rangle = \prod_{k<k_f} a^\dagger_k|0\rangle. \tag{6.4.2}$$

Exactly as in Chapter 3 we can introduce, for $k < k_f$, the b_k operators. Thus the Fermi gas ground state becomes the vacuum for the new types of particles whose creation operators are $a^\dagger_k\,(k > k_f)$ and $b^\dagger_k\,(k < k_f)$.

The Hamiltonian is as in Chapter 4 except that we must remember that the summation is not only over $\mathbf{k}$ but also over α, β, γ and δ, namely

$$H_0 = \sum_k \left(\frac{k^2}{2m} - E_f\right) a^\dagger_k a_k, \quad \text{where } E_f = \frac{k_f^2}{2m}.$$

We can further simplify the notation by using $\mathbf{h}$, $\mathbf{h}'$, $\mathbf{h}''$, etc., to denote momentum states below the Fermi sea ($h < k_f$) and $\mathbf{p}$, $\mathbf{p}'$, $\mathbf{p}''$, etc., to denote momentum states above the Fermi sea ($p > k_f$). Thus we could write instead of (6.4.2)

$$|g\rangle = \Pi a^\dagger_h|0\rangle. \tag{6.4.3}$$

6.5. Perturbation series

As explained in Chapter 4, we now consider the nucleons to interact adiabatically and, as discussed previously, they may be expected to form the perturbed ground state, whose energy is given by the Goldstone formula (cf. Chapter 4, (4.3.5)),

$$E = E_0 + \langle g | V + V \frac{Q}{E_0 - H_0} V + \ldots | g \rangle_L. \tag{6.5.1}$$

The complete Hamiltonian for our assumed two-body interaction is

$$H = H_0 + \frac{1}{2} \sum_{k_1 k_2 k_3 k_4} \langle k_1 k_2 | v | k_3 k_4 \rangle a_{k_1}^\dagger a_{k_2}^\dagger a_{k_4} a_{k_3}. \tag{6.5.2}$$

We remind the reader that the summation over the k's implies also summation over the different nucleon states. It is also best at this stage to mention that the matrix elements $\langle k_1 k_2 | v | k_3 k_4 \rangle$ have as yet only a formal significance. Since our assumed inter-nucleon interaction may have a hard core, these matrix elements can be actually infinite. The treatment of the hard core will be considered in section 6.8.

(i) Fig. 6.1 (ii)

The first-order term (in V) of (6.5.1) gives rise to two diagrams (direct and exchange) (Figs. 6.1 (i) and (ii)). The application of the rules already discussed in detail in Chapter 4 gives the following contributions to $\Delta E \equiv E - E_0$:

(i) Direct: $$\Delta E = \tfrac{1}{2} \Sigma \langle hh' | v | hh' \rangle \tag{6.5.3}$$

and

(ii) Exchange: $$\Delta E = -\tfrac{1}{2} \Sigma \langle hh' | v | h'h \rangle. \tag{6.5.4}$$

The second-order term in (6.5.1) gives rise to two diagrams and contributions (Figs. 6.2 (i) and (ii)).

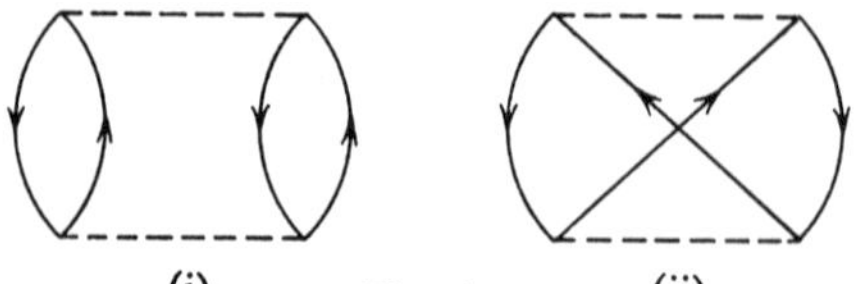

(i) Fig. 6.2 (ii)

up the contributions of the infinite subset consisting of all the direct ladder diagrams of the perturbation series. Thus in the low density limit when we take account of only the ladder diagrams (including the exchange ones), we obtain

$$\Delta E = \tfrac{1}{2}\Sigma\left[\langle hh'|t\left(\frac{h^2+h'^2}{2m}\right)|hh'\rangle - \langle hh'|t\left(\frac{h^2+h'^2}{2m}\right)|h'h\rangle\right]. \quad (6.6.5)$$

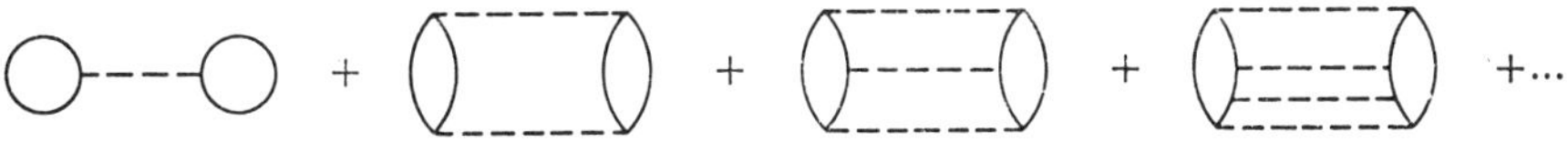

Fig. 6.4

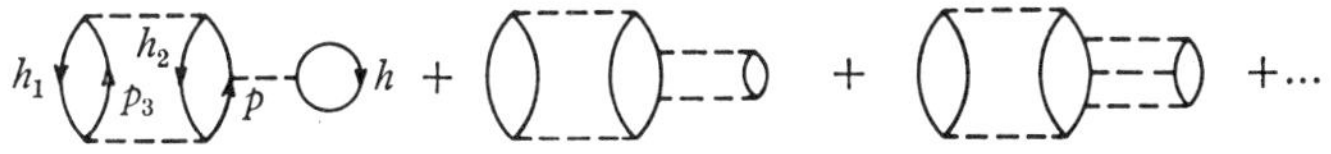

Fig. 6.5

This particular choice of ϵ has enabled us to sum the direct ladder diagrams shown in Fig. 6.4 but other choices would enable us to 'sum up' other subsets of diagrams. For example, to 'sum up' the infinite subset represented by Fig. 6.5, we will have to choose

$$\epsilon = \frac{h_1^2+h_2^2+h^2-p_3^2}{2m}.$$

For, bearing in mind the rules for writing down matrix elements, the t-matrix equation for this subset should read

$$\langle ph|t|ph\rangle = \langle ph|v|ph\rangle + \sum_{p'p''} \frac{\langle ph|v|p'p''\rangle\langle p'p''|t|ph\rangle}{\dfrac{h_1^2+h_2^2+h^2}{2m} - \dfrac{p_3^2+p'^2+p''^2}{2m}}. \quad (6.6.6)$$

6.7. Calculation of matrix elements of *t*

In an infinite nuclear medium at densities such as those that occur at the centres of finite nuclei, two nucleons coming together interact several times (because of the strong inter-nucleon forces) before separating. We saw how to take account of this by means of one single effective t interaction so that the expression for ΔE to first order in t might be a good approximation for a low density system. In this section we outline the procedure for calculating matrix elements of the t matrix.

Consideration of spin states is trivial. There is a separate t matrix equation for each distinct spin state with the appropriate part of the inter-nucleon potential. We avoid labelling the spin states so as not to complicate the notation. To evaluate $\langle \mathbf{k}_1 \mathbf{k}_2 | t | \mathbf{k}_1 \mathbf{k}_2 \rangle$, it is convenient to introduce the total and the relative momentum of the two particles, $2\mathbf{P}$ and $\mathbf{k}$ given by

$$2\mathbf{P} = \mathbf{k}_1 + \mathbf{k}_2, \tag{6.7.1}$$

$$\mathbf{k} = \tfrac{1}{2}(\mathbf{k}_1 - \mathbf{k}_2). \tag{6.7.2}$$

Since the interaction conserves total momentum, a typical t-matrix equation becomes

$$\langle \mathbf{k}' | t(\mathbf{P}) | \mathbf{k} \rangle = \langle \mathbf{k}' | v | \mathbf{k} \rangle + \sum_{\mathbf{k}''} \frac{\langle \mathbf{k}' | v | \mathbf{k}'' \rangle \langle \mathbf{k}'' | t(\mathbf{P}) | \mathbf{k} \rangle f(\mathbf{P}, \mathbf{k})}{\epsilon - \dfrac{1}{2m} |\mathbf{P} + \mathbf{k}''|^2 - \dfrac{1}{2m} |\mathbf{P} - \mathbf{k}''|^2}, \tag{6.7.3}$$

where $\mathbf{k}$, $\mathbf{k}''$ and $\mathbf{k}'$ are the relative momenta of the two particles in the initial, intermediate, and final states. ϵ denotes the initial kinetic energy of the nucleons, namely

$$\frac{|\mathbf{P} + \mathbf{k}|^2 + |\mathbf{P} - \mathbf{k}|^2}{2m},$$

and the matrix element $\langle \mathbf{k}' | t(\mathbf{P}) | \mathbf{k} \rangle$ introduced in (6.7.3) denotes simply

$$\langle \mathbf{P} + \mathbf{k}', \mathbf{P} - \mathbf{k}' | t | \mathbf{P} + \mathbf{k}, \mathbf{P} - \mathbf{k} \rangle.$$

Finally, we define f in (6.7.3) by

$$\begin{aligned} f(\mathbf{k}'', \mathbf{P}) &= 1 \quad \text{if} \quad |\mathbf{P} \pm \mathbf{k}''| > k_f \\ &= 0 \qquad \text{otherwise,} \end{aligned}$$

in order to take care of the Exclusion Principle in the summation.

This integral equation is not solved directly, but rather in terms of a 'wave matrix' M defined by

$$t = vM = v + v(Q/d)vM, \tag{6.7.4}$$

from which it follows immediately, that

$$M = 1 + (Q/d)vM. \tag{6.7.5}$$

Written out in full, (6.7.5) becomes

$$\langle \mathbf{r} | M | \mathbf{k} \rangle = \langle \mathbf{r} | \mathbf{k} \rangle + \int G(\mathbf{r}, \mathbf{r}') v(\mathbf{r}') \langle \mathbf{r}' | M | \mathbf{k} \rangle \, d\mathbf{r}', \tag{6.7.6}$$

where

$$G(\mathbf{r},\mathbf{r}') = \sum_{\mathbf{k}} f(\mathbf{k},\mathbf{P}) \frac{e^{i\mathbf{k}.(\mathbf{r}-\mathbf{r}')}}{\epsilon - \dfrac{P^2}{m} - \dfrac{k^2}{m}}. \tag{6.7.7}$$

It is convenient to denote $\langle \mathbf{r}|M|\mathbf{k}\rangle$ by $\psi_{\mathbf{k}}(\mathbf{r})$. This, as well as the $G(\mathbf{r},\mathbf{r}')$, depends on the total momentum $2\mathbf{P}$ in addition to its dependence on the initial energy ϵ. We omitted these labels in order to avoid further complication of the notation. The integral equation

$$\psi_{\mathbf{k}}(\mathbf{r}) = e^{i\mathbf{k}.\mathbf{r}} + \int G(\mathbf{r},\mathbf{r}')\, v(\mathbf{r}')\, \psi_{\mathbf{k}}(\mathbf{r}')\, d\mathbf{r}' \tag{6.7.8}$$

differs from the usual scattering equation (see problem P.1 (i) of Chapter 1) because of the presence of the factor f in G, which prevents the vanishing of the energy denominator. Also the 'scattered wave' decreases more rapidly than r^{-1} at large distances and G is not isotropic. Equation (6.7.8) is known as the Bethe–Goldstone equation.

The solution of (6.7.8) may be obtained as usual by partial wave analysis, though in general we obtain a set of coupled equations. We shall not write down the explicit equations, but refer the reader to the discussion by Werner (1959). However, we shall consider a simplification due to Brueckner & Gammel (1958), who remove the coupling due to f by averaging f over all directions of the total momentum $2\mathbf{P}$.

If θ is the angle between the directions of $\mathbf{P}$ and $\mathbf{k}$, then f is unity if

$$\cos\theta < (P^2 + k^2 - k_f^2)/kP,$$

and otherwise $f = 0$ (see Fig. 6.6). Averaging over the angle θ we obtain

$$\begin{aligned} \bar{f}(\mathbf{P},\mathbf{k}) &= 0, \quad \text{if} \quad P^2 + k^2 < k_f^2 \\ &= 1, \quad \text{if} \quad |\mathbf{P}-\mathbf{k}| > k_f \\ &= \frac{P^2 + k^2 - k_f^2}{kP} \end{aligned}$$

in all other cases.

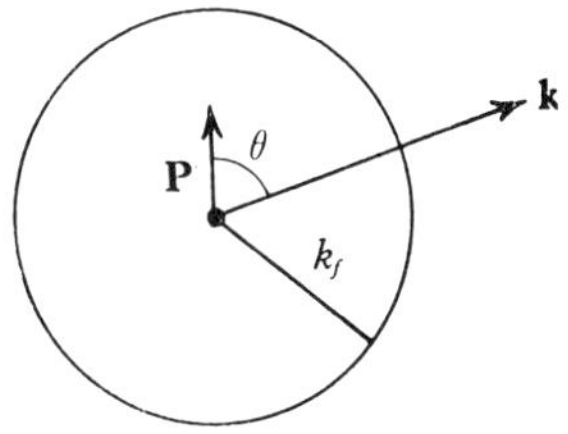

Fig. 6.6

This averaging procedure still leaves the coupling due to the tensor force in the triplet states. It therefore gives rise to coupled integral equations in these states but only to single integral equations for the singlet states. It is simply the treatment of the hard core in the inter-nucleon potential which gives rise to any difficulty in the solution of these integral equations (see the following section). Once they are solved, t-matrix elements are given by

$$\begin{aligned} \langle \mathbf{k}|t|\mathbf{k}\rangle &= \langle \mathbf{k}|vM|\mathbf{k}\rangle \\ &= \sum_{\mathbf{r}'\mathbf{r}} \langle \mathbf{k}|r'\rangle \langle \mathbf{r}'|v|\mathbf{r}\rangle \langle \mathbf{r}|M|\mathbf{k}\rangle \\ &\equiv \int e^{i\mathbf{k}.\mathbf{r}} v(\mathbf{r})\, \psi_{\mathbf{k}}(\mathbf{r})\, d\mathbf{r}, \end{aligned} \tag{6.7.9}$$

and hence the expression for ΔE may be found from (6.6.5).

6.8. Treatment of hard core

The method of solving integral equations such as (6.7.8) will be illustrated by considering the s-wave part of this equation. Since the method is the same for higher angular momentum components, and the coupled equations, we shall not write down the complete set of equations.

Let $\phi(r)/r$ be the s-wave part of $\psi_{\mathbf{k}}(\mathbf{r})$ and

$$\frac{\phi_0(r)}{r} = \frac{\sin kr}{kr}$$

be the s-wave part of $e^{i\mathbf{k}.\mathbf{r}}$.

The s-wave part of (6.7.8) then becomes

$$\phi(r) = \phi_0(r) + \int_0^\infty dr'\, G_0(r,r')\, v(r')\, \phi(r'), \tag{6.8.1}$$

with
$$G_0(r,r') = \frac{2}{\pi}\int dk \frac{\bar{f}(k,P) \sin kr \sin kr'}{\epsilon - P^2/m - k^2/m}. \tag{6.8.2}$$

If $r = c$ is the radius of the hard core in the potential,

$$\phi(r) = 0 \quad \text{at} \quad r = c.$$

Assume then that $v\phi$ has a δ-function contribution at $r = c$, that is

$$v\phi = \lambda\delta(r-c). \tag{6.8.3}$$

Equation (6.8.1) then becomes

$$\phi(r) = \phi_0(r) + \lambda\, G_0(r, c) + \int_c^\infty dr'\, G_0(r, r')\, v(r')\, \phi(r'), \quad (6.8.4)$$

with $0 = \phi(c) = \phi_0(c) + \lambda\, G_0(c, c) + \int_c^\infty dr'\, G(c, r')\, v(r')\, \phi(r')$, (6.8.5)

which determines λ. Thus

$$\phi(r) = \left\{\phi_0(r) - \phi_0(c)\frac{G_0(r, c)}{G_0(c, c)}\right\}$$
$$+ \int_c^\infty dr' \left\{G_0(r, r') - \frac{G_0(r, c)\, G_0(c, r')}{G_0(c, c)}\right\} v(r')\, \phi(r'). \quad (6.8.6)$$

This procedure of altering the inhomogeneous term and the Green function of the integral equation has removed the difficulties (infinities) associated with the hard core to give us integral equations which can be solved readily. From these solutions, t-matrix elements may be obtained as

$$\langle k|t|k\rangle = \int_0^\infty \phi_0\, v\phi\, dr$$
$$= \phi_0(c)\, \lambda + \int_c^\infty dr\, \phi_0(r)\, v(r)\, \phi(r)$$
$$= -\frac{\phi_0^2(c)}{G(c, c)} + \int_c^\infty dr \left\{\phi_0(r) - \phi_0(c)\frac{G(c, r)}{G(c, c)}\right\} v(r)\, \phi(r). \quad (6.8.7)$$

We have implicitly assumed that $\phi v = 0$ for $r < c$. This is not an exact result but only an approximation whose usefulness has to be checked at the end of the calculation.

When Brueckner & Gammel used this simple theory they obtained a value ~ 35 MeV for the binding energy per nucleon of nuclear matter! We can thus conclude that nuclear matter may not be regarded simply as a low-density system. Other terms in the perturbation series give rise to important contributions. Brueckner & Gammel therefore developed their theory to take account of additional terms which seem important physically and we shall outline their arguments in the following sections.

6.9. Propagator modification

By now we are very familiar with the use of the independent-particle model or the Hartree–Fock method as a first approximation in a many-body problem. In this model, as we have seen, the essential idea is to consider the motion of one particle in the average field produced by the motions of the other particles. Total energies and other properties are then expressed in terms of the single-particle energies and wave functions thus obtained. This model works so long as the inter-particle forces are not too strong to cause local correlations and condensations. For systems with strong two-particle forces, one has to calculate two-body interactions more exactly and to do so one uses an independent pair model.

In this model, one tries to consider the motion of a pair of particles in the average field produced by the rest of the particles of the system. Then one expresses energies and other properties in terms of the interaction energy of a pair of particles, pair wave functions and so on. For the method to be successful, three-body and higher-order clusterings must be small. For nuclear matter, since inter-particle distances and ranges of the inter-nucleon potentials are of similar order at the equilibrium density, it seems plausible that, while two-body correlations will be very important, many (more than two)-body correlations might be small. Thus the independent pair model may be a good scheme for nuclear matter. Indeed, the result (6.6.5) may be regarded as a first attempt at constructing an independent pair model. In section 6.6 we considered the motion (scattering) of a pair of nucleons in nuclear matter. But the other nucleons were taken account of only in a passive way, in the sense that we merely allowed for their presence by preventing the two nucleons under consideration from scattering into states already occupied by other nucleons. However, we can take account of the interaction of the other particles with these two nucleons in a simple way by altering the energy denominator of (6.6.1). As an example, consider the third-order diagrams shown in Fig. 6.7. Their contribution, according to the rules of Chapter 4, is

$$\frac{1}{2}\sum\frac{\langle h_1 h_2|v|pp'\rangle\langle pp'|v|h_1 h_2\rangle}{[(h_1^2+h_2^2-p^2-p'^2)^2]/4m^2}\,V(p),$$

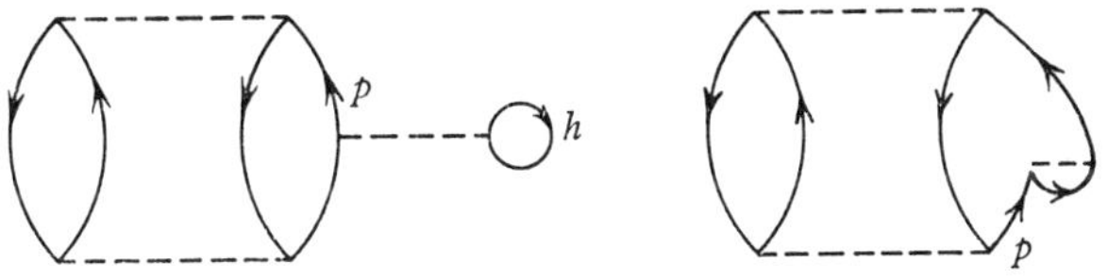

Fig. 6.7

where $$V(k) = \sum_h [\langle kh|v|kh\rangle - \langle kh|v|hk\rangle] \tag{6.9.1}$$

is reminiscent of the effective one-body potential of the Hartree–Fock theory.

If we consider the second-order diagram in Fig. 6.8 (from which Fig. 6.7 is derived by having 'self-energy insertions' (see chapter 10) attached to the line labelled p) and use for the energy denominator

$$k^2/2m + V(k) = \epsilon(k), \tag{6.9.2}$$

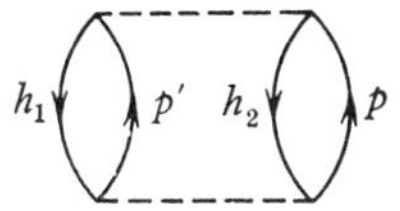

Fig. 6.8

rather than simply the kinetic energy, the other rules remaining unchanged, we obtain a contribution

$$\frac{1}{2}\sum \frac{\langle h_1 h_2|v|pp'\rangle\langle pp'|v|h_1 h_2\rangle}{\left[\dfrac{h_1^2+h_2^2}{2m} + V(h_1) + V(h_2) - \dfrac{p^2+p'^2}{2m} - V(p) - V(p')\right]}. \tag{6.9.3}$$

Expanding the denominator and retaining only the first power of V, it is clear that this new interpretation of the second-order diagram of Fig. 6.8 takes into account automatically, not only the contributions from Figs. 6.7 and 6.8, but also those from the three diagrams obtained from Fig. 6.8 by having the 'self-energy insertions' in the other three solid lines labelled h_1, h_2 and p'. Physically we have gone a stage further with the independent pair model and allowance has now been made (if somewhat imperfectly) for the nuclear binding of the pair of particles as they propagate and scatter in the medium.

6.10. Brueckner–Gammel equations

Brueckner & Gammel combined the propagator modification of section 6.9 with the replacement of the v interactions by the t matrix. This brings in a self-consistency condition as shown below.

It is convenient to distinguish t matrices calculated using simply the kinetic energy from those calculated with modified energy denominators, and we shall do so henceforth by calling the latter K matrices.

To return to the self-consistency condition, we first observe that $\epsilon(k)$ as given by (6.9.2) is such that the effective single-particle potential $V(k)$ is to be calculated using the K matrices, via the equation

$$V(k) = \Sigma\,[\langle kh|K|kh\rangle - \langle kh|K|hk\rangle], \tag{6.10.1}$$

which is a modification of (6.9.1) obtained by replacing v by K. Clearly, this is a new self-consistency condition since it implies that, to calculate $V(k)$, we need $\epsilon(k)$ and vice versa.

The actual self-consistent calculations involve further considerations and approximations and the equations are not as simple as (6.10.1) may lead one to expect. To see this in detail, we write down the K-matrix equation explicitly as

$$K(\epsilon) = v + v(Q/d)\,K(\epsilon) \tag{6.10.2}$$

with

$$\frac{Q}{d} \equiv \Sigma \frac{|pp'\rangle\langle pp'|}{\epsilon - \epsilon(p) - \epsilon(p')}, \tag{6.10.3}$$

where we have shown explicitly the dependence of K on ϵ. Thus, when we wish to calculate $\langle h_1 h_2|K|h_1 h_2\rangle$, we choose

$$\epsilon = \epsilon(h_1) + \epsilon(h_2),$$

with $\epsilon(h_1)$ and $\epsilon(h_2)$ defined by equations (6.9.2) and (6.10.1). The parameter for K in (6.10.1) is $\epsilon(k) + \epsilon(h)$ and all seems well. But the trouble is that the solution of (6.10.2) also demands knowledge of the $\epsilon(p)$'s.

To see the origin of the difficulty, we notice first that $\epsilon(k)$ has been interpreted as an effective one-particle energy. Thus, $V(k)$ has been considered as an average potential which one particle experiences as a result of interactions with the rest of the particles and there seems a unique prescription for calculating it for $k < k_f$. However, for $k > k_f$, no unique prescription for the calculation of $V(k)$ exists. This is because the calculation of $V(p)$ involves that of $\langle ph|K|ph\rangle$, and, just as in section 6.6, what has to go in the energy denominator depends on the exact position of the p line on the diagram, that is on the history of the particle which has come to occupy the previously unoccupied **p** level.

Brueckner & Gammel approximate the energy denominator to make $\epsilon(p)$ depend only on some average $\epsilon - \Delta$ (to be discussed below) and p. Thus their approximate self-consistent equations reduce to

$$\epsilon(h) = \frac{h^2}{2m} + V(h), \tag{6.10.4}$$

where

$$V(h) = \Sigma\,[\langle hh'|K(\epsilon(h)+\epsilon(h'))|hh'\rangle - \langle hh'|K(\epsilon(h)+\epsilon(h'))|h'h\rangle], \tag{6.10.5}$$

$$K(\epsilon) = v + v(Q/d)\,K(\epsilon), \tag{6.10.6}$$

$$\frac{Q}{d} = \sum_{pp'} \frac{|pp'\rangle\langle pp'|}{\epsilon - \epsilon(p,\epsilon) - \epsilon(p',\epsilon)} \tag{6.10.7}$$

and

$$\epsilon(p,\epsilon) = \frac{p^2}{2m} + \sum_h [\langle ph|K(\epsilon-\Delta)|ph\rangle - \langle ph|K(\epsilon-\Delta)|hp\rangle]. \tag{6.10.8}$$

We reiterate that Δ is an average excitation energy for a particle.

The energy shift is then given, to first-order in K, as

$$\Delta E = \frac{1}{2}\sum_{hh'} \begin{Bmatrix} \langle hh'|K(\epsilon(h)+\epsilon(h'))|hh'\rangle \\ -\langle hh'|K(\epsilon(h)+\epsilon(h')|h'h\rangle \end{Bmatrix}, \tag{6.10.9}$$

which is the analogue of equation (6.6.5) in the simple t-matrix theory.

6.11. Calculation of nuclear matter parameters

6.11.1. *Methods*

The solution of the K-matrix equation is along the lines that were employed for the solution of the t-matrix equations of section 6.7. Therefore we shall not enter into the details here but we shall merely outline the self-consistency procedure. To begin with, the K-matrix equation is solved using the kinetic energies for $\epsilon(h)$ and $\epsilon(p,\epsilon)$, namely

$$\epsilon(h) = h^2/2m, \quad \text{and} \quad \epsilon(p,\epsilon) = p^2/2m.$$

From the K-matrix elements thus obtained, new values are calculated for $\epsilon(h)$ and $\epsilon(p)$. These are now used to solve for the K matrices again, the process being repeated until approximate self-consistency is obtained. In this iterative procedure, a new coupling of partial waves (in addition to that due to the Exclusion Principle) would arise because of the energy denominators. This is circumvented by spherical averaging, exactly as in section 6.6.

The above method for the calculation of nuclear matter properties is, on the one hand, complicated because of the self-consistency condition, and on the other, open to some objection on the grounds that $\epsilon(k)$ for $k > k_f$ was evaluated by a somewhat arbitrary procedure. Formally, Brueckner & Gammel have shown how this difficulty can be overcome, but no practical application of their general method has so far been made. There have been, however, two attempts to simplify the calculations of nuclear matter parameters and we now turn to consider these briefly.

Moszkowski & Scott (1960) separate the inter-nucleon interaction into a short- and a long-range part

$$v = v_s + v_l.$$

v_s involves the hard core difficulty but, since it is short-ranged, the t matrix corresponding to it, say t_s, can be calculated to a first approximation as if the nucleons were interacting in free space, rather than in the nuclear medium. Let us call this approximation t_s^F. Since v_l involves no hard-core difficulties, it can be treated by ordinary perturbation theory. The difference between t_s^N, the t matrix for v_s in nuclear matter, and t_s^F, can now be calculated in terms of $\epsilon(k)$, the energy of nucleons in nuclear matter.

Bethe, Brandow & Petschek (1963) have developed a method which does not separate the potential into long- and short-range parts and these authors believe that their method should be more accurate than that of Moszkowski & Scott. They also emphasize that $V(p)$ has nothing to do with the actual potential energy felt by a nucleon of momentum p in moving through nuclear matter. Since the only purpose of the $V(k)$'s was to produce some numbers in the energy denominators of the K-matrix equations, these authors develop a 'reference spectrum method' by which rather simpler values than the self-consistent ones used by Brueckner & Gammel could be used in the energy denominators. Without going into the details of these methods, we proceed to summarize the results of recent calculations of the basic nuclear parameters.

6.11.2. *Results*

The earliest calculation of the nuclear matter parameters with a realistic two-body potential and transcending ordinary perturbation theory was that of Brueckner & Gammel (1958). With the best two-body interactions available at that time they obtained the values 15–18 MeV for the binding energy per nucleon of nuclear matter, in reasonable agreement with the empirical value of 16 MeV.

Subsequently two new inter-nucleon potentials have been proposed.

Brueckner & Masterson (1962), with a somewhat cruder form of the Brueckner–Gammel theory, have used the potential proposed by Lassila, Hull, Ruppel, McDonald & Breit (1962), while Razavy (1963) has used the potential proposed by Hamada & Johnston (1962) in a reference spectrum method (without self-consistency). Both calculations give the binding energy per nucleon in nuclear matter as of the order of 8 MeV at a density corresponding to $r_0 \sim 1{\cdot}3$ F. This is an unsatisfactory state of affairs, for the theory of nuclear matter as a many-body system seems sufficiently well developed for us to anticipate better agreement with experiment. In the next section, therefore, we examine possible ways out of this apparent dilemma.

6.12. Nuclear interactions and correlations

The theory of nuclear matter as developed so far attempts to predict bulk properties starting with given two-nucleon potentials. It may be that the assumption that the nuclear force is derivable from a potential is suspect. It may also be that three- or more-body forces are important. However, before drawing such far-reaching conclusions, we should explore other avenues in an attempt to reconcile the calculated value of about 8 MeV/particle with the experimental value.

6.12.1. *Soft-core interaction*

It has been suggested that the short-range repulsion between nucleons should not be treated as a hard core. Wong (1964) has considered a 'soft' repulsion with a Yukawa shape appropriate to the exchange of a very heavy meson with Compton wavelength of about 0·2 F. At the time of writing, no realistic 'soft' potential satisfying at once the two-nucleon scattering data and yielding a reasonable value for the nuclear binding energy exists. But it appears probable from the work of Wong that a suitable 'soft' force can in fact be found. Other ways of changing the two-nucleon force to 'fit' the nuclear matter-binding energy, without altering the fit to the two-nucleon scattering data may also be found, for example via a velocity-dependent interaction.

6.12.2. *Singularities in the K-matrix and two-body correlations*

In our exposition, we did not pause to ask whether the K-matrix equations had any non-singular solutions. The situation obtaining in other many-body problems makes us suspect that if the two-body force has a residual attractive term then bound pair states may be formed giving rise to the well-known gap in the single-particle energy spectrum (see Chapter 7). Empirically Bohr, Mottelson & Pines (1958) observe that the first intrinsic excited states of heavy deformed even-even nuclei lie at about 1 MeV above the ground state while the spacing between single-nucleon excited states is expected to be only about $\frac{1}{4}$ MeV or so. We might therefore inquire whether this means that nuclear matter is a superfluid (cf. Chapters 7 and 8) whose ground-state properties cannot be studied by perturbation theory.

Attractive two-body interactions imply a singularity in the K-matrix equation. This is best illustrated by considering a separable potential (Bell & Squires, 1961). The matrix element of the potential between two states in which the relative momentum of the two nucleons is $2\mathbf{k}$ and $2\mathbf{k}'$ is assumed to be given by

$$\langle \mathbf{k}'|v|\mathbf{k}\rangle = (\lambda/\Omega)\, v(\mathbf{k})\, v(\mathbf{k}'). \tag{6.12.1}$$

λ is a parameter to specify the strength of the interaction and $v(\mathbf{k})$ is assumed to be real and positive so that λ is negative for attractive forces and positive for repulsive forces.

If $2\mathbf{P}$ denotes the total momentum of the two nucleons, the K-matrix equation (6.10.2) may be written

$$\langle \mathbf{k}'|K(\mathbf{P})|\mathbf{k}\rangle = (\lambda/\Omega)\, v(\mathbf{k}')\, v(\mathbf{k}) + \frac{\lambda}{\Omega}\sum_{\mathbf{k}''} \frac{v(\mathbf{k}')\, v(\mathbf{k}'')\, f(\mathbf{k}'', \mathbf{P})\, \langle \mathbf{k}''|K(\mathbf{P})|\mathbf{k}\rangle}{\epsilon(\mathbf{k}+\mathbf{P})+\epsilon(\mathbf{k}-\mathbf{P})-\epsilon(\mathbf{k}''+\mathbf{P})-\epsilon(\mathbf{k}''-\mathbf{P})}. \tag{6.12.2}$$

The solution of this equation may be verified to be

$$\langle \mathbf{k}'|K(\mathbf{P})|\mathbf{k}\rangle = \frac{\Omega^{-1}\lambda v(\mathbf{k}')\, v(\mathbf{k})}{1+\lambda C(\mathbf{k},\mathbf{P})}, \tag{6.12.3}$$

with

$$C(\mathbf{k},\mathbf{P}) = \frac{1}{\Omega}\sum_{\mathbf{k}''} \frac{v^2(\mathbf{k}'')f(\mathbf{k}'',\mathbf{P})}{\epsilon(\mathbf{k}''+\mathbf{P})+\epsilon(\mathbf{k}''-\mathbf{P})-\epsilon(\mathbf{k}+\mathbf{P})-\epsilon(\mathbf{k}-\mathbf{P})}. \tag{6.12.4}$$

If we make the reasonable assumption that $\epsilon(p) > \epsilon(h)$ for all p and h, then the integrand in (6.12.4) is positive (we convert summation to integration in the usual manner) and it is not difficult to see that $C \to \infty$ as $k \to k_f$ and $P \to 0$.

It thus follows, since $C > 1$ for some region, that the series expansion of (6.12.3) in powers of λ is divergent in this region, however weak and non-singular the potential is.

Another consequence is that for $\lambda < 0$ the energy denominator in (6.12.3) can vanish for some $\mathbf{k}$ and $\mathbf{P}$ unless of course λ is larger than the maximum value of C^{-1}. This would make the integrals of the Brueckner–Gammel theory singular.

The existence of K-matrix singularities can also be seen in the light of the theory of integral equations. The K-matrix equation for $\mathbf{P} = 0$ may be written

$$K|\mathbf{h}\rangle = v|\mathbf{h}\rangle + \sum_{\mathbf{p}} \frac{v|\mathbf{p}\rangle\langle \mathbf{p}|K|\mathbf{h}\rangle}{2\epsilon(\mathbf{h})-2\epsilon(\mathbf{p})}. \tag{6.12.5}$$

This is an inhomogeneous integral equation. If we consider the homogeneous integral equation

$$K|\mathbf{h}\rangle = \mu \sum \frac{v|\mathbf{p}\rangle\langle \mathbf{p}|K|\mathbf{h}\rangle}{2\epsilon(\mathbf{h})-2\epsilon(\mathbf{p})}, \tag{6.12.6}$$

(6.12.5) will have a non-singular solution if and only if there is no eigenvalue μ of (6.12.6) equal to 1. Further, if and only if none of the eigenvalues μ are less than 1 in magnitude will there be a series expansion for K in powers of v. These are results of the standard theory of integral equations. Emery (1959) showed that as $k \to k_f$ there is always one eigenvalue that tends to zero, demonstrating that the series for K is divergent. Also, in the effective-mass approximation, in which the free-particle kinetic energy is modified to give $\epsilon(k)$ by multiplication by m/m^*, there will be no non-singular solutions of (6.12.5) in the triplet state if $m^* > 0{\cdot}75m$ and in the singlet state if $m^* > 0{\cdot}56m$ according to Emery's calculations (using the potential of Brueckner & Gammel).

In the singlet state the Brueckner–Gammel equations themselves gave a singularity for $\langle \mathbf{h}|K|\mathbf{h}\rangle$ for $h = 1{\cdot}468F^{-1}$ with $k_f = 1{\cdot}5F^{-1}$ (see Bell & Squires, 1961) if $\epsilon(k)$ was taken as continuous across the Fermi surface. The singularity disappeared when an energy gap was assumed across the Fermi surface of magnitude of about 0·4 MeV. Since the singularity, when it occurs, is so near to k_f, the numerical methods used for the solution of the Brueckner–Gammel equation will not reveal it. Also, the gaps in $\epsilon(k)$ across the Fermi surface assumed in the Brueckner–Gammel approximation, are roughly of the order of magnitude necessary to make the singularity disappear.

Emery & Sessler (1960) have calculated the energy gap of nuclear matter taking $k_f = 1{\cdot}4F^{-1}$ and using an effective mass $m^* = m$. They then obtained an energy gap of 0·1 MeV, but this turns out to be sensitive to the choice of m^* and disappears if $m^* = 0{\cdot}75m$. Even if one assumes a larger gap $\epsilon \sim 1$ MeV, corresponding to the observed gaps in large finite nuclei, the effect on the average energy per nucleon is going to be very small. Taking E_f, the Fermi energy, to be about 45 MeV for $k_f \sim 1{\cdot}5F^{-1}$, the effect on the energy per particle will be $\frac{3}{8}\epsilon^2/E_f$, which is about 0·008 MeV for $\epsilon \sim 1$ MeV, the $\frac{3}{8}$ factor being the probability of finding two nucleons in an even state.

Thus the conclusion seems to be that, while the pairing of nucleons might be of great importance in considering the excited states of large nuclei, this same pairing has very little effect on the binding energy/particle in nuclear matter. A useful discussion of this and related aspects of the problem has been given by Baker (1964).

6.12.3. *Estimate of three-body correlations*

We started with the expression for the ground-state energy of the many-nucleon system expressed formally by means of the Goldstone formula as an infinite series in ascending powers of the inter-nucleon potential. This series was certainly not convergent for a strong inter-nucleon force. It then turned out that, in any order of the perturbation series, there were terms proportional to different powers of the density. Use of the t matrix then enabled us to sum an infinite subseries of terms having the same density dependence in the original Goldstone result. In particular, the Brueckner–Gammel theory included the contributions of every term in the Goldstone series proportional to the density, ρ. The next logical step ought to be the investigation of the contributions of all the terms of the perturbation series that are proportional to ρ^2 and so on. In the language of Feynman graphs, the low-density theory of Brueckner & Gammel took account of all the graphs with two hole lines. An improved version should really include *all* graphs with three hole lines and so on. Physically the Brueckner–Gammel theory takes careful account of two-body correlations and thus goes beyond a single-particle approximation. The next step is a careful consideration of three-body correlations. A first and important step in this direction has been taken by Bethe (1965; cf. Faddeev, 1961). Three-body correlations contribute only about 3 to 6% of the binding energy as calculated by inclusion solely of two-body correlations. While more work with realistic forces will have to be carried out, we can conclude that the Goldstone result, if rearranged into a series in powers of the density (or the number of interacting particles), appears to be rapidly convergent. Thus the Brueckner–Gammel theory can be considered reasonably well established and the fit with nuclear binding data used to choose between competing phenomenological two-nucleon potentials.

6.13. Single-particle and rearrangement energies

We conclude this chapter by establishing an important theorem due to Hugenholtz & Van Hove (1958) and discussing its relevance to the calculation of the nuclear matter parameters.

We consider an N-particle system and denote by $E_0(N)$ the energy of the system at equilibrium density. If we now add a particle to the system, then assuming it occupies the lowest available single-particle level, we may expect the resulting $N+1$ particle system to be in its ground state, with energy

$$E_0(N+1) = E_0(N)+E_f', \tag{6.13.1}$$

where E_f', the energy of the particle added, is nearly equal to $\epsilon(k_f)$ the energy of the particle in the highest occupied level.

This equation, written in the form appropriate for $N \to \infty$ is

$$E'_f = \frac{dE_0}{dN} = \frac{d}{dN}\left(\frac{E_0}{N}N\right) = \frac{E_0}{N} + \frac{N}{\Omega}\frac{d}{d\rho}\left(\frac{E_0}{N}\right)$$

$$= \frac{E_0}{N} + \rho\frac{d}{d\rho}\left(\frac{E_0}{N}\right). \quad (6.13.2)$$

Now, for a system at equilibrium, $(d/d\rho)(E_0/N) = 0$. Thus we have established that the mean energy per particle of a system at equilibrium is equal to the energy of the particle of the system occupying the highest single-particle level for, in the limit of a large system, this latter energy equals the lowest energy with which we may add a particle to the system. Similar arguments will also show that the mean energy per particle equals the minimum energy required to remove a particle from the system. This latter energy is often termed the separation energy. We emphasize that the result just established is true for a system describable within an independent-particle model framework. A typical example for which this result is true is the Hartree–Fock theory.

In the Brueckner–Gammel theory

$$\epsilon(k_f) = \frac{k_f^2}{2m} + \sum_h \langle hk_f | K(\epsilon(k_f)) + \epsilon(h)) | hk_f \rangle$$

$$- \sum_h \langle hk_f | K(\epsilon(k_f) + \epsilon(h)) | k_f h \rangle, \quad (6.13.3)$$

while

$$E_0 = \sum_h \frac{h^2}{2m} + \frac{1}{2}\sum_{hh'} [\langle hh' | K(\epsilon(h) + \epsilon(h')) | hh' \rangle$$

$$- \langle hh' | K(\epsilon(h) + \epsilon(h')) | h'h \rangle]. \quad (6.13.4)$$

Then we have

$$E'_f = \frac{dE_0}{dN} = \frac{dE_0}{dk_f}\frac{dk_f}{dN} = \frac{\pi^2}{2\Omega k_f^2}\frac{dE_0}{dk_f}$$

$$= \epsilon(k_f) + \frac{1}{2}\sum_{hh'} \langle hh' | \frac{\partial k_f}{\partial N}\frac{\partial}{\partial k_f}\{K(\epsilon(h) + \epsilon(h'))\} | hh' - h'h \rangle$$

$$= \epsilon(k_f) + R, \quad (6.13.5)$$

where this equation defines the rearrangement energy R. Thus the single-particle energy at the Fermi surface is not equal to the minimum energy required to remove a particle from an N particle system at equilibrium density to form an $N-1$ particle system at equilibrium density. The difference R is the energy involved when the $N-1$ particles, which formed part of the N particle system at equilibrium, readjust themselves to be at equilibrium. R is non-zero because of the dependence of the K matrix on density (k_f). This dependence is due both to the change in the Exclusion Principle as well as a change in the energy denominator and may be exhibited as follows:

$$K = v + v(Q/d)K, \tag{6.13.6}$$

and therefore we have

$$\delta K = K\frac{\delta Q}{d}K + KQ\delta\left(\frac{1}{d}\right)K \tag{6.13.7}$$

to first order. Brueckner & Goldman (1960) in this way estimated that the rearrangement energy was of the order of 10 MeV.

This discussion makes us point out once again that the $V(p)$ of the Brueckner–Gammel theory has little in common with the actual potential energy felt by a nucleon of arbitrary energy added to nuclear matter. If the particle added has energy larger than E_f', then the unperturbed state so formed is degenerate with states which have other particles excited out of the Fermi sea at the expense of the energy of the added particle. Thus the state has a finite lifetime. This can be interpreted in terms of an imaginary part to the (optical model) potential felt by the added nucleon while the $\epsilon(p)$'s of the Brueckner–Gammel theory were explicitly real. However it is not appropriate to pursue this point further here.

Problems

P.6 (i). Neutrons and protons are considered as two states of the single particle, the nucleon, by the formal device of introducing the double-valued operator, isotopic spin and saying that a neutron is a nucleon with isotopic spin $+\frac{1}{2}$ and that a proton is a nucleon with isotopic spin $-\frac{1}{2}$. This gives rise to isotopic spin matrices (just as the ordinary spin gives rise to Pauli spin matrices)

$$\tau_x = \begin{pmatrix} 0 & 1 \\ 1 & 0 \end{pmatrix}, \quad \tau_y = \begin{pmatrix} 0 & -i \\ i & 0 \end{pmatrix} \quad \text{and} \quad \tau_z = \begin{pmatrix} 1 & 0 \\ 0 & -1 \end{pmatrix}$$

so that

$$\tau_z \lambda_N = \lambda_N, \quad \tau_z \lambda_P = -\lambda_P,$$

where $\lambda_N = \begin{pmatrix} 1 \\ 0 \end{pmatrix}$ and $\lambda_P = \begin{pmatrix} 0 \\ 1 \end{pmatrix}$ are the isotopic spin functions.

Show that a two-nucleon interaction of the form

$$V(r) = V_1(r) + V_2(r)\,\boldsymbol{\tau}_1 . \boldsymbol{\tau}_2$$

is independent of the charge of the interacting particles.

P.6 (ii). In a non-interacting Fermi gas state of nucleons, show that the probability of finding 2 nucleons with relative momentum k is

$$\frac{24k^2}{k_f^3}\left(1 - \frac{3}{2}\frac{k}{k_f} + \frac{k^3}{2k_f^3}\right),$$

where k_f is the Fermi momentum.

P.6 (iii). Assume that the internucleon interaction is $V_0 e^{-\mu r}/\mu r$ in even states and is zero in odd states.

Show that, to first order, the energy per particle is

$$\frac{3}{5}\frac{k_f^2}{2m} + \frac{V_0}{16\pi}\left(\frac{k_f}{\mu}\right)^3\left\{1 + \frac{6\mu^2}{k_f^2}\left(\frac{3}{8} - \frac{\mu}{16k_f} - \frac{\mu}{2k_f}\tan^{-1}\frac{2k_f}{\mu}\right)\right.$$
$$\left. + \left(3 + \frac{\mu^2}{4k_f^2}\right)\frac{\mu^2}{16k_f^2}\ln\left(1 + \frac{4k_f^2}{\mu^2}\right)\right\},$$

and that the interaction energy of a particle with momentum k is

$$V(k) = \frac{V_0}{\pi}\left\{\frac{k_f^3}{\mu^3} + \frac{3k_f}{2k}\left[\left(\frac{1}{4} + \frac{\mu^2}{k_f^2} - \frac{\mu^2}{k^2}\right)\ln\frac{\mu^2 + (k+k_f)^2}{\mu^2 + (k-k_f)^2}\right.\right.$$
$$\left.\left. + \frac{2k}{k_f} - \frac{2k\mu}{k_f^2}\tan^{-1}\frac{2k\mu}{\mu^2 + k^2 - k_f^2}\right]\right\}.$$

CHAPTER 7

SUPERCONDUCTIVITY

7.1. Introduction

In Chapter 5, we dealt with the Sommerfeld model of a metal, including the Coulombic interactions between electrons. It emerged that the single-particle energy spectrum resembled the non-interacting case in the sense that there were states available for particle excitation immediately outside the Fermi surface. Systems with this property are said to be normal. In the theory of superconductivity, the effect of the vibrating ionic lattice on the electrons is crucial, and must therefore be included at the outset. The level spectrum is then no longer normal. But first we shall survey briefly some of the more important observed properties of superconductors. While doing so, we shall take the opportunity to draw a number of highly important qualitative conclusions. These will then be used eventually as a basis for a quantitative explanation of the observations enumerated below.

7.2. Summary of observed properties

7.2.1. *Laws of similarity*

We will confine our attention initially to the so-called 'soft' superconductors (for a precise classification, see section 7.2.4 below), and then there are laws of similarity enabling results to be expressed in terms of reduced variables. Some of these will emerge below. The point to be made here is that such behaviour indicates that, as a first approximation at least, a theory of superconductivity should be possible, without invoking the specific properties of any particular metal. It is this theory with which we shall be concerned. For precisely the same reason, of course, one cannot expect the theory to account for the individual properties peculiar to any metal.

7.2.2. *Vanishing of d.c. resistance*

Below some critical temperature, T_c, usually of the order of a few degrees, a superconductor possesses that property from which it gets its name—its electrical resistivity vanishes. Above T_c the metal is normal. To understand the general implication of this, consider a current-carrying state for a normal metal. The appropriate model is a displaced Fermi sphere, smeared by the effect of temperature and whatever interactions are present. It is schematically drawn in Fig. 7.1. What happens when the field is turned off is well known. The effect of crystal defects and lattice vibrations is that the electrons are scattered individually in such a way that there is a net transfer of electrons from the leading to the trailing edge of the Fermi sphere. Soon, after a characteristic relaxation time, the zero-current state, given by Fig. 7.2, is effectively restored.

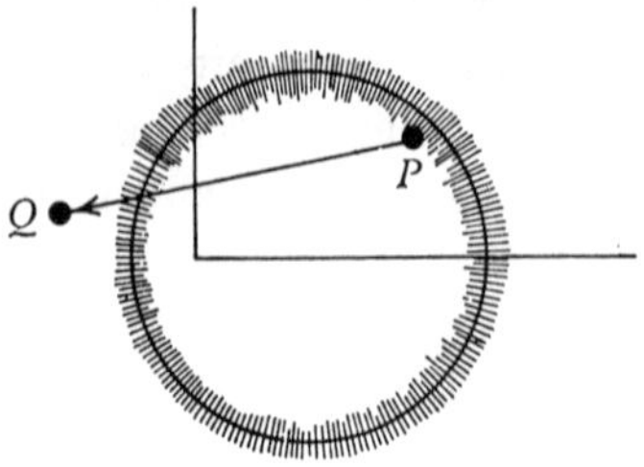

Fig. 7.1. Current carrying state for normal metal.

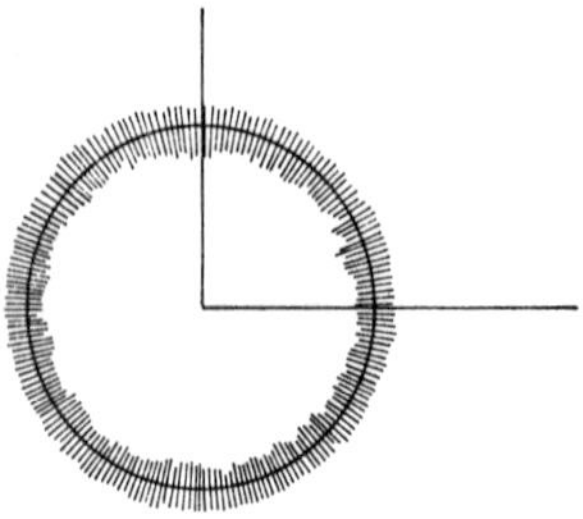

Fig. 7.2. Zero-current state.

However, in the case of a superconductor, a current-carrying state such as that indicated in Fig. 7.1 is stable in zero applied field. In other words, typical single-particle transitions of the type $P \rightarrow Q$ indicated in the diagram are no longer favoured energetically. This must be because the particle at P is bound in momentum space to the other particles.

Generally speaking, we will be concerned with the case when no current is flowing and we arrive at the following conclusions concerning this superconducting ground state. The particles are bound together in some manner in momentum space. Furthermore, single-particle transitions are inhibited by an energy gap (the degree of binding).

It is important at this stage to emphasize that there is an essential difference in character between the energy gap in a superconductor and that in an insulator. We have already indicated in the argument immediately above that the position of the gap in momentum space is wholly dependent on the electronic distribution. Nevertheless, we wish to stress again here that, whereas the gaps in an insulator are determined by the lattice only, the gap in a superconductor is generated by the electrons themselves and moves with the electrons. This is crucial in understanding the difference between superconductors and insulators.

Generally speaking, we shall be concerned in this chapter with describing fundamentally the superconducting ground state, and those excited states which carry no current.

7.2.3. *Low-temperature specific heat*

We are dealing here with that particular contribution to the observed specific heat from the valence electrons alone. The specific heat indicates the effective number of degrees of freedom available to absorb a given small amount of energy. Thus, while an ideal classical gas has a constant specific heat, that of the ideal Fermi gas is proportional to T. The latter is thus lower than the classical value for low enough T and this reflects the fact that while, in the classical gas, all states are available for excitation, only those in an energy width $k_B T$ about the Fermi surface are available in the degenerate case. For a superconductor at low temperature, the specific heat is found experimentally to be reduced even further. Indeed the low-temperature behaviour ($T < T_c$) is found to be represented by the formula

$$\frac{C_v}{\gamma T_c} = a\,e^{-bT_c/T}, \tag{7.2.1}$$

where γT is the low-temperature electronic specific heat of the normal metal. The parameters a and b, themselves weakly temperature dependent, are about 9 and 1·5 for all superconductors. The implication is that there is an extreme scarcity of available states for small energy increments, thus confirming our suspicions of the existence of an energy gap in the single-particle excitation spectrum. More details are exhibited in Fig. 7.3. Above T_c, the

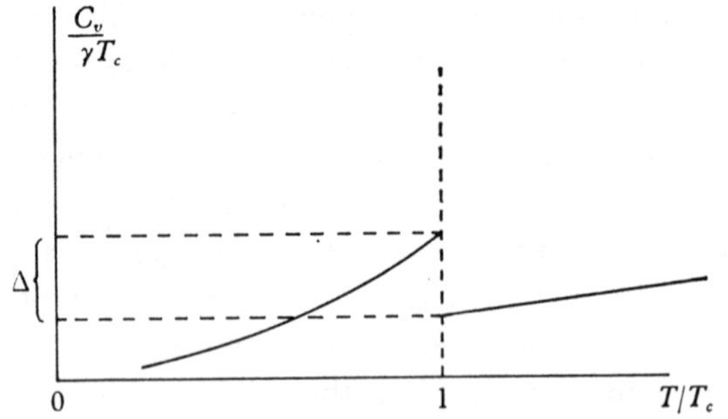

Fig. 7.3. Specific heat curve of typical superconductor.

specific heat reverts to its normal behaviour while at T_c itself there is a second-order phase transition (constant entropy, constant volume, no latent heat) with a discontinuity in the specific heat as indicated. The size of the discontinuity Δ is about $2{\cdot}5\gamma T_c$ for all superconductors. Our philosophy will be that once we understand the properties of the energy gap, an understanding of the transition in terms of a co-operative effect should follow.

7.2.4. *Penetration depth and coherence length*

It is well known that superconductivity can be destroyed by the application of a magnetic field greater than some critical value H_c. Below H_c, the field is expelled from the specimen, except for a thin, temperature-dependent surface layer, having thickness λ_L which is typically of order 10^{-5} cm. This exhibition of (nearly) perfect diamagnetism, the Meissner effect, gives rise to some of the most difficult theoretical problems and we shall not deal with it fully here.

However, it is vital that we stress at this point the idea of a coherence length ξ, of order 10^{-4} cm, over which strong correlations exist in a superconductor. This is, of course, different from the penetration depth, λ_L, referred to above.

This idea of coherence length, introduced into the theory by Pippard (1953), is so basic that we shall briefly summarize here the evidence which led him to invent the concept. The three points on which he focussed attention were:

(i) The sharpness of the transition from the normal to the superconducting state.

(ii) The fact that the penetration depth is very insensitive to the application of a magnetic field comparable with the critical field.

(iii) The existence of a large surface energy at a boundary between normal and superconducting phases in the presence of the critical field.

Pippard then noted that point (i) suggested that the phenomenon was one of co-operative behaviour in which local fluctuations (e.g. persistence of local order above the transition temperature) were more or less removed. Point (ii) likewise suggested, from the fact that a field equal to the critical H_c may produce a change of only 1 %, that the disturbing effect of the field is not confined simply to the penetration layer, but is distributed over a greater thickness of order 10^{-4} cm. Or again, if we write the surface energy in (iii) in the form $l \times H_c^2/8\pi$, the length l is found to be of the order of 10^{-4} cm.

An alternative order of magnitude estimate of the coherence length can be given from the Uncertainty Principle, though it does, in a sense, anticipate some of the later arguments of the Bardeen–Cooper–Schrieffer (B.C.S., 1957) theory. We argue that the superconducting electrons must be correlated over a distance Δl related to an uncertainty Δp in momentum through

$$\Delta l \sim \frac{h}{\Delta p}.$$

Now, Δp must involve the energy associated with the critical temperature T_c, and since, as we shall see, only electrons near the Fermi surface are involved, the velocity is v_f. Hence

$$\Delta l \sim \frac{hv_f}{kT_c}.$$

Taking T_c typically as a few degrees, and v_f corresponding to a Fermi energy $\sim$ 5 eV, we find again $\Delta l \sim 10^{-4}$ cm. Obviously, to sharpen up the concept, we need a detailed description of the superconducting wave function (cf. section 7.4).

At this point, we can give a more precise classification of superconductors. If the penetration depth λ_L is small compared with the coherence length ξ, then we speak of 'soft' or type I superconductors. When $\lambda_L/\xi \gg 1$, the physical behaviour is very different and such superconductors are referred to as type II (hard). The intermediate régime when $\lambda_L/\xi \sim 1$ is more difficult and must await a discussion of the Ginzburg–Landau theory (cf. section 7.9).

At this point, it is worthwhile to consider the basic difficulty which has beset the formulation of a microscopic theory of superconductivity for many years. It is concerned with the minute difference in the free energies of the superconducting and normal states of a metal. Thus, if F_s and F_n are these free energies in volume Ω, then we may write

$$\frac{H_c^2}{8\pi} = \frac{F_n - F_s}{\Omega}. \tag{7.2.2}$$

The observed variation of H_c with T is as shown in Fig. 7.4. It is found that H_{c0} is a few hundred oersteds, leading to a value of order 10^{-8} eV per atom for the energy difference at $T = 0$ between the normal and superconducting states.

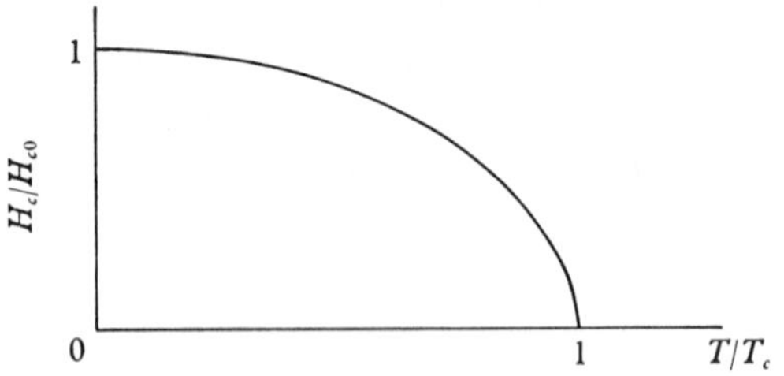

Fig. 7.4. Variation of critical field H_c with T.

When we contrast this with the fact that, as we saw in Chapter 5, the Fermi energy is typically 5 eV, the exchange energy about 2 or 3 eV, while correlation energies are of the order of 1 eV (and usually only accounted for to moderate accuracy by the methods of Chapter 5), the success of the microscopic theory given below is truly remarkable.

This 10^{-8} eV energy can again be interpreted in terms of an energy gap in the single-particle excitation spectrum in the following way. The energy supplied to destroy the gap at the transition temperature T_c is of order $k_B T_c$ per electron. But, by thinking of the normal Fermion system just above T_c, the electrons effectively taking part in the transition are confined to a narrow region of width of order $k_B T_c$ about the Fermi surface, where the density of states is ρ, say. Hence, the total energy change in the transition is of order $(k_B T_c)(\rho k_B T_c)$ which is typically of order 10^{-8} eV.

7.2.5. *Isotope effect*

It has been found that in many super conductors, T_c varies with average isotopic mass M. More precisely, it is found that for any given metal to a first approximation (for a summary of the data, see Garland, 1963)

$$T_c M^{\frac{1}{2}} = \text{const.} \tag{7.2.3}$$

Since, as the arguments of the previous paragraph indicate, T_c turns out to be proportional to H_c, a relation similar to (7.2.3) holds between H_c and M.

The importance of the isotope effect is as follows. We have seen a need to produce binding of the electrons in momentum space, thus producing an energy gap at the Fermi surface. Such binding can only be produced by attractive interactions, the largely repulsive screened Coulomb interaction being ineffective for this purpose (see, however, Luttinger, 1966). Furthermore, the bare minimum requirement on such attractive interactions is that they bind those particles near the Fermi surface, the others then having to overcome an energy gap because of the normal Fermi distribution alone, to make single-particle transitions. Until the isotope effect was discovered, the physical origin of this behaviour was not conclusively demonstrated (though Fröhlich (1950) had recognized the importance of the lattice interaction and theoretically predicted the isotope effect). Afterwards, however, people were highly motivated to seek an attractive component between physical electrons by taking a proper account of the lattice. This search proved to be successful. Theoretical investigation of the coupled electron-phonon system shows that the phonon variables can be effectively eliminated, while the modification to the screened Coulomb forces between electrons is such as to produce attractive interactions for pairs of particles sufficiently near the Fermi surface. Fröhlich's derivation of this interaction is summarized in Appendix 7A.1. A different well-known discussion is that of Bardeen & Pines (1955). From the point of view of the present chapter, it is not important that such calculations differ in their details. The B.C.S. theory, below, incorporates only the crudest features of the interaction, which all reasonable results have in common. Actually, all time-independent interactions (such as those mentioned above) are incorrect in principle as the following elementary discussion makes clear. In refinements of the B.C.S. theory, such time-dependence is incorporated as an essential feature of the problem (see, for example, Schrieffer, 1964; Rickayzen, 1965).

The simple physical picture often given is of an electron polarizing the lattice through which it is moving. The lattice deforms in an attempt to produce a compensating positive charge around any given electron. Thus, as the electron moves it leaves behind it a positively charged wake which, before the lattice relaxes, is capable of attracting another electron. The two electrons are thus indirectly attractively coupled.

If ω is the zero-point oscillator frequency of an ion, the normal electron gas is perturbed by a lattice with energy $\hbar\omega \sim 10^{-1}$ eV per ion, so this must be a measure of the effective width at the Fermi surface for which interactions are attractive. It should be noted that the phonons we have been considering are virtual. They are excited by the passage of electrons through a lattice even at absolute zero. At higher temperatures, of course, the thermal phonons soon dominate, the latter in contrast being responsible (in part) for resistance to current flow.

7.3. Pairing hypothesis

Let us now incorporate the features of the above discussion into the following assumption. We suppose that the mechanism responsible for superconductivity is an attractive interaction between pairs of particles, each occupying states having energy within some energy shell of width 2δ at the Fermi surface. In the case of metals exhibiting the isotope effect, we replace δ by $\hbar\omega$.

The above assumption now gives us a criterion for constructing a potential for the superconducting phase. However, it needs to be emphasized that we are still faced with a many-body Schrödinger equation for which there is no general method of solution. Experience in the many-body problem so far has indicated that each potential requires initial intuitive insight to ascertain what the important physical features are, before suitable many-body solutions can be found. The remainder of this section will be devoted to isolating what is believed to be the fundamental physical feature appropriate to the above assumption.

Primarily we are interested in the attractive interactions between particles near the Fermi surface. Let us, then, examine the elementary scattering processes to which they give rise. We will classify the latter according to the total momentum **k** of the pair. The question we will ask ourselves is this. How many pairs in our energy shell are there for any given **k**? The answer may be obtained as follows. Construct two spherical shells of radius k_f such that the vector joining their centres is **k**. Let the left one (see Fig. 7.5) denote the Fermi sphere while the right one is a geometrical construction only. The pairs of particles then, each near the Fermi surface which sum to give **k** are just diametrically opposite points on the intersection of the two shells. For vanishingly small width δ, the measure of the number of pairs of this type is, in general, the half circumference of the circle of intersection of the spheres. However, if $\mathbf{k} = 0$, a different situation arises. The spheres are now superposed and the corresponding number of pairs has the measure of half the surface area of the Fermi sphere. One sees in this way that those pairs near the Fermi surface with total momenta zero are infinitely more numerous than those with any given non-zero total momentum.

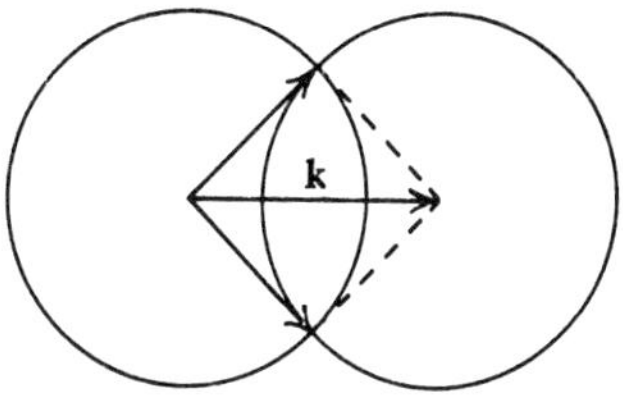

Fig. 7.5

For the preceding reason, we will now focus attention on that particularly numerous subset of pairs with total momentum zero and see if a study of these can lead us to an explanation of the existence of the superconducting ground state.

7.3.1. *Cooper pairs*

It will now be shown that pairs of the above kind with zero total momentum are likely to bind together to produce the desired energy gap. There is no *a priori* reason why they should, for each electron has a Fermi energy of about 5 eV. However, it is to be remembered that we are not considering two electrons in free space but in a medium of like particles and it turns out that the very Exclusion Principle which produces the 5 eV kinetic energy, operates in such a way as to produce the desired binding.

Let us consider, to begin with, a set of independent Fermions. Then we select two of these at the Fermi surface, with total momentum zero, and allow them to interact, the other particles remaining entirely passive, neither interacting with each other or with the two selected particles. Their sole role is to occupy the states below the Fermi surface. The two special particles now interact and scatter repeatedly in accordance with the conservation conditions. In particular, momentum is conserved, so the pair must scatter into

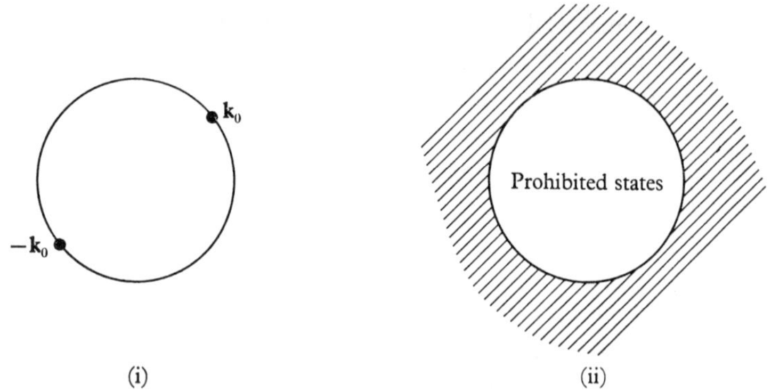

Fig. 7.6. Effect of interactions on two-particle states.

other states with net zero momentum. The only proviso is that the particles cannot enter the already fully occupied states below the Fermi surface. After infinite time (cf. section 4.5) a stationary state is achieved. The $t = -\infty$ and $t = 0$ states are indicated schematically in parts (i) and (ii) respectively of Fig. 7.6. We have the usual smearing due to interaction, but with only two particles involved.

The unperturbed energy of the pair was $2E_f$, where E_f is the usual Fermi energy $\hbar^2 k_f^2/2m$. To ascertain whether binding occurs, we have to examine the ground state of the Schrödinger equation

$$\left(-\frac{\hbar^2}{2m}\nabla_1^2 - \frac{\hbar^2}{2m}\nabla_2^2 + v(\mathbf{r}_1, \mathbf{r}_2)\right)\psi = E\psi; \tag{7.3.1}$$

where v simulates the superconducting interaction. We must clearly show that the corresponding eigenvalue is less than $2E_f$. To solve this equation, we go through the standard procedure of expanding ψ in terms of the unperturbed ($v = 0$) excited states. In accordance with our discussion above, these are just the two-particle plane-wave products with individual momenta **k** and $-\mathbf{k}$, given by $\Omega^{-1} e^{i\mathbf{k}.(\mathbf{r}_1-\mathbf{r}_2)}$ with $k > k_f$, and eigenvalues $E_{\mathbf{k}} = \hbar^2 k^2/m$.

Now the introduction of the two spin states is not essential to the understanding of this particular problem. We will, therefore, suppose that the momentum states are singly filled. Then the perturbed wave function given by (7.3.1) may be written

$$\psi = \sum_{k'>k_f} C_{\mathbf{k}'} \frac{e^{i\mathbf{k}'.(\mathbf{r}_1-\mathbf{r}_2)}}{\Omega}, \tag{7.3.2}$$

where the $C_{\mathbf{k}}$'s are constants to be determined. Substitution of (7.3.2) into (7.3.1) then gives

$$\sum_{k'>k_f} C_{\mathbf{k}'}(E_{\mathbf{k}'} - E + v(\mathbf{r}_1, \mathbf{r}_2)) \frac{e^{i\mathbf{k}'.(\mathbf{r}_1-\mathbf{r}_2)}}{\Omega} = 0, \tag{7.3.3}$$

or, multiplying by $e^{-i\mathbf{k}.(\mathbf{r}_1-\mathbf{r}_2)}$ and integrating over both space variables,

$$(E_{\mathbf{k}} - E) C_{\mathbf{k}} = -\sum_{k'>k_f} C_{\mathbf{k}'} v(\mathbf{k}, -\mathbf{k}, \mathbf{k}', -\mathbf{k}') \tag{7.3.4}$$

where

$$v(\mathbf{k}, -\mathbf{k}, \mathbf{k}', -\mathbf{k}') = \iint \frac{e^{-i\mathbf{k}.\mathbf{r}_1}}{\sqrt{\Omega}} \frac{e^{i\mathbf{k}.\mathbf{r}_2}}{\sqrt{\Omega}} v(\mathbf{r}_1, \mathbf{r}_2) \frac{e^{i\mathbf{k}'.\mathbf{r}_1}}{\sqrt{\Omega}} \frac{e^{-i\mathbf{k}'.\mathbf{r}_2}}{\sqrt{\Omega}} d\mathbf{r}_1 d\mathbf{r}_2. \tag{7.3.5}$$

Our next task is to choose the latter in accordance with the assumption concerning the interaction responsible for superconductivity. We take the Hermitian form

$$v(\mathbf{k}, -\mathbf{k}, \mathbf{k}', -\mathbf{k}') = \begin{cases} -v/\Omega & \text{for} \quad E_f < E_k, E_{k'} < E_f + \delta, \\ 0 & \text{otherwise}, \end{cases} \tag{7.3.6}$$

thus restricting the scattering states (the shaded region of Fig. 7.6 (ii)) to the energy range $(E_f, E_f + \delta)$. The following points will also be noted. The energy shell is not placed symmetrically about the Fermi level because of the inability of the given two particles to penetrate into the fully occupied sea. Within the shell, the matrix elements are chosen (as simply as possible) to be constant. Our approximation is clearly to replace the proper physical matrix element by some average value in the shell. The choice of the minus sign reflects the fact that the interaction is attractive, while the introduction of the Ω factor is to make the v on the right-hand side of order unity (see (7.3.5)).

We now substitute for v from (7.3.6) into (7.3.4), and we find that

$$C_{\mathbf{k}} = \frac{v}{E_k - E} \times \frac{1}{\Omega} \sum_{E_f < E_{k'} < E_f + \delta} C_{\mathbf{k}'}. \tag{7.3.7}$$

In general, the solution of such a secular equation is difficult, but in this particular case the calculation of the eigenvalues is relatively easy. The $C_{\mathbf{k}}$'s can be entirely eliminated by summing each side over $\mathbf{k}$ values in the energy range $(E_f, E_f + \delta)$ and then cancelling the factor $\sum_{E_f < E_k < E_f + \delta} C_{\mathbf{k}}$ common to both sides. We are left with the equation

$$\frac{1}{v} = \frac{1}{\Omega} \sum_{E_f < E_k < E_f + \delta} \frac{1}{E_k - E} \tag{7.3.8}$$

to be solved for the eigenvalues E.

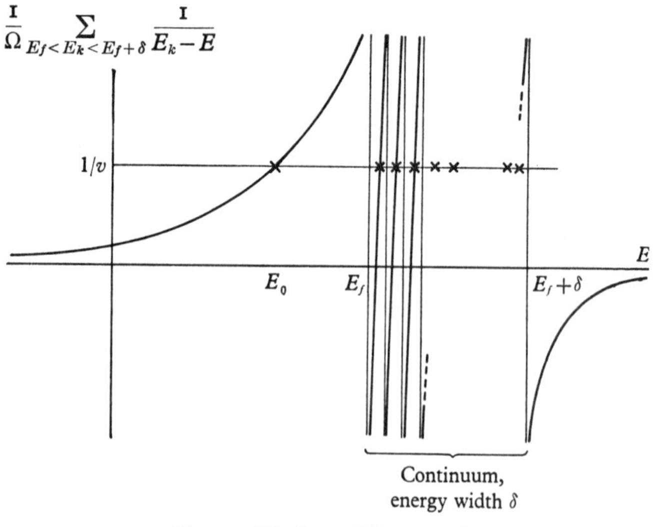

Fig. 7.7. Binding of Cooper pairs.

The roots of (7.3.8) are now explored by the well-known graphical method of plotting out both sides as functions of E, and examining where the curves intersect. The left-hand side, for $v > 0$, is just a straight line in the upper half-plane parallel to the E-axis. Turning to the right-hand side of (7.3.8), it will be observed that as E runs from E_f to $E_f + \delta$, the function fluctuates with great rapidity between plus and minus infinity. Each time it does so, it takes on the value

$1/v$ once. Thus, in this range, we get a continuum of solutions of (7.3.8). The important feature is, however, that there is a portion of the curve below E_f which separates from the continuum (cf. the plasmon discussion of Chapter 5) and decreases monotonically to zero as $E \to -\infty$. This branch of the curve takes the value $1/v$ at some volume-independent energy below the Fermi level. This corresponds to the low-lying bound state we have been seeking.

7.3.2. *Magnitude of binding energy*

Once this low-lying level E_0 has been shown to exist, the evaluation of the binding energy $E_f - E_0$ is not difficult. Applying (7.3.8) specifically to the ground state, we introduce a free-particle density of states factor ρ by means of the usual replacement,

$$\frac{1}{\Omega}\Sigma \to \int d\epsilon\, \rho(\epsilon), \tag{7.3.9}$$

to obtain

$$\frac{1}{v} = \int_{E_f}^{E_f+\delta} \frac{\rho(\epsilon)}{\epsilon - E_0}\, d\epsilon. \tag{7.3.10}$$

There are no divergency difficulties associated with this integral, since the denominator in the integrand is positive definite over the range of integration. The explicit free-particle expression for ρ may be used in (7.3.10) but since we are later forced to use an (accurate) expedient when dealing with the density of states appropriate to normal physical electrons, we shall use the same device here. The point to be made is that $(\epsilon - E_0)^{-1}$ is infinite not far below the range of integration, dropping to moderate values as ϵ increases (and indeed once one reaches $\epsilon = E_f + \delta$, as may be ascertained using (7.3.11) below). Thus in the range $(E_f, E_f + \delta)$, we expect $(\epsilon - E_0)^{-1}$ to vary much more rapidly than the density of states function ρ. Since δ is small, we approximate $\rho(\epsilon)$ by its value at the Fermi surface, perform the resulting elementary logarithmic integral and obtain without difficulty a binding energy given by

$$E_f - E_0 = \frac{\delta}{e^{1/\rho v} - 1} \sim \delta\, e^{-1/\rho v}, \tag{7.3.11}$$

the latter step following if $e^{1/\rho v} \gg 1$. If we recall that in metals, experiment indicates that $E_f - E_0 \sim 10^{-4}\,\text{eV}$ while $\delta \sim 10^{-1}\,\text{eV}$, the quantity $e^{1/\rho v}$ is certainly large in this case. Let us note how the exclusion of the pair from the Fermi sphere is essential to produce binding, for if k_f vanishes so does ρ and therefore so does (7.3.11).

We will pursue this example no further. The important point has been demonstrated. The very numerous class of pairs of particles near the Fermi surface, with total momentum zero, appear to be capable of condensing into bound pairs when the superconducting interaction is introduced. The binding energy per pair is volume-independent and is of the form (7.3.11). It should be noted that, although the latter is very small, the functional form of it is such that it cannot be expanded in a power series in the interaction parameter v, and thus in any many-body generalization of the above method, perturbation theory would not be easy to apply. For this reason, variation methods are most often used.

On the basis of the evidence that has been accumulated above, we now make the following basic assumption on which a many-body theory will be founded. The many-body behaviour of the particles in the energy shell at the Fermi surface, interacting attractively and thus producing the superconducting behaviour, is such that the particles with opposite momenta and opposite spins condense into bound pairs.

We will say no more about the assumption of opposite momenta. The opposite spin requirement has not been touched upon, however. Experimentally, in metals, once we agree to couple particles with opposite momenta, the spin choice is clear. The coupling of parallel spins would not be expected to alter the usual Pauli spin paramagnetism of the normal electrons. On the other hand, binding of pairs at the Fermi surface would greatly inhibit spin flipping as the latter would require the destruction of bound pairs. Observed Knight-shifts, though not yet understood in detail, are nevertheless generally in favour of opposite spin coupling. Theoretically, also, this seems reasonable. Other things being equal, the effect of exchange is to favour antiparallel spin alignment for attractive forces. (A simple calculation of the expectation value of the Hamiltonian for two interacting particles with respect to determinantal states corresponding to parallel and antiparallel spin cases is quite instructive by way of illustration.)

We have seen how the basic properties of superconductors lead, rather naturally, to the hypothesis of a new type of phonon mediated interaction between electrons, as exemplified by the Fröhlich coupling of Appendix 7A.2. The theory may then be formulated in terms of bound electron pairs, as originally supposed by Cooper (1956), and the binding energy, estimated crudely, has the correct order of magnitude. We are therefore in a position to construct a full many-body theory of the superconducting state, and this will be the major concern of the remainder of this chapter.

7.4. Bardeen–Cooper–Schrieffer (B.C.S.) theory

In order to focus attention on the essentials of the problem, and, at first, to minimize complicated detail, we will proceed by making reasonable assumptions. These are conveniently expressed via a simplified Hamiltonian, which we now consider.

7.4.1. *Reduced Hamiltonian*

The fundamental problem is that of finding the ground state of a set of Fermions in which both screened Coulomb and superconducting interactions are present. The discussion of the two-body problem given above strongly suggests that, in describing the main features of the superconducting phase, it is the attractive interactions which lead to pairing, and not the screened Coulomb repulsions, which will play the crucial role, and for this reason we shall ignore the latter completely.

Our final decision in formulating the problem concerns the precise choice of the many-body superconducting interaction, and for this we are guided by the concluding assumption of the previous section. Our physical picture is of a gas of independent Fermions, except for bound pairs near the Fermi surface. These pairs may scatter against each other, always conserving total momentum and individual spin, but scattering between members of different pairs is precluded as this would mean the breaking of pairing bonds.

We thus arrive at the following reformulation of the many-body problem. We start from a set of independent Fermions described by the Hamiltonian

$$H_0 = \Sigma \epsilon_K a_K^\dagger a_K. \tag{7.4.1}$$

Introducing the convention that K represents a single-particle state with momentum $\mathbf{k}$ and spin $\uparrow$ and $-K$ a state $-\mathbf{k}$, $\downarrow$, then the interaction

$$V = \frac{1}{2}\sum V_{KK'} a_K^\dagger a_{-K}^\dagger a_{-K'} a_{K'}, \tag{7.4.2}$$

applied at time $t = -\infty$, say, is constructed in accordance with our previous discussion. The summation in (7.4.2) is over E_k and $E_{k'}$, restricted to lie in the energy shell of thickness 2δ about the Fermi surface. The matrix elements $V_{KK'}$ as written above are to be thought of as negative (attractive). We are thus faced with investigating the wave function which evolves after infinite time and which is a stationary state of the reduced Hamiltonian

$$H = H_0 + V. \tag{7.4.3}$$

The situation is depicted in Fig. 7.8.

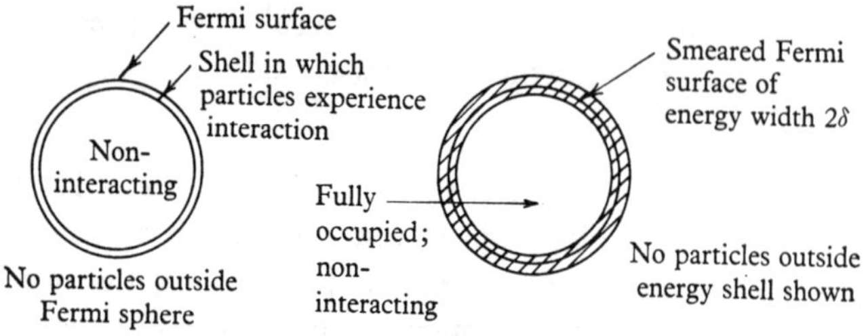

(i) $t = -\infty$. (ii) $t = 0$.

Fig. 7.8

There is a strong resemblance to the Cooper two-particle problem. There, only two particles, K and $-K$, were selected at the Fermi surface and their scattering in a shell at the Fermi limit studied. Here, however, many such pairs are allowed to scatter. This is a much more complicated situation because, whereas previously we knew that our chosen pair could freely scatter in the energy shell, now such scattering is inhibited by the possibility of the occupation of such states by other particles. The problem is truly a many-body one.

7.4.2. *B.C.S. wave function*

At this stage, it is possible to go ahead without any further approximations and solve for the ground-state wave function of (7.4.3) to $O(1/N)$ accuracy. The wave function thus obtained is that used by Bardeen, Cooper & Schrieffer (1957). In the present account of the theory, we have chosen to adopt the philosophy of B.C.S. in their original paper, where they regard the wave function as a very convenient trial function to be used variationally. We introduce then the trial variational wave function

$$|\psi\rangle = \prod_{K>0} (u_K + v_K a_K^\dagger a_{-K}^\dagger)|0\rangle, \tag{7.4.4}$$

where u_K and v_K are to be taken as real and such that

$$u_K = u_{-K} = u_{\mathbf{k}}, \quad v_K = -v_{-K} = v_{\mathbf{k}} \tag{7.4.5}$$

and finally, for what turns out to be normalization purposes,

$$u_{\mathbf{k}}^2 + v_{\mathbf{k}}^2 = 1. \tag{7.4.6}$$

If we recall the convention that $K = \mathbf{k}, \uparrow$ and $-K = -\mathbf{k}, \downarrow$, then we see that the restriction $K > 0$ implies, in conjunction with (7.4.5) and the anticommutation relations for the a's, that the product in (7.4.4) is over all distinct terms $(u_K + v_K a_K^\dagger a_{-K}^\dagger)$.

It must be agreed that the physical significance and utility of the choice (7.4.4) are far from obvious. However, we shall proceed to examine its detailed properties and consequences immediately, and its physical status will then become clear.

7.4.3. *Properties of trial wave function*

We first observe that there is no ambiguity in the definition (7.4.4) since it is readily seen that distinct pairs of the

$$(u_K + v_K a_K^\dagger a_{-K}^\dagger)$$

factors commute.

Secondly, any candidate for a trial wave function must correlate in a special way those pairs of particles with total momentum and total spin zero. The form (7.4.4) clearly reveals such special treatment. This in itself, of course, is hardly sufficient recommendation, but we will see later that the correlation is achieved in a particularly satisfactory way. Furthermore, the determinantal wave function of states below the Fermi sea is contained in (7.4.4), as we see by choosing

$$\left.\begin{matrix} u_K = 0 \\ v_K = 1 \end{matrix}\right\} (k < k_f), \qquad \left.\begin{matrix} u_K = 1 \\ v_K = 0 \end{matrix}\right\} (k > k_f). \tag{7.4.7}$$

Substituting these values directly into (7.4.4), we find

$$\prod_{\substack{K>0 \\ k<k_f}} a_K^\dagger a_{-K}^\dagger |0\rangle = \prod_{k<k_f} a_K^\dagger |0\rangle. \tag{7.4.8}$$

We eventually use u_K and v_K as variational parameters so it is always comforting if this simple solution is, if necessary, available to us.

A special feature of (7.4.4) which we must consider at this point is that it does not correspond to some prescribed and fixed number of particles. We have

$$|\psi\rangle = (\Pi u_K)\left(1+\Sigma\frac{v_K}{u_K}a_K^\dagger a_{-K}^\dagger+\Sigma\frac{v_K v_L}{u_K u_L}a_K^\dagger a_{-K}^\dagger a_L^\dagger a_{-L}^\dagger+\ldots\right)|0\rangle$$
$$= (\Pi u_K)(|\phi_0\rangle+|\phi_2\rangle+|\phi_4\rangle+\ldots), \qquad (7.4.9)$$

where $|\phi_i\rangle$ denotes a (non-normalized) wave function for i particles. The question may reasonably be asked why one might choose a wave function with the property that it is not an eigenstate of the number operator. The answer is that it is for computational convenience. By this device, in a way that will become apparent, one is essentially using variationally ϕ_N of the above series, for which particle number is strictly conserved, but avoiding the mathematical difficulties inherent in calculations of this kind. The physical nature of $|\phi_N\rangle$, itself, is clarified if we write it (see Blatt, 1964) as

$$|\phi_N\rangle = \left(\Sigma\frac{v_K}{u_K}a_K^\dagger a_{-K}^\dagger\right)^{N/2}|0\rangle.$$

Thus, it represents $N/2$ correlated electron pairs. The particular amplitude for any 2×2 plane-wave determinantal state, with indices K and $-K$, is given by v_K/u_K.

Finally, by observing that if $K \neq L$, then $u_K+v_K a_K^\dagger a_{-K}^\dagger$ commutes not only with $u_L+v_L a_L^\dagger a_{-L}^\dagger$ but also with its conjugate, we see that

$$\langle\psi|\psi\rangle = \langle 0|\,\Pi(u_K+v_K a_{-K}a_K)(u_K+v_K a_K^\dagger a_{-K}^\dagger)\,|0\rangle$$
$$= \Pi(u_K^2+v_K^2) = 1, \qquad (7.4.10)$$

the final step following from the restriction (7.4.6). The B.C.S. wave function is thus conveniently normalized.

7.4.4. *Vacuum property*

There exists a rather simple set of Fermion operators for which the B.C.S. wave function is the vacuum state. More precisely, let us define the operators

$$\alpha_K = u_K a_K - v_K a_{-K}^\dagger, \quad \alpha_K^\dagger = u_K a_K^\dagger - v_K a_{-K}, \qquad (7.4.11)$$

and so, by (7.4.5)

$$\alpha_{-K} = u_K a_{-K} + v_K a_K^\dagger, \quad \alpha_{-K}^\dagger = u_K a_{-K}^\dagger + v_K a_K. \quad (7.4.12)$$

The α's thus defined are Fermion operators, that is

$$\left.\begin{aligned} \alpha_K^\dagger \alpha_L + \alpha_L \alpha_K^\dagger &= \delta_{KL}, \\ \alpha_K^\dagger \alpha_L^\dagger + \alpha_L^\dagger \alpha_K^\dagger &= 0, \\ \alpha_K \alpha_L + \alpha_L \alpha_K &= 0, \end{aligned}\right\}, \quad (7.4.13)$$

as may be verified by direct substitution for the α's using (7.4.11), subsequent use of the anticommutation relations satisfied by the a's and the requirement (7.4.6) on the u's and v's.

Then, we have the property that $|\psi\rangle$ is a vacuum state for the α's:

$$\alpha_K |\psi\rangle = 0 \quad \text{(all } K\text{)}. \quad (7.4.14)$$

This is proved by writing the left-hand side of (7.4.14) entirely in terms of the a and $a^\dagger$ operators. Splitting the calculation into two parts, we have

$$\begin{aligned} a_K \prod_{L>0} (u_L + v_L a_L^\dagger a_{-L}^\dagger)|0\rangle &= a_K(u_K + v_K a_K^\dagger a_{-K}^\dagger) \\ &\qquad \times \prod_{\substack{L>0 \\ L\neq K}} (u_L + v_L a_L^\dagger a_{-L}^\dagger)|0\rangle \\ &= (u_K a_K + v_K[1 - a_K^\dagger a_K] a_{-K}^\dagger) \\ &\qquad \times \prod_{\substack{L>0 \\ L\neq K}} (u_L + v_L a_L^\dagger a_{-L}^\dagger)|0\rangle \\ &= (u_K a_K + v_K a_{-K}^\dagger + v_K a_K^\dagger a_{-K}^\dagger a_K) \\ &\qquad \times \prod_{\substack{L>0 \\ L\neq K}} (u_L + v_L a_L^\dagger a_{-L}^\dagger)|0\rangle \\ &= v_K a_{-K}^\dagger \prod_{\substack{L>0 \\ L\neq K}} (u_L + v_L a_L^\dagger a_{-L}^\dagger)|0\rangle, \end{aligned} \quad (7.4.15)$$

two terms in the penultimate line disappearing because a_K commutes with all terms in the product and so may be brought through to operate on the vacuum $|0\rangle$ to give zero. A very similar calculation gives

$$a_{-K}^\dagger \prod_{L>0} (u_L + v_L a_L^\dagger a_{-L}^\dagger)|0\rangle = u_K a_{-K}^\dagger \prod_{\substack{L>0 \\ L\neq K}} (u_L + v_L a_L^\dagger a_{-L}^\dagger)|0\rangle. \quad (7.4.16)$$

Multiplying (7.4.15) by u_K, (7.4.16) by v_K and subtracting, gives us then the desired result (7.4.14).

It should be noted that, if we make the determinantal choice (7.4.7) of the u's and v's, the α's become the b's of equations (3.9.5) The eventual choice of the u's and v's appropriate to the superconducting state will correspond closely to that of (7.4.7), because the energy involved in the transition to the normal state is so small.

7.4.5. *Excited states*

The Fermion relations (7.4.13) and the vacuum property (7.4.14) enable us to write down a set of single-particle excited states of the system. This may be done by analogy with the particle-hole description discussed in Chapter 3. The normalized excited states may be written

$$\alpha^\dagger_{K_1}\alpha^\dagger_{K_2}\dots\alpha^\dagger_{K_n}\alpha^\dagger_{K'_1}\alpha^\dagger_{K'_2}\dots\alpha^\dagger_{K'_n}|\psi\rangle, \tag{7.4.17}$$

where the α_K's ($\alpha_{K'}$'s) have momenta of sizes greater (less) than k_f. The restriction on the momenta in (7.4.17) is to ensure that these excited states are in one–one correspondence with those of the non-interacting system.

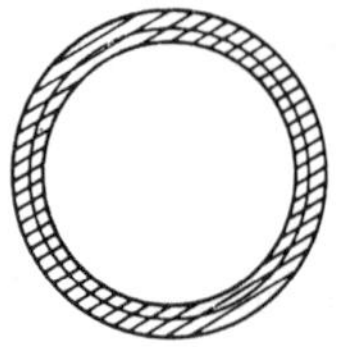

(i) Superconducting ground state.

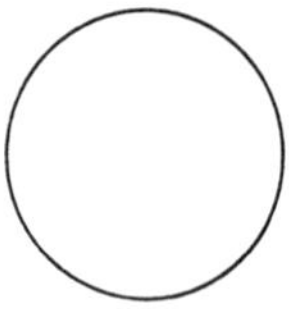

(ii) Reference normal ground state (no holes, no particles).

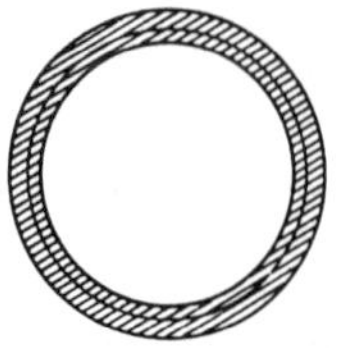

(iii) Superconducting excited state. The change from (i) should be in the shading, but this is clearly an awkward model.

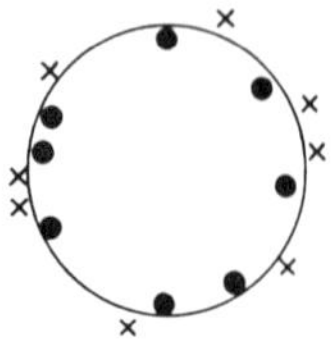

(iv) Reference normal excited state. Equal numbers of particles and holes. Depicts the superconducting state quite clearly.

Fig. 7.9

This one–one correspondence gives us a convenient way of depicting all the single-particle excitations of the B.C.S. wave function. We can, of course, think of actual **k**-space and a smeared

Fermi surface as indicated in Fig. 7.8 (ii). However, it is difficult to know how to display excited states. As shown in Fig. 7.9, the problem is solved if we use the corresponding normal state for reference purposes.

Thus, for example, some of the simpler excited states are drawn in reference **k**-space in Fig. 7.10. The simplest form (consistent with normal-state particle conservation) is shown in Fig. 7.10 (i). This is the state $\alpha_K^\dagger \alpha_{K'}^\dagger|\psi\rangle$ and is obtained from the ground state by the breaking up of the pairing K', $-K'$. A slightly more complicated

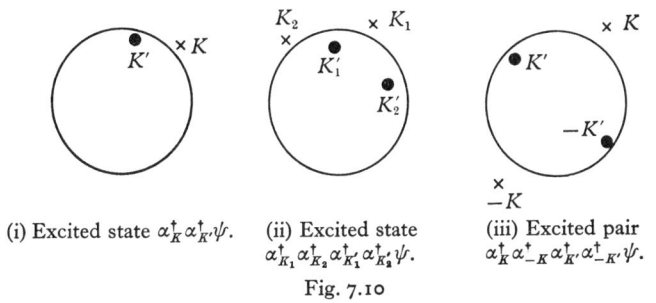

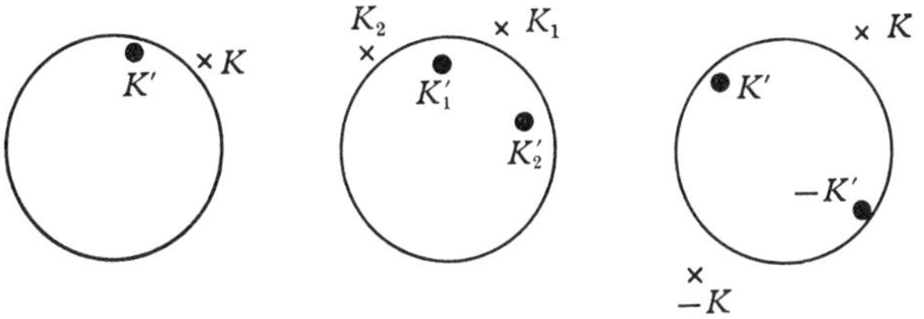

(i) Excited state $\alpha_K^\dagger \alpha_{K'}^\dagger \psi$. (ii) Excited state $\alpha_{K_1}^\dagger \alpha_{K_2}^\dagger \alpha_{K_1'}^\dagger \alpha_{K_2'}^\dagger \psi$. (iii) Excited pair $\alpha_K^\dagger \alpha_{-K}^\dagger \alpha_{K'}^\dagger \alpha_{-K'}^\dagger \psi$.

Fig. 7.10

excited state $\alpha_{K_1}^\dagger \alpha_{K_2}^\dagger \alpha_{K_1'}^\dagger \alpha_{K_2'}^\dagger|\psi\rangle$ is shown in Fig. 7.10 (ii). In general, the formation of this state from the ground state requires the breaking of two bonds, those between K_1', $-K_1'$ and K_2', $-K_2'$. There is a special case, however, when $K_1' = -K_2'$. In this case, the two pairs indicated below the Fermi surface are not distinct. For example, Fig. 7.10 (iii) corresponds to an 'excited pair'—the pair characterized by index K' having scattered to create a pair characterized by index K.

7.4.6. *Momentum distribution*

So far we have derived a number of properties which are quite obviously of considerable interest and utility, on the assumption that the wave function (7.4.4) will be useful for describing the ground state. We now give an important reason why this is so.

First, let us remark that while the introduction of the α's of (7.4.11), having the property (7.4.14), gives great insight into the physics, we obtain a bonus at the same time. Very often, calculations of averages, with respect to the B.C.S. wave function, of operators defined in terms of a's and $a^\dagger$'s can be simplified by expressing the latter in terms of α's and $\alpha^\dagger$'s. It is thus convenient to have available such expressions. These are readily obtained by inverting (7.4.11) to give

$$a_K = u_K \alpha_K + v_K \alpha^\dagger_{-K}, \quad a^\dagger_K = u_K \alpha^\dagger_K + v_K \alpha_{-K}, \tag{7.4.18}$$

and

$$a_{-K} = u_K \alpha_{-K} - v_K \alpha^\dagger_K, \quad a^\dagger_{-K} = u_K \alpha^\dagger_{-K} - v_K \alpha_K. \tag{7.4.19}$$

A good illustration of this method is the evaluation of the momentum distribution function $\langle\psi|a^\dagger_K a_K|\psi\rangle$, a problem to which we now turn.

Let us first consider the function $a_K|\psi\rangle$. Use of (7.4.19) and the vacuum property (7.4.14) gives the result

$$a_K|\psi\rangle = u_K \alpha_K|\psi\rangle + v_K \alpha^\dagger_{-K}|\psi\rangle = v_K \alpha^\dagger_{-K}|\psi\rangle. \tag{7.4.20}$$

Hence, recalling the Hermiticity properties, the momentum distribution is given by

$$\langle\psi|a^\dagger_K a_K|\psi\rangle = v^2_K. \tag{7.4.21}$$

At this stage, the significant point about this result is that the probability of occupation of the state K depends simply on v_K. Furthermore, the only constraints on the v_K's are expressed through (7.4.5) and (7.4.6). We may thus regard the v_K's with $K > 0$ as independent variables, the v_K's with $K < 0$ and all the u_K's being functions of them. Thus, apart from the fact implied by (7.4.21) and (7.4.5) that states K and $-K$ are equally occupied, the occupation of all states can be freely and independently varied. This observation implies that $|\psi\rangle$ will be a powerful variational wave function.

A simple related result can also be derived in passing. We recall that the momentum distribution is given by defining an operator $a^\dagger_K a_K$ which tells us the number of particles in state K. One can also ask what the probability is of simultaneous occupation of any two states. Clearly the generalization of $a^\dagger_K a_K$ is

$$a^\dagger_K a_K a^\dagger_L a_L = a^\dagger_K a^\dagger_L a_L a_K$$

and so the probability of simultaneous occupation of states K and L is just

$$\langle\psi|a^\dagger_K a^\dagger_L a_L a_K|\psi\rangle = (a_L a_K \psi, a_L a_K \psi), \tag{7.4.22}$$

where, by (7.4.18) and (7.4.20),

$$\begin{aligned} a_L a_K|\psi\rangle &= (u_L \alpha_L + v_L \alpha^\dagger_{-L}).v_K \alpha^\dagger_{-K}|\psi\rangle \\ &= v_K v_L \alpha^\dagger_{-L} \alpha^\dagger_{-K}|\psi\rangle \quad (K \neq -L). \end{aligned} \tag{7.4.23}$$

Thus, provided $K \neq -L$, that is the states concerned do not constitute a coupled pair, we have the Hartree-like result

$$\langle\psi|a_K^\dagger a_L^\dagger a_L a_K|\psi\rangle = v_K^2 v_L^2 = \langle\psi|a_K^\dagger a_K|\psi\rangle\langle\psi|a_L^\dagger a_L|\psi\rangle. \quad (7.4.24)$$

7.4.7. *Variational programme*

It is now assumed that the reader is convinced of the advantages of the B.C.S. wave function. What has not yet been discussed is just how, in view of the fact that the function does not describe a fixed number of particles in the sense indicated in (7.4.9), a variational calculation can be carried out, and what the final results mean.

Let us briefly recall the conventional variation procedure. We choose a normalized N-particle wave function, that is an eigenstate of the total number operator $N_{\text{op.}}$. The energy expression is then computed and minimized.

In the present B.C.S. case, there is no formal difficulty in calculating the energy. In terms of (7.4.9), it is simply

$$E \equiv \langle\psi|H|\psi\rangle = (\Pi u_K)^2 \sum_{n=0}^{\infty} \langle\phi_{2n}|H|\phi_{2n}\rangle \quad (7.4.25)$$

since it can be shown that scalar products between wave functions of different particle number are zero. However, we have seen that generally $|\psi\rangle$ is not an eigenstate of the number operator. We can nevertheless insist on the weaker requirement that the average of $N_{\text{op.}}$ with respect to $|\psi\rangle$ is N, that is

$$\bar{N} \equiv \langle\psi|N_{\text{op.}}|\psi\rangle = N. \quad (7.4.26)$$

The problem presented by the minimization of (7.4.25) subject to (7.4.26) being satisfied is a standard one. We consider the absolute minimization of

$$E - \mu\bar{N} \equiv \langle\psi|H - \mu N_{\text{op.}}|\psi\rangle, \quad (7.4.27)$$

where μ is a Lagrange parameter, as yet unspecified. On finding $|\psi(\mu)\rangle$, substitution in (7.4.26) gives us an equation from which μ may be determined.

But the subsidiary condition (7.4.26) alone is not sufficient to ensure that the calculation discussed in the previous paragraph is in the slightest degree related to a conventional calculation. We would like to show, of course, that what we have done is to have 'almost' satisfied $\quad N_{\text{op.}}|\Psi\rangle = N|\Psi\rangle.$

A proper way of investigating this matter is clearly by evaluating the mean square deviation

$$\Delta^2 N_{\text{op.}} \equiv \langle \psi | N_{\text{op.}}^2 | \psi \rangle - N^2 \qquad (7.4.28)$$

and showing this to be unimportant.

If the latter is true (and we will show it to be), then we are essentially using as variational function the member ϕ_N of the series (7.4.9). There will be a slight spread, of course, reflecting the non-vanishing of (7.4.28), which implies that other ϕ_i's, with i near to N, enter into the series. But those ϕ_i's, far removed from the neighbourhood of $i = N$, will enter with only very small weighting factors.

The variational procedure we will adopt may be summarized as follows. The optimum u_K and v_K occurring in the B.C.S. wave function are calculated by taking as Hamiltonian H that specified by (7.4.3) and expected to correspond to a superconductor, and minimizing the energy (7.4.25) subject to the restriction (7.4.26). To check that the calculation is valid, (7.4.28) will be evaluated and its effect shown to be small. Having obtained the ground state, the single-particle excitations are obtained using the techniques of section 7.4.5. The knowledge of the single-particle energy spectrum enables us later to investigate the low-temperature thermodynamics.

Finally, it will be noted that (7.4.27) displays the quantity

$$\mathscr{H} = H - \mu N_{\text{op.}}, \qquad (7.4.29)$$

and this is generally a more convenient Hamiltonian to use than H itself. This is an essentially trivial matter involving a re-definition of the zero for the single-particle energies. In future we always discuss the properties of $\mathscr{H}$.

7.5. Zero-temperature form of B.C.S. theory

For reference purposes, let us write down explicitly the Hamiltonian (7.4.29). It is

$$\mathscr{H} = \Sigma \epsilon_K a_K^\dagger a_K + \tfrac{1}{2} \Sigma V_{KK'} a_K^\dagger a_{-K}^\dagger a_{-K'} a_{K'}, \qquad (7.5.1)$$

where it is understood that the sums involving $V_{KK'}$ are restricted to states near the Fermi surface, as previously discussed, and we have introduced the notation

$$\epsilon_K = \mathrm{E}_K - \mu. \qquad (7.5.2)$$

Our first job is to calculate $\epsilon = \langle\psi|\mathscr{H}|\psi\rangle$ and minimize for the optimum u_K and v_K. The resulting Euler equation is the B.C.S. integral equation.

7.5.1. *B.C.S. integral equation*

The direct procedure is to use $|\psi\rangle$ as given by (7.4.4) and $\mathscr{H}$ as given by (7.5.1) to find ϵ. We, however, will use the canonical transformations (7.4.18) and (7.4.19) and the vacuum property (7.4.14), not only because it is probably easier, but also because it establishes the essential equivalence between the B.C.S. approach and the later work of Bogoliubov (1958; see also Valatin, 1958).

Thus, let us use Wick's theorem to rewrite $\mathscr{H}$ in normal product form with respect to the α's. We have

$$\begin{aligned} a_K^\dagger a_{-K}^\dagger a_{-K'} a_{K'} = {} & N_\alpha(a_K^\dagger a_{-K}^\dagger a_{-K'} a_{K'}) + N_\alpha(\underbrace{a_K^\dagger a_{-K}^\dagger}_{} a_{-K'} a_{K'}) \\ & + N_\alpha(a_K^\dagger a_{-K}^\dagger \underbrace{a_{-K'} a_{K'}}_{}) + N_\alpha(\underbrace{a_K^\dagger a_{-K}^\dagger a_{-K'}}_{} a_{K'}) \\ & + N_\alpha(\underbrace{a_K^\dagger a_{-K}^\dagger a_{-K'} a_{K'}}_{}) + N_\alpha(a_K^\dagger \underbrace{a_{-K}^\dagger a_{-K'}}_{} a_{K'}) \\ & + N_\alpha(a_K^\dagger \underbrace{a_{-K}^\dagger a_{-K'} a_{K'}}_{}) + N_\alpha(\underbrace{a_K^\dagger a_{-K}^\dagger a_{-K'}}_{} a_{K'})\ [\text{with } a_{-K}^\dagger \text{ contracted with } a_{K'}] \\ & + N_\alpha(\underbrace{a_K^\dagger \underbrace{a_{-K}^\dagger a_{-K'}}_{} a_{K'}}_{}) + N_\alpha(\underbrace{a_K^\dagger a_{-K}^\dagger}_{} \underbrace{a_{-K'} a_{K'}}_{}) \end{aligned}$$

and the various terms are easily evaluated explicitly. For example,

$$\begin{aligned} N_\alpha(\underbrace{a_K^\dagger a_{-K}^\dagger}_{} a_{-K'} a_{K'}) &= \langle\psi|(u_K \alpha_K + v_K \alpha_{-K}^\dagger)(u_K \alpha_{-K} - v_K \alpha_K^\dagger)|\psi\rangle \\ &\quad \times N[(u_{K'} \alpha_{-K'} - v_{K'} \alpha_{K'}^\dagger)(u_{K'} \alpha_{K'} + v_{K'} \alpha_{-K'}^\dagger)] \\ &= -u_K v_K . (u_{K'}^2 \alpha_{-K'} \alpha_{K'} - u_{K'} v_{K'} \alpha_{-K'}^\dagger \alpha_{-K'} \\ &\qquad - u_{K'} v_{K'} \alpha_{K'}^\dagger \alpha_{K'} - v_{K'}^2 \alpha_{K'}^\dagger \alpha_{-K'}^\dagger), \end{aligned}$$

and the other terms are evaluated similarly. On collecting terms, we find

$$\mathscr{H} = \epsilon + \mathscr{H}_{11} + \mathscr{H}_{20} + \mathscr{H}_{\text{int.}}, \tag{7.5.3}$$

where

$$\epsilon = \Sigma \epsilon_K v_K^2 + \tfrac{1}{2} \Sigma V_{KK'} u_K v_K u_{K'} v_{K'}, \tag{7.5.4}$$

$$\mathscr{H}_{11} = \Sigma \epsilon_K (u_K^2 - v_K^2) \alpha_K^\dagger \alpha_K - 2 \Sigma V_{KK'} u_K v_K u_{K'} v_{K'} \alpha_K^\dagger \alpha_K, \tag{7.5.5}$$

$$\begin{aligned} \mathscr{H}_{20} = {} & \Sigma \epsilon_K u_K v_K (\alpha_K^\dagger \alpha_{-K}^\dagger + \alpha_{-K} \alpha_K) \\ & + \tfrac{1}{2} \Sigma V_{KK'} (u_K^2 - v_K^2) u_{K'} v_{K'} (\alpha_K^\dagger \alpha_{-K}^\dagger + \alpha_{-K} \alpha_K) \end{aligned} \tag{7.5.6}$$

and

$$\begin{aligned} \mathscr{H}_{\text{int.}} = \Sigma\{ & \text{weighted terms of type } \alpha^\dagger\alpha^\dagger\alpha^\dagger\alpha^\dagger,\ \alpha^\dagger\alpha^\dagger\alpha^\dagger\alpha, \\ & \alpha^\dagger\alpha^\dagger\alpha\alpha,\ \alpha^\dagger\alpha\alpha\alpha,\ \alpha\alpha\alpha\alpha\}. \end{aligned} \tag{7.5.7}$$

Expression (7.5.7) arises from the term displayed above which has no contractions, while $\mathcal{E}$ comes, as usual, from the three fully contracted members. (In (7.5.4), some terms of type $V_{KK} v_K^4$ which are a factor Ω smaller than the corresponding $\epsilon_K v_K^2$ have been dropped.)

We thus turn to the minimization of $\mathcal{E}$ as given by (7.5.4) subject to the restrictions (7.4.5) and (7.4.6). The latter conditions are immediately satisfied by the substitutions

$$u_K = \cos\theta_K, \quad v_K = \sin\theta_K, \quad \theta_K = -\theta_{-K} = \theta_k. \tag{7.5.8}$$

Then we may write (7.5.4) in the form

$$\mathcal{E} = \Sigma\epsilon_K \sin^2\theta_K + \tfrac{1}{8}\Sigma V_{KK'} \sin 2\theta_K \sin 2\theta_{K'}, \tag{7.5.9}$$

from which we find

$$\delta\mathcal{E} = \Sigma\epsilon_K \sin 2\theta_K\, \delta\theta_K + \tfrac{1}{2}\Sigma V_{KK'} \cos 2\theta_K \sin 2\theta_{K'}\, \delta\theta_K. \tag{7.5.10}$$

The antisymmetry of θ_K leads to no complications, for on allowing independent variations in (7.5.10) we obtain the integral equation

$$\tan 2\theta_K = -\frac{1}{2\epsilon_K} \sum_{K'} V_{KK'} \sin 2\theta_{K'}, \tag{7.5.11}$$

and this equation automatically has a solution of the correct symmetry. This depends on the property $V_{K-K'} = 0$ expressing the spin conservation requirement, for it is readily seen that if θ_K^0 is any solution, then θ_K^0 $(K > 0)$, $-\theta_{-K}^0$ $(K < 0)$ is also a solution.

There are other ways of expressing (7.5.11). For example, we may write it in the form

$$\Delta_K = -\frac{1}{2} \sum_{K'} V_{KK'} \frac{\Delta_{K'}}{E_{K'}}, \tag{7.5.12}$$

where

$$\tan 2\theta_K = \frac{\Delta_K}{\epsilon_K} \tag{7.5.13}$$

and

$$E_K = \sqrt{(\epsilon_K^2 + \Delta_K^2)}, \tag{7.5.14}$$

the interrelations between the various quantities being summarized in Fig. 7.11. If we write

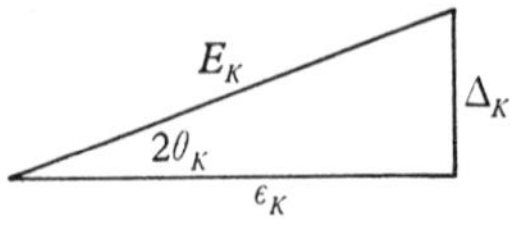

Fig. 7.11

$$\epsilon_K = \epsilon_{-K} = \epsilon_k, \tag{7.5.15}$$

the antisymmetry of θ_K leads to the antisymmetry of Δ_K:

$$\Delta_K = -\Delta_{-K} = \Delta_k. \tag{7.5.16}$$

Finally, spinless forms of the integral equation can be written down. On putting

$$V_{KK'} = V_{-K-K'} = V_{kk'}; \quad V_{K-K'} = V_{-KK'} = 0, \qquad (7.5.17)$$

(7.5.11), for example, reduces to

$$\tan 2\theta_k = -\frac{1}{2\epsilon_k} \sum_{k'} V_{kk'} \sin 2\theta_{k'}. \qquad (7.5.18)$$

It should be noted from (7.5.12) that $\Delta_K = 0$ is always a solution. This may be rewritten $u_K v_K = 0$ and is thus the normal Fermi sphere solution. The point is, however, that for a superconductor there are energetically lower solutions and these will now be investigated.

7.5.2. *Solution for averaged potential*

Let us now explicitly write down an interaction simulating that of a superconductor. We choose for computational convenience

$$V_{KK'} = \begin{cases} -V/\Omega & \text{if } K, K' \text{ have same spin and } |\epsilon_K|, |\epsilon_{K'}| < \delta, \\ 0 & \text{otherwise.} \end{cases} \qquad (7.5.19)$$

This should be compared with the choice (7.3.6) made when first discussing the concept of a Cooper pair. Many of the subsequent comments made there are relevant here and will not be repeated. One point needs discussion however. The quantity μ is so far no more than a Lagrange multiplier. How can we be sure that $|\epsilon_K| < \delta$ actually defines a shell around the Fermi surface? The answer is that a simple calculation for non-interacting particles gives $\mu = E_f$. For interacting particles in which only small amounts of energy are involved (certainly the case here), μ differs little from E_f and so $|\epsilon_K| < \delta$ defines a shell which covers the Fermi surface. Such a choice of energy shell circumvents a mathematical difficulty without affecting the physics.

Putting (7.5.19) in (7.5.12) gives the results

$$\Delta_K = 0 \quad (|\epsilon_K| > \delta), \qquad (7.5.20)$$

and

$$\Delta_K = \frac{V}{2}\frac{1}{\Omega} \sum_{\substack{|\epsilon_{K'}| < \delta \\ K', K \text{ same spin}}} \frac{\Delta_{K'}}{E_{K'}} \quad (|\epsilon_K| < \delta). \qquad (7.5.21)$$

The latter is solved by observing that $\Delta_K = \Delta$, independent of K, and whence, cancelling out the $\Delta = 0$ solution, we find

$$\frac{2}{V} = \frac{1}{\Omega} \sum_{|\epsilon_K|<\delta} \frac{1}{(\sqrt{\epsilon_K^2+\Delta^2})}, \tag{7.5.22}$$

where the summation in (7.5.22) is over singly occupied states. Once more reasoning as in the Cooper argument (cf. (7.3.8) and (7.3.10)) we obtain

$$\frac{2}{V} = \rho \int_{-\delta}^{\delta} \frac{d\epsilon}{\sqrt{(\epsilon^2+\Delta^2)}} \tag{7.5.23}$$

and direct integration gives

$$\Delta = \delta \operatorname{cosech} \frac{1}{\rho V} \sim 2\delta e^{-1/\rho V}. \tag{7.5.24}$$

Summarizing, our properly antisymmetrized solution is

$$\left.\begin{array}{ll} \Delta_K & (K>0), \\ -\Delta_{-K} & (K<0), \end{array}\right\} \tag{7.5.25}$$

where $\Delta_K = 0$ for $|\epsilon_K| > \delta$, while $\Delta_K = \Delta$ as given by (7.5.24) when $|\epsilon_K| < \delta$. We shall show shortly that (7.5.25) is indeed energetically below the plane-wave determinantal solution. For the present, we conclude this section by showing that our solution is almost an eigenfunction of the number operator by investigating the mean square deviation (7.4.28).

Writing the number operator in terms of the α's using (7.4.18) and (7.4.19), we have

$$N_{\text{op.}} = \sum_K [(u_K^2 - v_K^2)\alpha_K^\dagger \alpha_K + u_K v_K(\alpha_K^\dagger \alpha_{-K}^\dagger + \alpha_{-K}\alpha_K) + v_K^2]. \tag{7.5.26}$$

Hence

$$\langle\psi|N_{\text{op.}}|\psi\rangle = \sum_K v_K^2, \tag{7.5.27}$$

and equating the latter with N defines μ. The average of $N_{\text{op.}}^2$ is only slightly more difficult to compute and is

$$\langle\psi|N_{\text{op.}}^2|\psi\rangle = \sum_K u_K^2 v_K^2 + \sum_{K,L} v_K^2 v_L^2, \tag{7.5.28}$$

and so

$$\Delta^2 N_{\text{op.}} \equiv \langle\psi|N_{\text{op.}}^2|\psi\rangle - \langle\psi|N_{\text{op.}}|\psi\rangle^2 = \sum_K u_K^2 v_K^2. \tag{7.5.29}$$

Now from (7.5.8) and Fig. 7.11 we have the useful relations

$$2u_K v_K = \frac{\Delta_K}{E_K}, \quad u_K^2 - v_K^2 = \frac{\epsilon_K}{E_K}, \tag{7.5.30}$$

and appeal to the first of these gives

$$\Delta^2 N_{\text{op.}} = \frac{1}{4}\sum_K \frac{\Delta_K^2}{E_K^2} = \tfrac{1}{4}\Omega(2\rho)\int_{-\delta}^{\delta} \frac{\Delta^2}{\epsilon^2+\Delta^2}\,d\epsilon = \Omega\rho\Delta\tan^{-1}\frac{\delta}{\Delta}. \tag{7.5.31}$$

We note in particular the rigorous result that

$$\lim_{N\to\infty} \frac{\Delta^2 N_{\text{op.}}}{N^2} = 0, \tag{7.5.32}$$

and the remarks of section 7.4.7 will be recalled.

7.5.3. *Ground-state properties*

(*a*) *Total energy*. The total energy expression (7.5.9) reduces, on using the optimum condition (7.5.11), to

$$\mathcal{E} = \sum_K \epsilon_K\{\sin^2\theta_K - \tfrac{1}{4}\sin 2\theta_K \tan 2\theta_K\}. \tag{7.5.33}$$

Employing the solution (7.5.25) appropriate to the potential (7.5.19) we obtain, after a somewhat tedious but straightforward calculation, for the weak coupling case

$$\mathcal{E}(\delta) = \mathcal{E}(0) - \tfrac{1}{2}\Omega\rho\Delta^2, \tag{7.5.34}$$

which shows that our solution is energetically lower than the plane wave determinantal state having energy $\mathcal{E}(0)$. Since Δ typically is of order 10^{-4} e.v., the total energy is of the order anticipated in section 7.2.4.

(*b*) *Isotope effect*. It is of interest to note that we have here an example of the isotope effect. For the critical field $H_{\text{co.}}$ required to destroy the superconducting phase is given by (cf. (7.2.2))

$$\frac{H_{\text{co.}}^2}{8\pi} = \frac{\mathcal{E}(0)-\mathcal{E}(\delta)}{\Omega} = \tfrac{1}{2}\rho\Delta^2 = 2\rho\delta^2 e^{-2/\rho V}. \tag{7.5.35}$$

Denoting the ionic mass by M and putting

$$\delta = \hbar\omega \propto \frac{1}{M^{\frac{1}{2}}}, \tag{7.5.36}$$

we obtain the relation

$$H_{\text{co.}} M^{\frac{1}{2}} = \text{const.} \tag{7.5.37}$$

The proportionality indicated above simply relates an oscillator's frequency to its mass and arises naturally in theories of lattice vibrations (see, for example, (8.2.10), p. 265).

(*c*) *Momentum distribution.* A result of some interest is a precise expression for the smearing of the Fermi surface brought about by the interaction. This is given by the momentum distribution function (7.4.21). Using (7.5.8) and Fig. 7.11 once more, we obtain

$$\langle\psi|a_K^\dagger a_K|\psi\rangle = v_K^2 = \frac{1}{2}\left(1-\frac{\epsilon_K}{E_K}\right) = \frac{1}{2}\left(1-\frac{\epsilon_K-\mu}{\sqrt{[(\epsilon_K-\mu)^2+\Delta_K^2]}}\right), \tag{7.5.38}$$

a plot of which is given for the solution (7.5.25) in Fig. 7.12.

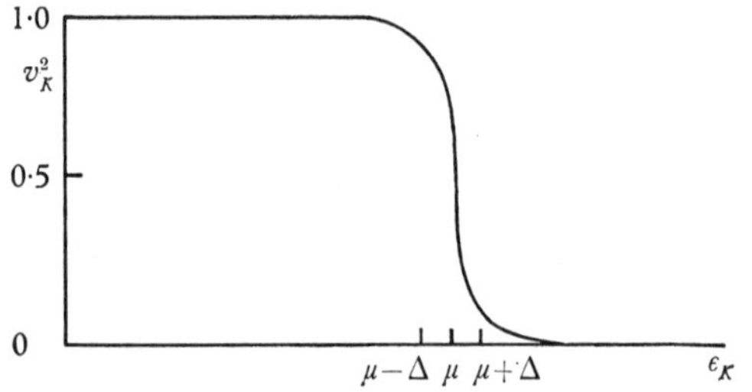

Fig. 7.12. Momentum distribution.

The continuity across the Fermi surface should be noted, in contradistinction to that obtained for electrons interacting via Coulomb forces only, in Chapter 5.

7.5.4. *Excited states*

(*a*) *Single-particle excitations.* In accordance with our previous discussion (see Fig. 7.9 and its interpretation in the text) the simplest kind of excited state $\alpha_K^\dagger \alpha_{K'}^\dagger|\psi\rangle$ has excitation energy

$$(\alpha_K^\dagger \alpha_{K'}^\dagger\psi, \mathscr{H}\alpha_K^\dagger \alpha_{K'}^\dagger\psi)-(\psi,\mathscr{H}\psi). \tag{7.5.39}$$

As a first step in the calculation of (7.5.39), we express $\mathscr{H}$, as given by (7.5.3)–(7.5.7), in simpler form using the optimum relationship (7.5.11).

The first remarkable fact is that $\mathscr{H}_{20}$ vanishes identically. This establishes the essential equivalence of the B.C.S. theory with that of the Bogoliubov method. The latter involves the perturbation problem posed by regarding $\mathscr{H}_{\text{int.}}$ as a perturbation in (7.5.3). If

$\mathscr{H}_{20}$ is non-zero, vanishing energy denominators are obtained. To avoid such 'dangerous' denominators it is necessary to set $\mathscr{H}_{20} = 0$, and the latter is equivalent to minimization of the energy by the B.C.S. method.

The next fact is that $\mathscr{H}_{11}$ reduces without difficulty to the form

$$\mathscr{H}_{11} = \sum_K E_K \alpha_K^\dagger \alpha_K, \tag{7.5.40}$$

where $E_K = \sqrt{(\epsilon_K^2 + \Delta_K^2)}$, as given by (7.5.14). Thus, it follows that

$$\mathscr{H} = \mathsf{E} + \mathscr{H}_{11} + \mathscr{H}_{\text{int.}}, \tag{7.5.41}$$

where $\mathscr{H}_{11}$ is given by (7.5.40), and the excitation energy (7.5.39) readily reduces to

$$\begin{aligned}(\alpha_K^\dagger \alpha_{K'}^\dagger \psi, [\mathscr{H}_{11} + \mathscr{H}_{\text{int.}}]\, \alpha_K^\dagger \alpha_{K'}^\dagger \psi) \\ = E_K + E_{K'} + (\alpha_K^\dagger \alpha_{K'}^\dagger \psi, \mathscr{H}_{\text{int.}}\, \alpha_K^\dagger \alpha_{K'}^\dagger \psi).\end{aligned} \tag{7.5.42}$$

If, for the moment, we suppose the last term vanishes as K, K' approach the Fermi surface (the former from above, the latter from below), then since each of E_K, $E_{K'}$ tends to Δ, we find an energy gap of size 2Δ between the ground state and the nearest single-particle excited state, as we anticipated.

It remains to evaluate the final term of (7.5.42). Using the Hamiltonian (7.5.1) and the transformations (7.4.18) and (7.4.19), we can write down an explicit expression for $\mathscr{H}_{\text{int.}}$ in terms of the α's, from which the required expectation value can be calculated. The following method, however, turns out to be much simpler to carry through.

We begin by asking what kind of terms in $\mathscr{H}_{\text{int.}}$ can give non-zero contributions to $(\alpha_K^\dagger \alpha_{K'}^\dagger \psi, \mathscr{H}_{\text{int.}}\, \alpha_K^\dagger \alpha_{K'}^\dagger \psi)$. Recalling the particle conservation requirement, these certainly must be proportional to operators of the type $\alpha_P^\dagger \alpha_Q^\dagger \alpha_R \alpha_S$. But we can go further. Unless (P, Q), (R, S) and (K, K') are all equal (in some order), the average still vanishes. There is thus essentially only one operator of the type sought, namely $\alpha_K^\dagger \alpha_K \alpha_{K'}^\dagger \alpha_{K'}$. If we now inspect the potential energy terms in (7.5.1), expressed in terms of the α's, and remember that K' is below and K above the Fermi surface, we find very few contributions of the type sought. Their sum total is

$$-4V_{KK'} u_K v_K u_{K'} v_{K'} \alpha_K^\dagger \alpha_K \alpha_{K'}^\dagger \alpha_{K'} \tag{7.5.43}$$

and thus

$$\begin{aligned}(\alpha_K^\dagger \alpha_{K'}^\dagger \psi, \mathscr{H}_{\text{int.}}\, \alpha_K^\dagger \alpha_{K'}^\dagger \psi) &= 4V_{KK'} u_K v_K u_{K'} v_{K'} \\ &= V_{KK'} \sin 2\theta_K \sin 2\theta_{K'}.\end{aligned} \tag{7.5.44}$$

We can now justify the neglect of this term in (7.5.42) for K, K' near the Fermi surface, since, under these conditions,

$$V_{KK'} = O(\Omega^{-1}),$$

while E_K and $E_{K'}$ remain of order unity.

Finally, let us note that for the total momentum operator, we have

$$\sum_{\text{All } K} \mathbf{k} a_K^\dagger a_K = \sum_{\text{All } K} \mathbf{k} \alpha_K^\dagger \alpha_K. \tag{7.5.45}$$

The significance of this is that K is more than merely a convenient reference parameter relating the system to its non-interacting counterpart; it represents the momentum and spin of a quasi-particle excitation. A somewhat similar situation will be discussed in a little more detail in section 8.3.7(*c*).

(*b*) *Current-carrying states.* Having demonstrated the existence of an energy gap, the metastability of current-carrying states is readily understood. Fig. 7.13 illustrates in **k**-space (see Fig. 7.9

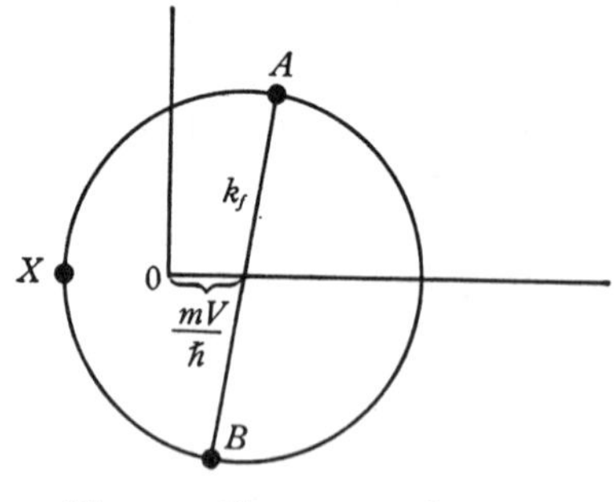

Fig. 7.13. Current-carrying state.

and the corresponding discussion) such a state, where the drift velocity is V and the electronic mass is m. The highly correlated pairs are given by diametrically opposite points near the Fermi surface as A and B. If, for any reason, the bond between a pair is broken, for the drift velocity V attained in practice, it is energetically more favourable for the pair to revert to its correlated state than to seek lower kinetic energies in the neighbourhood of X.

To be more precise from the above point of view, the decrease in kinetic energy in placing A and B near X is

$$\frac{\hbar^2}{2m} OA^2 + \frac{\hbar^2}{2m} OB^2 - 2\frac{\hbar^2}{2m} OX^2 = 2\hbar k_f V, \tag{7.5.46}$$

a result independent of the particular pair chosen. On the other hand, the increase in the energy of the system on breaking the bond is 2Δ and thus the pair prefers to remain correlated provided

$$V < \frac{\Delta}{\hbar k_f} \sim \frac{10^{-4} \times 10^{-12}}{10^{-19}} \text{cm/sec} = 10^3 \text{cm/sec}, \qquad (7.5.47)$$

typical values having been inserted for Δ and k_f.

7.6. Elevated temperature form of B.C.S. theory

Let us begin by summarizing those aspects of the zero-temperature situation which will be of importance to us in the following. The Hamiltonian is

$$\mathscr{H} = \epsilon + \sum_K E_K \alpha_K^\dagger \alpha_K + \mathscr{H}_{\text{int.}} \qquad (7.6.1)$$

where from (7.5.9) the ground-state energy is given by

$$\epsilon = \sum_K \epsilon_K \sin^2\theta_K + \frac{1}{8} \sum_{KK'} V_{KK'} \sin 2\theta_K \sin 2\theta_{K'}, \qquad (7.6.2)$$

and using (7.5.5)

$$E_K = \epsilon_K \cos 2\theta_K - \frac{1}{2} \sum_{K'} V_{KK'} \sin 2\theta_K \sin 2\theta_{K'}. \qquad (7.6.3)$$

The ground-state wave function $|\psi\rangle$, given by (7.4.4), is the vacuum state for the α's and the single-particle excited states are Fermi-like, of type $\alpha_K^\dagger \alpha_{K'}^\dagger |\psi\rangle$, where one of the indices, say K, lies above the Fermi surface, while the other lies below.

The situation is as in Fig. 7.10 (i) in reference **k**-space. The expectation value of $\mathscr{H}_{\text{int.}}$ with respect to such an excited state is required and is (by (7.5.44))

$$(\alpha_K^\dagger \alpha_{K'}^\dagger \psi, \mathscr{H}_{\text{int.}} \alpha_K^\dagger \alpha_{K'}^\dagger \psi) = V_{KK'} \sin 2\theta_K \sin 2\theta_{K'}. \quad (7.6.4)$$

A point to notice about these formulae is that the B.C.S. integral equation (7.5.11) has not been used. This is important, since the non-zero temperature theory requires a more general procedure, based on a minimization of the free energy.

7.6.1. *Elevated temperature B.C.S. integral equation*

We find it convenient, in accordance with the ideas developed in Chapter 3, to regard the B.C.S. ground state $|\psi\rangle$ as a no-particle system from which Fermions are excited in pairs as in Fig. 7.10 (i). A general excited state consists of many such pairs, while at any given temperature T, the state of the system is expressed as a probability distribution over all possible excited states. Let us suppose therefore that at temperature T, f_K is a probability of occupation of the state K (in reference **k**-space).

Our immediate purpose is to write down an expression for the Helmholtz free energy

$$F = U - TS, \tag{7.6.5}$$

where U is the internal energy and S is the entropy. For the case of Fermions, the latter takes the standard form (see, for example, Wilson, 1958)

$$S = -k_B \sum_K \{f_K \log f_K + (1 - f_K) \log (1 - f_K)\}. \tag{7.6.6}$$

It remains therefore to evaluate the internal energy U.

The latter task is performed by appropriately weighting the various contributions isolated with the help of (7.6.2), (7.6.3) and (7.6.4). Specifically, we have the total kinetic energy as the sum of the vacuum kinetic energy, plus that due to Fermions created, that is

$$\text{Kinetic energy} = \sum_K \epsilon_K \sin^2 \theta_K + \sum_K \epsilon_K \cos 2\theta_K f_K. \tag{7.6.7}$$

Similarly, the total potential energy is formed from three contributions, which arise via vacuum-vacuum, vacuum-Fermion and Fermion-Fermion interactions Thus we may write

$$\text{Potential energy} = \frac{1}{8} \sum_{KK'} V_{KK'} \sin 2\theta_K \sin 2\theta_{K'} + \sum_K f_K \times \left\{ -\frac{1}{2} \sum_{K'} V_{KK'} \sin 2\theta_K \sin 2\theta_{K'} \right\} + \frac{1}{2} \sum V_{KK'} \sin 2\theta_K \sin 2\theta_{K'} f_K f_{K'}. \tag{7.6.8}$$

Adding (7.6.7) and (7.6.8), we obtain, after a little rearrangement, the total internal energy

$$U = \frac{1}{2} \sum_K \epsilon_K - \frac{1}{2} \sum_K \epsilon_K (1 - 2f_K) \cos 2\theta_K + \frac{1}{8} \sum_{KK'} V_{KK'} \sin 2\theta_K \sin 2\theta_{K'} (1 - 2f_K)(1 - 2f_{K'}). \tag{7.6.9}$$

Our procedure now follows the zero-temperature case closely. With S and U given by (7.6.6) and (7.6.9), F is varied with respect to θ_K and f_K. Varying first θ_K, we obtain, in a straightforward manner,

$$\tan 2\theta_K = -\frac{1}{2\epsilon_K} \sum_{K'} V_{KK'} \sin 2\theta_{K'} (1 - 2f_{K'}), \tag{7.6.10}$$

or, putting (see Fig. 7.14)

$$\tan 2\theta_K = \frac{\Delta_K}{\epsilon_K}, \quad E_K = \sqrt{(\epsilon_K^2 + \Delta_K^2)} \tag{7.6.11}$$

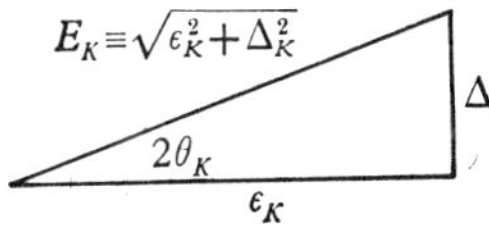

Fig. 7.14

we can write (7.6.10) in the alternative form

$$\Delta_K = -\frac{1}{2}\sum_{K'} V_{KK'} \frac{\Delta_{K'}}{E_{K'}}(1-2f_K). \tag{7.6.12}$$

[We should note that E_K introduced in (7.6.11) is not that of section 7.5 (see remarks below).]

On the other hand, variation with respect to f_K gives

$$\epsilon_K \cos 2\theta_K - \frac{1}{2}\sum_{K'} V_{KK'} \sin 2\theta_K \sin 2\theta_{K'}(1-2f_{K'}) + \frac{1}{\beta}\log\frac{f_K}{1-f_K} = 0, \tag{7.6.13}$$

where the notation $\beta = 1/k_B T$ has been used. The Euler equations (7.6.10) and (7.6.13), can be thrown into more convenient and transparent forms by eliminating the explicit dependence on $V_{KK'}$. Substitution of (7.6.10) in (7.6.11) gives

$$-\frac{1}{\beta}\log\frac{f_K}{1-f_K} = \epsilon_K \cos 2\theta_K + \sin 2\theta_K(\epsilon_K \tan 2\theta_K) = \frac{\epsilon_K}{\cos 2\theta_K} = E_K, \tag{7.6.14}$$

and hence

$$f_K = \frac{1}{e^{\beta E_K}+1}. \tag{7.6.15}$$

The latter, on substitution in (7.6.12), gives

$$\Delta_K = -\frac{1}{2}\sum_{K'} V_{KK'} \frac{\Delta_{K'}}{E_{K'}} \tanh \tfrac{1}{2}\beta E_{K'}, \tag{7.6.16}$$

which is the elevated temperature B.C.S. integral equation.

The function f_K as given by (7.6.15) is clearly of Fermi–Dirac form. It may be sketched as in Fig. 7.15. When $\Delta_K = 0$, we have $E_K = \epsilon_K$ and we obtain the temperature-dependent distribution appropriate to independent particles, but in a particle-hole description rather than in the usual form. When $T \to 0$, the area under the curve contracts to zero in order to give no particles and no holes as required for the ground state.

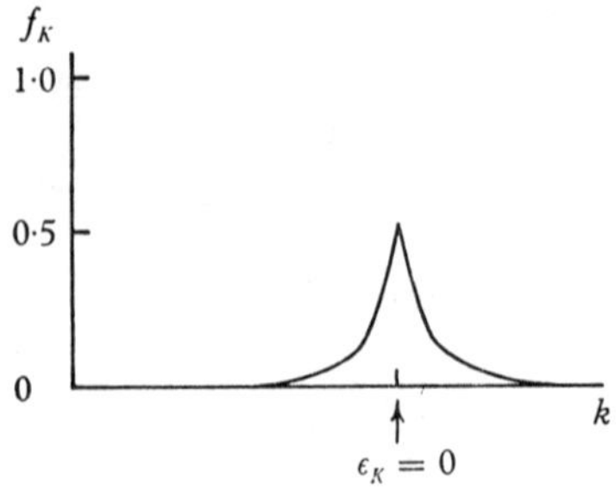

Fig. 7.15. Distribution function in particle-hole description.

Strictly speaking, Δ_K, θ_K and E_K should carry an index T because of their temperature dependence, this index having been omitted for notational convenience. When $T = 0$, they revert to their values of section 7.5.

Finally, we observe that E_K and Δ_K play the same roles here as they did previously. For the energy of interaction of an additional Fermion with index K, created from our thermally smeared interacting system, is given by

$$\left\{\epsilon_K \cos 2\theta_K - \frac{1}{2}\sum_{K'} V_{KK'} \sin 2\theta_K \sin 2\theta_{K'}\right\} + \sum_{K'} V_{KK'} \sin 2\theta_K \sin 2\theta_{K'} f_{K'}, \quad (7.6.17)$$

where use has been made of (7.6.3) and (7.6.4). This expression simplifies to

$$\epsilon_K \cos 2\theta_K - \frac{1}{2}\sum_{K'} V_{KK'} \sin 2\theta_K \sin 2\theta_{K'}(1 - 2f_{K'}) = E_K(\beta), \quad (7.6.18)$$

the latter step following from (7.6.10). Thus $E_K(T)$ is the energy required to create a Fermion, and so the energy required to create a particle-hole pair at the Fermi surface is just $2\Delta_K(T)$, where K is chosen at the Fermi surface.

7.6.2. *Solution for averaged potential*

Once more we solve the integral equation when the potential takes the simple form

$$V_{KK'} = \begin{cases} -V/\Omega & \text{if } K, K' \text{ have same spin and } |\epsilon_K|, |\epsilon_{K'}| < \delta, \\ 0 & \text{otherwise.} \end{cases} \quad (7.6.19)$$

Putting (7.6.19) in (7.6.16), we have, as before

$$\Delta_K = 0 \quad (|\epsilon_K| > \delta), \qquad \Delta_K = \Delta \quad (|\epsilon_K| < \delta), \tag{7.6.20}$$

where Δ, independent of K, is either zero (the normal solution) or is given by

$$\frac{2}{V} = \frac{1}{\Omega} \sum_{|\epsilon_{K'}|<\delta} \frac{1}{\sqrt{(\epsilon_K^2+\Delta^2)}} \tanh \tfrac{1}{2}\beta \sqrt{(\epsilon_K^2+\Delta^2)}$$

$$= \rho \int_{-\delta}^{\delta} \frac{d\epsilon}{\sqrt{(\epsilon^2+\Delta^2)}} \tanh \tfrac{1}{2}\beta \sqrt{(\epsilon^2+\Delta^2)}, \tag{7.6.21}$$

where the summation is over singly occupied states.

The following points are to be noted. As $\beta \to \infty$, we regain the zero-temperature situation and, as we have seen, it is possible to solve (7.6.21) for a non-zero Δ given by (7.5.24). On the other hand, for any given Δ, as β decreases from infinity, the right-hand side of (7.6.21) decreases monotonically until finally, for β sufficiently small, equation (7.6.21) can never be satisfied, and the normal alternative solution must be chosen. The critical temperature above which the superconducting state cannot occur is given by putting $\Delta(\beta) = 0$ when $\beta = \beta_c$ in (7.6.21). We find

$$\frac{1}{\rho V} = \int_0^{\delta} \frac{d\epsilon}{\epsilon} \tanh \tfrac{1}{2}\beta_c \epsilon = \int_0^{\beta_c\delta} \frac{dx}{x} \tanh \tfrac{1}{2}x. \tag{7.6.22}$$

The latter is a function of $\beta_c \delta$ only and for superconductors this is large. Equation (7.6.22) may be solved (see Appendix 7A.1) for $\beta_c \delta$ large, to give a critical temperature defined by

$$\frac{1}{\beta_c} = 1{\cdot}14\delta\, e^{-1/\rho V}. \tag{7.6.23}$$

Generally, it is possible in the $\beta_c \delta \gg 1$ limit to draw an absolute curve of $\Delta(\beta)/\Delta(\infty)$ against β_c/β as shown in Fig. 7.16. The details are again given in Appendix 7A.1. The curve clearly exhibits a co-operative effect. Qualitatively speaking, the thermal creation of a particle K excludes from the scattering states available to a Cooper pair, both K and $-K$, resulting in a reduced energy of interaction.

A number of interesting conclusions can be drawn from (7.6.23). In the first place, putting $\delta = \hbar\omega$, we obtain the isotope effect in the form

$$\beta_c \omega = \text{const.} \tag{7.6.24}$$

Also, recalling the energy gap formula (7.5.24) for zero temperatures and weak coupling, one finds a universal ratio for all B.C.S. superconductors, namely

$$\frac{2\Delta(\infty)}{k_B T_C} = 3{\cdot}52, \tag{7.6.25}$$

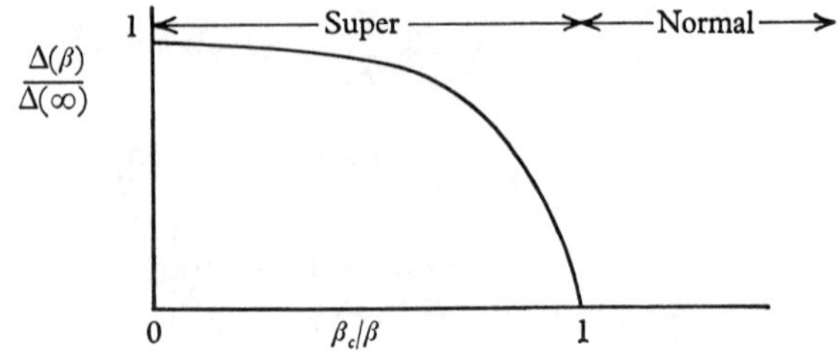

Fig. 7.16. $\Delta(\beta)/\Delta(\infty)$ against β_c/β.

while use of (7.5.35) for the critical field $H_{\text{co.}}$ at zero temperature gives a further absolute ratio

$$\frac{\gamma T_c^2}{H_{\text{co.}}^2} = 0{\cdot}17, \tag{7.6.26}$$

where $\gamma = \frac{2}{3}\pi^2 \rho k_B^2$ is the zero temperature limit of dC_v/dT for the corresponding normal electron gas. These results have been touched on in a qualitative way in our preliminary discussion in section 7.2.

7.6.3. *Critical field and specific heat*

The free energy (7.6.5) may be readily calculated using (7.6.20) and in this way we can investigate other thermodynamic properties. For example, Fig. 7.4 is obtained for the critical field. For details the reader is referred to the original B.C.S. paper.

Turning to the specific heat C, it is possible to write this in a physically revealing form as follows. We have

$$C = T\frac{d}{dT}(S/\Omega) = -\frac{\beta}{\Omega}\frac{dS}{d\beta}. \tag{7.6.27}$$

Using the formula (7.6.6) for S and recalling (7.6.15) and (7.6.11) we can write the latter in the form

$$C = \frac{k_B \beta^2}{\Omega} \sum_K f_K(1-f_K)\left[E_K^2 + \frac{\beta}{2}\frac{\partial}{\partial\beta}\Delta_K^2\right]. \tag{7.6.28}$$

The first term in the square brackets gives a contribution continuous for all temperatures. Above the critical temperature, this is the only contribution, since Δ_K is identically zero. On the other hand, the second term is discontinuous as a function of temperature as we go through the critical temperature (see Fig. 7.16), the fall in the specific heat as the system becomes normal being

$$\frac{k_B \beta_c^3}{2\Omega} \sum_K \left(\frac{\partial}{\partial \beta} \Delta_K^2\right)_{\beta=\beta_c} \sim 2{\cdot}43\gamma T_c. \qquad (7.6.29)$$

In this way, we see the general features of the specific heat curve in Fig. 7.3 emerging, and in fact closer analysis shows that if

$$2{\cdot}5 < T_c/T < 6,$$

equation (7.2.1) is essentially satisfied.

7.7. Collective excitations and flux quantization

We have discussed so far essentially the ground state of a superconductor, plus the excited states which can be reached by single-particle excitations.

We should now stress that collective, as well as single-particle excitations, exist in superconductors. While these do not affect the properties discussed quantitatively in earlier sections of this chapter, it is now known that a description of the collective excitations is necessary before any gauge invariant theory of the electromagnetic properties of superconductors can be constructed. This would lead us into a whole field, and too far away from the main theme of the volume. We must reluctantly therefore simply refer the reader to the discussions of Anderson (1958), Rickayzen (1959) and Prange (1963) for full details.

What we shall do, however, is to show that the Meissner effect implies flux quantization. The quantum of flux, involving, as we shall see, the total charge of a Cooper pair, gives a striking verification of the essential feature of the B.C.S. theory.

What we wish to do is to construct the free energies F_n and F_s of the normal and superconducting states as functions of the flux Φ. Then, to distinguish the states, we must insist, because of the Meissner effect, that

$$\frac{\partial F_s}{\partial \Phi} = 0.$$

To do this, we calculate first the energy levels of independent electrons in, to simplify the geometry, a hollow cylinder. The result (Byers & Yang, 1961) is given by

$$E_n = \frac{1}{2m}\left[p_r^2+p_z^2+\frac{\hbar^2}{r^2}\left(n+\frac{e}{ch}\Phi\right)^2\right], \tag{7.7.1}$$

where p_r and p_z are the momenta in the radial and z-directions, Φ is the flux and n an integer.

If we now compute the partition function and the free energy, we find these are essentially independent of Φ. But we have so far missed out the crucial property of the superconducting state, namely the pairing of electrons. Thinking now in terms of the level spectrum of these pairs, we can see that the average energy of states n and $-n$ increases from the zero field value by a term proportional to Φ^2. Thus, from the partition function, the free energy increases away from $\Phi = 0$, quadratically with Φ. This is already in marked contrast to the normal state. It is readily shown that such pairing of states n and $-n$ is also energetically favourable around $\Phi = ch/e$, or indeed any positive or negative multiple of this. In each case, the free energy increases as we go away from these multiples of ch/e.

But more care is required when Φ is $\frac{1}{2}(ch/e)$ or $(m+\frac{1}{2})(ch/e)$. In this case, $n+(e/ch)\,\Phi$ in (7.7.1) is $n+\frac{1}{2}$, and we can pair electrons occupying states with $n = 0$ and $n = -1$ and so on. The energy per particle then remains the same as for $\Phi = 0$, but again increases as we go away from the points $(m+\frac{1}{2})(ch/e)$. This time the increase is, for $m = 0$, proportional to

$$(\Phi-\tfrac{1}{2}(ch/e))^2.$$

Thus, we see that the free energy has maxima with respect to Φ at values $ch/2e$ or any multiples of it. Therefore, the existence of the Meissner effect then implies flux quantization, with $ch/2e$ as the unit.

This argument has taken no account of the actual size of the wave function of the B.C.S. pair in zero magnetic field. It turns out that this reduces the flux unit slightly below $ch/2e$.

7.8. Anderson theory of dirty superconductors

The theory so far discussed has considered only the interaction between the electrons and the lattice field. In reality, of course, some impurities will be present in all superconductors, and will limit the electronic mean free path. It might seem, at first sight, that when the mean free path becomes too short, the uncertainty in the electronic energy will be comparable with, or greater than, the energy gap, and thus inhibit the superconducting behaviour.

This circumstance is not found experimentally. Rather surprisingly, the transition temperature of superconductors is not dramatically sensitive to impurity concentration. The Anderson (1959) theory of such so-called 'dirty' superconductors gives us the basic reason why this should be so.

The starting-point of the theory is the construction of the one-electron eigenfunctions in the presence of the scattering centres. If we represent the complete set of Bloch functions in the pure superconductor by $\phi_{\mathbf{k}}$, then formally the exact eigenfunctions when the impurities are present may be written

$$\psi_n = \Sigma A_{n\mathbf{k}}\,\phi_{\mathbf{k}}. \tag{7.8.1}$$

If the Bloch functions are normalized, then for a normalized wave function ψ_n we must have

$$\sum_{\mathbf{k}} A_{n\mathbf{k}}^2 = 1. \tag{7.8.2}$$

We next observe that if the impurity centres are non-magnetic, so that time-reversal symmetry arguments apply, then the wave function

$$\psi_{-n} = \Sigma A_{n\mathbf{k}}^*\,\phi_{-\mathbf{k}} \tag{7.8.3}$$

is also an eigenfunction, with the same eigenvalue as (7.8.1). The natural generalization of the B.C.S. pairing is then obtained by coupling the states (7.8.1) and (7.8.3).

The complete set of eigenfunctions ψ_n may now be used to define creation and annihilation operators, $C_n^\dagger$ and C_n. Then, in terms of these, the important part of the Hamiltonian which gives rise to an energy gap has the form

$$H = -\sum_{nn'} V_{nn'}\,C_n^\dagger\,C_{-n}^\dagger\,C_{-n'}\,C_{n'}, \tag{7.8.4}$$

where $V_{nn'}$ can be related to the matrix elements $V_{kk'}$ calculated with respect to Bloch states.

Once $V_{nn'}$ is known, the B.C.S. derivation of the energy gap for pure superconductors follows through again. When the impurity concentration is large, $V_{nn'}$ represents essentially an average of $V_{kk'}$ over the Fermi surface. We anticipate therefore that, when the Anderson theory becomes valid, the effective interaction and the energy gap will be considerably less anisotropic than for pure superconductors.

In dirty superconductors, the relaxation time τ is often of the order of 10^{-14} sec. Then the uncertainty in the electronic energy, $h\tau^{-1}$ is of order 0·1 eV. The Anderson theory, very roughly, is applicable when this uncertainty $h\tau^{-1}$ is of the order of the energy gap. This is equivalent to the condition that the electronic mean free path is less than the coherence length.

7.9. Ginzburg–Landau theory

It would hardly be appropriate to end this chapter without some discussion of the basic features of hard (type II) superconductors on which so much research is taking place nowadays. We will see that inherent in such a description is the need to allow a spacially varying superfluid density, and the highly successful Ginzburg–Landau theory was designed to meet such a situation.

We shall be content, below, to give the original phenomenological discussion of Ginzburg & Landau (1950), but we should remark that Gor'kov (1959), using a Green function formulation of the B.C.S. method, was able to provide a microscopic basis for the theory.

The starting-point is to define a function $\Psi(\mathbf{r})$ such that $|\Psi(\mathbf{r})|^2$ is the density of superconducting electrons at position $\mathbf{r}$. Thus Ψ is a kind of effective wave function but in the Gor'kov derivation, for instance, Ψ turned out to be proportional to the local energy gap parameter $\Delta(\mathbf{r})$. This is a very reasonable result since the ability to create superconducting electrons is directly related to the size of Δ.

In terms of Ψ, the total free energy of the superconducting state in the presence of a magnetic field $H(\mathbf{r})$ is written plausibly in the form

$$F \equiv \int d\mathbf{r}\, F_H^s(\mathbf{r}) = \int d\mathbf{r} \left\{ F_0^s(\mathbf{r}) + \frac{1}{2m} \left| -i\hbar\nabla\Psi - \frac{e^*}{c}\mathbf{A}(\mathbf{r})\Psi \right|^2 + \frac{H^2(\mathbf{r})}{8\pi} \right\}. \tag{7.9.1}$$

Here e^* is the charge on one of the 'particles' constituting the superfluid. Since these are now known to be Cooper pairs, $e^* = 2e$. The term F_0^s is written

$$F_0^s = \frac{H_c^2(T)}{8\pi}\left(1 - 2\left|\frac{\Psi}{\Psi_0}\right|^2 + \left|\frac{\Psi}{\Psi_0}\right|^4\right), \tag{7.9.2}$$

where Ψ_0 denotes the zero field value of Ψ. The right-hand side of (7.9.2) may be regarded as a truncated power series in Ψ and is thus most appropriate near the transition point. The coefficients are chosen to give the expression the properties (i) $F_0^s = H_c^2(T)/8\pi$ when $\Psi = 0$, and (ii) $\partial F_0^s/\partial\Psi = 0 = F_0^s$ when $\Psi = \Psi_0$.

Before proceeding further, it is convenient to follow Abrikosov (1957) by using reduced variables. Thus we define the following dimensionless quantities:

$$\frac{\Psi}{\Psi_0} = \psi, \quad \frac{\mathbf{A}}{\sqrt{(2)}\,H_c(T)} = \mathbf{a}, \quad \frac{\mathbf{H}}{\sqrt{(2)}\,H_c(T)} = \mathbf{h}, \quad K = \frac{\sqrt{(2)}\,e^*}{\hbar c} H_c(T)\,\lambda_L^2. \tag{7.9.3}$$

In the definition of K, the fundamental length λ_L, given by

$$\lambda_L^2 = \frac{mc^2}{4\pi e^{*2}|\Psi_0|^2}, \tag{7.9.4}$$

is a convenient unit of length to use in rewriting the free energy as

$$\frac{F}{H_c^2(T)/4\pi} = \int d\mathbf{r}\{\tfrac{1}{2} - |\psi|^2 + \tfrac{1}{2}|\psi|^4 + |(i/K)\nabla\psi + \mathbf{a}\psi|^2 + h^2(\mathbf{r})\}. \tag{7.9.5}$$

The next step is to vary ψ and $\mathbf{a}$ in (7.9.5). The respective Euler equations are

$$[(i/K)\nabla + \mathbf{a}]^2\psi = \psi - |\psi|^2\psi \tag{7.9.6}$$

and (since $\mathbf{h} = \operatorname{curl}\mathbf{a}$)

$$-\operatorname{curl}\operatorname{curl}\mathbf{a} = |\psi|^2\mathbf{a} + (i/2K)(\psi^*\nabla\psi - \psi\nabla\psi^*); \tag{7.9.7}$$

(7.9.6) and (7.9.7) are the Ginzburg–Landau equations. For solutions ψ of these equations, (7.9.5) becomes

$$\frac{F}{H_c^2(T)/4\pi} = \int d\mathbf{r}\{\tfrac{1}{2} - \tfrac{1}{2}|\psi|^4 + h^2(\mathbf{r})\}. \tag{7.9.8}$$

The quantities $H_c(T)$, λ_L and K are the three essential parameters of the theory. $H_c(T)$, as defined above, is the thermodynamic bulk critical field and is given by

$$H_c^2(T)/8\pi = -\int_0^\infty M\,dH. \qquad (7.9.9)$$

For type I superconductors, (7.9.9) is hardly necessary, for $H_c(T)$ is then the usual critical field, above which $M = 0$, and below which we have the property $M = -H/4\pi$ of a perfect diamagnet. In type II materials, the $M(H)$ relationship is more complicated (cf. Fig. 7.19 and comments thereon in the text) and (7.9.9) is required in its generality.

7.9.1. *Superconductor surface*

To interpret λ_L, let us consider the case of a superconductor occupying the half-space $x > 0$ and with a small applied field $\mathbf{h}_0$ in the region $x < 0$. Inside the superconductor, the field and vector

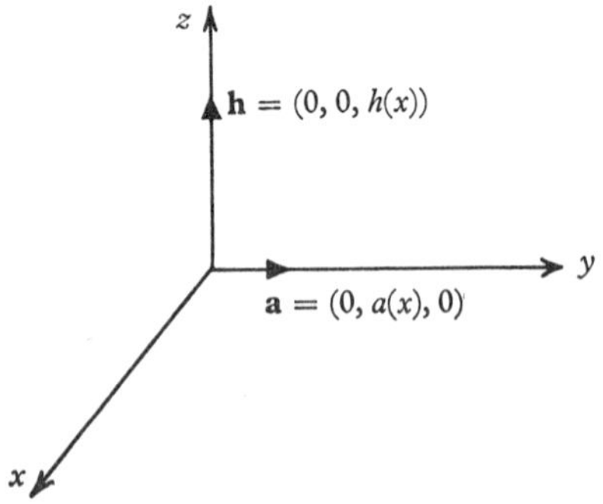

Fig. 7.17

potential may be described as shown in Fig. 7.17 where $a'(x) = h(x)$. The Ginzburg–Landau equations (7.9.6) and (7.9.7) then reduce to

$$-\frac{1}{K^2}\psi'' + a^2\psi = \psi - |\psi|^2\psi, \qquad (7.9.10)$$

and (neglecting terms in ψ quadratic in h_0)

$$a'' = a. \qquad (7.9.11)$$

Equation (7.9.11) has the elementary solution $a = -h_0 e^{-x}$, $h = h_0 e^{-x}$. Thus, $\lambda_L(=1)$ is a measure of the field penetration into the superconductor; it is, in fact, the usual London penetration depth. The explicit expression for a may now be used in (7.9.10) to obtain information about ψ. Expanding the latter as a power series in h_0^2 and discarding h_0^4 terms in (7.9.10), elementary methods yield the result (Douglass & Falicov, 1964)

$$\psi = 1 - \frac{K}{\sqrt{(2)}\,(2-K^2)}\left[e^{-\sqrt{(2)}\,Kx} - \frac{K}{\sqrt{2}}e^{-2x}\right]h_0^2, \qquad (7.9.12)$$

where we have used the boundary condition that $\psi'(x) = 0$ when $x = 0$.

The dimensionless parameter K, the only one to appear explicitly in (7.9.6) and (7.9.7), can be numerically estimated from (7.9.4) by a number of methods (Goodman, 1962). It is found that usually for soft superconductors, K_0 (the value of K when $T = 0$) is of the order of 0·1 at 0 °K, while the corresponding results for hard superconductors are characteristically an order of magnitude higher than this.

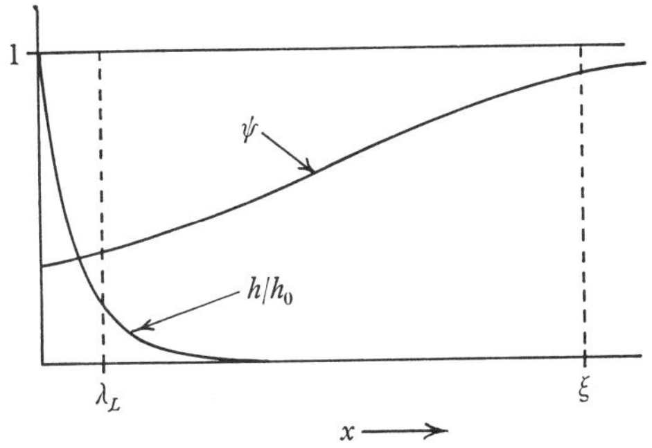

Fig. 7.18. ψ obeys (7.9.12) when h_0 is small.

When $K \ll 1$, the first exponential obviously dominates the second term in (7.9.12) and ψ increases to its asymptotic value of unity in a distance of order $[\sqrt{(2)}\,K]^{-1}$. This should, then, be a measure of the coherence length ξ. Thus we expect $K \simeq \lambda_L/\xi$, and in Gor'kov's calculation this result is almost exactly obtained.

When $K \gtrsim 1$, a rather different physical situation obtains, and to understand this, it is first necessary to introduce in a qualitative way the concept of surface energy (see, for example, F. London, 1954). The exclusion of an external field, H_0, from a superconductor raises the energy per unit volume by an amount $H_0^2/8\pi$. If this were the only consideration, the material would striate into superconducting regions of width $d_s < \lambda_L$, separated from each other by normal regions of width $d_n \ll \lambda_L$. Since $d_n/d_s \ll 1$, maximum advantage is taken of Cooper pairing, while $d_s < \lambda_L$ means (see Fig. 7.18) that the field would fairly approximate to H_0 everywhere. The reason that this does not usually happen is that there exists a positive surface energy. A crude estimate of this is obtained by observing (Fig. 7.18) that the boundary reduces the superfluid volume of the metal by a factor of order ξ per unit area, whereas the effective volume not accessible to the magnetic field is obtained by reducing the metallic volume by a factor λ_L per unit area. The surface energy per unit area is thus a factor of order $\xi - \lambda_L$ times $H_0^2/8\pi$. When $K \ll 1$, $\xi - \lambda_L > 0$ and we have a positive surface energy. However, when $K \gtrsim 1$, $\xi - \lambda_L < 0$ and an instability of the kind described above sets in.

Abrikosov has shown that the Ginzburg–Landau equations can be used to put the above discussion on a quantitative basis. He finds that the transition value of K is $1/\sqrt{2}$ (and thereby furnishes a precise distinction between type I and type II behaviour) and that above this value the striations take the form of vortices. A description of the latter is given in the following section.

7.10. Abrikosov's theory of type II superconductors

The observed magnetization versus field curves for typical types I and II superconductors are shown in Fig. 7.19. In the first case, there is the usual single transition field H_c, but in type II materials, it will be seen that there are two transitions, one at H_{c1}, corresponding to an initial penetration of flux and the other at H_{c2} when the penetration is complete. Actually, it has recently been found (Saint-James & de Gennes, 1963) that because of a very persistent superconducting sheath near the surface, an extremely weak diamagnetic effect can obtain up to some maximum value H_{c3}. We shall not enter into this latter effect here, but will be concerned with exhibiting the essential nature of the region $H_{c1} < H < H_{c2}$. If the two curves in Fig. 7.19 enclose the same area, then because of (7.9.9), H_c will be the thermodynamic critical field for the type II case.

It is clear that above H_{c1}, the striations referred to at the end of the previous section have occurred, and we must re-examine the Ginzburg–Landau equations bearing in mind this new physical effect. Now, for example, the situation depicted in Fig. 7.18 is no longer appropriate. It is convenient to examine the two limiting cases when H_0 is first near H_{c2} and then H_{c1}.

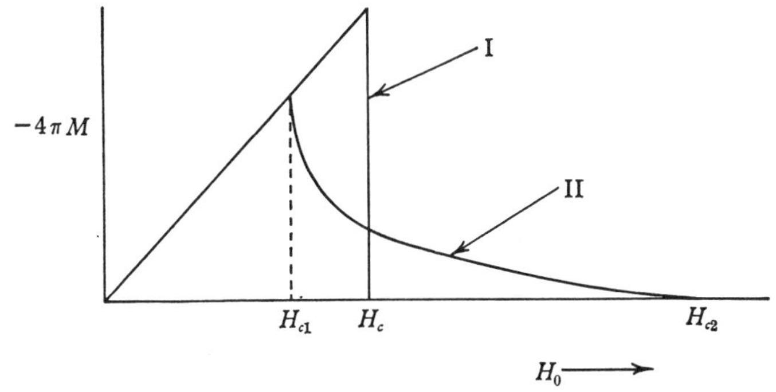

Fig. 7.19

(*a*) $H_0 \lesssim H_{c2}$. The starting-point in this case is to suppose that, as a zeroth approximation, we can write (Fig. 7.17) $h = h_0$, where h_0 is a constant. Thus, to within a trivial constant which serves to locate the *yz*-plane, $a = h_0 x$. We also know that now ψ is a small quantity. Hence, to begin with, we neglect the second Ginzburg–Landau equation which is homogeneous and quadratic in ψ and solve the first, omitting the cubic $|\psi|^2\psi$ term.

If we suppose ψ is independent of y, then once more (7.9.10) is obtained, where now, however, $a = h_0 x$. Thus, we find a Schrödinger equation appropriate to a simple harmonic oscillator; for bounded solutions, $h_0 = K/(2n+1)$. The highest possible value of h_0 is thus K (when $n = 0$), which implies (recall equation (7.9.3) and Fig. 7.19) that

$$H_{c2} = \sqrt{(2)}\, KH_c. \tag{7.10.1}$$

The corresponding eigenfunction is proportional to $\exp(-\frac{1}{2}K^2x^2)$.

The above analysis is not yet, however, satisfactory, the ψ we have found being highly localized in the x-direction. This has no physical significance and the resolution of this difficulty is to note that there is a whole family

$$\psi_n = e^{ikny}\, e^{-\frac{1}{2}K^2[x-(kn/K^2)]^2}, \tag{7.10.2}$$

satisfying (7.9.6) (with $|\psi|^2\psi$ omitted), and for which (7.10.1) is appropriate. Thus

$$\psi = \sum_{n=-\infty}^{\infty} c_n \psi_n \tag{7.10.3}$$

is an allowed solution if the c's are periodic in n, thus giving ψ periods of k/K^2 and $2\pi/k$ in x and y respectively. At this point with $H = H_{c2}$, k is quite arbitrary; its value will be fixed when the situation with H just below H_{c2} is resolved. (Then k becomes an available energy minimization parameter.) This latter problem is solved in considerable detail by Abrikosov, and we shall be content here to summarize, rather briefly, his method.

The next step is to include quadratic terms in ψ by substituting (7.10.3) into (7.9.7). This leads to revised values of h and a given by

$$h = h_0 - \frac{1}{2K}|\psi|^2, \tag{7.10.4}$$

and

$$a = h_0 x - \frac{1}{2K}\int^x |\psi^2|\, dx. \tag{7.10.5}$$

Then account is taken of third-order terms by returning, with (7.10.4) and (7.10.5) to (7.9.6) with $|\psi|^2\psi$ now included. This leads to the result

$$\frac{K-h_0}{K}\overline{|\psi|^2} + \left(\frac{1}{2K^2} - 1\right)\overline{|\psi|^4} = 0, \tag{7.10.6}$$

where a bar denotes a macroscopic average.

With these results, we can now compute macroscopic variables of interest. For the magnetic induction, (7.10.4) gives

$$b = \bar{h} = h_0 - \frac{(K-h_0)}{(2K^2-1)\beta}; \quad \beta = \frac{\overline{|\psi|^4}}{(\overline{|\psi|^2})^2}, \tag{7.10.7}$$

while the free energy, given by (7.9.8) is proportional to

$$f_1 = \tfrac{1}{2} + b^2 - \frac{(K-b)^2}{1+(2K^2-1)\beta}. \tag{7.10.8}$$

We see, then, that for given b, to minimize f_1 we must make β as small as possible.

So far, the results have been independent of the specific choice of the c's in (7.10.3). At this stage, however, because we are now concerned with the minimization of β, as defined in (7.10.7), a choice of c_n's (periodic in n) must be made. Strictly speaking, this should be on energy grounds—a complicated problem. But it would appear that the simplest choice, when all the c's are equal, is best. Under these circumstances

$$\beta = \frac{k}{K\sqrt{(2\pi)}}\left\{\sum_n \exp\left(-\frac{k^2}{2K^2}n^2\right)\right\}^2 \qquad (7.10.9)$$

and the minimum β is 1·18 when $k = K\sqrt{(2\pi)}$. Then, writing $b = h+4\pi m$, for external fields just below H_{c2}, (7.10.7) gives

$$-4\pi m = \frac{1}{1{\cdot}18(2K^2-1)}(h_{c2}-h_0). \qquad (7.10.10)$$

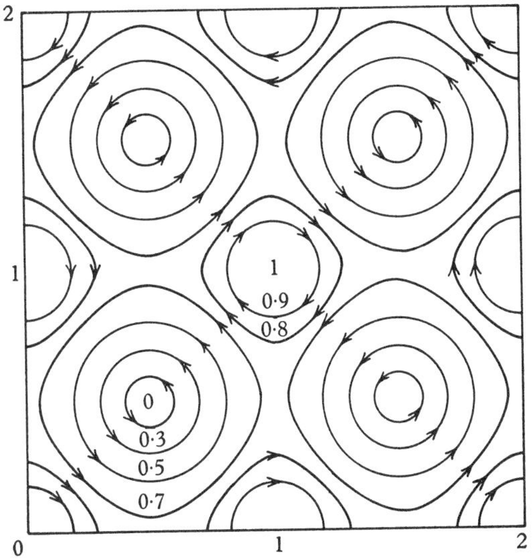

Fig. 7.20

Both in its first power dependence on $(h_{c2}-h_0)$ (see Fig. 7.19), and in its proportionality constant, (7.10.10) is found to be in agreement with experiment. The constant $|\psi|^2$ contours in the xy-plane, appropriate to the solution thus calculated, are shown (suitably normalized) in Fig. 7.20. The arrows indicate the current directions. It is clear that we have a configuration of vortices.

(*b*) *H_0 just greater than H_{c1}*. As H_0 decreases from the situation described above near H_{c2}, the assumption is that the vortices persist, though becoming more and more separated from each other, in qualitative agreement with the magnetization curve shown in

Fig. 7.19. In the limit $H_0 = H_{c1}$, we may consider a completely isolated vortex, at least as a first approximation.

Then, directing **a** perpendicular to the radius vector and writing $\psi = fe^{i\theta}$, (7.9.6) and (7.9.7) reduce to

$$-\frac{1}{K^2 r}\frac{d}{dr}\left(r\frac{df}{dr}\right) + q^2 f = f - f^3, \qquad (7.10.11)$$

$$\frac{d}{dr}\left[\frac{1}{r}\frac{d}{dr}(rq)\right] = qf^2, \qquad (7.10.12)$$

where $q = |\mathbf{a} - K^{-1}\nabla\theta|$. The field is given by

$$h = -\frac{1}{r}\frac{d}{dr}(rq). \qquad (7.10.13)$$

Equations (7.10.11) and (7.10.12) are to be solved under the boundary conditions that f (everywhere finite) $\to 1$ as $r \to \infty$, while $q \sim |K^{-1}\nabla\theta| = 1/Kr$ as $r \to 0$, and $\to 0$ as $r \to \infty$. This final condition means that at large r, $a \sim 1/Kr$ and so the magnetic induction per vortex line is

$$b = \int h\,dS = \oint \mathbf{a}.d\mathbf{l} = \frac{1}{Kr}.2\pi r = \frac{2\pi}{K}. \qquad (7.10.14)$$

The problem is now well defined, but only when $K \gg 1$ does further analytical progress seem possible. In the following, therefore, we will consider only this case. For $r < 1$, we may write q to leading order as $1/Kr$. On inserting this in (7.10.11), and solving in series, we find

$$\left.\begin{aligned} f^2 &= 1 - \frac{1}{K^2r^2} + \ldots \quad (Kr > 1), \\ f &= \text{const. } r + \ldots \quad (Kr < 1). \end{aligned}\right\} \qquad (7.10.15)$$

On returning to (7.9.8) and rearranging a little using (7.10.13) and (7.10.14), the energy per unit length of filament can be written

$$\epsilon = \pi \int_0^\infty \left[(1 - f^4)r - f^2\frac{d}{dr}(r^2q^2)\right] dr. \qquad (7.10.16)$$

Replacing the range of integration by its most important part, which, in view of the above discussion, is $(1/K, 1)$, integration of (7.10.16) gives, in leading order, $(2\pi/K^2)\ln K$. Abrikosov obtained the next order term by numerical methods and found that a more accurate expression is $(2\pi/K^2)(\ln K+0{\cdot}081)$. For a transition from the perfectly diamagnetic state to that with vortex structure, $\epsilon < 2h_{c1}b$. Thus, the transition occurs at the external field value

$$h_{c1} = \frac{1}{2K}(\ln K+0{\cdot}081) \quad (K \gg 1), \qquad (7.10.17)$$

a result that is in reasonable agreement with experiment. Abrikosov goes on to obtain the interaction energies between vortices, thus the equilibrium filament density, and thereby the details of Fig. 7.19 immediately above H_{c1}.

Problems (see also p. 448)

P.7(i). In the B.C.S. wave function, suppose the u's and v's are complex but satisfying

$$u_K = u_{-K}, \quad v_K = -v_{-K}, \quad |u_K|^2+|v_K|^2 = 1. \qquad (1)$$

Furthermore, let us *not* approximate $\mathscr{H}$ by the reduced Hamiltonian of (7.4.1), instead, writing

$$\mathscr{H} = \Sigma\epsilon_K a_K^\dagger a_K + \tfrac{1}{2}\Sigma V(K_1K_2K_1'K_2')\, a_{K_1}^\dagger a_{K_2}^\dagger a_{K_2'} a_{K_1'}. \qquad (2)$$

Following the general procedure of section 7.5.1, introduce the transformations (7.4.18) and (7.4.19) and rewrite $\mathscr{H}$ in the form

$$\mathscr{H} = \epsilon + \Sigma E_K \alpha_K^\dagger \alpha_K + H_{\text{int.}}, \qquad (3)$$

where

$$\epsilon = \sum_K \left\{\epsilon_K + \frac{1}{2}\sum_{K'}[V(KK'KK')-V(KK'K'K)]\,|v_{K'}|^2\right\}|v_K|^2$$
$$+\frac{1}{2}\sum_{KK'} V(K-KK'-K')\,u_K v_K u_{K'} v_{K'}, \qquad (4)$$

$$E_K = \sqrt{(\xi_K^2+|\Delta_K|^2)}, \qquad (5)$$

$$\xi_K = \epsilon_K + \sum_{K'}[V(KK'KK')-V(KK'K'K)]\,|v_{K'}|^2, \qquad (6)$$

$$|v_K|^2 = \frac{1}{2}\left(1-\frac{\xi_K}{E_K}\right), \quad |u_K|^2 = \frac{1}{2}\left(1+\frac{\xi_K}{E_K}\right), \qquad (7)$$

$$H_{\text{int.}} = +\tfrac{1}{2}\Sigma V(K_1K_2K_1'K_2')\, N_\alpha(a_{K_1}^\dagger a_{K_2}^\dagger a_{K_2'} a_{K_1'}), \qquad (8)$$

and where Δ_K satisfies the equation

$$\Delta_K = -\frac{1}{2}\sum_{K'} V(K-KK'-K')\frac{\Delta_{K'}}{E_{K'}}. \qquad (9)$$

CHAPTER 8

MANY-BOSON SYSTEMS

8.1. Introduction

The problem we shall deal with in this chapter is basically to find solutions of the Schrödinger equation which are symmetric with respect to interchange of any pair of co-ordinates for N interacting particles in a volume Ω. This symmetry, of course, is the feature characterizing a Boson assembly.

As in Chapters 5 and 7, our prime concern will be the calculation of the elementary excitation spectra of the various systems considered. When discussing the associated quasi-particles (cf. Chapter 1, section 1.11), we can make a definite statement about their statistics. This is because the Bosons forming the assembly carry an angular momentum which is an integral multiple of $\hbar$. Thus there is no way in which a given excitation can acquire an angular momentum of an odd multiple of $\frac{1}{2}\hbar$. Hence, the quasi-particles of a many-Boson system are themselves Bosons. (One observation relevant to section 8.2 below is that phonon theory applies to atoms or ions which may individually be Fermions, for example, solid ^{3}He. The statistics of the bare particles in that case is inconsequential.) The connection with the low-temperature thermodynamics when only low-lying states are important is now straightforward. For a given excitation spectrum $E(\mathbf{k})$, the occupation of the **k**th state is

$$n_{\mathbf{k}} = \{e^{\beta E(\mathbf{k})} - 1\}^{-1}, \quad \beta = \frac{1}{k_B T}, \tag{8.1.1}$$

and the treatment is then standard. We shall see, in addition, that other important physical properties follow from the shape of $E(\mathbf{k})$.

From the practical point of view it is necessary to distinguish between strong and weak coupling of the original particles. In the latter case one imagines a fluid which can be related (however tortuously) to a system of N independent Bosons. Work on liquid helium proceeds on this assumption and much of this chapter, in particular section 8.3, is concerned with this problem. On the other hand, in the strong coupling case one has a solid, for which it makes no sense to regard a gas of independent particles as the zeroth order approximation. Instead one must start from the other extreme of an array of static particles perturbed with a little kinetic energy. This leads us to the theory of lattice vibrations discussed in section 8.2. It should be remarked that formally the latter theory is very similar to those of black-body radiation and spin-waves, the quasi-particles of the elastic vibrations being phonons, of the electromagnetic field being photons and of the spin-waves being magnons. There are, of course, differences, but not sufficient to warrant separate detailed treatment here. For further information on these topics the reader may consult Heitler (1954) and Mattis (1965) respectively.

8.2. Phonons

The theory of lattice vibrations has been presented in a number of texts (e.g. Born & Huang, 1954; Ziman, 1960). We shall therefore minimize detail in our account, but nevertheless the theory is helpful in the present context as it provides possibly the simplest example of the reduction of a system of strongly interacting particles to one of weakly coupled quasi-particles.

8.2.1. *Collective co-ordinates and harmonic approximation*

To focus attention on aspects related to many-body theory generally, rather than on those pertaining to the individual study of lattice dynamics, we deal only with the special case of a Bravais lattice in which all the N atoms are like. Then, as we consider each atom vibrating about a site in a regular crystal lattice, the mere assumption of stable vibrations enables us to say a good deal.

Thus, by suitable choice of (collective) co-ordinates, sufficiently small vibrations can be described in classical mechanics by $3N$ independent simple harmonic oscillators, one for each degree of freedom. In the quantum description, one might then expect a parallel decomposition into independent quantum oscillators, and this, indeed turns out to be so. In either case, we redescribe the system by normal co-ordinates which, for sufficiently small displacements, vibrate independently.

We denote the lattice points by $\mathbf{a}, \mathbf{b}, \ldots, \mathbf{m}, \ldots$ and the actual atomic positions by $\mathbf{r}_a, \mathbf{r}_b, \ldots, \mathbf{r}_m, \ldots$. The displacement of the **m**th atom from its site is then

$$\mathbf{u}_m = \mathbf{r}_m - \mathbf{m}. \tag{8.2.1}$$

For small displacements, the total potential energy can be expanded in a Taylor series in the displacements, and formally written

$$V(\mathbf{r}_a, \ldots, \mathbf{r}_m, \ldots) = V_0 + V_1 + V_2 + V_3 + \ldots. \tag{8.2.2}$$

As usual, V_0 is an unimportant constant, V_1 is zero, and, provided V_3 and higher terms can be neglected, we may write the total Hamiltonian as

$$H = \frac{1}{2M} \sum_{\mathbf{m}} \mathbf{p}_{\mathbf{m}}^2 + \frac{1}{2} \sum_{\substack{\mathbf{mn} \\ ij}} A_{\mathbf{mn}}^{ij} u_{\mathbf{m}}^i u_{\mathbf{n}}^j, \tag{8.2.3}$$

where u^i denotes the ith component of $\mathbf{u}$,

$$A_{\mathbf{mn}}^{ij} = \left(\frac{\partial^2 V}{\partial u_{\mathbf{m}}^i \, \partial u_{\mathbf{n}}^j} \right)_0, \tag{8.2.4}$$

and M is the atomic mass.

The subscript in (8.2.4) denotes evaluation of the derivative at the equilibrium position. $A_{\mathbf{mn}}^{ij}$ in (8.2.3) satisfies the usual physical conditions for translational invariance, stability, etc.

In the harmonic approximation defined by (8.2.3), the essential problem, classical or quantum, is to diagonalize the potential energy under unitary transformation. The kinetic energy retains its simple form, in the new variables, under such a transformation, and the problem is then trivially solved.

8.2.2. *Classical analysis*

Writing $\mathbf{p}_l = M\dot{\mathbf{u}}_l$ in (8.2.3), the classical equations of motion

$$M\ddot{u}_{\mathbf{m}}^i = -\sum_{\mathbf{n}j} A_{\mathbf{mn}}^{ij} u_{\mathbf{n}}^j \tag{8.2.5}$$

may be solved by a Fourier analysis into progressive waves:

$$\mathbf{u}_{\mathbf{m}} = \frac{1}{\sqrt{N}} \sum_{\mathbf{q}} \mathbf{U}_{\mathbf{q}} e^{-i\mathbf{q}.\mathbf{m}}, \quad \mathbf{U}_{\mathbf{q}} = \frac{1}{\sqrt{N}} \sum_{\mathbf{m}} \mathbf{u}_{\mathbf{m}} e^{i\mathbf{q}.\mathbf{m}}. \tag{8.2.6}$$

On substituting for $\mathbf{u}_{\mathbf{m}}$ in (8.2.5) and comparing components, we find

$$M\ddot{U}_{\mathbf{q}}^i = -\Lambda_{\mathbf{q}} U_{\mathbf{q}}^i, \tag{8.2.7}$$

where $\Lambda_{\mathbf{q}}$ is a 3×3 matrix whose elements are

$$\Lambda_{\mathbf{q}}^{ij} = \sum_{\mathbf{n}} A_{\mathbf{on}}^{ij} e^{-i\mathbf{q}.\mathbf{n}}, \tag{8.2.8}$$

the suffix **o** appearing in the latter because of translational invariance.

Assuming that the $U_{\mathbf{q}}$'s have the form $\mathbf{e}_{\mathbf{q}} e^{i\omega_{\mathbf{q}}t}$, we obtain a decomposition into $3N$ independent linear oscillator equations

$$\ddot{U}_{\mathbf{q}}^{i} = -(\omega_{\mathbf{q}}^{i})^2 U_{\mathbf{q}}^{i}, \tag{8.2.9}$$

where the frequencies $\omega_{\mathbf{q}}^{i}$ and polarization vectors $\mathbf{e}_{\mathbf{q}}^{i}$ are obtained from the eigenequation

$$\Lambda_{\mathbf{q}} \mathbf{e}_{\mathbf{q}}^{i} = M(\omega_{\mathbf{q}}^{i})^2 \mathbf{e}_{\mathbf{q}}^{i}. \tag{8.2.10}$$

Elementary excitation spectrum. As might be expected from the above discussion, the dispersion functions $\omega_{\mathbf{q}}$ represent the elementary excitation spectra of the system (the quantum derivation is considered in the next section). Since the detailed nature of these dispersion curves has been discussed many times (Ziman 1960; Cochran, 1963) we shall summarize briefly here the main points:

(i) The dispersion relation is periodic: that is

$$\omega_{(\mathbf{q}+\mathbf{G})}^{i} = \omega_{\mathbf{q}}^{i}, \tag{8.2.11}$$

where **G** is any vector of the reciprocal lattice, namely a vector such that $e^{i\mathbf{G}.\mathbf{m}} = 1$, where **m** is any direct lattice vector. Thus it is sufficient to define $\omega_{\mathbf{q}}$ in a unit cell of the reciprocal lattice, or in the first Brillouin zone.

(ii) $\omega_{\mathbf{q}}^{i}$ is linear in q for small q. This is a characteristic feature of lattice vibrations, the proportionality constant being the group velocity of propagation of long waves through the lattice, or the speed of sound for the branch of the spectrum under consideration. This feature is crucially different from the zero coupling case, where the energies of the elementary excitations are $\propto q^2$, for small q. This linear law follows, for all q, for an elastic continuum, and illustrates only that the atomic character of the solid is not important for wavelengths large compared with the lattice spacing.

(iii) That $\omega_{\mathbf{q}}^{i}$ curve with greatest slope at the origin is the longitudinal mode, while the other two are the transverse modes.

8.2.3. *Quantum analysis*

Having established the basic method used in classical mechanics, it is a relatively simple matter to carry out the quantum mechanical analysis. We merely return to the harmonic Hamiltonian (8.2.3), and replace the p's by $(\hbar/i)\nabla$. As in the classical case, the object is then to reduce the many-body equation to a set of independent harmonic oscillators.

Motivated by the success of the change of variable (8.2.6), which yielded an eigenvalue equation for the vectors $\mathbf{U_q}$, we perform a Fourier analysis of the displacements once more. Each of the new variables is symmetric with respect to the original co-ordinates $\mathbf{r}_i$ defined by (8.2.1). Any wave function constructed from the $\mathbf{U_q}$ has, therefore, the correct symmetry and we can thus forget about the symmetry requirement in the subsequent analysis. The corresponding momentum operators are readily written down. We have

$$\frac{\hbar}{i}\frac{\partial}{\partial u^i_{\mathbf{m}}} = \frac{\hbar}{i}\sum_{\mathbf{n}j}\frac{\partial U^j_{\mathbf{n}}}{\partial u^i_{\mathbf{m}}}\frac{\partial}{\partial U^j_{\mathbf{n}}} = \frac{\hbar}{i}\frac{1}{\sqrt{N}}\sum_{\mathbf{n}} e^{i\mathbf{n}.\mathbf{m}}\frac{\partial}{\partial U^i_{\mathbf{n}}}, \tag{8.2.12}$$

and so $$\mathbf{p_m} = \frac{1}{\sqrt{N}}\sum_{\mathbf{n}} e^{i\mathbf{n}.\mathbf{m}}\mathbf{P_n}, \quad \mathbf{P_n} = \frac{1}{\sqrt{N}}\sum_{\mathbf{m}} e^{-i\mathbf{m}.\mathbf{n}}\mathbf{p_m}. \tag{8.2.13}$$

Since $\mathbf{u}^*_{\mathbf{l}} = \mathbf{u_l}$, $\mathbf{p}^*_{\mathbf{l}} = -\mathbf{p_l}$ and the reality of the eigenvalues requires similar relations for the $\mathbf{U}$'s and $\mathbf{P}$'s, we have

$$\mathbf{U}^*_{\mathbf{m}} = \mathbf{U}_{-\mathbf{m}}, \quad \mathbf{P}^*_{\mathbf{m}} = \mathbf{P}_{-\mathbf{m}}. \tag{8.2.14}$$

In the new variables, then, we have a kinetic energy of

$$\begin{aligned}\frac{1}{2M}\sum_{\mathbf{m}}\left[\frac{1}{\sqrt{N}}\sum_{\mathbf{n}} e^{i\mathbf{n}.\mathbf{m}}\mathbf{P_n}\right]^2 &= \frac{1}{2MN}\sum_{\mathbf{mnl}} e^{i\mathbf{n}.\mathbf{m}}e^{i\mathbf{l}.\mathbf{m}}\mathbf{P_n}.\mathbf{P_l} \\ &= \frac{1}{2M}\sum_{\mathbf{nl}}\delta_{\mathbf{n},-\mathbf{l}}\mathbf{P_n}.\mathbf{P_l} \\ &= \frac{1}{2M}\sum_{\mathbf{n}}\mathbf{P_n}.\mathbf{P_{-n}},\end{aligned} \tag{8.2.15}$$

while the potential energy becomes

$$\begin{aligned}&\frac{1}{2}\sum_{\mathbf{m}i\mathbf{n}j} A^{ij}_{\mathbf{mn}}\left(\frac{1}{\sqrt{N}}\sum_{\mathbf{q}} U^i_{\mathbf{q}}e^{-i\mathbf{q}.\mathbf{m}}\right)\left(\frac{1}{\sqrt{N}}\sum_{\mathbf{r}} U^j_{\mathbf{r}}e^{-i\mathbf{r}.\mathbf{n}}\right) \\ &= \frac{1}{2N}\sum_{ij\mathbf{qr}}\left(\sum_{\mathbf{mn}} A^{ij}_{\mathbf{mn}}e^{-i(\mathbf{q}.\mathbf{m}+\mathbf{r}.\mathbf{n})}\right)U^i_{\mathbf{q}}U^j_{\mathbf{r}} \\ &= \frac{1}{2N}\sum_{ij\mathbf{qr}}\left(\sum_{\mathbf{n}} e^{-i(\mathbf{q}+\mathbf{r}).\mathbf{n}}\right)\left(\sum_{\mathbf{h}} A^{ij}_{\mathbf{ho}}e^{-i\mathbf{q}.\mathbf{h}}\right)U^i_{\mathbf{q}}U^j_{\mathbf{r}},\end{aligned}$$

where we have used translational invariance again. Thus on using (8.2.8) the potential energy becomes

$$\frac{1}{2N}\sum_{ij\mathbf{q}\mathbf{r}} N\delta_{\mathbf{q},-\mathbf{r}}\Lambda^{ij}_{\mathbf{q}}U^i_{\mathbf{q}}U^j_{\mathbf{r}} = \frac{1}{2}\sum_{ij\mathbf{q}}\Lambda^{ij}_{\mathbf{q}}U^i_{\mathbf{q}}U^j_{-\mathbf{q}}. \tag{8.2.16}$$

Combining (8.2.15) and (8.2.16) we now have

$$H = \sum_{\mathbf{q}} H_{\mathbf{q}}, \tag{8.2.17}$$

where
$$H_{\mathbf{q}} = \frac{1}{2M}P_{\mathbf{q}}P^*_{\mathbf{q}} + \frac{1}{2}\sum_{ij}\Lambda^{ij}_{\mathbf{q}}U^i_{\mathbf{q}}U^{j*}_{\mathbf{q}}. \tag{8.2.18}$$

We have again achieved our aim of reducing the many-body Hamiltonian to a sum of single-particle Hamiltonians. The classical analogue is the reduction of (8.2.5) to the form (8.2.7).

(*a*) *Diagonalization of Hamiltonian.* The final objective is to decompose the motion into a form analogous to that given classically in (8.2.9). The method of doing this as before is to diagonalize the potential energy under unitary transformation, the kinetic energy remaining invariant in form. Let us denote the new variables by $\mathfrak{U}_{\mathbf{q}}$ and the transformation matrix by $Q_{\mathbf{q}}$, so that

$$\mathbf{U}_{\mathbf{q}} = Q_{\mathbf{q}}\mathfrak{U}_{\mathbf{q}}, \quad Q^{*T}_{\mathbf{q}}Q_{\mathbf{q}} = I, \tag{8.2.19}$$

the superscript T denoting the transpose, and I the unit matrix. Then (8.2.10) tells us that the potential energy in (8.2.18) must take the form

$$\tfrac{1}{2}\Sigma M(\omega^i_{\mathbf{q}})^2\,\mathfrak{U}^i_{\mathbf{q}}\mathfrak{U}^{i*}_{\mathbf{q}}, \tag{8.2.20}$$

and that the columns of $Q_{\mathbf{q}}$ are built up from the polarization vectors $\mathbf{e}^i_{\mathbf{q}}$ $(i=1,2,3)$. This final remark means $Q^{ij}=e^{ij}$, both denoting the ith component of $\mathbf{e}^j$.

On the other hand, the new momentum $\mathfrak{P}_{\mathbf{q}}$ is obtained by the same process as was used in (8.2.12). We find $\mathbf{P}_{\mathbf{q}} = Q^*_{\mathbf{q}}\mathfrak{P}_{\mathbf{q}}$ (an unsurprising result if we remember that position and momentum are canonically conjugate) and so the kinetic energy is unaltered in form since

$$\begin{aligned}\mathbf{P}^T_{\mathbf{q}}\mathbf{P}^*_{\mathbf{q}} = (Q^*_{\mathbf{q}}\mathfrak{P}_{\mathbf{q}})^T(Q^*_{\mathbf{q}}\mathfrak{P}_{\mathbf{q}})^* = \mathfrak{P}^T_{\mathbf{q}}Q^{*T}_{\mathbf{q}}Q_{\mathbf{q}}\mathfrak{P}^*_{\mathbf{q}} = \mathfrak{P}^T_{\mathbf{q}}I\mathfrak{P}^*_{\mathbf{q}}\\ = \mathfrak{P}^T_{\mathbf{q}}\mathfrak{P}^*_{\mathbf{q}}.\end{aligned} \tag{8.2.21}$$

The result of (8.2.19) and (8.2.21) is to allow (8.2.18) to be written

$$H_{\mathbf{q}} = \sum_{i=1}^{3} H^i_{\mathbf{q}}, \tag{8.2.22}$$

where
$$H_{\mathbf{q}}^{i} = \frac{1}{2M}\mathfrak{P}_{\mathbf{q}}^{i}\mathfrak{P}_{\mathbf{q}}^{i*} + \tfrac{1}{2}M(\omega_{\mathbf{q}}^{i})^{2}\mathfrak{U}_{\mathbf{q}}^{i}\mathfrak{U}_{\mathbf{q}}^{i*}. \tag{8.2.23}$$

The overall simplification is to express the original total Hamiltonian again as the sum of Hamiltonians corresponding to $3N$ uncoupled one-dimensional oscillators of frequencies $\omega_{\mathbf{q}}^{i}$, namely
$$H = \sum_{\mathbf{q}i} H_{\mathbf{q}}^{i}. \tag{8.2.24}$$

(*b*) *Phonon creation and annihilation operators.* An even more convenient reduction is possible if we introduce operators which 'factorize' (8.2.23) above. These are (Dirac, 1958)
$$\left.\begin{aligned} a_{\mathbf{q}}^{i} &= \frac{1}{(2\hbar M\omega_{\mathbf{q}}^{i})^{\frac{1}{2}}}\mathfrak{P}_{\mathbf{q}}^{i} - i\left(\frac{M\omega_{\mathbf{q}}^{i}}{2\hbar}\right)^{\frac{1}{2}}\mathfrak{U}_{\mathbf{q}}^{i*},\\ a_{\mathbf{q}}^{i*} &= \frac{1}{(2\hbar M\omega_{\mathbf{q}}^{i})^{\frac{1}{2}}}\mathfrak{P}_{\mathbf{q}}^{i*} + i\left(\frac{M\omega_{\mathbf{q}}^{i}}{2\hbar}\right)^{\frac{1}{2}}\mathfrak{U}_{\mathbf{q}}^{i}.\end{aligned}\right\} \tag{8.2.25}$$

The inverse relationships are also useful and are easily written down if we remember that (8.2.14), (8.2.19) and $\mathbf{P}_{\mathbf{q}} = Q_{\mathbf{q}}^{*}\mathfrak{P}_{\mathbf{q}}$ imply
$$\mathfrak{U}_{\mathbf{q}}^{*} = \mathfrak{U}_{-\mathbf{q}}, \quad \mathfrak{P}_{\mathbf{q}}^{*} = \mathfrak{P}_{-\mathbf{q}}, \tag{8.2.26}$$
and $\omega_{\mathbf{q}}^{i}$ is even in $\mathbf{q}$, as may be seen mathematically by appeal to (8.2.8) and (8.2.10). (Physically, of course, one expects reversal of direction of a progressive wave to leave the frequency unaltered.) The resulting inverses are
$$\left.\begin{aligned} \mathfrak{U}_{\mathbf{q}}^{i} &= -i\left(\frac{\hbar}{2M\omega_{\mathbf{q}}^{i}}\right)^{\frac{1}{2}}(a_{\mathbf{q}}^{i*} - a_{-\mathbf{q}}^{i}),\\ \mathfrak{P}_{\mathbf{q}}^{i} &= \left(\frac{\hbar M\omega_{\mathbf{q}}^{i}}{2}\right)^{\frac{1}{2}}(a_{\mathbf{q}}^{i} + a_{-\mathbf{q}}^{i*}).\end{aligned}\right\} \tag{8.2.27}$$

It is a straightforward matter to rewrite (8.2.23) as
$$H_{\mathbf{q}}^{i} = \tfrac{1}{2}\hbar\omega_{\mathbf{q}}^{i}(a_{\mathbf{q}}^{i}a_{\mathbf{q}}^{i*} + a_{\mathbf{q}}^{i*}a_{\mathbf{q}}^{i}), \tag{8.2.28}$$
and to verify the Boson commutation relations
$$[a_{\mathbf{q}}^{i}, a_{\mathbf{p}}^{j*}] = \delta_{\mathbf{qp}}\delta_{ij}, \tag{8.2.29}$$
all other commutators vanishing. Combining (8.2.23), (8.2.24), (8.2.28) and (8.2.29) we finally write the full Hamiltonian in the simple form
$$H = \sum_{\mathbf{q},i}\hbar\omega_{\mathbf{q}}^{i}(a_{\mathbf{q}}^{i*}a_{\mathbf{q}}^{i} + \tfrac{1}{2}). \tag{8.2.30}$$

The problem has now been reduced to the elementary situation discussed in section 8.1 (cf. also Chapter 1, section 1.11). Equation (8.2.30) represents a set of independent quasi-particles—phonons. In the ground state, no phonons are present and the total energy is

$$\frac{1}{2}\sum_{\mathbf{q},i} \hbar\omega_{\mathbf{q}}^{i},$$

the familiar zero-point energy of the oscillators. If a phonon of wave vector $\mathbf{q}$ and polarization i is created, the corresponding excitation energy is

$$E_{\mathbf{q}} = \hbar\omega_{\mathbf{q}}^{i}.$$

Thus, the dispersion curve $\omega_{\mathbf{q}}$ represents the excitation spectrum.

In the harmonic approximation the phonons are independent quasi-particles. But, of course, in practice, the neglected anharmonic terms, V_3, etc., ... of (8.2.2) lead to a phonon-phonon interaction, which we shall discuss briefly in section 8.2.4. First, however, we turn to the consideration of a very powerful technique, the theory of which is largely due to Van Hove (1954, 1961), which allows the direct measurement of elementary excitation spectra by inelastic neutron scattering. An amplified recent account is that of Kittel (1963).

In anticipation of these discussions, we will summarize the effect of transforming our original p's and u's to a's and a^*'s (via the P's and U's and $\mathfrak{P}$'s and $\mathfrak{U}$'s). Using (8.2.6), (8.2.19) and (8.2.27), we find

$$u_{\mathbf{m}}^{i} = -i\sum_{\mathbf{q},j}\left(\frac{\hbar}{2MN\omega_{\mathbf{q}}^{j}}\right)^{\frac{1}{2}} e_{\mathbf{q}}^{ij}\, e^{-i\mathbf{q}\,.\,\mathbf{m}}(a_{\mathbf{q}}^{j*} - a_{-\mathbf{q}}^{j}), \qquad (8.2.31)$$

where the e's appear as a result of the remark following (8.2.20). It should be noted that, despite outward appearances, $u_{\mathbf{m}}^{i}$ is real, as may be directly tested by changing the sign of the summation variable $\mathbf{q}$, forming u_{m}^{i*} and using the facts that the e's are odd and the ω's even in $\mathbf{q}$. The Heisenberg form of (8.2.31) is (recall (4.7.2) and (4.7.3))

$$u_{\mathbf{m}}^{i}(t) = -i\sum_{\mathbf{q},j}\left(\frac{\hbar}{2MN\omega_{\mathbf{q}}^{j}}\right)^{\frac{1}{2}} e_{\mathbf{q}}^{ij}[\exp\{-i(\mathbf{q}\,.\,\mathbf{m} - \omega_{\mathbf{q}}^{j}t)\}\, a_{\mathbf{q}}^{j*} + \exp\{i(\mathbf{q}\,.\,\mathbf{m} + \omega_{\mathbf{q}}^{j}t)\}\, a_{\mathbf{q}}^{j}], \qquad (8.2.32)$$

where, in changing the sign of the suffix in the second summation, we have once more used the symmetries of the e's and the ω's.

(*c*) *Van Hove correlation functions and neutron cross-sections.* We begin with the expression

$$\frac{d^2\sigma}{d\Omega\, d\epsilon} = \frac{k'}{k}\left(\frac{m}{2\pi\hbar^2}\right)^2 \sum_{i,f} p_i |\langle \mathbf{k}'f|V|\mathbf{k}i\rangle|^2 \delta(\omega + \epsilon_i - \epsilon_f) \quad (8.2.33)$$

for the differential cross-section per unit solid angle and unit outgoing energy range of scattered neutrons. The explanation of the various terms are as follows. The composite neutron-target wave function before impact is $|i\mathbf{k}\rangle = |i\rangle e^{i\mathbf{k}.\mathbf{r}}$ while afterwards it is $|f\mathbf{k}'\rangle = |f\rangle e^{i\mathbf{k}'.\mathbf{r}}$. Born approximation is valid (Breit, 1947), and so the usual 'golden rule' applies with matrix element $\langle \mathbf{k}'f|V|\mathbf{k}i\rangle$, where V is the potential seen by the neutron. The delta function expresses the requirement that the energy loss $\omega = (\hbar^2/2m)(k^2 - k'^2)$ of the neutron is the energy gain $\epsilon_f - \epsilon_i$ of the target. The p's are Boltzmann factors specifying the thermal distribution of the initial states, while the factors outside the summation arise from density of final states and incident flux factors. (See, for example, Schiff (1955), Chapter 8.)

Now let us use the identity

$$\delta(\omega) = (2\pi)^{-1}\int_{-\infty}^{\infty} e^{-i\omega t}\, dt$$

and (confining ourselves to non-magnetic materials) the Fermi pseudopotential form

$$V = (2\pi\hbar^2/m)\, b \sum_{\mathbf{m}} \delta(\mathbf{r} - \mathbf{r}_{\mathbf{m}}). \quad (8.2.34)$$

(In general, b should carry a suffix $\mathbf{m}$, even for crystals of one chemical element, since it may be spin- or isotope-dependent. The appropriate modification to meet this situation will be indicated presently.) Then (8.2.33) becomes

$$\frac{d^2\sigma}{d\Omega\, d\epsilon} = b^2 \frac{k'}{k} \sum_{i,f,\mathbf{m},\mathbf{n}} p_i \langle i|\exp\{-i\mathbf{K}.\mathbf{r}_{\mathbf{m}}\}|f\rangle\langle f|\exp\{i\mathbf{K}.\mathbf{r}_{\mathbf{n}}\}|i\rangle$$
$$\times \frac{1}{2\pi}\int_{-\infty}^{\infty} \exp\{-i(\omega + \epsilon_i - \epsilon_f)t\}\, dt, \quad (8.2.35)$$

where $\mathbf{K} = \mathbf{k} - \mathbf{k}'$, the momentum loss of the neutron.

The next step is to introduce Heisenberg time-dependent operators (cf. (4.7.2), for example). Observing that

$$\langle f|\exp\{i\mathbf{K}.\mathbf{r}_{\mathbf{n}}(t)\}|i\rangle = \langle f|\exp\{i\mathbf{K}.\mathbf{r}_{\mathbf{n}}\}(t)|i\rangle$$
$$= \langle f|\exp\{i\mathbf{K}.\mathbf{r}_{\mathbf{n}}\}|i\rangle \exp\{i(\epsilon_f - \epsilon_i)t\}, \quad (8.2.36)$$

we may drop the $e^{-i(\epsilon_i-\epsilon_f)t}$ from (8.2.35) and replace $\mathbf{r}_\mathbf{n}$ by $\mathbf{r}_\mathbf{n}(t)$. Then, summing over f gives

$$\frac{d^2\sigma}{d\Omega\, d\epsilon} = Nb^2 \frac{k'}{k} S(\mathbf{K}, \omega), \tag{8.2.37}$$

where

$$S(\mathbf{K}, \omega) = \frac{1}{2\pi}\int_{-\infty}^{\infty} dt \exp\{-i\omega t\} \frac{1}{N} \sum_{\mathbf{m},\mathbf{n}} \langle \exp\{-i\mathbf{K}.\mathbf{r}_\mathbf{m}(0)\} \times \exp\{i\mathbf{K}.\mathbf{r}_\mathbf{n}(t)\}\rangle_T, \tag{8.2.38}$$

the thermodynamic average being defined by the relationship

$$\langle O\rangle_T = \sum_i p_i \langle i|O|i\rangle. \tag{8.2.39}$$

The function $S(\mathbf{K}, \omega)$ is of great utility, as we will see. It has, also, general properties of a rather basic kind. Let us, then, before moving on to the practical problem of the energy spectrum, pause and discuss some of these latter properties, as they will be of considerable subsequent use to us.

First, from (8.2.38), the Fourier transform of $S(\mathbf{K}, \omega)$ is‡

$$\begin{aligned} S(\mathbf{K}, t) &= \frac{1}{N}\sum_{\mathbf{m},\mathbf{n}} \langle \exp\{-i\mathbf{K}.\mathbf{r}_\mathbf{m}(0)\} \exp\{i\mathbf{K}.\mathbf{r}_\mathbf{n}(t)\}\rangle_T \\ &= \frac{1}{N}\langle \rho_\mathbf{K}(0)\rho_{-\mathbf{K}}(t)\rangle_T, \end{aligned} \tag{8.2.40}$$

where ρ_K is the usual Fourier transform of the number density (see, for example, (5.5.2) or below). The transform from $\mathbf{K}$ to $\mathbf{r}$-space can also be performed if we wish. Using (8.2.40), we have

$$S(\mathbf{r}, t) = \frac{1}{N}\left\langle \sum_{\mathbf{m},\mathbf{n}} \frac{1}{(2\pi)^3}\int \exp\{-i\mathbf{K}.\mathbf{r}\} \exp\{-i\mathbf{K}.\mathbf{r}_\mathbf{m}(0)\} \times \exp\{i\mathbf{K}.\mathbf{r}_\mathbf{n}(t)\}\, d\mathbf{K}\right\rangle_T. \tag{8.2.41}$$

We must be careful with this integral because, when $t \neq 0$, the exponents are not additive. Nevertheless, the convolution theorem for Fourier transforms of operator products enables us to write it as

$$S(\mathbf{r}, t) = \frac{1}{N}\left\langle \sum_{\mathbf{m},\mathbf{n}} \int d\mathbf{r}'\, \delta(\mathbf{r}+\mathbf{r}_\mathbf{m}(0)-\mathbf{r}')\, \delta(\mathbf{r}'-\mathbf{r}_\mathbf{n}(t))\right\rangle_T. \tag{8.2.42}$$

‡ We follow Van Hove and depart here from the standard inversion procedure of writing each element in $(\mathbf{K}, \omega)$ space with a 2π in the denominator. The purpose is to make $S(\mathbf{r}, t)$ of (8.2.42) asymptotically equal to the number density.

Of especial interest are the values of (8.2.40) and (8.2.42) at $t = 0$. In the former case, we have

$$S(\mathbf{K}) \equiv \int_{-\infty}^{\infty} S(\mathbf{K}, \omega)\, d\omega = \frac{1}{N}\langle \rho_{\mathbf{K}}\rho_{-\mathbf{K}}\rangle_T, \qquad (8.2.43)$$

and this is the structure factor introduced in section 5.6, while in the latter case, the position operators commute and we obtain

$$S(\mathbf{r}) \equiv \int S(\mathbf{K})\, e^{i\mathbf{K}.\mathbf{r}}\, d\mathbf{K} = \frac{1}{N}\sum_{\mathbf{m},\mathbf{n}} \langle \delta(\mathbf{r}-\mathbf{r}_{\mathbf{m}}+\mathbf{r}_{\mathbf{n}})\rangle_T. \qquad (8.2.44)$$

The $\mathbf{m} = \mathbf{n}$ terms contribute a delta function at the origin and the remainder defines the usual asymptotically normalized pair distribution function, $g(r)$. Specifically (8.2.44) becomes

$$S(\mathbf{r}) = \delta(\mathbf{r}) + \rho g(\mathbf{r}). \qquad (8.2.45)$$

Finally, let us observe that under certain conditions (8.2.37) can be integrated with respect to energy. When the incident neutron energy is large compared with $\hbar\omega$, for fixed scattering direction, $\hbar$K does not depend on the final neutron energy. Thus, using (8.2.43) and (8.2.45), the differential cross-section per unit solid angle is

$$\frac{d\sigma}{d\Omega} = \int \frac{d^2\sigma}{d\Omega\, d\mathcal{E}}\, d\mathcal{E} = Nb^2\hbar\frac{k'}{k}\int_{-\infty}^{\infty} S(\mathbf{K}, \omega)\, d\omega = Nb^2\hbar\frac{k'}{k} S(\mathbf{K}), \qquad (8.2.46)$$

where, by (8.2.45),

$$S(\mathbf{K}) = 1 + \rho\int g(\mathbf{r})\, e^{i\mathbf{K}.\mathbf{r}}\, d\mathbf{r} = 1 + \rho\int [g(\mathbf{r}) - 1]\, e^{i\mathbf{K}.\mathbf{r}}\, d\mathbf{r}. \qquad (8.2.47)$$

The last step, here, follows since the Fourier transform of unity contributes an unimportant delta function at $\mathbf{K} = 0$. In liquids (see below) it is to be preferred since (by construction in (8.2.45)) $g(r) \to 1$ as $r \to \infty$. Equations (8.2.46) and (8.2.47) are used in practice for the experimental determination of $g(\mathbf{r})$.

It should be emphasized that the above discussion is applicable to any array of scattering centres—not just solids, and in particular a number of these results will be applied to liquids later in this chapter. It is also true that the formalism holds for types of radiation other than neutrons (X-rays, for instance). However, neutrons of wavelength comparable to interatomic distances in solids and of energies commensurate with typical excitation energies are conveniently produced by reactors. In this way, a significant region of $(\mathbf{K}, \omega)$ space can be explored experimentally.

In the following, we develop the formalism further for the specific case of solids and show how it may be used in the experimental determination of excitation spectra.

(*d*) $S(\mathbf{K}, \omega)$ *in solids*. The important point about (8.2.37) is that the right-hand side contains the rather basic function $S(\mathbf{K}, \omega)$, which is amenable to systematic theoretical investigation, as we shall see below, while the left-hand side is directly accessible to experiment. Actually, in practice, when the b's carry suffices (recall remarks following (8.2.34)), (8.2.37) remains essentially the same, b being replaced by a mean value $\langle b \rangle$, and this quantity (the coherent part), is distinguishable from a virtually isotropic background (the incoherent part) which is proportional to $\langle b^2 \rangle - \langle b \rangle^2$. Methods of separating these two components from the raw data exist (see, for example, Egelstaff (1963)), but this need not be done for the purpose of determining excitation spectra, as the one-phonon processes in the coherent part which describe these excitations are clearly visible in the composite picture.

Thus, let us consider, in detail, the part of the integrand in (8.2.41), given by

$$
\begin{aligned}
&\langle \exp\{-i\mathbf{K}.\mathbf{r}_{\mathrm{m}}(0)\} \exp\{i\mathbf{K}.\mathbf{r}_{\mathrm{n}}(t)\}\rangle_T \\
&\quad = \exp\{-i\mathbf{K}.(\mathbf{m}-\mathbf{n})\} \langle \exp\{-i\mathbf{K}.\mathbf{u}_{\mathrm{m}}(0)\} \exp\{i\mathbf{K}.\mathbf{u}_{\mathrm{n}}(t)\}\rangle_T \\
&\quad = \exp\{-i\mathbf{K}.(\mathbf{m}-\mathbf{n})\} \langle \exp\{-i\mathbf{K}.[\mathbf{u}_{\mathrm{m}}(0)-\mathbf{u}_{\mathrm{n}}(t)]\}\rangle_T \\
&\qquad\qquad \times \exp\{\tfrac{1}{2}[\mathbf{K}.\mathbf{u}_{\mathrm{m}}(0), \mathbf{K}.\mathbf{u}_{\mathrm{n}}(t)]\} \\
&\quad = \exp\{-i\mathbf{K}.(\mathbf{m}-\mathbf{n})\} \exp\{-\tfrac{1}{2}\langle[\mathbf{K}.\mathbf{u}_{\mathrm{m}}(0)-\mathbf{K}.\mathbf{u}_{\mathrm{n}}(t)]^2\rangle\}_T \\
&\qquad\qquad \times \exp\{\tfrac{1}{2}[\mathbf{K}.\mathbf{u}_{\mathrm{m}}(0), \mathbf{K}.\mathbf{u}_{\mathrm{n}}(t)]\},
\end{aligned}
\tag{8.2.48}
$$

where each step in this calculation requires a little explanation. The first step arises on using (8.2.1), the idea being to focus attention on the $\mathbf{u}(t)$'s which are defined by (8.2.32). The next stage results from a theorem (Messiah, 1961, p. 442) that if operators A and B both commute with their commutator, then $e^A e^B = e^{(A+B)} e^{\frac{1}{2}[A,B]}$. If $A = -i\mathbf{K}.\mathbf{u}_{\mathrm{m}}(0)$ and $B = i\mathbf{K}.\mathbf{u}_{\mathrm{n}}(t)$, the conditions are more than fulfilled. In fact the commutator is readily seen to be a simple function (see (8.2.29) and the subsequent remark). The final step in (8.2.48) is obtained using another result given by Messiah (p. 449) which states that, for a single oscillator,

$$\langle e^{i\xi}\rangle_T = \exp\{-\tfrac{1}{2}\langle\xi^2\rangle_T\}$$

provided ξ is a linear combination of the appropriate a and a^*. Since we are, after all, concerned with $3N$ such oscillators, this result is evidently applicable in (8.2.48).

Clearly, it is now necessary to evaluate the exponents in (8.2.48). In order to do so, we use (8.2.32) to find (Glauber, 1955)

$$\begin{aligned}\langle u^i_{\mathbf{m}}(0)\,u^j_{\mathbf{n}}(t)\rangle_T = \sum_{\mathbf{q},k}\frac{\hbar}{2MN\omega^k_{\mathbf{q}}}e^{ik}_{\mathbf{q}}\,e^{jk}_{\mathbf{q}}\\ \times[\exp\{-i[\mathbf{q}.(\mathbf{m}-\mathbf{n})+\omega^k_{\mathbf{q}}t]\}\langle a^{k*}_{\mathbf{q}}\,a^k_{\mathbf{q}}\rangle_T\\ +\exp\{i[\mathbf{q}.(\mathbf{m}-\mathbf{n})+\omega^k_{\mathbf{q}}t]\}\langle a^k_{\mathbf{q}}\,a^{k*}_{\mathbf{q}}\rangle_T].\end{aligned} \quad (8.2.49)$$

Here, because of (8.2.29) and (8.2.39)

$$\langle a^{k*}_{\mathbf{q}}\,a^k_{\mathbf{q}}\rangle_T = n^k_{\mathbf{q}}, \quad \langle a^k_{\mathbf{q}}\,a^{k*}_{\mathbf{q}}\rangle_T = n^k_{\mathbf{q}}+1, \quad (8.2.50)$$

where

$$n^k_{\mathbf{q}} = [\exp\{\hbar\omega^k_{\mathbf{q}}/k_B T\}-1]^{-1}. \quad (8.2.51)$$

If there were no correlation between the displacements at different lattice sites, such u-products as appear in (8.2.49) would average to zero.

This result needs to be generalized a little further. Writing $\mathbf{K} = (K^1, K^2, K^3)$, and invoking (8.2.49), we obtain

$$\begin{aligned}&\langle\{\mathbf{K}.\mathbf{u}_{\mathbf{m}}(0)\}\{\mathbf{K}.\mathbf{u}_{\mathbf{n}}(t)\}\rangle_T\\ &= \sum_{i,j}K^iK^j\langle\mathbf{u}^{i*}_{\mathbf{m}}(0)\,\mathbf{u}^j_{\mathbf{n}}(t)\rangle_T\\ &= \sum_{\mathbf{q},k}\frac{\hbar}{2MN\omega^k_{\mathbf{q}}}(\mathbf{K}.\mathbf{e}^k_{\mathbf{q}})^2[\exp\{-i[\mathbf{q}.(\mathbf{m}-\mathbf{n})+\omega^k_{\mathbf{q}}t]\}n^k_{\mathbf{q}}\\ &\qquad +\exp\{i[\mathbf{q}.(\mathbf{m}-\mathbf{n})+\omega^k_{\mathbf{q}}t]\}(n^k_{\mathbf{q}}+1)].\end{aligned} \quad (8.2.52)$$

This equation can now be used to evaluate (8.2.48).

Taking the commutator first and utilizing the fact that it is clearly a c-number, we have

$$\begin{aligned}[\mathbf{K}.\mathbf{u}_{\mathbf{m}}(0), \mathbf{K}.\mathbf{u}_{\mathbf{n}}(t)] = \langle\{\mathbf{K}.\mathbf{u}_{\mathbf{m}}(0)\}\{\mathbf{K}.\mathbf{u}_{\mathbf{n}}(t)\}\rangle_T\\ -\langle\{\mathbf{K}.\mathbf{u}_{\mathbf{n}}(t)\}\{\mathbf{K}.\mathbf{u}_{\mathbf{m}}(0)\}\rangle_T,\end{aligned} \quad (8.2.53)$$

while the other unknown exponent is

$$\langle\{\mathbf{K}.\mathbf{u}_{\mathrm{m}}(0)-\mathbf{K}.\mathbf{u}_{\mathrm{n}}(t)\}^2\rangle_T$$
$$=\langle\{\mathbf{K}.\mathbf{u}_{\mathrm{m}}(0)\}^2\rangle_T-\langle\{\mathbf{K}.\mathbf{u}_{\mathrm{m}}(0)\}\{\mathbf{K}.\mathbf{u}_{\mathrm{n}}(t)\}\rangle_T$$
$$-\langle\{\mathbf{K}.\mathbf{u}_{\mathrm{n}}(t)\}\{\mathbf{K}.\mathbf{u}_{\mathrm{m}}(0)\}\rangle_T+\langle\{\mathbf{K}.\mathbf{u}_{\mathrm{n}}(t)\}^2\rangle_T. \quad (8.2.54)$$

The final term here is actually time independent, as explicit substitution for $\mathbf{u}_{\mathrm{n}}(t)$ in the form $e^{iH_0t}\mathbf{u}_{\mathrm{n}}e^{-iH_0t}$ shows. Thus we can replace the t there by zero. Also, because of (8.2.52), terms one and four in (8.2.54) are equal and independent of their lattice position suffices. Hence, substitution of (8.2.53) and (8.2.54) into (8.2.48) gives

$$\langle\exp\{-i\mathbf{K}.\mathbf{r}_{\mathrm{m}}(0)\}\exp\{i\mathbf{K}.\mathbf{r}_{\mathrm{n}}(t)\}\rangle_T$$
$$=\exp\{-i\mathbf{K}.(\mathbf{m}-\mathbf{n})\}\exp\{-2W\}\exp\{\langle[\mathbf{K}.\mathbf{u}_{\mathrm{m}}(0)][\mathbf{K}.\mathbf{u}_{\mathrm{n}}(t)]\rangle_T\}, \quad (8.2.55)$$

where

$$2W=\langle\{\mathbf{K}.\mathbf{u}_{\mathrm{m}}(0)\}^2\rangle_T=\sum_{\mathbf{q},k}\frac{\hbar}{2MN\omega_{\mathbf{q}}^k}(\mathbf{K}.\mathbf{e}_{\mathbf{q}}^k)^2(2n_{\mathbf{q}}^k+1). \quad (8.2.56)$$

The term e^{-2W} depends on temperature and is called the Debye–Waller factor.

Now the fluctuating function (8.2.52) will be small compared with (8.2.56). Thus, it is proper to expand the final exponential term in (8.2.55). On applying (8.2.38), the various orders in the expansion give rise to contributions to $S(\mathbf{K},\omega)$ which have clear physical interpretations. Below it will be sufficient for present purposes to consider the first two orders only.

In zeroth order, we find

$$S_0(\mathbf{K},\omega)=\frac{1}{2\pi}\int_{-\infty}^{\infty}dt\,e^{-i\omega t}\sum_{\mathbf{m},\mathbf{n}}e^{-i\mathbf{K}.(\mathbf{m}-\mathbf{n})}e^{-2W}. \quad (8.2.57)$$

Integration over time yields $2\pi\delta(\omega)$, implying elastic scattering, while use of the identity (Ziman, 1960)

$$\sum_{\mathbf{m},\mathbf{n}}e^{-i\mathbf{K}.(\mathbf{m}-\mathbf{n})}=(2\pi)^3\Omega\sum_{\mathbf{G}}\delta(\mathbf{K}-\mathbf{G}), \quad (8.2.58)$$

appropriate to the present case of a monatomic lattice, reveals a Bragg scattering law. The sole effect of temperature is to diminish the heights of the peaks from their rigid lattice value by a factor e^{-2W}.

In first order, we have

$$S_1(\mathbf{K},\omega) = \frac{1}{2\pi}\int_{-\infty}^{\infty} dt\, e^{-i\omega t} \sum_{\mathbf{m},\mathbf{n}} e^{-i\mathbf{K}.(\mathbf{m}-\mathbf{n})} e^{-2W} \langle\{\mathbf{K}.\mathbf{u}_\mathbf{m}(0)\}\{\mathbf{K}.\mathbf{u}_\mathbf{n}(t)\}\rangle_T, \tag{8.2.59}$$

where the thermal average is given by (8.2.44). Thus, once more integrating over time and applying (8.2.51), we obtain

$$S_1(\mathbf{K},\omega) = e^{-2W} \sum_{\mathbf{q},k} \frac{\hbar}{2MN\omega_\mathbf{q}^k} (\mathbf{K}.\mathbf{e}_\mathbf{q}^k)^2 [\delta(\omega+\omega_\mathbf{q}^k)\,\delta(\mathbf{K}+\mathbf{q}-\mathbf{G})\, n_\mathbf{q}^k + \delta(\omega-\omega_\mathbf{q}^k)\,\delta(\mathbf{K}-\mathbf{q}+\mathbf{G})\,(n_\mathbf{q}^k+1)]. \tag{8.2.60}$$

The interpretation of this result in terms of phonon creation and annihilation is clear. We see that the contribution to the differential cross-section (8.2.37) of the second part of (8.2.60) arises from those neutron energy losses $\omega = \omega_\mathbf{q}^k$ compatible with corresponding momentum losses of $\mathbf{q}-\mathbf{G}$. Similarly, the first part is explained in terms of phonon destruction.

Clearly, the higher terms in $S(\mathbf{K},\omega)$ represent multiphonon processes but we will not pursue the theory further in detail. The point we wish to reiterate is that it provides us, through (8.2.37), with a rather direct means of observing excitation spectra. We do not enter into practical details, but the essential feature is that for fixed $\mathbf{K}$ (and, thus, fixed scattering angle) a plot of the inelastic spectrum versus ω reveals one or possibly more sharp peaks, due to the one-phonon processes of (8.2.60), against a continuous multi-phonon and incoherent background. Further information is given by Van Hove and co-workers (1954, 1961) and Vols. I and II of *Inelastic Scattering of Neutrons from Solids and Liquids* (International Atomic Energy Agency, Vienna, 1963). The similarity with the determination of plasmon spectra using charged particle probes (Chapter 5, section 12) will be noted.

8.2.4. *Phonon-phonon interactions*

As we have seen, the harmonic approximation leads to a system of independent phonons. For many situations, and in particular for low-temperature thermodynamics when few phonons are excited, the approximation is a good one and very properly the subject has received much attention. On the other hand, there are certain situations when the inclusion of anharmonicity is crucial. This is the case, for example, when the temperature is very high and many phonons are produced. There is nothing in the harmonic approximation, for instance, to account for thermal expansion and melting, the explanations for these having to be sought in the anharmonic terms. Another situation when such forces must be included is in transport theory. Heat current resistance in a perfect crystal depends on phonon-phonon scattering which is absent in the harmonic approximation. An infinite thermal conductivity is thus erroneously predicted. The situation can be improved, however, by taking $\sum_{3}^{\infty} V_n$ of (8.2.2) as a perturbation.

Let us begin by considering V_3, which may be written in the form

$$V_3 = \frac{1}{3!} \sum_{\substack{\mathbf{mnl} \\ ijk}} B^{ijk}_{\mathbf{mnl}}\, u^i_{\mathbf{m}}\, u^j_{\mathbf{n}}\, u^k_{\mathbf{l}}. \tag{8.2.61}$$

This term is the direct analogue of V_2 as defined through (8.2.2), (8.2.3) and (8.2.4) and is obtained by the same expansion technique.

Substituting for u from (8.2.31) we find

$$V_3 = \frac{(-i)^3}{3!}\left(\frac{\hbar}{2MN}\right)^{\frac{3}{2}} \sum_{\substack{\mathbf{m'n'l'} \\ i'j'k'}} \frac{B\binom{i'j'k'}{\mathbf{m'n'l'}}}{(\omega^{i'}_{\mathbf{m'}}\, \omega^{j'}_{\mathbf{n'}}\, \omega^{k'}_{\mathbf{l'}})^{\frac{1}{2}}}$$

$$\times (a^{i'*}_{\mathbf{m'}} - a^{i'}_{-\mathbf{m'}})(a^{j'*}_{\mathbf{n'}} - a^{j'}_{-\mathbf{n'}})(a^{k'*}_{\mathbf{l'}} - a^{k'}_{-\mathbf{l'}}), \tag{8.2.62}$$

where (once more, in accordance with the remark following (8.2.20), replacing Q by e)

$$B\binom{i'j'k'}{\mathbf{m'n'l'}} = \sum_{\substack{\mathbf{mnl} \\ ijk}} B^{ijk}_{\mathbf{mnl}}\, e^{ii'}_{\mathbf{m'}}\, e^{jj'}_{\mathbf{n'}}\, e^{kk'}_{\mathbf{l'}}\, e^{-i(\mathbf{m'.m}+\mathbf{n'.n}+\mathbf{l'.l})}. \tag{8.2.63}$$

There are two important points to stress concerning V_3. First, the periodicity of the lattice implies that (8.2.63) and thus (8.2.62) vanish unless the 'pseudo-momentum' conservation law

$$\mathbf{m'} + \mathbf{n'} + \mathbf{l'} = \mathbf{G} \tag{8.2.64}$$

is satisfied, where $\mathbf{G}$ is a reciprocal lattice vector. If $\mathbf{G} = 0$, we have a normal process; otherwise we have an Umklapp process (see

Ziman, 1964). Secondly, from (8.2.62), we see that if we multiply out the creation and annihilation operators, various triple products arise which can be represented graphically as in Fig. 8.1.

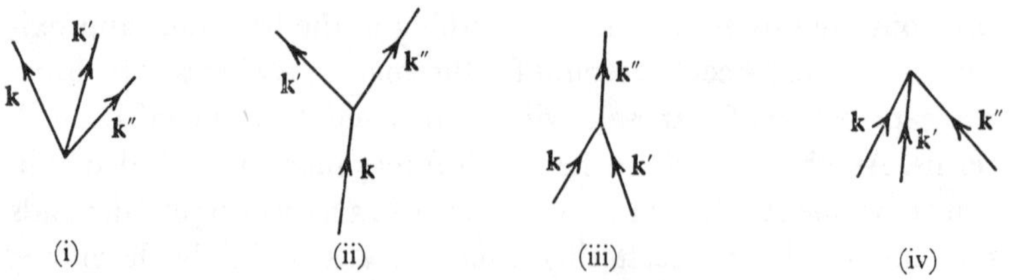

(i) Creation of three phonons. (ii) One phonon splits into two. (iii) Two phonons merge into one. (iv) Annihilation of three phonons.

Fig. 8.1. $\mathbf{k}+\mathbf{k}'+\mathbf{k}'' = \mathbf{G}$.

The above discussion is easily generalized to deal with $V_4, V_5, \ldots$. Pseudomomentum conservation laws similar to (8.2.64) and graphs analogous to those of Fig. 8.1 may be readily written down for the 4, 5, ...-phonon processes which thus arise. Thus, we are in a position to consider the totality of these effects by field-theoretical methods. For example, in Fig. 8.2 we have a graph of vertex order two, showing a phonon splitting into two, which later recombine. Self-energy bubbles like this can be included in any graph to raise vertex order, and it will become clear in Chapter 10 that the total effect of all such bubble insertions is to cause renormalization of the single-particle energies. This, of course, is already taken into account if one uses the observed $\omega_{\mathbf{q}}$ dispersion relationship. For systematic attacks on this problem the reader should consult Van Hove, Hugenholtz & Howland (1961), Carruthers (1961) and Kokkedee (1962, 1963).

Fig. 8.2

This concludes, then, our discussion of phonons in crystals. We turn next to the more difficult problem of phonon modes in fluids.

8.2.5. *Phonons in classical liquids*

The macroscopic theory of sound in gases and liquids is well-known. Sound is propagated by longitudinal density (or pressure) variations of long wavelength with speed $(dP/d\rho)^{\frac{1}{2}}$, where P is the pressure at density ρ. Because the concept of shear largely disappears in the case of fluids, there is nothing to correspond to the well-defined transverse modes that exist in a solid (certainly below some critical frequency).

We now address ourselves to the corresponding microscopic classical theory. This is readily understood in a qualitative way. Relative transverse displacements of neighbouring atoms invoke little in the way of forces tending to restore the configuration to its original position, and this implies that shear waves, as remarked above, are rather poorly defined entities. On the other hand, in longitudinal displacements, local departures from uniformity produce local pressure gradients. Any given atom is struck preferentially from the high-pressure side, thus acquiring a net momentum towards a nearby low-density region. The overall effect is to create a piling up of atoms in regions formerly of low density and vice versa. The process repeats itself and we have stable oscillations.

To put these arguments on a quantitative basis presents obstacles not met in the case of solids. There, one could expand the potential energy in terms of small displacement parameters. This approach is, of course, no longer open to us. Nevertheless, we know that sound waves occur in fluids and they correspond to density fluctuations. This suggests a line of attack. We write down, below, the microscopic expression for an elementary density fluctuation and show that it satisfies a classical simple harmonic equation within a certain (random phase) approximation. The method is that introduced by Bohm & Pines (1953) in their analysis of plasma waves in a classical electron gas. For simplicity, we assume initially that the interaction potential can be Fourier analysed, the essential features thereby being brought out. The extension to the physically more realistic case is indicated later.

(*a*) *Equation of motion of density fluctuations.* We take the Hamiltonian

$$H = \tfrac{1}{2}\Sigma M\mathbf{v}_i^2 + \sum_{i<j} V(\mathbf{r}_i - \mathbf{r}_j), \tag{8.2.65}$$

for which the equations of motion are

$$M\dot{\mathbf{v}}_i = -\sum_{j\neq i} \nabla_i(\mathbf{r}_i - \mathbf{r}_j) = -\frac{\mathbf{k}}{\Omega}\sum_{i\neq j}\sum_{\mathbf{k}} i\,V(\mathbf{k})\,e^{i\mathbf{k}.(\mathbf{r}_i-\mathbf{r}_j)}. \tag{8.2.66}$$

In accordance with the procedure outlined above, we write down the expression for the number density

$$\rho(\mathbf{r}) = \sum_i \delta(\mathbf{r}-\mathbf{r}_i) \tag{8.2.67}$$

and take its Fourier transform

$$\rho_{\mathbf{k}} = \int \rho(\mathbf{r})\, e^{-i\mathbf{k}.\mathbf{r}}\, d\mathbf{r} = \sum_i e^{-i\mathbf{k}.\mathbf{r}_i}. \tag{8.2.68}$$

Differentiation with respect to time then gives

$$\dot{\rho}_{\mathbf{k}} = -i \sum_i (\mathbf{k}.\mathbf{v}_i)\, e^{-i\mathbf{k}.\mathbf{r}_i} \tag{8.2.69}$$

and

$$\ddot{\rho}_{\mathbf{k}} = -\sum_i (\mathbf{k}.\mathbf{v}_i)^2 e^{-i\mathbf{k}.\mathbf{r}_i} - i \sum_i \mathbf{k}.\dot{\mathbf{v}}_i e^{-i\mathbf{k}.\mathbf{r}_i}. \tag{8.2.70}$$

The equation of motion (8.2.66) is now used to eliminate $\dot{\mathbf{v}}_i$ from (8.2.70) to obtain

$$\ddot{\rho}_{\mathbf{k}} = -\sum_i (\mathbf{k}.\mathbf{v}_i)^2 e^{-i\mathbf{k}.\mathbf{r}_i} - \frac{1}{M\Omega} \sum_{\mathbf{k}'} \mathbf{k}.\mathbf{k}'\, V(\mathbf{k}') \sum_i e^{i(\mathbf{k}'-\mathbf{k}).\mathbf{r}_i} \sum_j e^{-i\mathbf{k}'.\mathbf{r}_j}. \tag{8.2.71}$$

Now is the appropriate time to approximate. Clearly, the difficulty lies in the final part of (8.2.71). The simplification comes about by examining the various contributions to the sum over $\mathbf{k}'$. For a roughly random distribution of particle co-ordinates (as occurs in a fluid but not in a solid), provided $\mathbf{k} \neq \mathbf{0}$, the individual terms of $\rho_{\mathbf{k}}$ are scattered more or less randomly about zero. Thus $\rho_{\mathbf{k}}$ is a small quantity if $\mathbf{k} \neq \mathbf{0}$. But the final term of (8.2.71) contains products of the form $\rho_{\mathbf{k}'-\mathbf{k}}\rho_{\mathbf{k}'}$. These, in comparison with $\rho_{\mathbf{k}}$, are negligible, except for the special cases when $\mathbf{k}'-\mathbf{k}$ or $\mathbf{k}'$ vanish. The latter case does not matter, because of the presence of the multiplying factor $\mathbf{k}.\mathbf{k}'$ in the whole expression. We are, thus, left with the single term corresponding to $\mathbf{k}' = \mathbf{k}$ which must be retained in (8.2.71). This is the random phase approximation.

The first term of (8.2.71), present even when there are no interactions between the particles, can also be put in a more convenient form using the following results:

(i) For any fixed $\mathbf{k}$, there are many vectors $\mathbf{r}_i$ for which $e^{-i\mathbf{k}.\mathbf{r}_i}$ is equal to any chosen constant (of modulus not exceeding unity).

(ii) To such a subset of many $\mathbf{r}_i$, it is proper to assign the velocity distribution for the total system.

(iii) For any chosen velocity $\mathbf{v}_i$ within the subset, randomness means that all orientations θ_i with respect to $\mathbf{k}$ are equally likely. Thus we have

$$\Sigma' \mathbf{v}_i \cos^2 \theta_i e^{-i\mathbf{k}.\mathbf{r}_i} = \tfrac{1}{3}\overline{v^2}\Sigma' e^{-i\mathbf{k}.\mathbf{r}_i}, \tag{8.2.72}$$

where Σ' denotes a sum over the kind of subset indicated in (i), and $\overline{v^2}$ is the mean square velocity in the system. The factor $\frac{1}{3}$ arises from summing over angles for fixed velocity using (iii). Also in (8.2.72) we have replaced a sum of squares by the mean, multiplied by the number of terms, this being possible since the exponential factor is a constant for the subset. If we now sum (8.2.72) over all subsets, we finally obtain

$$\sum_i (\mathbf{k}.\mathbf{v}_i)^2 e^{-i\mathbf{k}.\mathbf{r}_i} = \tfrac{1}{3}k^2\overline{v^2}\rho_{\mathbf{k}}. \tag{8.2.73}$$

Combining (8.2.73) with the random phase approximation, (8.2.71) may now be written

$$\ddot{\rho}_{\mathbf{k}} = -\omega_{\mathbf{k}}^2 \rho_{\mathbf{k}}; \quad \omega_{\mathbf{k}}^2 = k^2\left[\tfrac{1}{3}\overline{v^2} + \frac{\rho}{M} V(\mathbf{k})\right]. \tag{8.2.74}$$

It should be emphasized that ρ in (8.2.74) is the macroscopic average density N/Ω, and is not to be confused with the local density of (8.2.67). Equation (8.2.74) is the oscillator equation which we had anticipated. At long wavelengths it yields the characteristic linear phonon-type spectrum

$$\omega_{\mathbf{k}} \sim ck; \quad c = \left[\tfrac{1}{3}\overline{v^2} + \frac{\rho}{M} V(0)\right]^{\frac{1}{2}}, \tag{8.2.75}$$

c being the velocity of sound.

Our result is in essential agreement with the classical derivation of c from the equation of state. There, perturbation theory for small V, not, however, precluding hard cores, gives

$$c^2 = \frac{1}{\beta M}\left[1 + \rho\int_0^\infty (1 - e^{-\beta V(r)})\, 4\pi r^2\, dr + \ldots\right]; \quad \beta = \frac{1}{k_B T}. \tag{8.2.76}$$

For soft cores, one can expand the exponential to obtain

$$c^2 = \frac{k_B T}{M} + \frac{\rho}{M} V(0) + \ldots, \tag{8.2.77}$$

in agreement with (8.2.75). From this, it would seem plausible that a more accurate analysis of (8.2.71) would lead to the higher-order terms in (8.2.77) necessary to obtain (8.2.76). This would then be a further example of the important technique used earlier (cf.

Chapters 5 and 6) of summing an infinite series of divergent terms to produce a convergent result. In this case one could suppose a given hard core were softened by some limiting technique, the terms evaluated and summed, and finally the hard core limit taken.

(*b*) *Limits of validity of classical theory.* This is as far as we shall go with the classical theory, except to inquire to what extent classical theory applies to real liquids. This may be decided by the following rough argument. Let us consider an atom (molecule) moving about in the liquid with a mean free path of order $\rho^{-\frac{1}{3}}$. Conditions for that particle are much as though it were confined to move in a cube of side $\rho^{-\frac{1}{3}}$, and hence, the energy levels should be spaced at about $h^2\rho^{\frac{2}{3}}/M$ apart. If this separation is very small with respect to $k_B T$, the energy levels can be supposed continuous and classical physics should apply. Generally speaking, this criterion holds with fair accuracy for most liquids. There are exceptions, however, these being the low-temperature liquids of light atoms and molecules such as hydrogen, neon and helium. The latter is the most spectacular case since it remains a liquid (at one atmosphere pressure) down to $T = 0$, strong quantum effects thus necessarily coming into play. The ^{3}He isotope yields the Fermion system dealt with in Chapter 5; ^{4}He corresponds to Bose statistics. The study of the latter problem occupies most of the remainder of this chapter.

(*c*) $S(\mathbf{K}, \omega)$ *in liquids.* As we saw in section 8.2.3 (*c*) the structure factor $S(\mathbf{K})$ of the liquid, measured by X-ray or neutron experiments, is related to the Van Hove correlation function $S(\mathbf{K}, \omega)$ by

$$S(\mathbf{K}) = \int_{-\infty}^{\infty} S(\mathbf{K}, \omega)\, d\omega. \tag{8.2.78}$$

Recently, information about the forces between atoms or ions in simple classical fluids like argon or sodium has been forthcoming (see Johnson, Hutchinson & March, 1964), and a problem for the future is the practical calculation of $S(\mathbf{K}, \omega)$ from the two-body potential curve $V(r)$. This will be very different from, and probably more difficult than, the case of solids (dealt with in section 8.2.3 (*d*)), where one was able to use expansions in terms of the small displacements from lattice sites. A general theory for soft core functions (which $V(r)$, here, is not!) is described in terms of Green functions in Chapter 10, sections 8–10 inclusive.

Nevertheless, certain general results for the energy (ω) moments of $S(\mathbf{K}, \omega)$ may be obtained, provided only that the two-body potential is velocity independent (though some doubts have been cast on this in real liquids by Randolph, 1964). Thus, classically, the moment theorem of $S(\mathbf{K}, \omega)$ derived in Chapter 5 becomes

$$\overline{\omega^2} = \int_{-\infty}^{\infty} \omega^2 S(\mathbf{K}, \omega)\, d\omega = \frac{k_B T K^2}{M} \tag{8.2.79}$$

and
$$\overline{\omega^4} = \int_{-\infty}^{\infty} \omega^4 S(\mathbf{K}, \omega)\, d\omega$$

$$= \frac{K^4 k_B T}{M^2}\left(3k_B T + \int d\mathbf{r}\, g(r)\left[\frac{1-\cos Kx}{K^2}\right]\frac{\partial^2 V}{\partial x^2}\right), \tag{8.2.80}$$

where $g(r)$ is the pair function in the fluid.

We shall not prove the latter result, but refer to the original work of Placzek (1952) or the simpler derivation of de Gennes (1959).

Furthermore, as follows from density fluctuation arguments,

$$S(0) = k_B T \rho K_T, \tag{8.2.81}$$

where K_T is the isothermal compressibility.

Until recently, it has been supposed that $S(K)$ had a Taylor expansion in K^2 of the form

$$S(K) = S(0) + s_1 K^2 + s_2 K^4 + \dots. \tag{8.2.82}$$

Such an expansion implies that $g(r)$ falls off exponentially at large r, but, more recently, Enderby, Gaskell & March (1965) have pointed out that:

(i) In liquids like argon, where Van der Waals forces operate, in the region away from the critical point the correlation function $g(r)-1$ has the same range as the forces, that is r^{-6}, and hence $S(K)$ contains a K^3 term (cf. 8.3.47 below). Thus $S(K)$ is not analytic at the origin, contrary to the form of (8.2.82). This appears to suggest that many of the classical models will not be good approximations, based as they are on functions of K^2 (cf. Kadanoff & Martin, 1963).

(ii) In liquids like sodium, which are metallic, the electron screening theory of Chapter 5, section 5.2, leads to

$$V(r) \sim \cos 2k_f r / r^3$$

(or, more generally, modified by a phase factor) and this reflects itself again in the form of the pair function $g(r)$. This time $S(K)$ is not analytic at $K = 2k_f$ and the form is (cf. section 5.2)

$$S(K) = S(2k_f) + \text{const.}\,(K - 2k_f) \ln |K - 2k_f|. \quad (8.2.83)$$

Having summarized these rather general arguments for classical fluids, we deal finally with a simple treatment due to Feynman & Cohen (1956), which leads to the results (8.2.79) and (8.2.81). It is intimately connected with Feynman's theory of liquid helium, as we shall see below. The basic idea is that when we are concerned with disturbances of long wavelength, the liquid may be treated as a compressible continuum. Then, as before, we can use the density fluctuations introduced earlier in this section as normal co-ordinates. Explicitly, if the number density $\rho(r)$ and its Fourier transform are defined as previously in (8.2.67) and (8.2.68), in random-phase approximation (cf. (8.2.74) and (8.2.75)), we have

$$E = \frac{1}{2} \sum_{\mathbf{K}} \frac{M}{NK^2} [\dot{\rho}_{\mathbf{K}} \dot{\rho}_{\mathbf{K}}^* + \omega_{\mathbf{K}}^2 \rho_{\mathbf{K}} \rho_{\mathbf{K}}^*]. \quad (8.2.84)$$

Now, as we saw in section 8.2.3 (*c*), $S(K)$ is just $1/N$ times the expectation value of $|\rho_{\mathbf{K}}|^2$. Since the average values of the potential and kinetic energies are equal for a harmonic oscillator, it follows that

$$S(K) = \langle E_{\mathbf{K}} \rangle / Mc^2, \quad (8.2.85)$$

where $\langle E_{\mathbf{K}} \rangle$ is the average energy of the oscillator representing sound of wave number **K**.

On quantizing (8.2.84), at temperature T, the oscillator representing phonons of wave number **K** may be in its nth excited state E_n with probability proportional to $e^{-E_n/k_B T}$. Thus with $\beta = 1/k_B T$, and using the Planck formula for the average energy $\langle E_{\mathbf{K}} \rangle$, namely

$$\langle E_{\mathbf{K}} \rangle = \tfrac{1}{2}\hbar\omega_{\mathbf{K}} + \frac{\hbar\omega_{\mathbf{K}}}{e^{\beta\hbar\omega_{\mathbf{K}}} - 1}$$

$$= \tfrac{1}{2}\hbar c K \coth \frac{\beta\hbar c K}{2}, \quad (8.2.86)$$

and hence

$$S(K) = \frac{\hbar K}{2Mc} \coth \frac{\beta\hbar c K}{2}. \quad (8.2.87)$$

We shall return to this result in section 8.3, but we note immediately that, for small x, coth $x \sim 1/x$ and hence from (8.2.87)

$$\operatorname*{Lt}_{K\to 0} S(K) = \frac{1}{\beta Mc^2}. \qquad (8.2.88)$$

Thus, relating the velocity of sound to the compressibility, we have (8.2.81).

Secondly, although the theory is a long wavelength approximation, we can consider the Van Hove correlation function by noting that in this model,

$$\rho_{\mathbf{K}}(t) = \rho_{\mathbf{K}}(0)\, e^{i\omega_{\mathbf{K}} t}, \qquad (8.2.89)$$

or (8.2.89) with $\omega_{\mathbf{K}}$ replaced by $-\omega_{\mathbf{K}}$, are solutions of the Heisenberg equations of motions (Jones & March, to be published). Hence, since

$$S(\mathbf{K}, t) = \frac{1}{N}\langle \rho_{\mathbf{K}}(0)\, \rho_{-\mathbf{K}}(t)\rangle, \qquad (8.2.90)$$

we can linearly combine the above solutions and generalize (8.2.87) to

$$S(\mathbf{K}, \omega) = \frac{\hbar K}{2Mc} \coth \frac{\beta\hbar cK}{2} [a\delta(\omega - cK) + b\delta(\omega + cK)], \quad (8.2.91)$$

where, to satisfy (8.2.78), we have $a+b = 1$. If we invoke detailed balance, that is $S(\mathbf{K}, \omega) = e^{\hbar\beta\omega} S(\mathbf{K}, -\omega)$, then

$$a = \frac{-e^{\beta\hbar\omega}}{1-e^{\beta\hbar\omega}}; \quad b = \frac{1}{1-e^{\beta\hbar\omega}}. \qquad (8.2.92)$$

If we take the second moment of (8.2.91) with respect to ω, then we find, in the limit $\hbar \to 0$ or $\beta \to 0$,

$$\int_{-\infty}^{\infty} \omega^2\, S(\mathbf{K}, \omega)\, d\omega = \frac{K^2}{\beta M}. \qquad (8.2.93)$$

Thus, we arrive at the result (8.2.79), under such classical conditions, though we should reiterate that the theory is valid only for small K. In particular it cannot be expected to yield the correct limiting form $S(K) \to 1$ as $K \to \infty$.

8.3. Liquid ^{4}He

We now come to 'weakly' coupled many-Boson theory as defined in section 8.1. Here, the discussion of the interacting system is linked with the behaviour of a set of quite independent Bosons. It is therefore necessary to examine the properties of the latter system before proceeding further.

8.3.1. *Ideal Bose–Einstein gas*

In this ideal case, the particles do not interact and the Schrödinger equation is

$$T\Phi \equiv -\frac{\hbar^2}{2M}\sum_1^N \nabla_i^2 \Phi = E\Phi. \tag{8.3.1}$$

The normalized solutions of the latter are readily written down. One has

$$\Phi = (N_1^! N_2^! \ldots /N^!)^{\frac{1}{2}} S\phi_{\mathbf{k}_1}(\mathbf{r}_1)\, \phi_{\mathbf{k}_2}(\mathbf{r}_2) \ldots \phi_{\mathbf{k}_N}(\mathbf{r}_N);$$

$$\phi_{\mathbf{k}}(\mathbf{r}) = \frac{1}{\sqrt{\Omega}} e^{i\mathbf{k}.\mathbf{r}}, \tag{8.3.2}$$

where S denotes that the product is symmetrized by summing over all distinct permutations of the $\mathbf{r}_i$ (or alternatively, the $\mathbf{k}_i$, as this amounts to the same thing), and N_i is the number of indices equal to $\mathbf{k}_i$. Correspondingly the energy E is a sum of single-particle energies $E_{\mathbf{k}} = \hbar^2 k^2/2M$, that is

$$E = \sum_{i=1}^N E_{\mathbf{k}_i}. \tag{8.3.3}$$

The ground state is that for which (8.3.3) is a minimum, namely that for which all $\mathbf{k}_i$ vanish. Every particle is static. Excited states are formed by allowing non-zero momentum indices, the excitation energy being the total energy as given by (8.3.3).

If $a_{\mathbf{k}}^\dagger$ and $a_{\mathbf{k}}$ are the second quantized creation and annihilation operators referred to the single-particle levels $\phi_{\mathbf{k}}(\mathbf{r})$, the kinetic energy operator is

$$T = \sum_{\mathbf{k}} E_{\mathbf{k}} a_{\mathbf{k}}^\dagger a_{\mathbf{k}}. \tag{8.3.4}$$

It is clear, of course, from what has already been said, that $E_{\mathbf{k}}$ is the elementary excitation spectrum for the system in the sense of section 1.11. To formalize the matter, so as to connect closely with

(1.11.1), all one has to do is to observe that since $E_0 = 0$, the total Hamiltonian as given by (8.3.4) is

$$T = \sum_{\mathbf{k} \neq 0} E_{\mathbf{k}} a_{\mathbf{k}}^{\dagger} a_{\mathbf{k}}. \tag{8.3.5}$$

Thus E and $H_{\text{q.p. int.}}$ of (1.11.1) are zero while $H_{\text{q.p.}}$ is given by the right-hand side of (8.3.5). A quasi-particle is an excited particle of the Bose gas and the quasi-particles may appear or disappear one at a time. The momenta and energies of elementary excitations are additive in the sense of section 1.11. The free-particle parabolic form of excitation spectrum is shown in Fig. 8.3. A point to note for

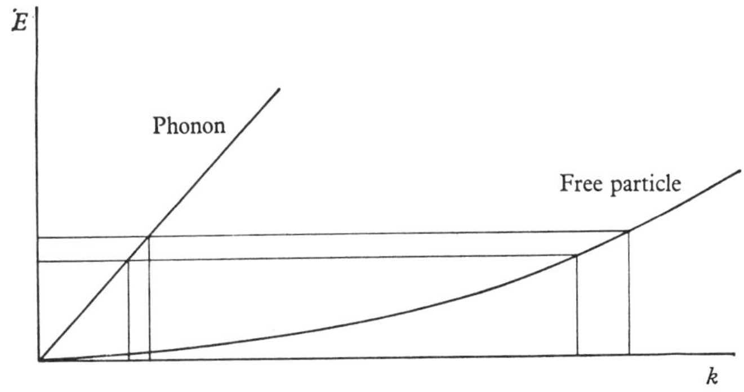

Fig. 8.3. Excitation spectrum for phonons and free particles.

future use is the abundance of low-lying excited states for any given energy range for such a single-particle spectrum, where the slope at the origin is zero. This is in contrast with the linear phonon case (cf. section 8.2) where the number of possible low-lying excited states is fewer.

It is also convenient to record at this stage, for future reference, the relation

$$a_{\mathbf{k}}^{\dagger} a_0 \Phi \propto \sum_i e^{i\mathbf{k}.\mathbf{r}_i} \Phi \quad (\mathbf{k} \neq 0), \tag{8.3.6}$$

which is readily verified to hold when Φ is given by (8.3.2). When $\Phi = 1$, the ground-state wave function, (8.3.6), of course, represents an elementary excitation of a particle from zero momentum to state **k**.

Low-temperature thermodynamics. At zero temperature all particles are in the zero-momentum state. It turns out that as the temperature is raised, this state is more and more depleted, until finally at some degeneracy temperature, T_0, it is no longer macroscopically occupied. This all seems intuitively reasonable. The matter is put on a quantitative basis using (8.1.1) to specify quasi-particle populations. In the standard way, the total number of quasi-particles is

$$N_{\text{q.p.}} = \sum_{\mathbf{k}} n_{\mathbf{k}} = \int_0^\infty \frac{g(E)}{e^{\beta E}-1}\,dE, \tag{8.3.7}$$

while the total energy is

$$E_{\text{q.p.}} = \sum_{\mathbf{k}} E_{\mathbf{k}} n_{\mathbf{k}} = \int_0^\infty \frac{Eg(E)}{e^{\beta E}-1}\,dE, \tag{8.3.8}$$

where $g(E)\,dE$ is the number of allowed states in the energy range $(E, E+dE)$. In the present case when $E_{\mathbf{k}} = \hbar^2k^2/2M$, we have

$$g(E) = \lambda E^{\frac{1}{2}}; \quad \lambda = \frac{2\pi}{h^3}\Omega(2M)^{\frac{3}{2}}. \tag{8.3.9}$$

Use of (8.3.9) in (8.3.7) enables one to write

$$N_{\text{q.p.}} = \lambda(k_B T)^{\frac{3}{2}} \int_0^\infty \frac{x^{\frac{1}{2}}\,dx}{e^x - 1}, \tag{8.3.10}$$

and the degeneracy temperature, T_0, is defined by $N_{\text{q.p.}} = N$. In terms of T_0, (8.3.10) becomes

$$N_{\text{q.p.}} \equiv N(T/T_0)^{\frac{3}{2}}, \tag{8.3.11}$$

and by simple subtraction the occupation of the ground state is

$$N - N_{\text{q.p.}} = N[1-(T/T_0)^{\frac{3}{2}}]. \tag{8.3.12}$$

This is shown pictorially in Fig. 8.4.

We see that as the temperature is lowered through T_0, the ground-state occupation becomes macroscopic. This piling up of particles in the zero-momentum state is the Bose–Einstein condensation phenomenon and the state with zero momentum is sometimes called the condensate.

Combining (8.3.9) with (8.3.8) we obtain

$$E_{\text{q.p.}} = \lambda(k_B T)^{\frac{5}{2}} \int_0^\infty \frac{x^{\frac{3}{2}}\,dx}{e^x - 1}. \tag{8.3.13}$$

The various thermodynamic properties are thus accessible and, for example, the specific heat law takes the form $C_v \sim T^{\frac{3}{2}}$. It should

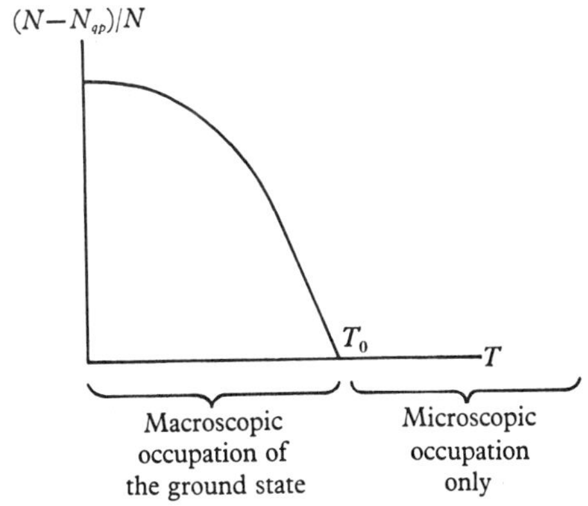

Fig. 8.4. Occupation of ground state for Bose gas.

be noted that the dependence of C_v on T gives information about the density of available low-lying energy states. For instance, if $E_{\mathbf{k}}$ had been of linear phonon form and thus with fewer available states (Fig. 8.3) the specific heat would have been of Debye T^3 form.

Above T_0, the simple concept of the creation of quasi-particles no longer applies, as no more quasi-particles can be created (from a now fully depleted condensate). The appropriate procedure is to replace (8.1.1) by

$$n_{\mathbf{k}} = \{e^{\beta[E(\mathbf{k})-\mu]} - 1\}^{-1}, \tag{8.3.14}$$

where μ, the chemical potential, is obtained by satisfying the normalization requirement

$$\sum_{\mathbf{k}} n_{\mathbf{k}} = N. \tag{8.3.15}$$

In this way the specific heat curve of Fig. 8.5 can be obtained by the usual standard techniques (cf. the treatment of Fermions in section 5.11.1).

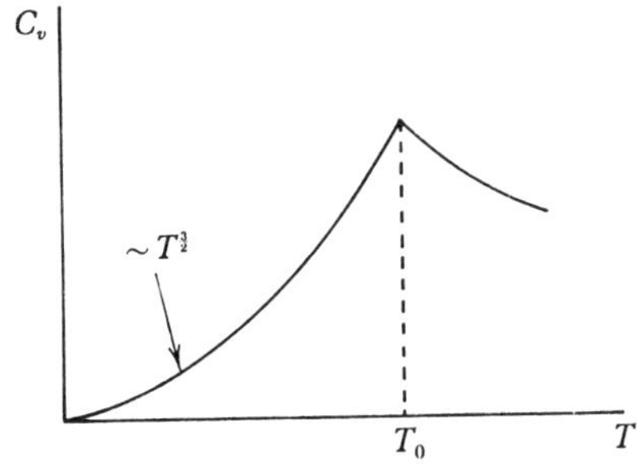

Fig. 8.5. Specific heat curve.

8.3.2. *Introduction of interactions*

We must now inquire what happens when interactions are introduced into the perfect Bose gas. The Hamiltonian then takes the form

$$H = -\frac{\hbar^2}{2M}\sum_i \nabla_i^2 + \sum_{i<j} V(\mathbf{r}_{ij}). \tag{8.3.16}$$

If V can be Fourier transformed, then (8.3.16) can be alternatively written

$$H = \sum_{\mathbf{k}} E_{\mathbf{k}} a_{\mathbf{k}}^{\dagger} a_{\mathbf{k}} + \tfrac{1}{2}\sum \langle \mathbf{k}_1 \mathbf{k}_2 | V | \mathbf{k}_1' \mathbf{k}_2' \rangle a_{\mathbf{k}_1}^{\dagger} a_{\mathbf{k}_2}^{\dagger} a_{\mathbf{k}_2'} a_{\mathbf{k}_1'} \tag{8.3.17}$$

and, as a matter of fact, even in the hard core case, at least under certain circumstances, the problem can be recast in this form (Huang, 1963).

The interaction we have switched on has the effect of exciting particles out of the condensate until, at infinite time, a steady-state balance is achieved. Then, for any chosen V, it is not clear how far the condensate is depleted. One expects, and assumes, that for sufficiently small perturbations, the condensate is macroscopically occupied by analogy with temperature excitation. Even for the large interactions between helium atoms in liquid He II, the evidence points to this (see, for example, Huang, 1963, p. 379 and Penrose & Onsager, 1956).

Thus, when the need arises below (in section 8.3.7) we shall take the occupation of the zero momentum state to be $O(N)$ and that of the others to be $O(1)$. It should be remarked that this is not the only possibility. Luban (1962) has drawn attention to the possible occurrence of a 'smeared' Bose–Einstein condensation such that in some neighbourhood of the origin in **k**-space, the average occupation of the **k**th state is $O(N^x)$, where x is a **k**-dependent number between 0 and 1. In so far as it has been investigated, such an assumption appears to have entirely reasonable consequences.

8.3.3. *Basic phenomenology*

At this point, it will be convenient to summarize some of the properties of helium.

(i) The potential curve for the interaction between He atoms is as shown in Fig. 8.6, in the form calculated by Slater & Kirkwood

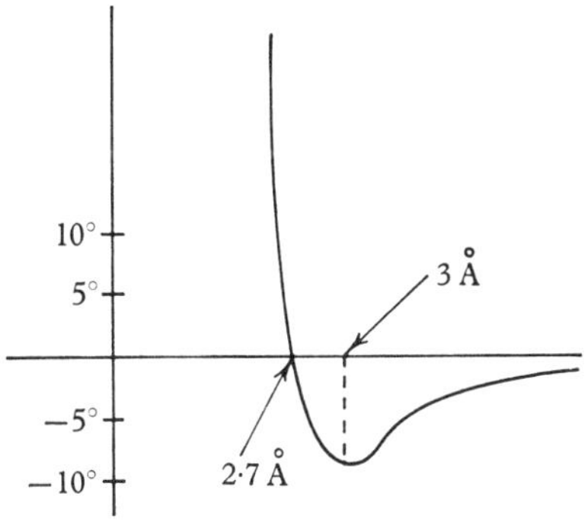

Fig. 8.6. Potential energy of interaction between He atoms.

(1931). Its essential characteristics are: (*a*) A hard repulsive core corresponding to an atomic diameter of about 2·7 Å. (*b*) A weakly attractive Van der Waals tail.

(ii) Helium gas liquefies at 4·2 °K and then remains liquid down to the lowest temperatures studied (for pressures up to ~ 25 atmospheres). The reasons for this exceptional behaviour are: (*a*) The small atomic mass, which results in a large zero-point energy. (*b*) The relatively weak Van der Waals attraction, which, combined with (*a*), makes any lattice arrangement unstable.

It turns out that other possible candidates do not fulfil both the necessary requirements. Thus, the other inert gases have weak Van der Waals interactions, but are too massive. Hydrogen, on the other hand, satisfies (*a*) but the intermolecular forces are too strong.

As a consequence of (*a*) and (*b*), liquid helium has a large specific volume (~ 46 Å^3/atom) corresponding to an average interatomic spacing of about 3·6 Å.

(iii) As the liquid is cooled, a remarkable transition takes place at 2·2 °K. This change is dramatically visible as the hitherto bubbling liquid suddenly becomes quiescent. The variation of the specific heat through the transition or λ-point, at temperature T_λ, is shown in Fig. 8.7. The transition is doubtlessly linked with the Bose–Einstein condensation of section 8.3.1, and, for example, the T_0 of that section, using the mass and density appropriate to ^{4}He, turns out to be 3·2°, compared with the measured T_λ of 2·2°. Furthermore, the Fermi liquid ^{3}He does not exhibit such a transition.

(*a*) *Liquid* He II. He I, the liquid at temperatures above T_λ, is a

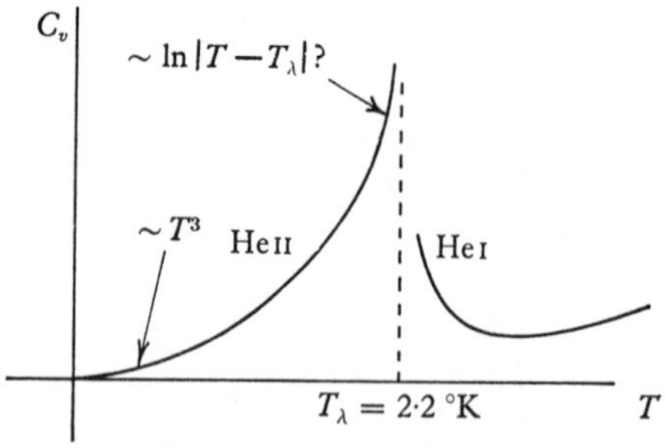

Fig. 8.7. Schematic dependence of C_v on T for liquid helium.

fairly normal liquid and our object below is to explain some of the low-temperature properties of the static (or at least very slowly moving) fluid He II formed below T_λ. By restricting ourselves in this way, we avoid two difficult and, as yet, unsettled problems; namely, the thermodynamics near the λ-point and the excitations exhibited by the fluid in motion (see Brout, 1963, and Feynman, 1955, for views on these matters). Also, it would be inappropriate to dwell at length on the various macroscopic properties of He II, and we shall simply supplement the above discussion with the following ruthless summary:

(i) He II exhibits superfluidity. This implies that there is no resistance to flow through narrow channels. If, on the other hand, the viscosity is measured by rotating a disc in the fluid, a (small) non-zero value is obtained.

(ii) He II has an exceptionally large thermal conductivity. The rate of heat transport is not governed by the usual temperature gradient theory and is so large that it can reasonably be described as a thermal superconductor (Lynton, 1959). It is this property which explains (iii) above. Thus, in He I, heat is carried away by the bubbling process just as in any other normal liquid near its boiling point. As soon as He II appears, this mechanism is no longer necessary.

(iii) Well-defined temperature waves may be propagated through the fluid. These are, in all essential respects, analogous to ordinary sound (pressure) waves. This phenomenon of second sound was predicted theoretically by Landau (see, for example, F. London, 1961) and subsequently confirmed experimentally by Peshkov (1946).

8.3.4. *Two-fluid model and Landau spectrum*

The above experimental facts can be explained using the two-fluid model of London, Tisza & Landau (see, for example, F. London, 1961). Fundamentally, we can view this model as based on the concept of elementary excitations and quasi-particles discussed in Chapter 1, section 1.11. We have already observed that as the temperature is increased from absolute zero, a quasi-particle 'gas' is generated. In this way, we are led to the idea of two interpenetrating fluids, a background component (which we shall see to be superfluid) and a normal component of quasi-particles.

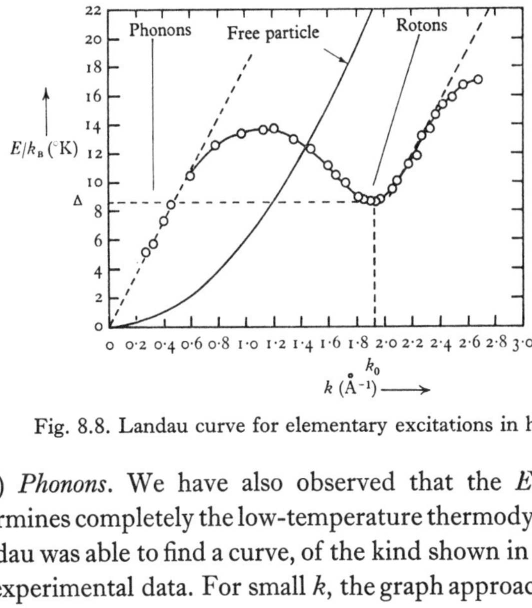

Fig. 8.8. Landau curve for elementary excitations in helium.

(*a*) *Phonons*. We have also observed that the $E(\mathbf{k})$ spectrum determines completely the low-temperature thermodynamics. Thus Landau was able to find a curve, of the kind shown in Fig. 8.8, to fit the experimental data. For small k, the graph approaches the linear form

$$E = ck, \tag{8.3.18}$$

where c is the velocity of sound in liquid helium. Clearly, in this region, the quasi-particles are just the phonons expected if one recalls the classical theory of section 8.2.5. Being lowest in energy, these phonons dominate at low temperatures and explain the observed Debye T^3 specific-heat law observed below about 0·5 °K.

(*b*) *Rotons*. Nevertheless, phonons alone are unable to explain the larger specific heats encountered as temperatures are raised to, say, about 1 °K. This led Landau to postulate a further kind of excitation, the roton, requiring a certain minimum energy Δ for its existence. Thus rotons can occur in significant numbers only above about 0·5 °K. At higher temperatures, the rotons largely dictate the thermodynamics (Landau & Lifshitz, 1958, p. 201). In the region of the roton minimum, the elementary excitation spectrum can be described by the equation

$$E = \Delta + \frac{(k-k_0)^2}{2\mu}. \tag{8.3.19}$$

From our preceding work in section 8.2.5, we might feel we have some intuitive understanding of phonons in the present context. We expect them to arise from a quantization of the elementary density fluctuations. However, in the case of the rotons, we have no theory to fall back on at this stage. In fact, this matter is not yet understood, though many important qualitative ideas have been suggested by Feynman (1955) and Feynman & Cohen (1956). To satisfy immediate curiosity, it should be stated that Feynman speculates that, in contrast with the phonon case, a roton arises from a localized disturbance, involving a few atoms only, creating a quantum vortex ring having a radius of about half an atomic spacing. Further shrinkage is impossible, and thus the classical property of vortex propagation through the background fluid (like a smoke-ring) does not obtain. This explains the small group velocity of rotons as given by dE/dk near $k = k_0$. It should be added that despite the name there is no reason to believe that rotons carry angular momentum. Fortunately, much progress can be made without detailed assumptions of the above kind.

For momenta above the roton region, the spectrum should be of parabolic single-particle type, somewhat analogous to that in the perfect Boson gas of section 8.3.1. If the momentum is high enough, as will become clear in Chapter 10, such excitations will be unstable, resulting in a decomposition into several excitations of lower momenta and energy. At the low temperatures of interest here, however, quasi-particles outside the phonon and roton regions are of little significance because of the comparatively large energies required to excite them.

Finally, let us remark that neutron diffraction experiments can nowadays pinpoint the Landau spectrum better than thermodynamic data. As was earlier emphasized, the Van Hove analysis of section 8.2.3 (*c*) was quite general, holding in particular for liquids. Using that formalism, Cohen & Feynman (1957) have shown that for slow neutrons and low temperatures, there is a result quite analogous to that for solids obtained in section 8.2.3 (*d*). This is that the scattering at fixed angle arises from the production or destruction of a single-excitation, the multiple processes contributing only to a diffuse background. (In addition, a simple argument shows that the above is the only scattering which takes place, elastic scattering in a liquid being virtually impossible.)

In fact, Fig. 8.8 is taken from the neutron diffraction measurements of Henshaw & Woods (1961). They find $c = 237$ m/sec, $\Delta/k_B = 8{\cdot}6(5)$ °K, $k_0 = 1{\cdot}91$ Å^{-1} and $\mu/M = 0{\cdot}16$, M being the mass of a ^{4}He atom. It will be noted that at the highest measured k, the curve has not yet attained an asymptotic form. Presumably it will do so at higher k. We shall see in section 8.3.6 (*d*), below, that the roton dip is associated very directly with the first peak in the $S(k)$ curve; probably the high k behaviour observed by Henshaw & Woods is connected with the second peak in $S(k)$.

8.3.5. *Landau spectrum and macroscopic properties*

As we have already remarked, Landau's spectrum satisfies the observed low-temperature thermodynamics. It is also consistent with other experimental results. The intention below is to confine our discussion to explaining qualitatively the remaining properties touched on in section 8.3.3, before going on to our main task of attempting to justify the Landau spectrum from first principles.

(*a*) *Second sound.* To begin with, let us interpret the abnormally high thermal conductivity in terms of the two-fluid model. When heat is applied to some part of the liquid, the density of quasi-particles in that location is increased. In an attempt to re-establish equilibrium, background fluid moves in and normal fluid (the quasi-particles) flows away. The thermal conductivity is limited only by the rate at which these flows can take place and this mechanism is much more efficient than the usual one in terms of thermal gradients. Similarly, second sound can be explained as a wave of varying quasi-particle number density. A local increase in temperature implies a corresponding local increase in quasi-particle density. Nearby background fluid flows in, leaving behind a region thereby enriched in normal fluid, etc. In this way, a thermal pulse travels through the liquid with little attentuation.

(*b*) *Viscosity.* The viscosity problem, also, can be understood in terms of the two-fluid model. A rotating object is immersed not only in the background fluid, but also in the normal component. It thus experiences a non-zero drag. On the other hand, as we will see shortly, the background fluid flows frictionlessly through narrow pipes provided the velocity is not too high. For this reason, in the context of the two-fluid model, the background, as distinct from the normal component, is referred to as the superfluid, even though the latter term is used to describe a property of He II as a whole.

(*c*) *Excitations and superfluidity.* The final question to settle here is the crucial one of why an excitation spectrum of Landau type implies superfluid behaviour. The original argument of Landau was one based on Galilean transformations between reference frames. The following form of it has been given by Beliaev (1959).

Let us suppose that the fluid contains, to begin with, a single quasi-particle with momentum $\mathbf{k}$. The group velocity is then

$$\mathbf{v}_g = \frac{\partial E}{\partial \mathbf{k}}. \tag{8.3.20}$$

Now let us suppose that we give every particle in the fluid a velocity $\mathbf{u}$ so that this becomes the macroscopic fluid velocity. The momentum of the fluid as a whole changes from $\mathbf{k}$ to

$$\mathbf{k}' = \mathbf{k} + M_t \mathbf{u}, \tag{8.3.21}$$

M_t being the total mass, while the new group velocity of the quasi-particle is

$$\mathbf{v}_g' = \frac{\partial E'}{\partial \mathbf{k}'} = \mathbf{v}_g + \mathbf{u}. \tag{8.3.22}$$

From (8.3.20) to (8.3.22), we have

$$\frac{\partial E'}{\partial \mathbf{k}} = \frac{\partial E'}{\partial \mathbf{k}'} = \frac{\partial E}{\partial \mathbf{k}} + \mathbf{u} \tag{8.3.23}$$

and, thus, integration with respect to $\mathbf{k}$ gives

$$E' = E(\mathbf{k}) + \mathbf{u}.\mathbf{k}, \tag{8.3.24}$$

for all $\mathbf{k}$.

The point to note about this result is that E' is greater than zero for all $\mathbf{k}$, if and only if

$$|\mathbf{u}| < E(\mathbf{k})/k \quad (\text{all } k). \tag{8.3.25}$$

The minimum value of $E(\mathbf{k})/k$ is given by the slope of the line OA in Fig. 8.9 and thus E' is greater than zero if the macroscopic fluid velocity is less than the slope of OA. If E' is greater than zero it implies that the liquid is energetically stable against the formation of elementary excitations. In other words, it costs energy to create

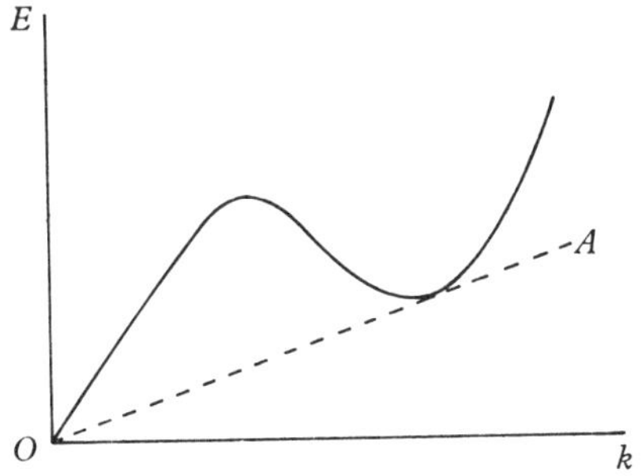

Fig. 8.9

quasi-particles and if no energy is available, no normal fluid will be formed. If u were too high, the liquid would be unstable with respect to the formation of negative energy excitations because of the valley in the $E(\mathbf{k})$ curve.

Actually, it seems that one cannot naïvely take over the Landau $E(\mathbf{k})$ curve determined from the thermodynamic data and apply it immediately to the fluid in motion. The reason for this is that the critical velocity, as determined by (8.3.25) and the Landau spectrum, should be Δ/k_0 (see Figs. 8.8 and 8.9). This gives a value a few hundred times higher than experiment. To account for the discrepancy, Onsager & Feynman have suggested that a further kind of excitation, the quantized vortex line, occurs in the flowing liquid. We will not pursue this matter here. It should be stressed, however, that the preceding argument indicates that superflow occurs for sufficiently low macroscopic velocities.

Because of the part played by the roton valley in determining the slope of the line OA, it should not be overlooked that it is primarily the (linear) shape of the phonon portion of the spectrum which determines that the liquid will be superfluid. To see this quite clearly, let us imagine for a moment that $E(\mathbf{k}) = k^2/2M$ as in the ideal case (Fig. 8.3). Then, in (8.3.24), one can always choose a $\mathbf{k}$ (say, as $-M\mathbf{u}$) so that $E' < 0$. Geometrically, all straight lines with non-zero slope through the origin intersect the parabola $k^2/2M$ at two distinct points. On the other hand, if $E(\mathbf{k}) = ck$ in (8.3.24), provided $u < c$, this possibility does not exist at small k. One may say that the interactions in He II bring about such a scarcity of low-lying excitations (compared with the perfect gas case; see Fig. 8.3) that superflow is possible.

As we have indicated, the behaviour of He II is largely determined by the Landau excitation spectrum. The determination of the latter is thus the central problem to which we will address ourselves. Two approaches have so far been used with success. The first considers the problem as it stands, without modification. Despite the many difficulties, Feynman (1955) has been able to make considerable progress and some of his work is considered in the following section. The alternative method is to consider simplified model systems which in some way resemble He II. There is the disadvantage of losing some physical features, but by suitable modification there is the possibility of carrying the mathematics further and in this way gaining greater insight. The most famous example of this kind is the soft-core attractive interaction problem introduced in the pioneering work of Bogoliubov (1947) (see also Bogoliubov & Zubarev, 1955). (This is considered in section 8.3.7.)

8.3.6. *Feynman theory*

We should begin by commenting that the Feynman theory is not entirely quantitative in nature, but there can be little doubt that it contains the essential physics of the situation.

The problem is to investigate the solutions of the Schrödinger equation with Hamiltonian (8.3.16), where the probability amplitude $\Psi(\mathbf{r}_1, \mathbf{r}_2, \ldots, \mathbf{r}_N)$ is symmetric with respect to interchange of any pair of co-ordinates. The corresponding energies are

$$E = \frac{\hbar^2}{2M} \sum_i \int |\nabla_i \Psi|^2 \, d\mathbf{r}_1 \ldots d\mathbf{r}_N + \sum_{i<j} \int V(r_{ij}) \, |\Psi|^2 \, d\mathbf{r}_1 \ldots d\mathbf{r}_N, \quad (8.3.26)$$

and we shall refer to this expression later.

The potential V is that shown in Fig. 8.6 and discussed in section 8.3.3. The effect of the strong repulsive core is to make it difficult for atoms to overlap to any degree; that is, Ψ for such configurations is practically zero. Provided this restriction is met, to a rough approximation there is little other preference for the positions of the atoms. Thus, looking over a few atoms at a time, the atoms are more or less uniformly spaced. We might alternatively say that the local macroscopic density is constant. The large volume per atom compared with core volume ensures that atoms can easily move around each other and the system thus changes without difficulty from one configuration of high probability to another.

A further point concerns the nodal behaviour of the eigenfunctions. The ground state, Φ, is characteristically without nodes and may be taken to be everywhere positive. Now the excited states are orthogonal to Φ (and to each other). This means that their real and imaginary parts must each take on both positive and negative values. We now turn to an analysis of such states.

(*a*) *Density fluctuations and excited state wave functions.* Approximate phonon wave functions may be written down with the help of section 8.2.5. We saw there that, to a first approximation, the classical density fluctuations $\rho_{\mathbf{k}}$, as given by (8.2.68), were the normal modes for a liquid (see (8.2.74)). This was proved, admittedly, assuming soft cores, but there is no reason to doubt its general validity. In the random phase approximation, then, the $\rho_{\mathbf{k}}$ may be regarded as oscillating independently. The first excited state of a simple harmonic oscillator is merely the co-ordinate multiplied by the ground-state Gaussian form. This indicates that in the present case the (unnormalized) phonon wave functions may be represented by

$$\Psi = \rho_{\mathbf{k}}\Phi = \sum_i e^{i\mathbf{k}.\mathbf{r}_i}\Phi. \tag{8.3.27}$$

It should be noted that we modify the exact ground-state wave function Φ, thereby maintaining the requirement, as valid in excited states as in the ground state, that atoms do not overlap. As it stands, (8.3.27) has the necessary symmetry property.

It will be recalled (see (8.3.6)), that in the non-interacting case, the single-particle excitations can be written in the form (8.3.27) with, of course, Φ taken to be constant. Nevertheless, we reiterate that, despite appearances, (8.3.27) describes phonon wave functions for small k. Quite apart from the method of derivation, one can understand the physical nature of a wave function such as (8.3.27) as follows. Considering the real part, let us draw schematically $\cos \mathbf{k}.\mathbf{r}$ as in Fig. 8.10 and indicate by circles the atomic positions in the ground state. As we have noted, the atoms never significantly overlap and are fairly evenly spaced. A typical situation is that shown

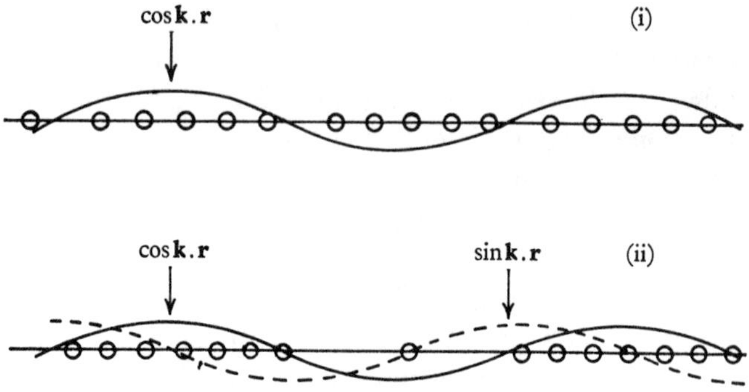

Fig. 8.10

in Fig. 8.10 (i). For small $\mathbf{k}$, there are many atoms per wavelength and because of the even distribution, $\sum_i \cos \mathbf{k}.\mathbf{r}_i$ will vanish or be exceedingly small. But there also exists a much smaller amplitude that the atoms in the ground state may be distributed as in Fig. 8.10 (ii). For such a case with many more positive terms than negative terms, $\sum_i \cos \mathbf{k}.\mathbf{r}_i$ will be large. The total effect, therefore, of multiplying Φ by $\sum_i \cos \mathbf{k}.\mathbf{r}_i$ is to give density fluctuations as described by Fig. 8.10 (ii), say, high probability at the expense of uniform distributions. This is also the effect of the exponential sum, for it will be seen that by symmetry, $\sum_i \sin \mathbf{k}.\mathbf{r}_i$ vanishes not only for the situation shown in Fig. 8.10 (i) but also for that of Fig. 8.10 (ii). This description will be valid until the wavelength becomes comparable with the mean inter-particle spacing (see below).

(*b*) *Phonon-like character of elementary excitations.* The fundamental problem now is to show that there are no elementary excitations of lower energy than the phonons, at sufficiently small momenta. This is necessary because it has been pointed out already (see section 8.3.5) that the existence of, say, a parabolic spectrum of free-particle type in this region would imply no superfluidity in He II. We have already touched on one example, however, of the way in which the statistics frustrate attempts to construct excited-state wave functions of low excitation energy. More explicitly, one might attempt to describe the excitation of a single particle moving through the loosely packed fluid, the other atoms participating in a back-flow round it. In an effective mass approximation, the wave function might be tentatively written $e^{i\mathbf{k}\cdot\mathbf{r}_1}\Phi$, where $\mathbf{r}_1$ is the co-ordinate of the excited particle and Φ, just as before, governs the relative distribution of all the particles. But this wave function is not satisfactory as it stands. It must be properly symmetrized. On doing this, we end up with the function (8.3.27), which represents phonons and not single-particle excitations for small k. In other words, there is a lower bound in energy and momentum below which (8.3.27) cannot represent single-particle excitations. As we have already indicated, there is a natural cut-off in the k range for which (8.3.27) can represent phonons.

The general reasoning is as follows. Let us try to construct an excited state of as low energy as possible, which is not of phonon type. It cannot correspond to a local macroscopic density variation as this can be described in terms of phonons. It must therefore correspond to a stirring (the terminology is Feynman's) of the atoms in the ground state so that there are no gross concentrations or deficiencies of atoms anywhere in the system. Now the particle distribution affects the potential energy only through the pair function (recall section 1.4.3) and the latter, from what has just been said, is little altered by stirring. Everything, therefore, depends on the kinetic energy term in (8.3.26), which in turn depends on grad Ψ in configuration space. We shall now show that the orthogonality of Ψ to Φ combines with the Bose statistics to produce such a fluctuation in grad Ψ in configuration space that the energy is inevitably raised by some non-zero amount above that of the ground state. (The minimum possible energy difference should be the characteristic energy Δ required to create a roton.) This means there will always be lower energy phonon states.

Since both the real and imaginary parts of Ψ separately satisfy the same Schrödinger equation, we may assume, without loss of generality, in the following, that Ψ is real. Then the orthogonality properties imply that Ψ has nodes and thus takes on a positive maximum and a negative minimum value. The next step is to show that the positions of these cannot be arbitrarily far separated in configuration space.

Let us suppose that Ψ takes on its maximum value for some configuration A, with atoms at positions $(\mathbf{a}_1, \mathbf{a}_2, \ldots, \mathbf{a}_N)$, so that

$$\Psi_{\text{max.}} = \Psi(\mathbf{a}_1, \mathbf{a}_2, \ldots, \mathbf{a}_N), \tag{8.3.28}$$

and its minimum for configuration B, with atoms at positions $(\mathbf{b}_1, \mathbf{b}_2, \ldots, \mathbf{b}_N)$, so that

$$\Psi_{\text{min.}} = \Psi(\mathbf{b}_1, \mathbf{b}_2, \ldots, \mathbf{b}_N). \tag{8.3.29}$$

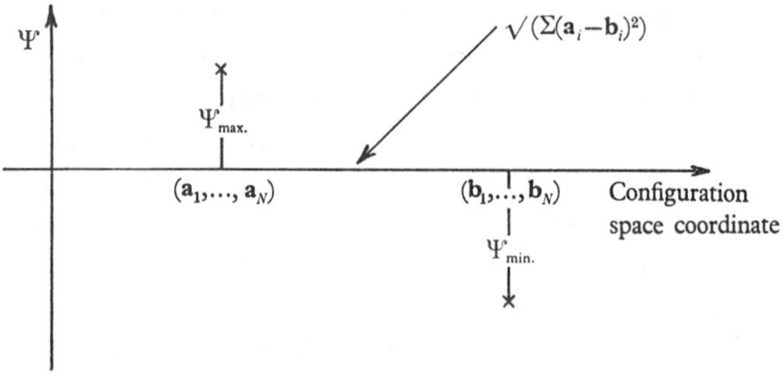

Fig. 8.11

The situation is shown schematically in Fig. 8.11. The configuration A (which, as has been stressed, has a constant local macroscopic density) may be depicted in real space as in Fig. 8.12, say. The same applies to B, and so the A and B sites intermingle in such a way that within about half an atomic spacing of an A site is a B site and vice versa. We can speak of adjacent A and B sites. Now for most configurations, $\sqrt{[\Sigma(\mathbf{a}_i - \mathbf{b}_i)^2]}$ will be very large, since the distance between at least one pair $\mathbf{a}_i$, $\mathbf{b}_i$ will be large. But for a set when $\mathbf{a}_i$ and $\mathbf{b}_i$ correspond to adjacent sites for all i, the separation in configuration space will be minimal. The requirement that Ψ oscillates between its maximum and minimum values in this distance places a lower bound on the kinetic energy density in this region. On the other hand, the ground state has no nodes and need not meet such a condition. One expects, therefore, a state of the kind here considered, to be always separated from the ground state by some minimum energy, Δ. The above analysis shows that if Fig. 8.12 is to represent the maximum and minimum amplitudes, the two configurations should more or less symmetrically interpenetrate without overlap so that A positions are about half an atomic spacing from the nearest B positions much as in Fig. 8.13.

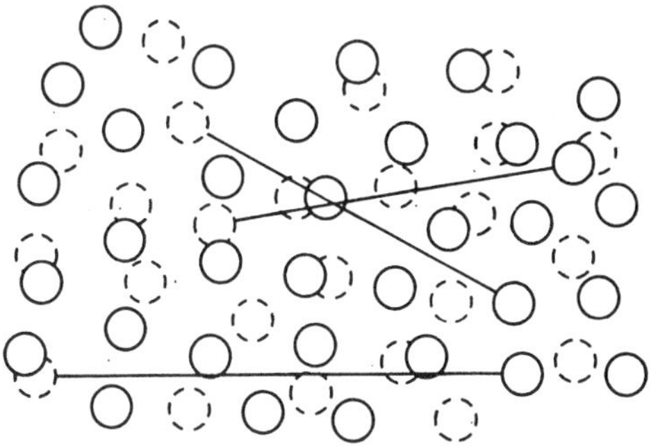

Fig. 8.12. Full circles are A sites, broken circles are B sites. The relative positions of the configurations are tentative. All one can be sure about is that within approximately half an atomic spacing of an A site there is a B site. The full lines denote examples of possible correlations of A sites with B sites of the non-optimal kind discussed in the text.

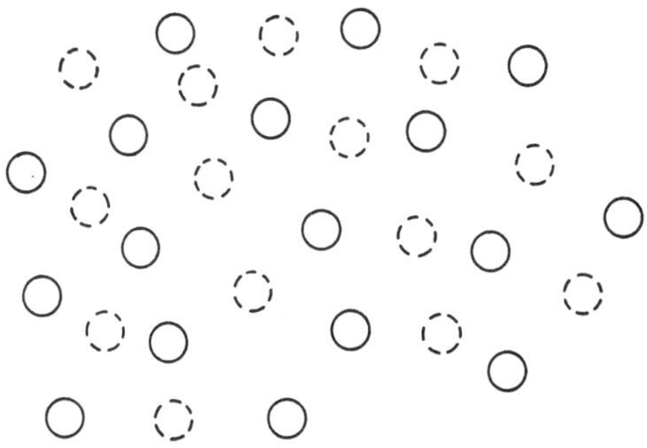

Fig. 8.13

(*c*) *Variational trial wave function.* The above kind of excitations describe the rotons. We now attempt to construct the $E(\mathbf{k})$ curve in this region by means of approximate trial variational wave functions formed in accordance with the above discussion. Let us suppose, to begin with, that we only consider those configurations in which every particle must be on an A site or a B site. By definition, those configurations when every A site is occupied and when every B site is occupied are the most likely to be observed. But it is also possible (though less probable) that all atoms except one are at A sites, the odd man out being at a B site. Less probable still, but nevertheless permissible, is that all but two A sites are occupied, and so on. To account for this, one writes

$$\Psi = F\Phi, \quad F = \sum_i f(\mathbf{r}_i), \tag{8.3.30}$$

the Φ being present to exclude the possibility of serious overlap of atoms. We expect that the above variation of probabilities will be accounted for if f always takes the value $+1$ on an A site and -1 on a B site. But if we examine this function, we see it is also reasonable for describing situations when an atom is on neither an A nor a B site. Thus we might expect f to vary smoothly between $+1$ and -1 and thus, for example, if an atom is nearly on an A site, its f value is nearly $+1$, if it is roughly midway between, its f value is approximately zero, etc.

The precise determination of f is now carried out using the variational principle. We take as Hamiltonian of the system,

$$H = -\frac{\hbar^2}{2M}\sum_i \nabla_i^2 + \sum_{i<j} V(r_{ij}) - E_0, \tag{8.3.31}$$

where E_0 is the ground-state energy, and thus we have

$$H\Phi = 0. \tag{8.3.32}$$

Defining $$\mathfrak{N} = \langle\Psi|H|\Psi\rangle, \quad \mathfrak{D} = \langle\Psi|\Psi\rangle, \tag{8.3.33}$$

the next step is to find those f's for which the expression

$$\epsilon = \mathfrak{N}/\mathfrak{D} \tag{8.3.34}$$

is stationary. For these optimum functions, the corresponding ϵ's give the excitation energies.

From (8.3.30), (8.3.31) and (8.3.32), we obtain the result

$$H\Psi = -(\hbar^2/2M)\sum_i (\Phi\nabla_i^2 F + 2\nabla_i\Phi . \nabla_i F)$$

$$= -(\hbar^2/2M)\Phi^{-1}\sum_i \nabla_i . (\Phi^2\nabla_i F),$$

from which explicit dependence on E_0 and the potential energy has been eliminated. Thus, we find

$$\mathfrak{N} = (\hbar^2/2M)\sum_i \int (\nabla_i F^*).(\nabla_i F)\,\Phi^2\,d\mathbf{r}_1\,d\mathbf{r}_2\ldots d\mathbf{r}_N$$

$$= (\hbar^2/2M)\sum_i \int (\nabla_i f^*(\mathbf{r}_i)).(\nabla_i f(\mathbf{r}_i))\,\Phi^2\,d\mathbf{r}_1\,d\mathbf{r}_2\ldots d\mathbf{r}_N$$

$$= \rho\hbar^2/2M \int \nabla f^*(\mathbf{r}).\nabla f(\mathbf{r})\,d\mathbf{r}, \qquad (8.3.35)$$

where we have used the reality of Φ and the fact that the diagonal element of the first-order density matrix is just the ordinary constant particle density ρ. Also, we have

$$\mathfrak{D} = \langle\Psi|\Psi\rangle = \langle\Phi|F^*F|\Phi\rangle$$

$$= \langle\Phi|\sum_{ij} f^*(\mathbf{r}_i)f(\mathbf{r}_j)\,|\Phi\rangle$$

$$= \rho\int f^*(\mathbf{r}_1)f(\mathbf{r}_2)\,S(\mathbf{r}_1-\mathbf{r}_2)\,d\mathbf{r}_1\,d\mathbf{r}_2, \qquad (8.3.36)$$

where $S(\mathbf{r})$ is the probability per unit volume that a particle is at position $\mathbf{r}$ relative to a chosen atom. It is thus the quantity defined by equation (8.2.45). The Euler equation for (8.3.35) is now readily written down. It is

$$-\frac{\hbar^2}{2M}\nabla^2 f(\mathbf{r}) = \epsilon\int S(\mathbf{r}-\mathbf{r}')f(\mathbf{r}')\,d\mathbf{r}'. \qquad (8.3.37)$$

We turn now to discuss the nature of the elementary excitations implied by this equation.

(*d*) *Structure factor and excitation spectrum.* It is easy to verify that the Euler equation (8.3.37) has solutions

$$f(\mathbf{r}) = e^{i\mathbf{k}.\mathbf{r}}, \qquad (8.3.38)$$

with corresponding energies

$$\epsilon(\mathbf{k}) = \hbar^2k^2/2MS(\mathbf{k}), \qquad (8.3.39)$$

where (cf. (8.2.44))

$$S(\mathbf{k}) = \int S(\mathbf{r})\,e^{i\mathbf{k}.\mathbf{r}}\,d\mathbf{r}. \qquad (8.3.40)$$

The formula (8.3.39) represents a genuine energy-momentum relationship, for, with f's given by (8.3.38), it is readily verified that the corresponding Ψ''s share the property of the true eigenstates that they are eigenfunctions of the momentum operator. Since this operator is, of course, Hermitian, it follows that any pair of wave functions drawn from either of the above two families are orthogonal provided only that they correspond to different momenta. *Inter alia*, it follows that the spectrum (8.3.39) lies above the exact curve.

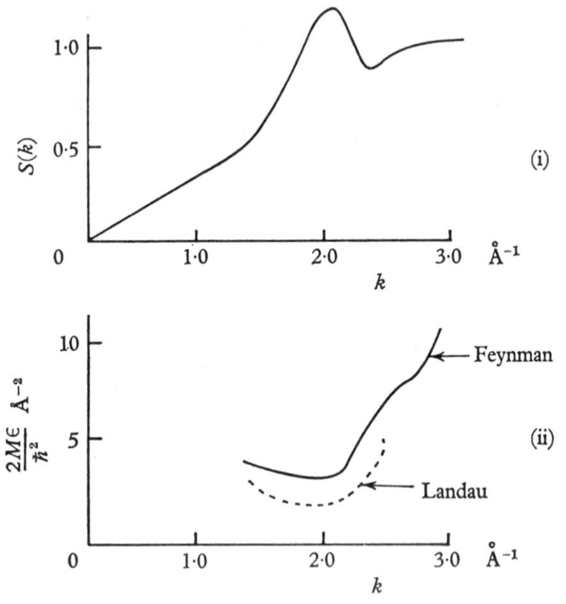

Fig. 8.14. (i) Schematic form of measured structure factor for liquid helium. (ii) Schematic form of Feynman excitation energy. Dotted curve shows Landau spectrum for comparison.

$S(k)$, as defined by (8.3.40), is the liquid structure factor which can be found experimentally, at least for not too small k. In Fig. 8.14 (i), we sketch $S(k)$ as measured by Beaumont & Reekie (1955). Because of the observed low-temperature T^3 specific heat law, we expect the experimentally unattainable low k part to extrapolate linearly through the origin, thus producing a phonon energy expression consistent with the Feynman discussion. The lower Fig. 8.14 (ii) gives $\epsilon(k)$ thus determined from (8.3.39). One should note that (8.3.38) implies that our wave function is once more of the type (8.3.27), though the domain of $\mathbf{k}$ is here in the roton range.

The general features of the experimentally observed curve are reproduced including the structure at high k found by Henshaw & Woods but the minimum roton energy is about a factor of two too high. Nevertheless, one feels that the essential physics is present and all that is required is a further refinement of the same basic approach. In this connexion, the concluding remarks of section 8.3.6 (*e*), below, should be noted.

We see that the roton dip in the $E(k)$ curve arises from the peak in $S(k)$. The latter, occurring at $k = 2\pi/a$, in turn arises from the local order in a liquid, namely the tendency for neighbouring atoms to be found at about their inter-atomic spacing a, from each other. This comfortably fits in with the analysis pertaining to Fig. 8.10. As k increases, the possibility of selecting phonon states in the manner indicated there, becomes more and more difficult because of the geometrical facts of life concerning the wavelength and the atomic radius. Eventually the simple phonon concept disappears. Around $k = 2\pi/a$, however, we have a situation for which the system has great preference (see Fig. 8.15), and this explains the relatively low excitation energies in this region. The sketch clearly shows that no density fluctuations are now involved.

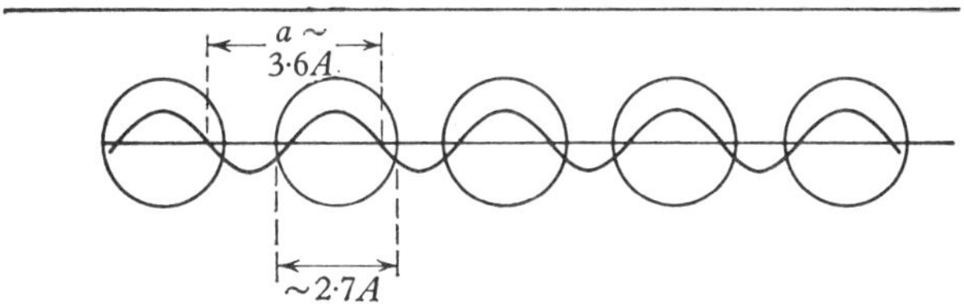

Fig. 8.15

At higher k still, a kink in the $E(k)$ curve reflects correlation between next-nearest neighbours—smaller in liquid helium than for other liquids because of the large specific volume and the zero-point energy. As mentioned previously, only at high k does (8.3.39) at last represent a single-particle excitation.

(*e*) *Alternative derivation of Feynman excitation spectrum.* The Feynman excitation spectrum, derived earlier by construction of an explicit wave function can be obtained from an argument which generalizes the form of $S(\mathbf{k}, \omega)$ given in section 8.2.5 (*c*). There, in equation (8.2.91), we found in the limit $T \to 0$,

$$S(\mathbf{k}, \omega) = S(k)\,\delta(\omega - ck). \tag{8.3.41}$$

Recalling that this is true for an elementary excitation which is simply a phonon, it suggests, for a general excitation $\epsilon(k)$, that we write

$$S(\mathbf{k}, \omega) = S(k)\,\delta(\omega - \epsilon(k)). \tag{8.3.42}$$

We now recall the moment theorem for classical fluids given by (8.2.79). The analogue at $T = 0$ is (cf. Chapter 5)

$$\int_{-\infty}^{\infty} \omega S(\mathbf{k}, \omega)\, d\omega = \frac{\hbar k^2}{2M}. \tag{8.3.43}$$

Substituting (8.3.42) into (8.3.43), we obtain immediately

$$\epsilon(k) = \frac{\hbar k^2}{2M\,S(k)}. \tag{8.3.44}$$

This is simply the Feynman result for a trial wave function $\rho_{\mathbf{k}}\Phi$.

The excitation spectrum thereby obtained agrees with experiment in the phonon region, but, as we have seen, there are quantitative differences in the roton part. Thus, at short wavelengths, it seems that the Van Hove function must be broadened by interaction between different excitations.

Feynman & Cohen gave arguments for improving the Feynman wave function by taking cognizance of the backflow round a given particle as it moves through the liquid and with the modified wave function, considerably better agreement was found.

(*f*) *Asymptotic form of radial distribution function.* One further result concerning the radial distribution function $g(r)$ follows in an elementary way from the Feynman structure factor for small k, namely

$$S(k) \sim \frac{\hbar k}{2Mc}, \tag{8.3.45}$$

where c is the sound velocity. Clearly, the small k behaviour of $S(k)$ has implications concerning the large r behaviour of $g(r)$, which, on transforming (8.2.47) and using the spherical symmetry of the liquid case, is given by

$$g(r) = 1 + \frac{1}{(2\pi)^3 \rho} \int_0^{\infty} \{S(k) - 1\} \frac{\sin kr}{kr}\, 4\pi k^2\, dk. \tag{8.3.46}$$

Now there is a precise result in Fourier transform theory (see, for example, Lighthill, 1958) that if $F(x)$ is a well-behaved function, with its derivatives, then

$$\int_0^\infty F(x)\sin xr\,dx = \frac{F(0)}{r} - \frac{F''(0)}{r^3} + \frac{F^{iv}(0)}{r^5} + \ldots. \qquad (8.3.47)$$

From (8.3.45), (8.3.46) and (8.3.47) it then follows immediately (Enderby, Gaskell & March, 1965) that

$$g(r) = 1 - \frac{\hbar}{2\pi^2\rho Mc} r^{-4} + \ldots. \qquad (8.3.48)$$

This derivation of (8.3.48) depends only on the Feynman result (8.3.45), plus the assumption that for the Boson fluid there are no points at which $S(k)$ or its derivatives have cusps or singularities. Such an assumption appears valid for Bose fluids, but not for Fermi fluids (cf. the result for $S(k)$ for non-interacting Fermions, in which case $S''(k)$ has a discontinuity at $k = 2k_f$, which profoundly changes the asymptotic form of $g(r)$ as given by (1.5.13)).

It is interesting to note that an explicit calculation of Huang (1960; see also Isihara & Yee, 1964) for a gas of hard sphere Bosons, also leads to the r^{-4} dependence for large r.

A final point worth stressing is the role of the attractive Van der Waals tail in the He–He pair potential $V(r)$ in determining the properties of liquid ^{4}He. It seems certain that the detailed nature of $V(r)$ determines the quantitative form of $S(k)$ and may also be responsible for the shape of the excitation curve toward the roton minimum. In this connexion, reference might be made to the conclusion of Enderby, Gaskell & March (1965) that for classical insulating fluids like Ar, with Van der Waals forces, there appears a term in $S(k)$ for small k, proportional to k^3, and whose magnitude depends linearly on the strength of the Van der Waals forces. On the other hand (Lee, Huang & Yang, 1957; Isihara & Yee, 1964) find for hard sphere Bosons of diameter a,

$$S(k) = k(k^2 + 16\pi a\rho)^{-\frac{1}{2}}, \qquad (8.3.49)$$

which also gives a k^3 term for small k. Furthermore, for Fermions, the statistics already yield a term $\propto k^3$, and hence it seems entirely possible that the k^3 term, correcting Feynman's result (8.3.45) for larger k, will depend both on statistics and on the full details of the force curve. It does not seem clear, at present, at $T = 0$, whether a term in k^2 follows the linear term in (8.3.44). The presence of such even terms does not affect the arguments concerning the asymptotic form of $g(r)$, as the even powers lead to terms falling off exponentially with distance, as in the Ornstein–Zernike theory (see, for example, Rosenfeld, 1951).

(*g*) *Wave functions for fluid in motion, and quantized vortex lines.* The Feynman theory, as discussed so far, has considered in detail the excited states of the fluid at rest.

We turn finally to consider the form that the wave functions would take for the fluid in motion. As before, we begin from the exact ground-state wave function Φ, and form

$$\Psi = \left[\exp\left\{i \sum_j s(\mathbf{r}_j)\right\}\right] \Phi. \tag{8.3.50}$$

The only assumption we make at this point is that $s(\mathbf{r})$ shall vary only slowly with position.

It is then possible to show that (8.3.50) represents the fluid in motion with a mean momentum density $\rho \mathbf{v}$, where

$$\mathbf{v} = (\hbar/M) \operatorname{grad} s. \tag{8.3.51}$$

In addition, if we make the requirement that s varies slowly more precise by insisting that grad s changes only by a fraction of itself over the interatomic spacing, then it turns out that this wave function is, to a good approximation, a solution of the wave equation. Furthermore, the wave functions thus generated are orthogonal to the ground state and to one another.

What we have not done so far is to impose the condition that Ψ describes an assembly of Bosons. Since Φ already has the desired characteristics, we must write down the condition that the modulus of the phase factor is invariant when we interchange atom positions. This means that if we draw a contour in the liquid formed by N atoms, and consider a displacement $\delta \mathbf{r}$ as required to take one atom into the position of its neighbour, then it follows that

$$\sum_{\text{contour}} \operatorname{grad} s(\mathbf{r}_j) \,.\, \delta \mathbf{r}$$

must be an integral multiple of 2π. But s is, as we have seen, essentially the velocity potential and hence we have

$$\oint \mathbf{v}.d\mathbf{r} = \frac{nh}{M}, \tag{8.3.52}$$

where n is an integer.

Thus, the wave function (8.3.50) describes a flow which clearly, from (8.3.51) is irrotational, that is

$$\operatorname{curl} \mathbf{v} = 0,$$

and also for which the circulation round any contour is quantized in units of h/M.

Nowhere, in this argument, have we shown that the wave function Ψ is, in any sense, the only possible choice to approximate to the fluid in motion. In fact, while no proof has been given as far as we are aware, it is believed that it is the only one which will at once describe flow, lead to irrotational motion and have the correct Bose symmetry. These statements should be qualified by remarking that little or nothing is known about the case when the velocity potential s varies rapidly.

Thus, the conclusion is that the theory points to the existence of a new excitation, the quantized vortex line (cf. p. 297). That this is more than academic in interest has been demonstrated experimentally, particularly by Vinen (1961). The idea of his experiment is that a quantum vortex having a circulation as given by (8.3.52), if it occurs around a thin oscillating wire, will deflect the wire in a direction perpendicular to the motion, and hence will gradually turn the plane of oscillation. The rate of rotation of the plane of oscillation will be proportional to the circulation, and hence will allow an experimental test of (8.3.52). While it would be out of place to go into further details here, the quantization of circulation is essentially confirmed by Vinen's experiment. The existence of free vortex lines, however, can hardly be said to be conclusively demonstrated.

The reader will have noticed the interesting similarity between the arguments for flux quantization in a superconductor given in Chapter 7 and the circulation quantization of a vortex.

8.3.7. *Bogoliubov model*

In the Feynman theory we saw that it was difficult to follow the mathematics through step by step. The advantage of the following treatment is that a fairly rigorous solution can be worked out, but at the price of having to consider a much simplified problem. Nevertheless, there is sufficient common ground with the Feynman theory to justify discussing the Bogoliubov model in some detail.

The method of second quantization is of great utility in many-Boson problems, but the difficulty in applying it to He II is that the hard core interaction potential between atoms has divergent matrix elements. This is the reason why we must modify the original problem. We assume we are now dealing with a set of interacting particles having convergent matrix elements and that the interaction can be taken as a perturbation. Specifically we consider

$$H = \sum_{\mathbf{k}} \epsilon_{\mathbf{k}} a_{\mathbf{k}}^{\dagger} a_{\mathbf{k}} + \frac{1}{2} \sum_{\mathbf{k}_1+\mathbf{k}_2=\mathbf{k}_1'+\mathbf{k}_2'} \langle \mathbf{k}_1 \mathbf{k}_2 | V | \mathbf{k}_1' \mathbf{k}_2' \rangle a_{\mathbf{k}_1}^{\dagger} a_{\mathbf{k}_2}^{\dagger} a_{\mathbf{k}_2'} a_{\mathbf{k}_1'}, \tag{8.3.53}$$

where the *a*'s are Boson operators.

The final comments of section 8.3.2 will be recalled. It will be assumed, in the calculation that follows, that the interaction is so weak that not only is the occupation of the condensate macroscopic, but, in addition, the number of excited particles is so small, that if N_0 is the average occupation of the zero-momentum state, N_0/N can as a first approximation, be replaced by unity. Because of the special status of the condensate, we rewrite (8.3.53), in order to emphasize its role, as follows:

$$H = \sum \epsilon_{\mathbf{k}} a_{\mathbf{k}}^{\dagger} a_{\mathbf{k}} + \tfrac{1}{2} V_0 a_0^{\dagger 2} a_0^2 + \sum_{\mathbf{k} \neq 0} (V_0 + V_{\mathbf{k}}) a_{\mathbf{k}}^{\dagger} a_{\mathbf{k}} a_0^{\dagger} a_0$$
$$+ \frac{1}{2} \sum_{\mathbf{k} \neq 0} V_{\mathbf{k}} (a_{\mathbf{k}}^{\dagger} a_{-\mathbf{k}}^{\dagger} a_0^2 + a_{\mathbf{k}} a_{-\mathbf{k}} a_0^{\dagger 2}) + V_R, \tag{8.3.54}$$

where the remaining potential energy terms, denoted by V_R, are such that, at most, one zero-momentum operator is involved. It is now supposed that V_R in (8.3.54) can be ignored in comparison with the other terms. The original argument of Bogoliubov was that, in order of magnitude (in a sense which will become apparent below), each of these is smaller than the terms displayed (albeit there are more of them) and might be plausibly dropped. This seems very reasonable physically, for if 'almost all' particles are in the condensate, the scattering of the very few promoted particles among themselves should be of secondary importance. Alternatively expressed, we suppose Bose–Einstein condensation is crucial and only those terms in which it explicitly participates are retained. We picture our simplified system as allowing only the excitation of pairs **k**, −**k** from the condensate and the re-entry of such pairs back into the condensate. The coupling between two sets of excited pairs in our approximation is zero. The parallel with the B.C.S. theory of Chapter 7 and the simplified reduced Hamiltonian is strong here.

(*a*) *Weak coupling limit.* We now simplify H further using our weak coupling assumptions. We have

$$N_{\text{op.}} = a_0^\dagger a_0 + \sum_{\mathbf{k} \neq 0} a_{\mathbf{k}}^\dagger a_{\mathbf{k}} \tag{8.3.55}$$

and hence

$$N_{\text{op.}}^2 = (a_0^\dagger a_0)^2 + \sum_{\mathbf{k} \neq 0} (a_{\mathbf{k}}^\dagger a_{\mathbf{k}} a_0^\dagger a_0 + a_0^\dagger a_0 a_{\mathbf{k}}^\dagger a_{\mathbf{k}}) + \dots \tag{8.3.56}$$

If we note that

$$(a_0^\dagger a_0)^2 = a_0^{\dagger 2} a_0^2 + a_0^\dagger a_0, \tag{8.3.57}$$

we can now use (8.3.56) in (8.3.54) to obtain

$$H = \tfrac{1}{2} V_0 (N_{\text{op.}}^2 - a_0^\dagger a_0) + \sum_{\mathbf{k} \neq 0} \epsilon_{\mathbf{k}} a_{\mathbf{k}}^\dagger a_{\mathbf{k}} + \sum_{\mathbf{k} \neq 0} V_{\mathbf{k}} a_{\mathbf{k}}^\dagger a_{\mathbf{k}} a_0^\dagger a_0 + \tfrac{1}{2} \Sigma V_{\mathbf{k}} (a_{\mathbf{k}}^\dagger a_{-\mathbf{k}}^\dagger a_0^2 + a_{\mathbf{k}} a_{-\mathbf{k}} a_0^{\dagger 2}). \tag{8.3.58}$$

The higher terms of (8.3.56) belong to the class of terms we have decided to ignore in H and thus are consistently omitted. If we look at the first part of (8.3.58), we see that for N-particle wave functions, $N_{\text{op.}}^2$ can be replaced by N^2 and in comparison, $a_0^\dagger a_0$, which counts particles in the condensate (of which there can be at most N) is negligible.

Our intention is to use a new set of operators

$$A_{\mathbf{k}} = a_{\mathbf{k}} a_0^\dagger / \sqrt{N}; \quad A_{\mathbf{k}}^\dagger = a_{\mathbf{k}}^\dagger a_0 / \sqrt{N} \quad (\mathbf{k} \neq 0) \tag{8.3.59}$$

describing the excitation of particles from the condensate. To this end, we note that for operations on N-body wave functions

$$a_{\mathbf{k}}^\dagger a_{\mathbf{k}} = \frac{a_{\mathbf{k}}^\dagger a_{\mathbf{k}}}{N} N_{\text{op.}} = \frac{a_{\mathbf{k}}^\dagger a_{\mathbf{k}}}{N} \left\{ a_0^\dagger a_0 + \sum_{\mathbf{l} \neq 0} a_{\mathbf{l}}^\dagger a_{\mathbf{l}} \right\} = \frac{a_{\mathbf{k}}^\dagger a_0 a_{\mathbf{k}} a_0^\dagger}{N} - \frac{a_{\mathbf{k}}^\dagger a_{\mathbf{k}}}{N} + \frac{1}{N} a_{\mathbf{k}}^\dagger a_{\mathbf{k}} \sum_{\mathbf{l} \neq 0} a_{\mathbf{l}}^\dagger a_{\mathbf{l}}. \tag{8.3.60}$$

The latter term belongs to the class we are neglecting in present approximation, while the second is negligible compared with the first because of the macroscopic occupation of the condensate. The net effect is that $a_{\mathbf{k}}^{\dagger} a_{\mathbf{k}}$ can be replaced by the first term of (8.3.60), which is just $A_{\mathbf{k}}^{\dagger} A_{\mathbf{k}}$. Likewise, the third term of (8.3.58) can be rewritten using

$$a_{\mathbf{k}}^{\dagger} a_{\mathbf{k}} a_0^{\dagger} a_0 = a_{\mathbf{k}}^{\dagger} a_0 a_{\mathbf{k}} a_0^{\dagger} - a_{\mathbf{k}}^{\dagger} a_{\mathbf{k}},$$

and the latter is negligible compared with the former. There is no problem in the case of the final part of (8.3.58) which, in totality, may thus be written

$$H = \tfrac{1}{2}N^2 V_0 + \sum_{\mathbf{k} \neq 0} (\epsilon_{\mathbf{k}} + NV_{\mathbf{k}}) A_{\mathbf{k}}^{\dagger} A_{\mathbf{k}} + \frac{1}{2} \sum_{\mathbf{k} \neq 0} NV_{\mathbf{k}} (A_{\mathbf{k}}^{\dagger} A_{-\mathbf{k}}^{\dagger} + A_{\mathbf{k}} A_{-\mathbf{k}}). \tag{8.3.61}$$

Within the present approximation, the A's and $A^{\dagger}$'s are Boson operators. This may be seen by observing that

$$A_{\mathbf{k}} A_{\mathbf{l}} - A_{\mathbf{l}} A_{\mathbf{k}} = 0 = A_{\mathbf{k}}^{\dagger} A_{\mathbf{l}}^{\dagger} - A_{\mathbf{l}}^{\dagger} A_{\mathbf{k}}^{\dagger} \quad (\mathbf{l}, \mathbf{k} \neq 0), \tag{8.3.62}$$

this following from the fact that the a's and $a^{\dagger}$'s satisfy similar relations. Also

$$\begin{aligned} A_{\mathbf{k}} A_{\mathbf{l}}^{\dagger} - A_{\mathbf{l}}^{\dagger} A_{\mathbf{k}} &= \frac{1}{N}(a_0^{\dagger} a_0 \delta_{\mathbf{kl}} - a_{\mathbf{l}}^{\dagger} a_{\mathbf{k}}) \\ &= \frac{N_{\text{op.}}}{N} \delta_{\mathbf{kl}} - \frac{1}{N} \sum_{\mathbf{m} \neq 0} a_{\mathbf{m}}^{\dagger} a_{\mathbf{m}} \delta_{\mathbf{kl}} - \frac{1}{N} a_{\mathbf{l}}^{\dagger} a_{\mathbf{k}} \quad (\mathbf{l}, \mathbf{k} \neq 0). \end{aligned} \tag{8.3.63}$$

Just as before, for relevant wave functions, the right-hand side of (8.3.63) can be replaced by $\delta_{\mathbf{kl}}$, and the A's are thus Boson operators.

It thus remains to investigate (8.3.61) for Boson A's. It will be recalled that a similar problem was solved in Chapter 7, section 7.5.1, for Fermions. Thus, the treatment here follows closely that given previously, the only modification being to allow for the different statistics. Let us, then, introduce the Bogoliubov transformation

$$\left.\begin{aligned} A_{\mathbf{k}} &= u_{\mathbf{k}} \alpha_{\mathbf{k}} + v_{\mathbf{k}} \alpha_{-\mathbf{k}}^{\dagger}, \\ A_{\mathbf{k}}^{\dagger} &= u_{\mathbf{k}} \alpha_{\mathbf{k}}^{\dagger} + v_{\mathbf{k}} \alpha_{-\mathbf{k}}, \end{aligned}\right\} \quad \left.\begin{aligned} \alpha_{\mathbf{k}} &= u_{\mathbf{k}} A_{\mathbf{k}} - v_{\mathbf{k}} A_{-\mathbf{k}}^{\dagger}, \\ \alpha_{\mathbf{k}}^{\dagger} &= u_{\mathbf{k}} A_{\mathbf{k}}^{\dagger} - v_{\mathbf{k}} A_{-\mathbf{k}}, \end{aligned}\right\} \tag{8.3.64}$$

where the $u_{\mathbf{k}}$, $v_{\mathbf{k}}$ are real and symmetric and satisfy

$$u_{\mathbf{k}}^2 - v_{\mathbf{k}}^2 = 1. \tag{8.3.65}$$

Thus defined, the α's are readily verified to be Boson operators also. Replacing the A's by α's in (8.3.61), we obtain

$$H = E + H_{11} + H_{20}, \tag{8.3.66}$$

where
$$E = \tfrac{1}{2}N^2V_0 + \sum_{\mathbf{k}\neq 0} [(\epsilon_{\mathbf{k}} + NV_{\mathbf{k}})\,v_{\mathbf{k}}^2 + NV_{\mathbf{k}}\,u_{\mathbf{k}}\,v_{\mathbf{k}}], \tag{8.3.67}$$

$$H_{11} = \sum_{\mathbf{k}\neq 0} [(\epsilon_{\mathbf{k}} + NV_{\mathbf{k}})\,(u_{\mathbf{k}}^2 + v_{\mathbf{k}}^2) + 2NV_{\mathbf{k}}\,u_{\mathbf{k}}\,v_{\mathbf{k}}]\,\alpha_{\mathbf{k}}^\dagger \alpha_{\mathbf{k}} \tag{8.3.68}$$

and

$$H_{20} = \sum_{\mathbf{k}\neq 0} [(\epsilon_{\mathbf{k}} + NV_{\mathbf{k}})\,u_{\mathbf{k}}\,v_{\mathbf{k}} + \tfrac{1}{2}NV_{\mathbf{k}}(u_{\mathbf{k}}^2 + v_{\mathbf{k}}^2)]\,(\alpha_{\mathbf{k}}^\dagger \alpha_{-\mathbf{k}}^\dagger + \alpha_{\mathbf{k}}\,\alpha_{-\mathbf{k}}). \tag{8.3.69}$$

Diagonalization is accomplished if H_{20} vanishes, that is if, for all $\mathbf{k}$,

$$2(\epsilon_{\mathbf{k}} + NV_{\mathbf{k}})\,u_{\mathbf{k}}\,v_{\mathbf{k}} + NV_{\mathbf{k}}(u_{\mathbf{k}}^2 + v_{\mathbf{k}}^2) = 0. \tag{8.3.70}$$

Writing
$$u_{\mathbf{k}} = \cosh\theta_k, \quad v_{\mathbf{k}} = \sinh\theta_k, \tag{8.3.71}$$

where θ_k is real and symmetric, the u's and v's satisfy the subsidiary requirements (see (8.3.65) and immediately before) and substitution in (8.3.70) gives

$$\tanh 2\theta_k = \frac{-NV_{\mathbf{k}}}{\epsilon_{\mathbf{k}} + NV_{\mathbf{k}}}. \tag{8.3.72}$$

Under this condition, substituting for

$$u_{\mathbf{k}}^2 + v_{\mathbf{k}}^2 = \cosh 2\theta_{\mathbf{k}} = \frac{\epsilon_{\mathbf{k}} + NV_{\mathbf{k}}}{\sqrt{[(\epsilon_{\mathbf{k}} + NV_{\mathbf{k}})^2 - N^2V_{\mathbf{k}}^2]}};$$

$$2u_{\mathbf{k}}\,v_{\mathbf{k}} = \sinh 2\theta_{\mathbf{k}} = \frac{-NV_{\mathbf{k}}}{\sqrt{[(\epsilon_{\mathbf{k}} + NV_{\mathbf{k}})^2 - N^2V_{\mathbf{k}}^2]}} \tag{8.3.73}$$

in (8.3.68) gives

$$H_{11} = \sum_{\mathbf{k}\neq 0} E_{\mathbf{k}}\,\alpha_{\mathbf{k}}^\dagger \alpha_{\mathbf{k}}, \quad E_{\mathbf{k}} = \sqrt{[(\epsilon_{\mathbf{k}} + NV_{\mathbf{k}})^2 - N^2V_{\mathbf{k}}^2]}. \tag{8.3.74}$$

Finally, since

$$v_{\mathbf{k}}^2 = \sinh^2\theta_{\mathbf{k}} = \tfrac{1}{2}(\cosh 2\theta_{\mathbf{k}} - 1) = \frac{1}{2}\left(\frac{\epsilon_{\mathbf{k}} + NV_{\mathbf{k}}}{E_{\mathbf{k}}} - 1\right), \tag{8.3.75}$$

one can somewhat simplify (8.3.67) to read

$$E = \tfrac{1}{2}N^2V_0 + \frac{1}{2}\sum_{\mathbf{k}\neq 0} (E_{\mathbf{k}} - \epsilon_{\mathbf{k}} - NV_{\mathbf{k}}). \tag{8.3.76}$$

It should be noted that, as in the B.C.S. scheme, if we had adopted the variational wave function Φ satisfying

$$\alpha_{\mathbf{k}}\,\Phi = 0 \quad \text{(all } \mathbf{k}\text{)} \tag{8.3.77}$$

and minimized

$$\langle\Phi|H|\Phi\rangle = E, \tag{8.3.78}$$

we would have obtained as Euler equation the result (8.3.72). Thus the two approaches are entirely equivalent.

At this stage, the problem is basically solved. What we have to do now is to analyse the above results and, where possible, make contact with our earlier work. It is, therefore, convenient to note at this stage that the $V_{\mathbf{k}}$'s used above are related to the $V(k)$'s of section 8.2.5 by the relationship

$$V_{\mathbf{k}} = (1/\Omega)\, V(k), \tag{8.3.79}$$

$V(k)$ being the Fourier transform of $V(r)$ and thus of order unity.

(*b*) *Ground state and sound velocity*. The most readily accessible quantity of physical interest is the sound velocity, c. This may be calculated from (8.3.76) and the equations

$$c = \sqrt{\left(\frac{1}{M}\frac{dP}{d\rho}\right)}, \quad P = \rho^2 \frac{d}{d\rho}\left(\frac{E}{N}\right), \tag{8.3.80}$$

where, as usual, $\rho = N/\Omega$, the average number density. The intermediate parameter P, is the pressure. Examination of the second term of (8.3.76) reveals it to be of second order in the potential and therefore negligible compared with the first, in the weak coupling limit. In this case, we obtain

$$c = \sqrt{\frac{\rho V(0)}{M}}, \tag{8.3.81}$$

which agrees with the classical result (8.2.77) for low enough temperatures. It should be remarked that (8.3.81) does not depend on the detailed analysis of part (*a*), above. For, if almost all the particles are in the condensate, by taking only the interactions between these, we obtain the leading order term $\frac{1}{2}N^2V_0$ in the energy.

Another point to which we should draw attention is the spatial form of the ground-state wave function. Huang (1963) has derived this in detail. We shall be content, here, to quote the final answer and note that it satisfies (8.3.77). This is in the spirit of the B.C.S. procedure (see sections 7.4.2 and 7.4.4). Huang shows that the (unnormalized) ground-state wave function is given by

$$\Phi = \Sigma' \left(\frac{v_{\mathbf{k}_1}}{u_{\mathbf{k}_1}}\right)^{n_{\mathbf{k}_1}} \left(\frac{v_{\mathbf{k}_2}}{u_{\mathbf{k}_2}}\right)^{n_{\mathbf{k}_2}} \cdots \left| n_0 \begin{matrix} n_{\mathbf{k}_1} n_{\mathbf{k}_2} \cdots \\ n_{-\mathbf{k}_1} n_{-\mathbf{k}_2} \cdots \end{matrix} \right\rangle, \tag{8.3.82}$$

where Σ' denotes summation over all possible N-particle wave functions of the type indicated for which $n_{\mathbf{k}} = n_{-\mathbf{k}}$ for all $\mathbf{k}$. The wave functions displayed in (8.3.82) are just those of independent Boson type discussed in section 8.3.1; by the notation, however, we have chosen to emphasize the special role of states of opposite momenta. The expression (8.3.82) should not, perhaps, surprise us, as it is the direct Boson analogue of the N-particle component (see (7.4.9) and subsequently) of the B.C.S. wave function. To verify (8.3.82), one notes, using (8.3.59) and the properties of the Boson creation and annihilation operators that for Φ given by (8.3.82)

$$u_{\mathbf{k}} A_{\mathbf{k}} \Phi = \Sigma' \left(\frac{v_{\mathbf{k}_1}}{u_{\mathbf{k}_1}}\right)^{n_{\mathbf{k}_1}} \left(\frac{v_{\mathbf{k}_2}}{u_{\mathbf{k}_2}}\right)^{n_{\mathbf{k}_2}} \cdots \frac{v_{\mathbf{k}}^{n_{\mathbf{k}}}}{u_{\mathbf{k}}^{n_{\mathbf{k}}-1}} \cdots \frac{\sqrt{(n_0+1)}\sqrt{n_{\mathbf{k}}}}{\sqrt{N}} \left| n_0+1 \begin{matrix} n_{\mathbf{k}_1} \ldots n_{\mathbf{k}}-1 \ldots \\ n_{-\mathbf{k}_1} \ldots n_{-\mathbf{k}} \ldots \end{matrix} \right\rangle \tag{8.3.83}$$

and

$$v_{\mathbf{k}} A_{-\mathbf{k}}^{\dagger} \Phi = \Sigma' \left(\frac{v_{\mathbf{k}_1}}{u_{\mathbf{k}_1}}\right)^{n_{\mathbf{k}_1}} \left(\frac{v_{\mathbf{k}_2}}{u_{\mathbf{k}_2}}\right)^{n_{\mathbf{k}_2}} \cdots \frac{v_{\mathbf{k}}^{n_{\mathbf{k}}+1}}{u_{\mathbf{k}}^{n_{\mathbf{k}}}} \cdots \frac{\sqrt{n_0}\sqrt{(n_{\mathbf{k}}+1)}}{\sqrt{N}} \left| n_0-1 \begin{matrix} n_{\mathbf{k}_1} \ldots n_{\mathbf{k}} \ldots \\ n_{-\mathbf{k}_1} \ldots n_{-\mathbf{k}}+1 \ldots \end{matrix} \right\rangle. \tag{8.3.84}$$

Consistent with our assumption that the occupation of the condensate is almost complete, we cancel the $\sqrt{(n_0+1)}$ of (8.3.83) and the $\sqrt{n_0}$ of (8.3.84) with the $\sqrt{N}$ factors. The resulting expressions are equal, as may be seen by replacing $n_{\mathbf{k}}$ by $n_{\mathbf{k}}+1$, $n_{-\mathbf{k}}$ by $n_{-\mathbf{k}}+1$ and, to conserve particle number, n_0 by n_0-2 in (8.3.83). But by definition of $\alpha_{\mathbf{k}}$ (see (8.3.64)), the equivalence of (8.3.83) and (8.3.84) means $\alpha_{\mathbf{k}} \Phi = 0$.

Finally let us note that the Bogoliubov theory, as here presented, should give the exact leading terms in an expansion in (the coupling constant premultiplying) V, but this will not be true for the density. For, on expanding the second part of (8.3.76) in powers of $\rho V(k)/\epsilon(k)$, the leading term is linear in ρ and quadratic in V. Thus, the second term, and, by inference, all those discarded in proceeding from (8.3.53) to (8.3.61), correct the first member of the series. For a review of the theory of the exact expansion in terms of the density, see Lieb (1963).

(*c*) *Excited states.* We have been able to reduce the system to a set of independent quasi-particle Bosons given by (8.3.74). The excited states are then of the kind $\alpha^\dagger_{\mathbf{k}_1}\alpha^\dagger_{\mathbf{k}_2}\ldots\Phi$, the associated excitation energy being $E_{\mathbf{k}_1}+E_{\mathbf{k}_2}+\ldots$. The simplest excitations are thus $\alpha^\dagger_{\mathbf{k}}\Phi$ with excitation energies $E_{\mathbf{k}}$. To show, then, that the latter is the excitation spectrum in the manner discussed in section 1.11, we have but to show that $\hbar\mathbf{k}$ represents the quasi-particle momentum as well as the original single-particle momentum. We proceed as in section 7.5.4 (*a*). It is necessary to prove that

$$\Sigma\mathbf{k}a^\dagger_{\mathbf{k}}a_{\mathbf{k}} = \Sigma\mathbf{k}\alpha^\dagger_{\mathbf{k}}\alpha_{\mathbf{k}}. \tag{8.3.85}$$

This is done by observing, first of all, that

$$\begin{aligned}\Sigma\mathbf{k}A^\dagger_{\mathbf{k}}A_{\mathbf{k}} &= \Sigma\mathbf{k}[u^2_{\mathbf{k}}\alpha^\dagger_{\mathbf{k}}\alpha_{\mathbf{k}}+v^2_{\mathbf{k}}+v^2_{\mathbf{k}}\alpha^\dagger_{-\mathbf{k}}\alpha_{-\mathbf{k}}+u_{\mathbf{k}}v_{\mathbf{k}}(\alpha^\dagger_{\mathbf{k}}\alpha^\dagger_{-\mathbf{k}}+\alpha_{-\mathbf{k}}\alpha_{\mathbf{k}})]\\ &= \Sigma\mathbf{k}\alpha^\dagger_{\mathbf{k}}\alpha_{\mathbf{k}},\end{aligned} \tag{8.3.86}$$

where symmetry has been exploited and (8.3.65) has been used. But it has already been shown (see (8.3.60) and subsequent comments) that $A^\dagger_{\mathbf{k}}A_{\mathbf{k}}$ can be replaced by $a^\dagger_{\mathbf{k}}a_{\mathbf{k}}$ and hence (8.3.85) follows.

Having thus proved that $E_{\mathbf{k}}$ is the elementary excitation spectrum, we go on to examine its detailed form. Clearly, this depends on the particular choice of $V_{\mathbf{k}}$ but, nevertheless, certain general observations can be made. For small k, using (8.3.74) we have the phonon-type spectrum

$$E_{\mathbf{k}} \sim \sqrt{[2\epsilon_{\mathbf{k}}\rho V(0)]} = \hbar k\sqrt{[\rho V(0)/M]} \tag{8.3.87}$$

corresponding to a sound velocity of $dE/d(\hbar k) = \sqrt{[\rho V(0)/M]}$, in agreement with (8.3.81). As seen earlier (section 8.3.5), a phonon-type spectrum is the essential feature which implies superflow. It should be noted that $V(0)$ must be positive for (8.3.87) to be meaningful. This implies that necessarily the interaction potential must contain a repulsive part. This is in contrast with the B.C.S. Fermion theory, where, to obtain a superfluid, it is necessary to introduce some attraction into the effective inter-electronic force.

Turning to the other extreme of large k, when $\epsilon_{\mathbf{k}} \gg \rho V(k)$, (8.3.74) gives

$$E_{\mathbf{k}} = \epsilon_{\mathbf{k}}+\rho V(k)+\ldots, \tag{8.3.88}$$

an expression with a dominant single-particle term. The precise intermediate form of $E_{\mathbf{k}}$ depends on the details of $V_{\mathbf{k}}$ and, in particular, it is not difficult to construct a variety of such functions yielding curves of Landau type each with a 'roton' minimum.

Finally, we shall show that the excited states $\alpha_{\mathbf{k}}^{\dagger}\Phi$, when expressed in configuration space co-ordinates, are just the functions (8.3.27) used by Feynman to describe phonon and roton states. To prove this, we look at the two factors of $\alpha_{\mathbf{k}}^{\dagger}\Phi$ as specified through (8.3.64). First, recalling (8.3.82), we have

$$v_{\mathbf{k}} A_{-\mathbf{k}} \Phi = \frac{v_{\mathbf{k}} a_{-\mathbf{k}} a_0^{\dagger}}{\sqrt{N}} \Sigma' \ldots \left(\frac{v_{\mathbf{k}}}{u_{\mathbf{k}}}\right)^{n_{\mathbf{k}}} \ldots \left| n_0 \begin{matrix} n_{\mathbf{k}_1} \ldots n_{\mathbf{k}} \ldots \\ n_{-\mathbf{k}_1} \ldots n_{-\mathbf{k}} \ldots \end{matrix} \right\rangle$$

$$= v_{\mathbf{k}} \Sigma' \ldots \left(\frac{v_{\mathbf{k}}}{u_{\mathbf{k}}}\right)^{n_{\mathbf{k}}} \ldots \frac{\sqrt{(n_0+1)}\sqrt{n_{\mathbf{k}}}}{\sqrt{N}} \left| n_0+1 \begin{matrix} n_{\mathbf{k}_1} \ldots n_{\mathbf{k}} \ldots \\ n_{-\mathbf{k}_1} \ldots n_{-\mathbf{k}}-1 \ldots \end{matrix} \right\rangle$$

$$= v_{\mathbf{k}} \frac{a_{\mathbf{k}}^{\dagger} a_0}{\sqrt{N}} \Sigma' \ldots \left(\frac{v_{\mathbf{k}}}{u_{\mathbf{k}}}\right)^{n_{\mathbf{k}}} \ldots \left| n_0+2 \begin{matrix} n_{\mathbf{k}_1} \ldots n_{\mathbf{k}}-1 \ldots \\ n_{-\mathbf{k}_1} \ldots n_{-\mathbf{k}}-1 \ldots \end{matrix} \right\rangle. \tag{8.3.89}$$

Now let us re-define our summations, replacing $n_{\mathbf{k}}-1$ by $n_{\mathbf{k}}$ and $n_{-\mathbf{k}}-1$ by $n_{-\mathbf{k}}$ and thus to conserve particle number, n_0 by n_0-2. The effect is to replace the above expression by

$$\frac{v_{\mathbf{k}} a_{\mathbf{k}}^{\dagger} a_0}{\sqrt{N}} \Sigma' \ldots \left(\frac{v_{\mathbf{k}}}{u_{\mathbf{k}}}\right)^{n_{\mathbf{k}}+1} \ldots \left| n_0 \begin{matrix} n_{\mathbf{k}_1} \ldots n_{\mathbf{k}} \ldots \\ n_{-\mathbf{k}_1} \ldots n_{-\mathbf{k}} \ldots \end{matrix} \right\rangle = \frac{v_{\mathbf{k}}^2}{u_{\mathbf{k}}} a_{\mathbf{k}}^{\dagger} a_0 \Phi. \tag{8.3.90}$$

Thus, we have

$$\alpha_{\mathbf{k}}^{\dagger}\Phi = u_{\mathbf{k}} A_{\mathbf{k}}^{\dagger}\Phi - v_{\mathbf{k}} A_{-\mathbf{k}}\Phi = \frac{1}{u_{\mathbf{k}}} a_{\mathbf{k}}^{\dagger} a_0 \Phi. \tag{8.3.91}$$

If we think of Φ as expanded in terms of the independent-particle states as in (8.3.82), one recalls the result (8.3.6) that the effect of $a_{\mathbf{k}}^{\dagger} a_0$ on any such state is to multiply it by $\sum_i e^{i\mathbf{k}.\mathbf{r}_i}$. Hence this is also the effect on Φ and one has

$$\alpha_{\mathbf{k}}^{\dagger}\Phi = \Sigma e^{i\mathbf{k}.\mathbf{r}_i}\Phi, \tag{8.3.92}$$

which were the excited-state wave functions of the more intuitive Feynman analysis.

(*d*) *Momentum distribution.* The momentum distribution is a property of some interest. It is given by

$$\langle \Phi | a_{\mathbf{k}}^{\dagger} a_{\mathbf{k}} | \Phi \rangle = v_{\mathbf{k}}^2 \quad (\mathbf{k} \neq 0), \tag{8.3.93}$$

and thus by (8.3.75). Without making too detailed assumptions concerning the functional form of $V_{\mathbf{k}}$, one can see that the graphical nature of this function is as shown in Fig. 8.16. The average

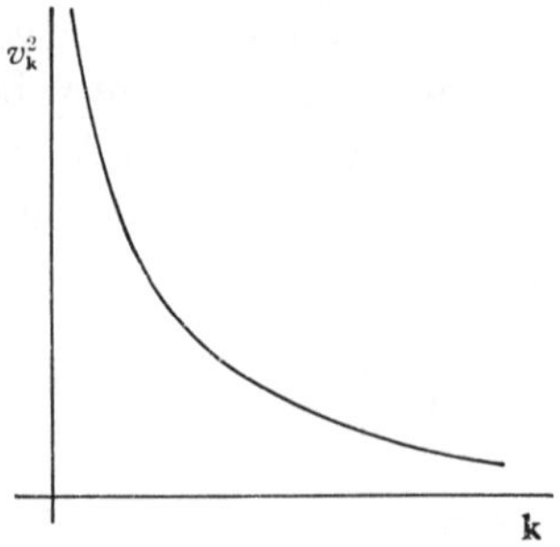

Fig. 8.16. Momentum distribution.

occupation of the zero-momentum state is obtained by simple subtraction as

$$N_0 = N - \sum_{\mathbf{k} \neq 0} v_{\mathbf{k}}^2. \tag{8.3.94}$$

The latter sum can be converted to Ω times a convergent integral. This may be shown straightforwardly by expanding $v_{\mathbf{k}}^2$ in the regions of possible divergency. We obtain quite easily

$$v_{\mathbf{k}}^2 = \frac{1}{2}\sqrt{\frac{\rho V(k)}{2\epsilon(k)}} + \dots \quad (\text{small } \mathbf{k}) \tag{8.3.95}$$

and

$$v_{\mathbf{k}}^2 = \left(\frac{\rho V(k)}{2\epsilon(k)}\right)^2 + \dots \quad (\text{large } \mathbf{k}). \tag{8.3.96}$$

Thus, we arrive at a necessary criterion for the applicability of the Bogoliubov method, namely that the parameters of the system are such that the depletion of the condensate, as given by the second term of (8.3.94), should be small compared with the total number of particles, N. An explicit example of the application of this result will be given in the treatment of the charged Boson gas in the following section.

8.4. Charged Boson gas

To conclude this chapter, we shall consider the Boson analogue of the high-density electron gas of Chapter 5, namely, the charged Boson gas moving in a neutralizing uniform background.

Interest in this problem was initially generated by Schafroth (1955; see also 1960) who was concerned with the superconducting properties of such a system, believing them to be related to superconductivity as it occurs in metals (Blatt, 1964).

A number of subsequent treatments have been given. The first due to Foldy (1961) and its later improvement by Girardeau (1962) show how the correlation energy in the high density limit depends on the density, and, in that sense, give the analogue of the series (5.3.22) for the correlation energy of an electron gas at high densities, discussed fully in Chapter 5. Other accounts, by Gaskell (1962) and Stephen (1962), use collective co-ordinate and elevated temperature approaches respectively. Below, we describe Foldy's work, as an illustration of the Bogoliubov formalism of the previous section.

8.4.1. *Foldy's treatment*

The results of section 8.3.7 apply almost without modification. The one proviso is that we must remember to allow for the neutralizing background of charge. The one time this concerns us is when considering the total energy. Clearly in (8.3.76), the first term, representing the self-energy of a uniform charge distribution, no longer occurs. Otherwise the formulae are as before, except that now we have $V_{\mathbf{k}} = 4\pi e^2/(\Omega k^2)$ appropriate to Coulomb interactions, and furthermore we take $\epsilon_{\mathbf{k}} = \hbar^2 k^2/2m$.

We begin by investigating the conditions under which the Bogoliubov method applies to this problem. With this purpose in mind (see the remarks following (8.3.96)) let us calculate the depletion of the condensate as given by the second term of (8.3.94). We have

$$N - N_0 = \frac{\Omega}{(2\pi)^3}\int_0^\infty v_{\mathbf{k}}^2\, 4\pi k^2\, dk \tag{8.4.1}$$

which, on using (8.3.74), (8.3.75) and our special form of $V_{\mathbf{k}}$, gives

$$\frac{N-N_0}{N} = \frac{r_s^{\frac{3}{4}}}{\pi 3^{\frac{1}{4}}}\int_0^\infty \left[\frac{\xi^4+2}{(\xi^4+4)^{\frac{1}{2}}} - \xi^2\right] d\xi = 0{\cdot}2114 r_s^{\frac{3}{4}}, \tag{8.4.2}$$

where r_s is the usual 'electron' separation parameter given by $(3\Omega/4\pi N)^{\frac{1}{3}} me^2/\hbar^2$. Since we have seen that the Bogoliubov method is applicable to the case when the depletion (8.4.2) is small, the results we derive below should hold in the high density limit $r_s \to 0$.

(*a*) *Ground-state energy*. Taking into account the modification noted above concerning the effect of the background charge, the total energy of the system is (cf. (8.3.76))

$$E = \frac{\Omega}{2(2\pi)^3}\int_0^\infty (E_{\mathbf{k}} - \epsilon_{\mathbf{k}} - NV_{\mathbf{k}})\, 4\pi k^2\, dk. \tag{8.4.3}$$

Once again we use (8.3.74) and the Coulombic form of V_k and find an energy per particle given by

$$E/N = r_s^{-\frac{3}{4}}(3^{\frac{1}{4}}/\pi)\int_0^\infty [\xi^2(\xi^4+4)^{\frac{1}{2}} - \xi^4 - 2]\, d\xi = -0{\cdot}803 r_s^{-\frac{3}{4}}. \tag{8.4.4}$$

This is clearly the leading term in a high density expansion. Since the independent particle contribution is zero (the kinetic energy being zero and the potential energy being cancelled as a result of the background charge), (8.4.4) is all correlation energy and is, therefore, necessarily negative.

(*b*) *Excitation energies*. Using (8.3.72), we have

$$E_{\mathbf{k}} = \hbar\left(\omega_p^2 + \frac{\hbar^2 k^4}{4m^2}\right)^{\frac{1}{2}}, \quad \omega_p = (4\pi\rho e^2/m)^{\frac{1}{2}}. \tag{8.4.5}$$

Thus, for small k, the excitation energies are like those for plasma oscillations, with a specific dispersion relation. The finite energy gap and the shape of the excitation spectrum implies that a charged Bose gas, at high densities, is both a superfluid and a superconductor. The Coulombic interactions are, of course, crucial for obtaining the energy gap, for, as we saw in (8.3.87), if $V(k)$ is finite as k tends to zero, a phonon-type spectrum is obtained. The situation is very reminiscent of the discussion of the collective modes of such systems (recall Fig. 5.8 of Chapter 5, where the $k = 0$ limit is independent of the statistics). In that case (cf. (8.2.74) and (8.2.75)), if $V(0)$ exists, one obtains zero sound with frequency proportional to wave index, and if V is Coulombic, one finds plasma oscillations, the excitation of which costs energy $\hbar\omega_p$.

It is also interesting that, from (8.4.5), there is a continuous change in the energy of the single-particle excitations, as k increases, from plasmon to free particle form.

Actually, there is a serious difficulty in this account which is not superficially obvious (Lieb & Sakakura, 1964). If one attempts to calculate the next order term in the energy series (8.4.4), the potential energy terms dropped in the Bogoliubov theory must be included.

When this is done, a divergent integral occurs, much as in the Gell-Mann & Brueckner calculation of Chapter 5. The reason for, and resolution of, this is rather as before and a further contribution to (8.4.4) of $(-\frac{1}{8})\ln r_s$ is obtained. This result was first obtained by Girardeau (1962).

In the low density limit, the arguments discussed in section 5.8 of Chapter 5 again apply, and the energy of the ground state is given by (5.8.4) (see also the comments immediately following this equation). Since the low-lying excitations are now phonon-like, it is clear that, at some transition density, the energy gap must disappear as the density is lowered.

Problems

P.8 (i). Let us suppose that with $\epsilon_{\mathbf{k}} = k^2/2M$, the energy spectrum, as given by (8.3.74), defines a Landau-type curve of the kind shown in Fig. 8.8, the phonon part being given by (8.3.18) and the roton part by (8.3.19). Show that in the phonon region

$$NV_{\mathbf{k}} \sim Mc^2, \tag{1}$$

and the next order correction term is negative, and that in the roton region

$$NV_{\mathbf{k}} = \frac{\epsilon_{\mathbf{k}_0}}{2}\left[\left(\frac{\Delta}{\epsilon_{\mathbf{k}_0}}\right)^2 - 1\right] - (k-k_0)\frac{\epsilon'_{\mathbf{k}_0}}{2}\left[\left(\frac{\Delta}{\epsilon_{\mathbf{k}_0}}\right)^2 + 1\right] + c_2(k-k_0)^2 + \dots, \tag{2}$$

where
$$\lim_{\Delta/\epsilon_{\mathbf{k}} \to 1} c_2 = \frac{1}{2}\left(\frac{1}{M} + \frac{1}{\mu}\right). \tag{3}$$

Thus, sketch two typical $NV_{\mathbf{k}}$ curves, one for $\Delta > \epsilon_{\mathbf{k}_0}$ and one for $\Delta < \epsilon_{\mathbf{k}_0}$, which give a Landau-type spectrum.

If $V(k)$ $(=\Omega V_k$, by (8.3.79)) is interpreted simply as the Fourier transform of some spherically symmetric potential $V(r)$, the leading order correction to (1) will be quadratic in k and with negative coefficient. Use (8.3.47) to show that this implies that $V(r)$ has the attractive r^{-6} tail characteristic of helium.

P.8 (ii). Show that an alternative derivation of the Feynman spectrum (8.3.39) from the wave function (8.3.30) is to require the latter to be an eigenfunction of momentum.

P.8 (iii). Investigate the energy spectrum (8.3.74) for the pathological interaction

$$V(k) = \begin{cases} V(0)\,(>0) & (k \leqslant k_0), \\ 0 & (k > k_0). \end{cases} \tag{4}$$

Thus show explicitly that, in general, not only a negative range in $V(k)$, but also a precipitous fall in this function can induce a 'roton' minimum.

Take the infinite k_0 limit in (4) and thus obtain a real space delta function interaction. Investigate in one dimension the excitation spectrum, total energy and ground-state depletion of such a system using (8.3.74), (8.3.76) and (8.3.94). The results so obtained should be compared with the exact results of Lieb (1963) and Lieb & Liniger (1963).

P.8 (iv). Apply the Debye model to (8.2.52) and (8.2.56). (In this approximation, isotropy and a single value for the sound velocity $c = \omega_q^k/q$ is assumed for longitudinal and transverse waves. Then $\sum_{\mathbf{q},k}$ is replaced by $\frac{1}{3}\int_0^{q_D} d\mathbf{q}$, where $q_D = (6\pi^2 N/\Omega)^{\frac{1}{3}}$. The characteristic Debye temperature θ_D which arises in this approximation is defined through $k_B\theta_D = \hbar q_D c$.)

Write $n_{\mathbf{q}}^k = 0$ in (8.2.52) to show that the Debye zero-point correlations are given by

$$\langle\{\mathbf{K}.\mathbf{u}_{\mathbf{m}}(0)\}\{\mathbf{K}.\mathbf{u}_{\mathbf{n}}(t)\}\rangle_{T=0}^{(D)}$$

$$= \frac{9\hbar^2|\mathbf{m}-\mathbf{n}|}{M4k_B\theta_D q_D^2}\left\{\frac{1-\exp\left[iq_D(|\mathbf{m}-\mathbf{n}|+ct)\right]}{|\mathbf{m}-\mathbf{n}|+ct} + \frac{1-\exp\left[-iq_D(|\mathbf{m}-\mathbf{n}|-ct)\right]}{|\mathbf{m}-\mathbf{n}|-ct}\right\}.$$

Prove, also, that (8.2.56) becomes

$$2W^{(D)} = \frac{9K^2}{4Mk_B\theta_D}\left\{1+4\left(\frac{T}{\theta_D}\right)\Phi\left(\frac{\theta_D}{T}\right)\right\},$$

where Φ is the function

$$\Phi(x) = \frac{1}{x}\int_0^x \frac{y\,dy}{e^y-1}$$

characteristic of the Debye theory.

CHAPTER 9

GRAND PARTITION FUNCTIONS

9.1. Introduction

So far we have been mostly concerned with the ground state and the low-lying excited states of a general many-body system. For studying the higher excited states, whose degeneracy will increase with the volume of the system under consideration, the methods employed earlier are not very suitable and it will prove necessary to invoke the full machinery of Statistical Mechanics. From the outset, we shall deal therefore with the calculation of the grand partition function Z_G.

Deduction of the bulk properties of the system from Z_G is a standard procedure (see, for example, Landau & Lifshitz, 1958) and in section 9.7 we shall immediately calculate thermodynamic properties from Z_G for a system with Coulomb interactions. Wider applications of the theory are, of course, possible, but we consider only the above example here. Further reference to elevated temperature theory will be made in Chapter 10.

Using analogous methods to those described in Chapter 4, it will prove possible to make a diagrammatic analysis of Z_G. An alternative procedure would have generalized the Bloch matrix approach of Chapter 1, sections 1.6 and 1.7 (in fact the Bloch equation is invoked also in the procedure presented in this chapter), and this method has been used, for example, by Montroll & Ward (1958).

9.2. Grand partition function

As usual, the Hamiltonian H for N spinless Fermions is given by (3.6.1) and (3.6.2), or alternatively, and more usefully for the present discussion, in terms of $a_{\mathbf{k}}$ and $a_{\mathbf{k}}^{\dagger}$ corresponding to a basic set of plane waves by (3.6.3) and (3.6.4). We remind the reader that the interaction $v(\mathbf{r}_1, \mathbf{r}_2)$ is being assumed to depend only on $|\mathbf{r}_1 - \mathbf{r}_2|$ and to have matrix elements between pairs of plane-wave states given by (1.3.6).

The grand partition function Z_G is now defined as

$$Z_G = \mathrm{Tr}\,[\exp\{-\beta(H-\mu N)\}]$$
$$= \mathrm{Tr}\,[\exp\{\alpha N-\beta H\}]. \tag{9.2.1}$$

Tr in (9.2.1) simply refers to the trace of the operator following it. N is the number operator $\Sigma a_{\mathbf{k}}^{\dagger} a_{\mathbf{k}}$ of Chapter 3 (3.4.2), μ is the chemical potential, β the inverse temperature (strictly $1/k_B T$), and $\alpha = \beta\mu$. Later in the chapter it will be convenient to use the fugacity z defined as e^{α}, in place of α.

We now introduce the distribution operator C, defined by

$$C = \exp\{-\beta(H-\mu N)\}. \tag{9.2.2}$$

C is a function of β and satisfies the Bloch equation (cf. Chapter 1, (1.6.3))

$$-\frac{\partial C}{\partial \beta} = (H-\mu N)\,C, \tag{9.2.3}$$

with the boundary condition

$$[C]_{\beta=0} = 1. \tag{9.2.4}$$

As already remarked in Chapter 1, (9.2.3) is very similar to the time-dependent Schrödinger equation with β replacing *it*, the formal solution of (9.2.3) with the boundary condition (9.2.4) being (9.2.2). Exploiting this analogy we shall obtain a perturbation expansion for C just as in Chapter 4. To keep the analogy as close as possible we shall often refer to β as a 'time', the inverted commas distinguishing it from the usual time variable.

9.2.1. *Perturbation series for distribution operator*

It will be convenient to write the modified Hamiltonian $H-\mu N$ in the form

$$H-\mu N = H_0-\mu N+V$$
$$= H_0'+V, \tag{9.2.5}$$

where V is as usual the total perturbation.

In analogy with Chapter 4, we introduce the 'time'-dependent operators of the interaction picture,

$$V(u) = \exp\{uH_0'\}\,V\exp\{-H_0'u\}, \tag{9.2.6}$$

the 'time' variable being, of course, u.

In terms of the $a_{\mathbf{k}}$'s we have

$$V(u) = \frac{1}{2}\sum_{\mathbf{k}'s} \langle \mathbf{k}_1 \mathbf{k}_2 | v | \mathbf{k}_3 \mathbf{k}_4 \rangle a^{+}_{\mathbf{k}_1}(u)\, a^{+}_{\mathbf{k}_2}(u)\, a_{\mathbf{k}_4}(u)\, a_{\mathbf{k}_3}(u), \quad (9.2.7)$$

where

$$a_{\mathbf{k}}(u) = a_{\mathbf{k}} \exp\{-(\epsilon_{\mathbf{k}} - \mu)\, u\} \quad (9.2.8)$$

and

$$a^{+}_{\mathbf{k}}(u) = a^{+}_{\mathbf{k}} \exp\{(\epsilon_{\mathbf{k}} - \mu)\, u\}. \quad (9.2.9)$$

Then

$$\frac{\partial}{\partial \beta}(\exp\{\beta H'_0\} C) = \exp\{\beta H'_0\}\frac{\partial C}{\partial \beta} + \exp\{\beta H'_0\} H'_0 C$$

$$= -\exp\{\beta H'_0\} VC$$

$$= -V(\beta)(\exp\{\beta H'_0\} C). \quad (9.2.10)$$

From (9.2.10) we obtain

$$\exp\{\beta H'_0\} C = 1 - \int_0^{\beta} V(u_1)\, du_1 + \int_0^{\beta} du_2 \int_0^{u_2} du_1\, V(u_2)\, V(u_1) + \ldots \quad (9.2.11)$$

and hence

$$C = \sum_0^{\infty} (-1)^r \int_{\beta > u_1 > u_2 > \ldots > u_r > 0} du_1 \ldots du_r \exp\{-\beta H'_0\} V(u_1) \ldots V(u_r). \quad (9.2.12)$$

Hence a formal expansion for Z_G, the trace of C, is

$$Z_G = \sum_0^{\infty} (-1)^r \int_{\beta > u_1 > \ldots > u_r > 0} du_r \ldots du_1 \times \mathrm{Tr}\,[\exp\{\alpha N - \beta H_0\}\, V(u_1) \ldots V(u_r)]. \quad (9.2.13)$$

9.2.2. *Grand partition function Z_G in terms of expectation values*

Let us denote by Z^0_G, the grand partition function for the non-interacting system, that is

$$Z^0_G = \mathrm{Tr}\,[\exp\{-\beta(H_0 - \mu N)\}]. \quad (9.2.14)$$

Now we define the expection value of an operator θ for non-interacting particles in thermodynamic equilibrium at temperature β as

$$\langle\langle \theta \rangle\rangle_0 = \frac{\mathrm{Tr}\,[\exp\{-\beta(H_0 - \mu N)\}\theta]}{Z^0_G}. \quad (9.2.15)$$

Using this definition, (9.2.13) may be rewritten

$$\frac{Z_G}{Z^0_G} = \sum_0^{\infty} (-1)^r \int_{\beta > u_1 > \ldots > u_r > 0} du_1 \ldots du_r \langle\langle V(u_1) \ldots V(u_r) \rangle\rangle_0. \quad (9.2.16)$$

9.3. Diagrammatic expansion

In Chapter 4, we obtained a diagrammatic perturbation series for the ground-state energy by writing a given time-ordered product of operators in terms of normal products and contractions. The normal products had expectation value zero in the ground state of the unperturbed system. Since the analogy of the ground-state expectation value is the average value defined by $\langle\langle\ \rangle\rangle_0$, we must generalize the normal-product definition so that this average value of the normal product of two operators is zero, if we are to exploit the analogy.

9.3.1. *Normal products*

We saw in Chapter 3, (3.4.1) that $N_{\mathbf{k}} = a_{\mathbf{k}}^{\dagger} a_{\mathbf{k}}$ is the number operator corresponding to the single-particle state $|\mathbf{k}\rangle$. Then its average value is

$$\langle\langle N_{\mathbf{k}}\rangle\rangle_0 = \frac{\mathrm{Tr}\left[\exp\{-\beta(H_0-\mu N)\}\, a_{\mathbf{k}}^{\dagger} a_{\mathbf{k}}\right]}{Z_G^0}$$

$$= \frac{\mathrm{Tr}\left[\Pi \exp\{(\alpha-\beta\epsilon_{\mathbf{k}'})\, a_{\mathbf{k}'}^{\dagger} a_{\mathbf{k}'}\}\, a_{\mathbf{k}}^{\dagger} a_{\mathbf{k}}\right]}{Z_G^0}$$

$$= \frac{\exp\{\alpha-\beta\epsilon_{\mathbf{k}}\}}{1+\exp\{\alpha-\beta\epsilon_{\mathbf{k}}\}} = \frac{1}{1+\exp\{\beta(\epsilon_{\mathbf{k}}-\mu)\}}$$

$$= f_{\mathbf{k}}, \tag{9.3.1}$$

where $f_{\mathbf{k}}$, the Fermi–Dirac function, is of course giving the probability of occupation of the level $|\mathbf{k}\rangle$.

We now define the normal product $N(a_{\mathbf{k}}^{\dagger} a_{\mathbf{k}})$ by the following equation:

$$N(a_{\mathbf{k}}^{\dagger} a_{\mathbf{k}}) = (1-f_{\mathbf{k}})\, a_{\mathbf{k}}^{\dagger} a_{\mathbf{k}} - f_{\mathbf{k}}\, a_{\mathbf{k}} a_{\mathbf{k}}^{\dagger}. \tag{9.3.2}$$

This definition reduces to that of Chapter 4 in the case of zero temperature, as we shall now show. As $\beta \to \infty$, $f_{\mathbf{k}} \to 1$ for all k with $\epsilon_{\mathbf{k}} < \mu$, and $f_{\mathbf{k}} \to 0$ for all k with $\epsilon_{\mathbf{k}} > \mu$. Thus $N(a_{\mathbf{k}}^{\dagger} a_{\mathbf{k}})$ becomes $a_{\mathbf{k}}^{\dagger} a_{\mathbf{k}}$ for $\epsilon_{\mathbf{k}} > \mu$, and $-a_{\mathbf{k}} a_{\mathbf{k}}^{\dagger}$ for $\epsilon_{\mathbf{k}} < \mu$, which is the desired result.

At finite temperature, the state $|\mathbf{k}\rangle$ behaves like an occupied level with probability $f_{\mathbf{k}}$ and an unoccupied level with probability $1-f_{\mathbf{k}}$.

From the definition of the normal product in (9.3.2), we obtain its average value as

$$\langle\langle N(a_{\mathbf{k}}^{\dagger} a_{\mathbf{k}})\rangle\rangle_0$$

$$= \frac{\mathrm{Tr}\left[\exp\{-\Sigma\beta(\epsilon_{\mathbf{k}'}-\mu)\, a_{\mathbf{k}'}^{\dagger} a_{\mathbf{k}'}\}\{(1-f_{\mathbf{k}})\, a_{\mathbf{k}}^{\dagger} a_{\mathbf{k}} - f_{\mathbf{k}} a_{\mathbf{k}} a_{\mathbf{k}}^{\dagger}\}\right]}{Z_G^0}$$

$$= \frac{\exp\{-\beta(\epsilon_{\mathbf{k}}-\mu)\}(1-f_{\mathbf{k}})}{1+\exp\{-\beta(\epsilon_{\mathbf{k}}-\mu)\}} - f_{\mathbf{k}}(1-f_{\mathbf{k}})$$

$$= 0,$$

the last step following trivially from (9.3.1).

9.3.2. *Contractions*

Contractions between two 'time'-dependent operators are defined, following again the procedure of Chapter 4, as the difference between their 'time'-ordered and normal products.

The non-vanishing contraction for the 'time'-dependent creation and annihilation operators are

$$\overbrace{a_{\mathbf{k}}^{\dagger}(u_2)\, a_{\mathbf{k}'}}(u_1) = T[a_{\mathbf{k}}^{\dagger}(u_2)\, a_{\mathbf{k}'}(u_1)] - N[a_{\mathbf{k}}^{\dagger}(u_2)\, a_{\mathbf{k}'}(u_1)]$$
$$= f_{\mathbf{k}}\, \delta_{\mathbf{k}\mathbf{k}'} \exp\{(\epsilon_{\mathbf{k}}-\mu)(u_2-u_1)\} \quad \text{if} \quad u_2 \geqslant u_1$$

and $$-(1-f_{\mathbf{k}})\, \delta_{\mathbf{k}\mathbf{k}'} \exp\{(\epsilon_{\mathbf{k}}-\mu)(u_2-u_1)\} \quad \text{if} \quad u_2 < u_1. \qquad (9.3.3)$$

9.3.3. *Graphical representation*

Wick's theorem can now be applied to the products appearing in (9.2.16) for the expansion of Z_G/Z_G^0. Each product is expressed in terms of contractions and normal products.

A typical non-vanishing nth-order term of (9.2.16) will contain:

(i) A factor $(-1)^n$.

(ii) Integrations over $u_1 \ldots u_n$ with $\beta > u_1 > u_2 > u_n > 0$.

(iii) n interactions at 'times' $u_n \ldots u_1$, each interaction giving a matrix element which is zero unless there is conservation of momentum in the interaction.

(iv) Non-vanishing contractions amongst the $4n$ creation and annihilation operators associated with the n interactions.

As in Chapter 4, every such term can be described by a graph if we represent:

(i) The matrix element of the interaction at u_1, namely

$$\langle \mathbf{k}_1 \mathbf{k}_2 | v | \mathbf{k}_3 \mathbf{k}_4 \rangle$$

by Fig. 9.1.

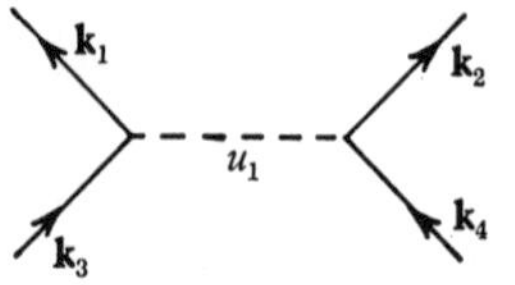

Fig. 9.1

Fig. 9.2

Fig. 9.3

(ii) The contraction $\overbrace{a_{\mathbf{k}}(u_2)\, a^{\dagger}_{\mathbf{k}}(u_1)}$ by Fig. 9.2 if $u_2 \geqslant u_1$ and by Fig. 9.3 if $u_2 < u_1$.

Then a typical graph corresponding to an nth-order term will consist of:

(i) n 'vertices' (horizontal dashed lines) corresponding to the n interactions at $u_n \ldots u_1$, drawn so that the vertex with label u_i is above the vertex with label u_j if $i < j$.

(ii) Directed solid lines, each labelled by a momentum index, and drawn so that (a) each solid line joins two vertices or a vertex to itself, (b) two directed lines go into each vertex and two come out.

A directed line with arrow pointing up is termed, as usual, a particle line and one with arrow pointing down a hole line. It will be seen that the directed lines form sets of closed loops.

9.3.4. *Rules for calculating contributions from graphs*

In section 9.3.3 we have just seen how to represent a term of the series (9.2.16) for Z_G/Z_G^0 by a graph. At this stage, we give the rules for obtaining the contribution to Z_G/Z_G^0 from a given graph with labels for the directed and interaction lines:

(i) Associate a matrix element $\frac{1}{2}\langle \mathbf{k}\mathbf{k}' | v | \mathbf{k}''\mathbf{k}''' \rangle$ with the interaction line (vertex) at which directed lines $\mathbf{k}''$, $\mathbf{k}'''$ enter the vertex and $\mathbf{k}, \mathbf{k}'$ leave.

(ii) Associate a factor $\exp\{(u-\omega)(\epsilon_{\mathbf{k}}-\mu)\}$ with directed line $\mathbf{k}$ starting at 'time' ω and ending at 'time' u.

(iii) Associate a factor f_k with line $\mathbf{k}$ for which the arrow points downwards.

(iv) Associate a factor $(1-f_{\mathbf{k}})$ with line $\mathbf{k}$ for which the arrow points upwards.

(v) Associate a factor $(-1)^{h+l}$, with a diagram in which there are l closed loops and h hole lines.

Then the contribution to Z_G/Z_G^0 is obtained by integrating the product of the factors thus obtained over the times $u_1 \ldots u_n$ as shown below:

$$\int_{\beta > u_1 > \ldots > u_n},$$

and summing over the different states $\mathbf{k}, \mathbf{k}'$, etc. Finally, for an nth-order diagram we must multiply the result by $(-1)^n$.

These rules are the same as those of Chapter 4 except for the statistical factors which, as we have seen already, occur since the state $\mathbf{k}$ behaves as an occupied level with probability $f_{\mathbf{k}}$ and an unoccupied level with probability $(1-f_{\mathbf{k}})$.

9.3.5. *Volume dependence of contributions*

It is important at this point to isolate the volume dependence of the various contributions, for reasons which we discussed fully in Chapter 4. To do so, we note that (i) every matrix element gives a factor proportional to $1/\Omega$, (ii) every summation over $\mathbf{k}$ gives a factor proportional to Ω.

Let us consider specifically a connected (in the sense of Chapter 4) graph which contributes to Z_G/Z_G^0, and which has n interaction lines and $2n$ directed lines.

The Ω dependence of its contribution will consist of the following factors: (i) $1/\Omega^n$ due to the n interaction lines, and (ii) $\Omega^{2n-(n-1)}$ due to summation over momentum states. The index is $2n-(n-1)$ as shown, since of $2n$ possible momentum summations, $n-1$ of these do not arise because of momentum conservation at the $(n-1)$ interactions. At the last interaction, momentum conservation is then included automatically. Thus, as in Chapter 4, we arrive at the important result that contributions from connected diagrams are proportional to Ω.

In general, of course, diagrams corresponding to terms in the expansion of Z_G/Z_G^0 consist of disconnected parts. By arguments very similar to those employed in Chapter 4, we can arrive at the exponential theorem; namely, that the contribution to Z_G/Z_G^0 from all diagrams is equal to the exponential of the contributions from all connected diagrams. Thus we arrive at the result, which must also follow on physical grounds, that the series for $\ln Z_G/Z_G^0$ consists of connected diagrams only—the contributions from which are all explicitly proportional to Ω. Thus we write

$$\ln Z_G = \ln Z_G^0 + \sum_0^\infty (-1)^r \int_{\beta > u_1 > u_2 > \ldots > u_r > 0} \times du_1 \ldots du_r \, \langle\langle V(u_1) \ldots V(u_r) \rangle\rangle_c, \quad (9.3.4)$$

where c indicates that only connected diagrams should be drawn.

9.4. Connection with ground-state theory

In ground-state perturbation theory the direct graphs arising in the first two orders were the following:

Fig. 9.4

The graphs corresponding to the first- and second-order terms in the expansion of $\ln Z_G$ are not only those of Fig. 9.4 (and those obtainable from them—the so-called exchange graphs) but also graphs such as those of Fig. 9.5.

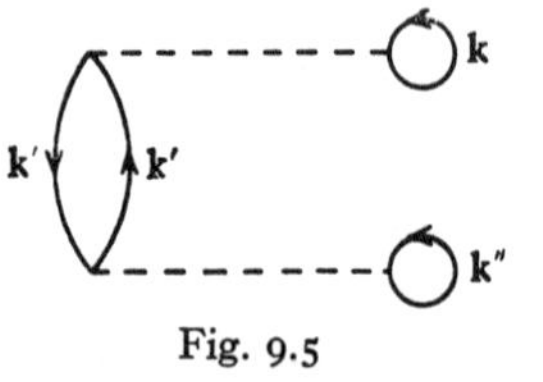

Fig. 9.5

In ground-state perturbation theory, such a graph cannot arise, since, either $k' < k_f$ and hence there cannot be a particle line of momentum k', or $k' > k_f$ and there cannot be a hole line of momentum k' (cf. p. 113).

In the present case, particle and hole lines with the same momentum can arise. The reason is again the fact that any single-particle level acts partly as an occupied and partly as an empty level. The contribution of the graph of Fig. 9.5 to $\ln Z_G$ is obtained according to the rules enunciated earlier as

$$(-1)^2 \sum_{\mathbf{kk'k''}} \langle \mathbf{k'k''}|v|\mathbf{k'k''}\rangle \langle \mathbf{kk'}|v|\mathbf{kk'}\rangle f_{\mathbf{k'}}(1-f_{\mathbf{k'}})f_{\mathbf{k}}f_{\mathbf{k''}}(-1)^{3+3}$$

$$\times \int_0^\beta du_2 \int_0^{u_2} du_1 \exp\{-(u_1-u_2)\epsilon_{\mathbf{k'}}\}\exp\{-(u_2-u_1)\epsilon_{\mathbf{k'}}\}$$

$$= \sum_{\mathbf{kk'k''}} f_{\mathbf{k}}f_{\mathbf{k'}}f_{\mathbf{k''}}(1-f_{\mathbf{k'}})\langle \mathbf{k'k''}|v|\mathbf{k'k''}\rangle \langle \mathbf{kk'}|v|\mathbf{kk'}\rangle.$$

This contribution is in general finite even in the limit of zero temperature, and therefore a note of caution is called for whenever we attempt to calculate zero-temperature properties from an expression for the same quantities at non-zero temperatures T by taking the limit $T \to 0$. The physics is not necessarily the same in both cases.

Thus, in the ground-state case, one calculates the properties of a system of a given number N of particles while in the elevated temperature case one has a fixed chemical potential and one calculates average properties of large numbers of possible configurations of an indefinite number of particles. All that one requires is that the average number of particles in these configurations is N. Going into details as to how one should take the limit $\beta \to \infty$ will take us too far away from our main theme here. We merely refer to the survey by Bloch (1963), to Kohn & Luttinger (1960), and to Luttinger & Ward (1960).

9.5. Alternative expansion for ln Z_G

Let us effect a slight generalization of the Hamiltonian with which we have worked so far and write

$$H^g = H_0 + gV,$$

where g is a coupling constant ultimately to be put equal to one to regain our original case. Then we generalize the grand partition function to Z_G^g given by

$$Z_G^g = \mathrm{Tr}\,[\exp\{\alpha N - \beta H_0 - \beta g V\}]. \qquad (9.5.1)$$

From (9.5.1), making use of the property that the trace of a product of operators is unaltered for any cyclic interchange of the operators, we can deduce that

$$g\frac{\partial Z_G^g}{\partial g} = -\beta \,\mathrm{Tr}\,[\exp\{(\alpha N - \beta H^g)\} gV]. \qquad (9.5.2)$$

An expansion for $g(\partial Z_G^g/\partial g)$ can now be carried out in a manner similar to that obtained in section 9.3 for Z_G. The expansion consists of all closed diagrams in which one interaction takes place at 'time' zero.

A general nth-order diagram will consist of several connected parts not joined to each other by any interaction line. The contributions from such disconnected parts are multiplicative. One of the connected parts of the diagram will contain the interaction at time zero. We shall call it the linked part of the graph. The other connected parts of the diagram are termed unlinked parts. These unlinked parts are components that occur in the expansion of Z_G. If we combine all the linked parts of all the diagrams arising in the expansion of $g(\partial Z_G^g/\partial g)$ with all the members occurring in the expansion of Z_G in all possible ways we shall obtain all the diagrams in the expansion of $g(\partial Z_G^g/\partial g)$.

Thus we can enormously reduce the number of diagrams we have to draw if we consider not $g(\partial Z_G^g/\partial g)$ but $(1/Z_G^g)(\partial Z_G^g/\partial g)$. In the expansion of the latter only connected diagrams having one interaction line at 'time' zero arise.

Hence we obtain the result that

$$\frac{1}{Z_G^g} g\frac{\partial Z_G^g}{\partial g} = g\frac{\partial}{\partial g}\ln Z_G^g$$

$$= -\beta \,\mathrm{Tr}\,[\exp\{\alpha N - \beta H^g\} gV]_c, \qquad (9.5.3)$$

where $[\ ]_c$ denotes that we need only consider connected diagrams. This is to be contrasted with (9.5.2), where the sum over all diagrams is, of course, implied.

Hence, by integration with respect to g,

$$\ln Z_G = \ln Z_G^0 - \beta \int_0^1 \frac{dg}{g} \mathrm{Tr}\,[\exp\{\alpha N - \beta H^g\} gV]_c, \quad (9.5.4)$$

where, of course Z_G is simply $[Z_G^g]_{g=1}$.

Rules for contributions to $\ln Z_G$ arising from the representation of (9.5.4) by graphs are trivially altered from those enunciated earlier. Therefore we shall not spell these out again.

9.6. Ring diagrams

Since it is entirely impracticable at present to calculate the contributions to $\ln Z_G$ given by all the graphs corresponding to (9.5.4), we content ourselves with the evaluation of contributions due to an important class of diagrams, the so-called ring diagrams of Chapter 5. In this way we shall establish the connexion between the present approach and that adopted in Chapters 4 and 5. We shall largely follow a method used by Thouless (1960).

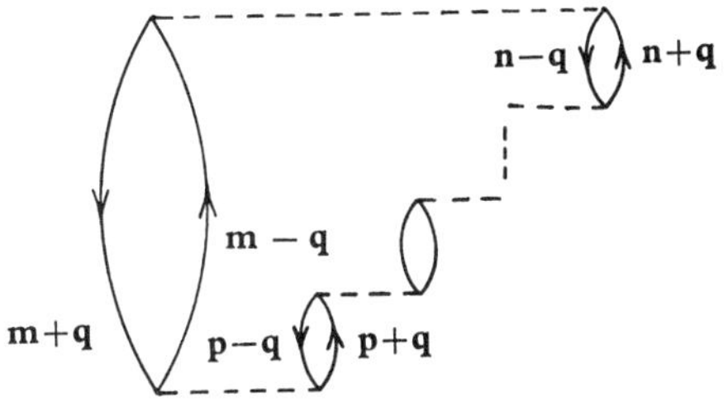

Fig. 9.6

A typical ring diagram is shown in Fig. 9.6. At each vertex, a particle-hole pair interact with the same transfer of momentum $2\mathbf{q}$, a typical matrix element being

$$\langle \mathbf{m}-\mathbf{q}, \mathbf{p}+\mathbf{q}|v|\mathbf{m}+\mathbf{q}, \mathbf{p}-\mathbf{q}\rangle = v(2\mathbf{q}).$$

If we 'remove' the last interaction, the diagram of Fig. 9.6 becomes that of Fig. 9.7.

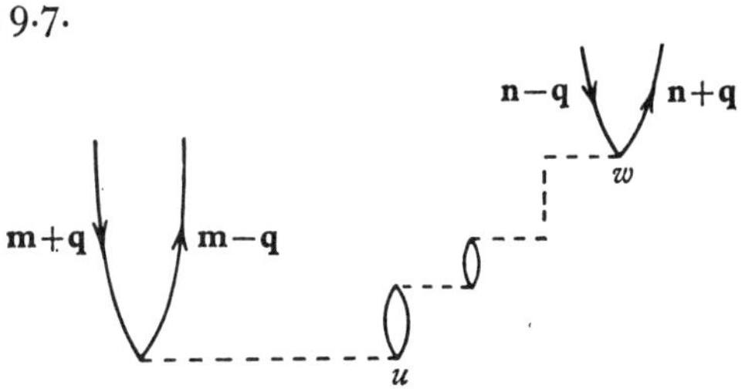

Fig. 9.7

The propagator (cf. section 9.3.4) for a particle of momentum $\mathbf{k}$ going from 'time' u to 'time' w is

$$\begin{aligned} S(\mathbf{k}, u-w) &= -f_{\mathbf{k}} \exp\{(\epsilon_{\mathbf{k}}-\mu)(u-w)\} \quad (u \geqslant w) \\ &= (1-f_{\mathbf{k}}) \exp\{(\epsilon_{\mathbf{k}}-\mu)(u-w)\} \quad (u < w). \end{aligned} \tag{9.6.1}$$

Fig. 9.7 describes the interaction of particle-hole pairs with the same transfer of momentum $2\mathbf{q}$ at each interaction. In particular the particle-hole pair at 'time' u has momenta $(\mathbf{m}+\mathbf{q}, \mathbf{m}-\mathbf{q})$. After several such interactions at time w, the particle-hole pair finds itself with momenta $(\mathbf{n}+\mathbf{q}, \mathbf{n}-\mathbf{q})$. This process can be described in terms of a propagator, just as we described a free particle by the propagator of (9.6.1). The new particle-hole propagator will be denoted by $R_{\mathrm{nm}}(\mathbf{q}, u-w)$.

$R_{\mathrm{nm}}(\mathbf{q}, u-w)$ will be given by the sum of ring diagrams such as that of Fig. 9.7. It is a function of $u-w$ only and will be periodic in t, satisfying

$$R_{\mathrm{nm}}(\mathbf{q}, t-\beta) = R_{\mathrm{nm}}(\mathbf{q}, t). \tag{9.6.2}$$

An integral equation for R_{nm} can now be written down. It is

$$\begin{aligned} R_{\mathrm{nm}}(\mathbf{q}, u-w) &= S(\mathbf{n}+\mathbf{q}, u-w)\, S(\mathbf{n}-\mathbf{q}, w-u)\, \delta_{\mathrm{nm}} \\ &\quad -\sum_p \int_0^\beta S(\mathbf{n}+\mathbf{q}, y-w)\, S(\mathbf{n}-\mathbf{q}, w-y)\, v(2\mathbf{q})\, R_{\mathrm{pm}}(\mathbf{q}, u-y)\, dy. \end{aligned} \tag{9.6.3}$$

The first term on the right-hand side of (9.6.3) arises from the non-interacting propagation of the particle-hole pair.

To solve the integral equation we expand $R_{\mathrm{nm}}(\mathbf{q}, u-w)$ in a Fourier series

$$R_{\mathrm{nm}}(\mathbf{q}, u-w) \sum_{-\infty}^{\infty} R_{\mathrm{nm}}(\mathbf{q}, \nu) \exp\{2\pi i(\nu/\beta)(u-w)\} \tag{9.6.4}$$

and substitute into (9.6.3). After some manipulation we obtain

$$R_{\mathrm{nm}}(\mathbf{q}, \nu) = \frac{\tanh(\frac{1}{2}\beta\epsilon'_{\mathbf{n}+\mathbf{q}}) - \tanh(\frac{1}{2}\beta\epsilon'_{\mathbf{n}-\mathbf{q}})}{2\beta(\epsilon'_{\mathbf{n}+\mathbf{q}}+\epsilon'_{\mathbf{n}-\mathbf{q}}) - 4\pi i\nu} \times \left[\delta_{\mathrm{nm}} + \beta \sum_p v(2\mathbf{q})\, R_{\mathrm{pm}}(\mathbf{q}, \nu)\right], \tag{9.6.5}$$

where

$$\epsilon'_{\mathbf{k}} = \epsilon_{\mathbf{k}} - \mu. \tag{9.6.6}$$

If we now define

$$X(\mathbf{q},\nu) = \sum_{\mathbf{n}} \tfrac{1}{2}\beta\, v(2\mathbf{q})\, [\beta(\epsilon_{\mathbf{n}+\mathbf{q}}-\epsilon_{\mathbf{n}-\mathbf{q}})-2\pi i\nu]^{-1} \times [\tanh(\tfrac{1}{2}\beta\epsilon'_{\mathbf{n}+\mathbf{q}})-\tanh(\tfrac{1}{2}\beta\epsilon'_{\mathbf{n}-\mathbf{q}})] \quad (9.6.7)$$

we find from (9.6.5) that

$$\beta \sum_{\mathbf{nm}} R_{\mathbf{nm}}(\mathbf{q},\nu)\, v(2\mathbf{q}) = X(\mathbf{q},\nu)-\beta X(\mathbf{q},\nu) \sum_{\mathbf{pm}} v(2\mathbf{q})\, R_{\mathbf{pm}}.$$

Thus we have immediately that

$$\beta \sum_{\mathbf{m}}\sum_{\mathbf{n}} R_{\mathbf{nm}}(\mathbf{q},\nu)\, v(2\mathbf{q}) = \frac{X(\mathbf{q},\nu)}{1+X(\mathbf{q},\nu)}. \quad (9.6.8)$$

A little thought now shows that

$$-\sum_{\mathbf{nm}} R_{\mathbf{nm}}(\mathbf{q},u-w)\, v(2\mathbf{q})$$

is the contribution from the sum of ring diagrams to $g(\partial/\partial g)(\ln Z_G^g)$. [In the closed ring diagram, one vertex has been singled out and each distinct ring diagram occurs a number of times equal to the number of its vertices.]

Hence the formal expression for the contribution to $\ln Z_G$ from the ring diagrams corresponding to momentum transfer $2\mathbf{q}$ is

$$-\ln(1+X(\mathbf{q},\nu)). \quad (9.6.9)$$

To avoid the discontinuity of $S(\mathbf{k},u-w)$ at $u=w$, we subtract the contribution of the first-order ring diagram from (9.6.9). Then we can write the final result as

$$[\ln Z_G]_R = \ln Z_G^0 - \frac{\Omega}{(2\pi)^3}\int \sum_{\nu=-\infty}^{\infty} \{\ln[1+X(\mathbf{q},\nu)]-X(\mathbf{q},\nu)\}\, d\mathbf{q}, \quad (9.6.10)$$

the subscript R denoting the fact that only the contributions from the ring diagrams of second- and higher-orders are included in $[\ln Z_G]_R$.

It will be useful in the physical applications of (9.6.10) to exhibit the dependence on Planck's constant explicitly. We have in fact worked so far with $\hbar = 1$, and the modification to (9.6.10) is then that $\Omega/(2\pi)^3$ becomes simply $\Omega/(2\pi\hbar)^3$.

9.7. Equation of state of electron gas

To apply (9.6.10) to physical systems, we must specify the interaction in order to calculate $X(\mathbf{q}, \nu)$. In this section, we shall deal with a gas of electrons, considered already at zero temperature in Chapter 5.

The physical quantity related most directly to $\ln Z_G$ is the equation of state, since (Landau & Lifshitz, 1958)

$$\frac{p\Omega}{k_B T} = \ln Z_G, \tag{9.7.1}$$

where p is the pressure. In fact, this relation gives the basic physical reason why, in our calculation of $\ln Z_G$, the series expansion had to yield solely terms proportional to Ω. Even the use of (9.7.1) with inclusion of only ring diagrams on the right-hand side leads to great complication, and we shall have to content ourselves with the limiting cases of high and low temperatures. Irrespective of these limits, we are, of course, working with an interaction whose Fourier transform $v(\mathbf{q})$ is given by

$$v(\mathbf{q}) = \left(\frac{2}{\pi}\right)^{\frac{1}{2}} \frac{\hbar^2 e^2}{q^2}. \tag{9.7.2}$$

We shall deal first with the classical Debye–Huckel limit.

9.7.1. *Debye–Huckel limit*

In this case, it turns out that the only term we need to consider, in the sum given in (9.6.10), is the $\nu = 0$ contribution. Furthermore, the important region is that of small momentum transfer, and, as $\mathbf{q} \to 0$,

$$\begin{aligned} X(\mathbf{q}, \nu) &\sim \left(\frac{2}{\pi}\right)^{\frac{1}{2}} \frac{\hbar^2 e^2}{q^2} \exp(\alpha) \hbar^{-3} \left(\frac{m}{\beta}\right)^{\frac{3}{2}} \beta \\ &= \left(\frac{2}{\pi}\right)^{\frac{1}{2}} \frac{1}{\hbar} \frac{e^2}{q^2} z \left(\frac{m}{\beta}\right)^{\frac{3}{2}} \beta, \end{aligned} \tag{9.7.3}$$

where we have written z for e^α.

Furthermore, $\ln Z_G^0$ becomes the usual result for an ideal Maxwell–Boltzmann gas, and is given by

$$\ln Z_G^0 = \frac{m^{\frac{3}{2}}}{\hbar} z\Omega(2\pi\beta)^{-\frac{3}{2}}. \tag{9.7.4}$$

Substituting these results in (9.6.10) we find

$$\beta p\Omega = \frac{m^{\frac{3}{2}}}{\hbar^3} z\Omega(2\pi\beta)^{-\frac{3}{2}} + \frac{\Omega\hbar^{-3}}{(2\pi)^2}\int_0^\infty \{xz - q^2 \ln(1 + xzq^{-2})\}\, dq, \quad (9.7.5)$$

where we have written

$$x = e^2(2/\Pi)^{\frac{1}{2}}\, \beta\hbar^{-1}\left(\frac{m}{\beta}\right)^{\frac{3}{2}}. \quad (9.7.6)$$

The integral over q is now straightforward, yielding the value $(\pi/3)(xz)^{\frac{3}{2}}$, and thus

$$\beta p\Omega = z\Omega\hbar^{-3}(2\pi\beta/m)^{-\frac{3}{2}} + \frac{\Omega\hbar^{-3}}{(2\pi)^3}\frac{\pi}{3}\left\{ze^2\left(\frac{2}{\pi}\right)^{\frac{1}{2}}\beta\hbar^{-1}\left(\frac{m}{\beta}\right)^{\frac{3}{2}}\right\}^{\frac{3}{2}}. \quad (9.7.7)$$

But, as usual, the fugacity z is related to ρ through

$$\rho = \frac{z}{\Omega}\frac{\partial}{\partial z}\ln Z_G$$

$$= \frac{z\hbar^{-3}}{(2\pi)^{\frac{3}{2}}(\beta/m)^{\frac{3}{2}}} + \frac{\pi\hbar^{-3}}{2(2\pi)^2}\left\{ze^2\left(\frac{2}{\pi}\right)^{\frac{1}{2}}\beta\hbar^{-1}\left(\frac{m}{\beta}\right)^{\frac{3}{2}}\right\}^{\frac{3}{2}}. \quad (9.7.8)$$

Evidently, as we switch off the interactions, i.e. let $e \to 0$, we obtain the usual perfect gas result

$$z_0 = \hbar^3\left(\frac{2\pi\beta}{m}\right)^{\frac{3}{2}}\rho. \quad (9.7.9)$$

If we now solve (9.7.8) for z, to $O(e^2)$, we obtain immediately

$$z = \hbar^3\left(\frac{2\pi\beta}{m}\right)^{\frac{3}{2}}\rho\{1 - \pi^{\frac{1}{2}}e^2\beta^2\rho^{\frac{1}{2}} + O(e^4)\}. \quad (9.7.10)$$

Finally we substitute (9.7.10) for z into (9.7.7) to obtain the final result

$$\beta p = \rho\{1 - (\tfrac{1}{3}\pi^{\frac{1}{2}})\, e^3\beta^{\frac{3}{2}}\rho^{\frac{1}{2}}\}. \quad (9.7.11)$$

This is the classical Debye–Huckel equation of state, and it is worthy of note that we can regard (9.7.11) as an expansion in $(e^2\beta\rho^{\frac{1}{3}})$, which is, in fact, the appropriate dimensionless expansion variable in this limit. Thus, we see that an expansion in e^2 is also, of necessity, a low-density expansion.

We shall now contrast this with the quantum-mechanical limit $\beta \to \infty$. In this case, obviously Planck's constant must come into the appropriate dimensionless grouping of variables and indeed the expansion parameter as $T \to 0$ (cf. Chapter 6) is $(me^2/\hbar^2\rho^{\frac{1}{3}})$. This time, the expansion in e^2 is also a high-density expansion. It is amusing that the ideal quantum gas of Fermions is the high density gas, whereas perfect classical gases exist only at low densities. It is, of course, the Exclusion Principle which dominates the Fermion problem. We turn finally then to illustrate these remarks by a calculation again based on the result (9.7.1) for the equation of state.

9.7.2. *Gell-Mann & Brueckner limit*

In this case, we must use the fact that as $q \to 0$, which again is due to the dominance of small momentum transfer as seen in Chapter 5,

$$X(\mathbf{q}, \nu) = 2m\hbar^{-3}(2\pi)^{-\frac{1}{2}} v(q) p_0[1 - \nu \tan^{-1}(1/\nu)], \qquad (9.7.12)$$

where $\nu = 2\pi mk/\beta q p_0$, and p_0 is defined by

$$z = e^{\mu\beta} = \exp\{\beta p_0^2/2m\}. \qquad (9.7.13)$$

Then, from (9.6.10), we have

$$\ln Z_G = \ln Z_G^0 + \tfrac{1}{2}\Omega(2\pi\hbar)^{-3} \int 4\pi q^2\, dq \sum_\nu \{2X(\mathbf{q}, \nu) - \ln[1 + 2X(\mathbf{q}, \nu)]\}, \qquad (9.7.14)$$

where the factor 2 multiplying $X(\mathbf{q}, \nu)$ has come in via spin degeneracy. As $\beta \to \infty$, i.e. $T \to 0$, ν becomes a continuous variable with $d\nu = 2\pi m/\beta q p_0$. Hence since

$$X(\mathbf{q}, \nu) = \left(\frac{2me^2 p_0}{\hbar^2 q^2 \pi}\right) R(\nu), \qquad (9.7.15)$$

with

$$R(\nu) = 1 - \nu \tan^{-1}(1/\nu). \qquad (9.7.16)$$

we find, putting $q = p_0 Q$,

$$\ln Z_G = \ln Z_G^0 + \Omega(2\pi\hbar)^{-3} \beta p_0^5 m^{-1} \times \int_0^\infty Q^3\, dQ \int_{-\infty}^{\infty} \left\{\frac{4me^2}{\hbar\pi Q^2 p_0} R(\nu) - \ln\left[1 + \frac{4me^2}{\pi\hbar p_0 Q^2} R(\nu)\right]\right\} d\nu. \qquad (9.7.17)$$

(*a*) *Internal energy*. We may now obtain the internal energy from the grand partition function in the usual way by differentiating $\ln Z_G$ with respect to β at constant z.

(*b*) *Density*. The density may be found by differentiating $\ln Z_G$

with respect to $\beta\mu = \ln z = (\beta p_0^2/2m)$ at fixed β. For this purpose we substitute

$$p_0 = \left(\frac{2m\mu}{\beta}\right)^{\frac{1}{2}} \tag{9.7.18}$$

in (9.7.17). After performing these differentiations and eliminating μ between the two equations, we can then obtain the ground-state energy/particle.

(*c*) *Correlation energy*. If we denote the double integral in (9.7.17) by I, then we find

$$\begin{aligned} \frac{E}{N} &= -\frac{1}{N}\frac{\partial \ln Z_G}{\partial \beta} \\ &= [2\rho m(2\pi\hbar)^3]^{-1}\left[\frac{8\pi p_0^5}{5} + 3I\left(\frac{2m\mu}{\beta}\right)^{\frac{3}{2}}\right] \end{aligned} \tag{9.7.19}$$

and the density is given by

$$\begin{aligned} \rho &= \frac{1}{\Omega}\frac{\partial \ln Z_G}{\partial \mu} \\ &= [2m(2\pi\hbar)^3]^{-1}\left[\frac{16m\pi p_0^3}{3} + \frac{5\beta}{\mu}(2m\mu/\beta)^{\frac{5}{2}} I\right]. \end{aligned} \tag{9.7.20}$$

Now suppose $p_0 = p_f + \delta p_0$, p_f being the usual Fermi momentum given by $p_f = \hbar(3\pi^2\rho)^{\frac{1}{3}}$. Then the change in the ground-state energy per particle due to Coulomb interactions is given by

$$E_{\text{correlation}} = \Delta\left(\frac{E}{N}\right) = \frac{[4\pi(\delta p_0/p_f) + \frac{3}{2}I]\,p_f^5}{m\rho(2\pi\hbar)^3}. \tag{9.7.21}$$

On the other hand, to satisfy (9.7.20) we must take

$$\frac{\delta p_0}{p_f} = -\frac{5}{8\pi}I, \tag{9.7.22}$$

and hence

$$E_{\text{correlation}} = -\frac{3Ip_f^2}{8\pi m}. \tag{9.7.23}$$

Recalling the definition of I, and using the results of Appendix 5A.1, this may be shown to be the Gell-Mann & Brueckner formula.

In summary, the ring integral formula (9.6.10) has been shown to reproduce, on the one hand, the well-established Debye–Huckel theory, and on the other the high density electron gas correlation energy. Although many workers have made enormous efforts to push such perturbative treatments to higher order, what we essentially require, but still lack, are accurate non-perturbative approximations to the solution of the Bloch equation. Unfortunately, such methods are only now being developed for the one-particle approximations, and it seems far ahead to the solution of the equivalent many-body integral equation.

Problems

P.9 (i). Calculate the free energy and the internal energy of a classical electron gas from the Debye–Huckel theory.

P.9 (ii). For a low-temperature electron gas, calculate the temperature dependence of the static dielectric constant $\epsilon(k, 0)$. How would this be relevant in the calculation of the specific heat of a dilute metallic alloy?

CHAPTER 10

GREEN FUNCTIONS

10.1. Introduction

The most unifying and powerful many-body method yet devised is that of the Green function. The one-body Green function, which will concern us initially in sections 10.2–10.7, gives directly the single-particle excitation spectrum, which has been seen in previous chapters to be so vitally important in explaining the low-temperature properties of systems. In addition, there are theorems which allow one to infer, once the Green functions are known, other physically interesting quantities such as, for example, the total energy and the momentum distribution.

The two-particle Green function gives, in particular, the collective modes, and we shall deal with this aspect in considerable detail in sections 10.8–10.12 below. But we should remark at this stage that it also arises in the theory of quasi-particle interactions. This leads by a more fundamental path (Landau, 1959) to the Landau theory of Fermi liquids dealt with from a phenomenological point of view in Chapter 5. Although it is a natural extension of the work of this chapter, we shall not consider this problem; the reader is referred to the detailed account by Nozières (1963).

The theory is most straightforwardly applied to normal Fermi systems but with suitable modification, to take into account certain anomalous Green functions, it can also be applied to condensed systems, both of Bose and Fermi types (see section 10.14). Until section 10.14 is reached, we shall develop the formalism with Fermions strictly in mind. Only then will the modifications required for Bosons be considered.

Furthermore, the Green function formalism is readily generalized to elevated temperatures and this matter will be taken up in section 10.13.

Very often, in practice, many-body Green functions are calculated by perturbation theory. This will, in general, be the tool we use. For this purpose, it is desirable to work in **k**-space rather than direct space. Green functions also have either time or energy arguments, and a word here concerning the roles of these is appropriate. To begin with, the theory is most naturally and easily described in terms of the former, and this state of affairs continues until we have derived the linked cluster theorem. After that, emphasis is shifted to the Green functions with energy arguments, our results being, thereby, very conveniently expressed in a form useful for practical computations.

Another important technique is the equation-of-motion method of Martin & Schwinger (1959). Here it is convenient to work in $(\mathbf{r}, t)$-space. This approach will be adopted in the elevated temperature description of section 10.13.

10.2. Definitions and generalities

Let us introduce the notation

$$a_{\mathbf{k}t} = e^{iHt} a_{\mathbf{k}} e^{-iHt}, \quad a^{\dagger}_{\mathbf{k}t} = (a_{\mathbf{k}t})^{\dagger} \qquad (10.2.1)$$

so that $a_{\mathbf{k}t}$ is the Heisenberg time-dependent operator associated with the Schrödinger time-independent $a_{\mathbf{k}}$ corresponding to a plane-wave basis. It is not to be confused with

$$a_{\mathbf{k}}(t) = e^{iH_0 t} a_{\mathbf{k}} e^{-iH_0 t}, \qquad (10.2.2)$$

introduced in (4.7.2) of Chapter 4, where H_0 is only the kinetic energy part of H. Then the single-particle Green function is defined by

$$G(\mathbf{k}, t_2 - t_1) = i\langle \Psi_n(0)|\, T\{a_{\mathbf{k}t_2}\, a^{\dagger}_{\mathbf{k}t_1}\}\, |\Psi_n(0)\rangle, \qquad (10.2.3)$$

where T is the chronological operator defined as before and $\Psi_n(0)$ is the Heisenberg time-independent exact normalized (hence the suffix n) ground state of the interacting N-particle system. Thus, we may write

$$H\Psi_n(0) = E\Psi_n(0); \quad \langle \Psi_n(0)|\Psi_n(0)\rangle = 1. \qquad (10.2.4)$$

In the theory of the single-particle Green function it is convenient to regard $\mathbf{k}$ as a momentum-spin index. On occasion, if there is no danger of ambiguity, we may use $\mathbf{k}$ for momentum only. For example, $k < k_f$ means that the state $\mathbf{k}$ lies below the Fermi level. The symbols k and $\mathbf{k}$, as used here, are quite unrelated to the auxiliary variable K which will later enter the theory rather as in chapter 4.

More explicitly, (10.2.3) is often written as

$$G(\mathbf{k}, t_2 - t_1) = G^{(r)}(\mathbf{k}, t_2 - t_1) + G^{(a)}(\mathbf{k}, t_2 - t_1), \qquad (10.2.5)$$

where

$$\left.\begin{aligned} G^{(r)}(\mathbf{k}, t_2-t_1) &= i\langle\Psi_n(0)|a_{\mathbf{k}t_2}\,a^{\dagger}_{\mathbf{k}t_1}|\Psi_n(0)\rangle \quad (t_2-t_1>0) \\ &= 0 \quad (t_2-t_1<0), \end{aligned}\right\} \tag{10.2.6}$$

and

$$\left.\begin{aligned} G^{(a)}(\mathbf{k}, t_2-t_1) &= 0 \quad (t_2-t_1>0) \\ &= -i\langle\Psi_n(0)|a^{\dagger}_{\mathbf{k}t_1}\,a_{\mathbf{k}t_2}|\Psi_n(0)\rangle \quad (t_2-t_1<0). \end{aligned}\right\} \tag{10.2.7}$$

$G^{(r)}$ and $G^{(a)}$ are called respectively the retarded and advanced parts of the Green function.

In writing G with a single time argument, we have already recognized that the right-hand side of (10.2.3) depends only on the time difference t_2-t_1. This is easily shown, for, if $t_2 > t_1$, (10.2.3) becomes

$$\begin{aligned} G(\mathbf{k}, t_2-t_1) &= i\langle\Psi_n(0)|\{e^{iHt_2}a_{\mathbf{k}}\,e^{-iHt_2}\}\{e^{iHt_1}a^{\dagger}_{\mathbf{k}}\,e^{-iHt_1}\}|\Psi_n(0)\rangle \\ &= i\,e^{iE(t_2-t_1)}\langle\Psi_n(0)|a_{\mathbf{k}}\,e^{-iH(t_2-t_1)}a^{\dagger}_{\mathbf{k}}|\Psi_n(0)\rangle \\ &= i\langle\Psi_n(0)|a_{\mathbf{k}}\,e^{-i(H-E)(t_2-t_1)}a^{\dagger}_{\mathbf{k}}|\Psi_n(0)\rangle. \end{aligned} \tag{10.2.8}$$

There is a similar argument for $t_2 < t_1$.

Let us make the following additional comments on (10.2.3). First of all, we saw above that it was necessary to be able to interchange the two $a_{\mathbf{k}t}$'s in order to describe both particle and hole propagation. The use of T rather than, say, P, to do this, is for mathematical convenience only. At a certain stage below, it is desirable to analyse a perturbation series and, as we have seen in Chapter 4, the T-operator is most convenient for this purpose. Secondly, the way in which the form (10.2.8) brings out the dependence on excitation energies should be noted. In order to study Green functions, a detailed knowledge of the ground state (in particular, the ground-state energy) is not a prerequisite.

The Fourier transform of $G(\mathbf{k}, t)$ with respect to time is also called (somewhat ambiguously) a single-particle Green function. The precise definitions are

$$G(\mathbf{k}, \omega) = \int_{-\infty}^{\infty} dt\, G(\mathbf{k}, t)\, e^{i\omega t} \tag{10.2.9}$$

and

$$G(\mathbf{k}, t) = \int_{-\infty}^{\infty} \frac{d\omega}{2\pi}\, G(\mathbf{k}, \omega)\, e^{-i\omega t}. \tag{10.2.10}$$

As was mentioned in section 10.1, in applications, $G(\mathbf{k}, \omega)$ is often the more useful form.

The unperturbed Green functions, denoted by zero suffices, are given by the special case $H = H_0$, when $|\Psi_n(0)\rangle$ becomes the independent-Fermion ground-state solution $|\Psi(-\infty)\rangle$, in the language of Chapter 4. We saw in that chapter that the G_0's played a very useful role in perturbation theory; they are even more important in the theory of many-body Green functions. $G_0(\mathbf{k}, t)$ has already been evaluated. For, on using the substitutions specified above, (10.2.3) reduces to the form indicated in (4.13.16) and (4.13.17)

Thus,

$$G_0(\mathbf{k}, t) = \begin{cases} \pm i e^{-i\epsilon_{\mathbf{k}} t} & (+ \text{ if } t > 0 \text{ and } k > k_f, \\ & - \text{ if } t < 0 \text{ and } k < k_f), \\ 0 & (\text{otherwise}). \end{cases} \tag{10.2.11}$$

The Fourier transform of this is

$$G_0(\mathbf{k}, \omega) = \frac{1}{\epsilon_{\mathbf{k}} - \omega \mp i0} \quad (k \gtrless k_f), \tag{10.2.12}$$

as may be verified by showing that the expression

$$\int_{-\infty}^{\infty} \frac{d\omega}{2\pi} \frac{e^{-i\omega t}}{\epsilon_{\mathbf{k}} - \omega \mp i0} \quad (k \gtrless k_f) \tag{10.2.13}$$

reduces to (10.2.11). When $t > 0$, we may evaluate (10.2.13) by using a semicircle in the upper half-plane and when $t < 0$, we complete the contour in the lower half-plane. The $\mp\, i0$ decides whether for any particular case, a pole does or does not contribute a residue. In this way we obtain (10.2.11).

These unperturbed Green functions are important because the propagators of interacting systems can be generated and calculated in terms of them by perturbation theory. For this purpose it is good enough to write (10.2.12) in the cruder form $G_0(\mathbf{k}, \omega) = (\epsilon_{\mathbf{k}} - \omega)^{-1}$.

10.3. Quasi-particles

So far, we have not discussed the physical reasons for introducing the Green function (10.2.3). We now proceed to do this.

Let us first write (10.2.3) in the form

$$G(\mathbf{k}, t) = i\langle \Psi_n(0)|\, e^{iHt} a_{\mathbf{k}}\, e^{-iHt} a_{\mathbf{k}}^{\dagger} |\Psi_n(0)\rangle \qquad (t > 0), \quad (10.3.1\,a)$$

$$G(\mathbf{k}, -t) = -i\langle \Psi_n(0)|\, e^{iHt} a_{\mathbf{k}}^{\dagger}\, e^{-iHt} a_{\mathbf{k}} |\Psi_n(0)\rangle \quad (t > 0), \quad (10.3.1\,b)$$

and begin by concentrating on (10.3.1 *a*). Clearly $a_{\mathbf{k}}^{\dagger}|\Psi_n(0)\rangle$ is the amplitude for the creation of a bare particle with index **k**. Next, $e^{-iHt} a_{\mathbf{k}}^{\dagger}|\Psi_n(0)\rangle$ represents the composite state after time t (this following since a formal solution of an equation such as (4.4.1) is of the form $\Phi(t) = e^{-iHt}\Phi(0)$). Then $a_{\mathbf{k}}\, e^{-iHt} a_{\mathbf{k}}^{\dagger}|\Psi_n(0)\rangle$ describes the amplitude for removing, after time t, the state added at time zero. Now if no particle had been inserted in the first place, the system would have evolved into the state $e^{-iHt}|\Psi_n(0)\rangle$. Thus, the probability amplitude for adding a bare particle at time zero, removing at time t and regaining the original many-body system is simply

$$\{e^{-iHt}|\Psi_n(0)\rangle\}^{\dagger}\{a_{\mathbf{k}}\, e^{-iHt} a_{\mathbf{k}}^{\dagger}|\Psi_n(0)\rangle\},$$

which is just (10.3.1 *a*), apart from a trivial factor of i.

Similarly, it is clear from (10.3.1 *b*) that for $t > 0$, $G(\mathbf{k}, -t)$ is the probability amplitude that the many-body system is not disturbed by the removal at time zero and subsequent creation at time t, of a particle having index **k**. In the case of Fermions, therefore, this corresponds to hole propagation. Because they are concerned with propagation of this kind, Green functions are also called propagators.

We have seen that (10.2.11) represents the above probability amplitudes in the case of non-interacting particles. These expressions are easily interpreted. The initial creation of a particle (or hole) is either allowed or forbidden because of obvious Exclusion Principle requirements. In the former case, the created particle (or hole) propagates in this state indefinitely without scattering, and its removal does not perturb the motion of the other particles.

In an interacting system, the above situation is altered as follows. First, the probability of placing a bare particle in the system is not simply zero or unity. It is, clearly, given in terms of the momentum distribution function, $n_{\mathbf{k}}$, by $1 - n_{\mathbf{k}}$, which is appreciable when $k - k_f$ is small, irrespective of its sign. Secondly, even if a particle is created, during its subsequent motion it interacts with the other particles in the system, thus further decreasing the amplitude of the Green function, as is clear from the discussion following (10.3.1).

However, under certain circumstances, to be described below, the attenuation is sufficiently slow that we can claim something akin to particle motion. We say that the whole phenomenon, that is the bare particle together with the polarized medium around it, corresponds to a quasi-particle. It is clear that for a well-defined quasi-particle, $k - k_f$ should be small. The exercise is elementary, in principle (see Abrikosov, Gorkov & Dzyaloshinski, 1963, p. 17, footnote) to show, for Fermions, that the statistics alone decide that to a first approximation, the quasi-particle lifetime (the $\Gamma_{\mathbf{k}}^{-1}$ of (10.3.3) below) is proportional to $(k - k_f)^{-2}$.

A weakly excited many-body state may be viewed as consisting of a distribution of such quasi-particles, small in number compared with the total number of bare particles composing the system. Thus any given region of polarization, constituting a quasi-particle, is well defined and quite separate, in general, from other such regions. Hence, to a first approximation, the system behaves like a dilute gas. Any individual quasi-particle is thought of as having a definite birth and subsequent decay, the totality of such processes being such as to maintain a constant average population. We have, of course, made use of such ideas in earlier parts of this book (particularly in Chapters 5, 7 and 8).

The usual discussion of the above effects, by complex variable theory applied to a spectral representation of (10.2.8), will be avoided here. Instead we take a more pragmatic approach by appealing to the results one finds on calculating the Green function by perturbation theory. For normal (i.e. non-superfluid) systems we shall see that our propagator can be written

$$G(\mathbf{k}, \omega) = \frac{1}{\epsilon_{\mathbf{k}} - \omega - M(\mathbf{k}, \omega)}, \tag{10.3.2}$$

where $M(\mathbf{k}, \omega)$ is a certain well-behaved function to be defined later, and called the Dyson irreducible or proper self-energy. The latter vanishes for non-interacting particles, as we know from (10.2.12). The assumption we shall make, and this is what is found in practice, is that (10.3.2) resembles (10.2.12) to the extent that for any given $k \gtrless k_f$, $G(\mathbf{k}, \omega)$ can be analytically continued into the lower (upper) half plane and has a simple pole near and below (above) the real axis at some point

$$\omega = \xi_{\mathbf{k}} = \epsilon_{\mathbf{k}} - i\Gamma_{\mathbf{k}} \quad (\Gamma_{\mathbf{k}} \gtrless 0) \tag{10.3.3}$$

say, satisfying the equation

$$\epsilon_{\mathbf{k}} - \omega - M(\mathbf{k}, \omega) = 0. \tag{10.3.4}$$

Although considerably more is known about the poles of G (Galitskii & Migdal, 1958; Nozières, 1963; Schrieffer, 1964), the above is sufficient for our present purposes.

In order to interpret the results above, it is desirable to transform into t-space. When $k > k_f$ and $t > 0$, we evaluate (10.2.10) by closing the contour using a semicircle in the lower half-plane. This results in a sum of terms arising from the various poles, the damping of each being given by the distance of the associated pole from the real ω-axis. Hence, that given by (10.3.3) will dominate and so, under these circumstances, $G(\mathbf{k}, \omega) \sim iz_{\mathbf{k}} \exp(-i\epsilon_{\mathbf{k}} t) \exp(-\Gamma_{\mathbf{k}} t)$. Here, $z_{\mathbf{k}}$ is minus the residue at $\xi_{\mathbf{k}}$, and is therefore determined by (10.3.2) to be

$$z_{\mathbf{k}} = \frac{1}{1 + \left(\dfrac{\partial M(\mathbf{k}, \omega)}{\partial \omega}\right)_{\omega = \xi_{\mathbf{k}}}}. \tag{10.3.5}$$

When $k > k_f$ and $t < 0$, one closes the contour in the upper half-plane. There are no poles near the real axis by hypothesis, all damping is severe and we may write $G(\mathbf{k}, \omega) \sim 0$. A similar discussion is possible for $k < k_f$ and we can summarize all the information thus gleaned, by the equation

$$G(\mathbf{k}, t) \sim \begin{cases} \left.\begin{matrix} 0 & (k < k_f) \\ iz_{\mathbf{k}} \exp(-i\xi_{\mathbf{k}} t) & (k > k_f) \end{matrix}\right\} & (t > 0), \\[2ex] \left.\begin{matrix} -iz_{\mathbf{k}} \exp(-i\xi_{\mathbf{k}} t) & (k < k_f) \\ 0 & (k > k_f) \end{matrix}\right\} & (t < 0). \end{cases} \tag{10.3.6}$$

Thus (10.3.6) shows that quasi-particle disturbances persist with energy $\epsilon_{\mathbf{k}}$ for a lifetime $\Gamma_{\mathbf{k}}^{-1}$. Actually, (10.3.4) provides the basis for a precise definition of a Fermi surface in momentum space, inside which one has quasi-holes and outside which one has quasi-particles in the above-defined sense.

The quantity $z_{\mathbf{k}_f}$ gives the discontinuity, at the Fermi level, of the momentum distribution function (recall Chapter 5, section 5.10.1). We shall not prove this here, but merely point out that this is suggested by (10.3.6). For, if the remainder terms are denoted by $g(\mathbf{k}, t)$, we have

$$G(\mathbf{k}_f+0, -0)-g(\mathbf{k}_f+0, -0) = -iz_{\mathbf{k}_f-0} \tag{10.3.7}$$

and
$$G(\mathbf{k}_f-0, -0)-g(\mathbf{k}_f-0, -0) = 0. \tag{10.3.8}$$

Thus, on assuming appropriate continuity properties for g and z, we have

$$z_{\mathbf{k}_f} = n_{\mathbf{k}_f+0} - n_{\mathbf{k}_f-0}, \tag{10.3.9}$$

since
$$G(\mathbf{k}, -0) = -i\langle\Psi_n(0)|a_{\mathbf{k}}^{\dagger}\, a_{\mathbf{k}}|\Psi_n(0)\rangle = -in_{\mathbf{k}}. \tag{10.3.10}$$

Equation (10.3.9) shows, in particular, that $z_{\mathbf{k}_f}$ is real and such that $0 \leqslant z_{\mathbf{k}_f} \leqslant 1$. The upper limit follows since for any $\mathbf{k}$, $0 \leqslant n_{\mathbf{k}} \leqslant 1$ by Pauli Principle requirements; and the lower limit since, if $z_{\mathbf{k}_f}$ were negative, the momentum distribution would be energetically unstable against particles falling back into the Fermi sea.

In practice, we calculate $M(\mathbf{k}, \omega)$ (below we shall do this by perturbation theory) and use it in (10.3.4) to find the pole of $G(\mathbf{k}, \omega)$, thereby obtaining the quasi-particle energy and lifetime. For well-defined quasi-particles, $\Gamma_{\mathbf{k}}$ is small and with this assumption we can solve (10.3.4) by a Taylor series in $\Gamma_{\mathbf{k}}$ to give

$$\mathcal{E}_{\mathbf{k}} - \epsilon_{\mathbf{k}} - M_0(\mathbf{k}, \epsilon_{\mathbf{k}}) = 0 \tag{10.3.11}$$

and
$$\Gamma_{\mathbf{k}} = \frac{M_1(k, \epsilon_{\mathbf{k}})}{1+\left(\dfrac{\partial M_0(k, \omega)}{\partial \omega}\right)_{\omega=\epsilon_{\mathbf{k}}}}, \tag{10.3.12}$$

where M_0 and M_1 are respectively the real and imaginary parts of M. We should note that in obtaining (10.3.11) and (10.3.12) we have not assumed the real displacement $\epsilon_{\mathbf{k}} - \mathcal{E}_{\mathbf{k}}$ of the pole from its non-interacting position to be small.

We see from the above discussion why we are interested in calculating the single-particle Green function. It gives us the quasi-particle energies and lifetimes. In concluding this section we would like to make the point that another important physical quantity, namely the ground-state energy, is given as a bonus once the above propagator is known, no separate and distinct calculation being necessary. We shall simply quote the final formula here:

$$E = \frac{i}{4\pi}\int \frac{2}{(2\pi)^3}\,d\mathbf{k}\int_C d\omega(\omega - \mathcal{E}_{\mathbf{k}})\,G(\mathbf{k}, \omega), \tag{10.3.13}$$

where the contour C consists of the real axis and the infinite semicircle in the upper half-plane.

With the importance of $G(\mathbf{k}, \omega)$ indicated by the above discussion, we show, in the next few sections, how to calculate it explicitly.

10.4. Green function and U matrix

Let us recall certain basic properties of the U matrix of Chapter 4 (see especially sections 4.4 and 4.5).

(i) We have
$$|\Psi(t)\rangle = U(t,t')|\Psi(t')\rangle, \tag{10.4.1}$$

from which it follows that

$$|\Psi(0)\rangle = U(0-\infty)|\Psi(-\infty)\rangle = U(0,\infty)|\Psi(\infty)\rangle. \tag{10.4.2}$$

Furthermore, by the appropriate limiting procedure,

$$|\Psi(\infty)\rangle = |\Psi(-\infty)\rangle = |g\rangle. \tag{10.4.3}$$

(ii) For conservative systems (and we always consider these)

$$U(t,t') = e^{iH_0t}e^{iH(t'-t)}e^{-iH_0t}. \tag{10.4.4}$$

(In general, this cannot be simplified further since operator exponents are not additive.) This enables us to relate the Heisenberg $a_{\mathbf{k}t}$ of (10.2.1) to the interaction $a_{\mathbf{k}}(t)$ of (10.2.2). We find

$$a_{\mathbf{k}t} = e^{iHt}e^{-iH_0t}a_{\mathbf{k}}(t)\,e^{iH_0t}e^{-iHt} = U(0,t)\,a_{\mathbf{k}}(t)\,U(t,0). \tag{10.4.5}$$

(iii) Thirdly, we note the composition law

$$U(t_1,t_2) = U(t_1,t_3)\,U(t_3,t_2). \tag{10.4.6}$$

We now use the above results to express G, as given by (10.2.3), in a form which is suitable for the application of perturbation theory. First, from (10.4.2) and (10.4.3),

$$|\Psi_n(0)\rangle = K^{-\frac{1}{2}}|\Psi(0)\rangle, \tag{10.4.7}$$

where K is a normalization factor. Requiring that

$$\langle\Psi_n(0)|\Psi_n(0)\rangle = 1,$$

we find

$$\begin{aligned} K = \langle\Psi(0)|\Psi(0)\rangle &= \langle\Psi(\infty)|\,U(\infty,0)\,U(0,-\infty)|\Psi(-\infty)\rangle \\ &= \langle g|\,U(\infty,-\infty)|g\rangle. \end{aligned} \tag{10.4.8}$$

10.4.1. *Green function in interaction picture*

Let us suppose, to begin with, that $t_2 > t_1$. Then, using (10.4.2), (10.4.3), (10.4.5) and (10.4.7) we can rewrite (10.2.3) as

$$G(\mathbf{k}, t_2 - t_1) = (i/K)\langle g|U(\infty, 0)\{U(0, t_2)\, a_k(t_2)\, U(t_2, 0)\}$$
$$\times\{U(0, t_1)\, a_{\mathbf{k}}^{\dagger}(t_1)\, U(t_1, 0)\}\, U(0, -\infty)|g\rangle, \quad (10.4.9)$$

and use of (10.4.6) and (10.4.8) then gives

$$G(\mathbf{k}, t_2 - t_1) = \frac{i\langle g|U(\infty, t_2)\, a_{\mathbf{k}}(t_2)\, U(t_2, t_1)\, a_{\mathbf{k}}^{\dagger}(t_1)\, U(t_1, -\infty)|g\rangle}{\langle g|U(\infty, -\infty)|g\rangle} \quad (t_2 > t_1). \quad (10.4.10)$$

Now let us recall the explicit form of U. It is (Chapter 4, (4.5.3) and (4.15.2))

$$U(t_1\, t_2) = 1 + \sum_{n=1}^{\infty} \frac{(-i)^n}{n!} \int_{t_2}^{t_1} dt'_n \ldots \int_{t_2}^{t_1} dt'_1 TV(t'_1) \ldots V(t'_n). \quad (10.4.11)$$

The brackets have been omitted from around the V-product since we intend to use T in a slightly more general context in the present chapter. Generally, let $\Omega(t)$ be a time-dependent operator. Then we will take

$$\{TV(t_1) \ldots V(t_n)\}\Omega(t) = T\{V(t_1) \ldots V(t_n)\Omega(t)\}. \quad (10.4.12)$$

Expressions such as that on the left-hand side of (10.4.12) did not occur previously and so, without ambiguity, one could write $TV(t_1) \ldots V(t_n) = T\{V(t_1) \ldots V(t_n)\}$. With the definition (10.4.12), we assert that for $t \geqslant t_2 > t_1$,

$$a_{\mathbf{k}}^{\dagger}(t)\, U(t_2, t_1) = U(t_2, t_1)\, a_{\mathbf{k}}^{\dagger}(t), \quad a_{\mathbf{k}}(t)\, U(t_2, t_1) = U(t_2, t_1)\, a_{\mathbf{k}}(t). \quad (10.4.13)$$

The proof follows by showing that, for this time sequence, $a_{\mathbf{k}}^{\dagger}(t)$ (and similarly $a_{\mathbf{k}}(t)$) commutes with every T-product in (10.4.11). We have

$$\{TV(t'_1) \ldots V(t'_n)\}\, a_{\mathbf{k}}^{\dagger}(t) \equiv T\{V(t'_1) \ldots V(t'_n)\, a_{\mathbf{k}}^{\dagger}(t)\}$$
$$= (-1)^{4n}\, T\{a_{\mathbf{k}}^{\dagger}(t)\, V(t'_1) \ldots V(t'_n)\}, \quad (10.4.14)$$

this last step following since each V involves products of four a operators and thus $a_{\mathbf{k}}^{\dagger}(t)$ required $4n$ interchanges to reach its new position. Now we observe that by definition (10.4.11), every t_i' lies between t_1 and t_2 and so is less than t. Thus the final expression in (10.4.14) can be rewritten as $a_{\mathbf{k}}^{\dagger}(t)\,T\{V(t_1')\ldots V(t_n')\}$ and this establishes (10.4.13).

Now let us use (10.4.13) in (10.4.10). We have, for $t_2 > t_1$,

$$\begin{aligned} &U(\infty, t_2)\,a_{\mathbf{k}}(t_2)\,U(t_2, t_1)\,a_{\mathbf{k}}^{\dagger}(t_1)\,U(t_1, -\infty) \\ &\quad = U(\infty, t_2)\,U(t_2, t_1)\,U(t_1, -\infty)\,a_{\mathbf{k}}(t_2)\,a_{\mathbf{k}}^{\dagger}(t_1) \\ &\quad = U(\infty, -\infty)\,a_{\mathbf{k}}(t_2)\,a_{\mathbf{k}}^{\dagger}(t_1), \end{aligned} \tag{10.4.15}$$

and thus (10.4.10) becomes

$$G(\mathbf{k}, t_2-t_1) = \frac{i\langle g|U(\infty, -\infty)\,a_{\mathbf{k}}(t_2)\,a_{\mathbf{k}}^{\dagger}(t_1)|g\rangle}{\langle g|U(\infty, -\infty)|g\rangle} \quad (t_2 > t_1). \tag{10.4.16}$$

By a similar calculation, we obtain for the other time sequence

$$G(\mathbf{k}, t_2-t_1) = -\frac{i\langle g|U(\infty, -\infty)\,a_{\mathbf{k}}^{\dagger}(t_1)\,a_{\mathbf{k}}(t_2)|g\rangle}{\langle g|U(\infty, -\infty)|g\rangle} \quad (t_1 > t_2). \tag{10.4.17}$$

In (10.4.16) and (10.4.17) we have now rewritten G in the interaction representation and can therefore use the graphical methods of Chapter 4.

Explicitly, we may write (10.4.16) and (10.4.17) for both time orders as

$$\begin{aligned} &G(\mathbf{k}, t_2-t_1) \\ &= \frac{i\sum_{n=0}^{\infty}\frac{(-i)^n}{n!}\int_{-\infty}^{\infty}dt_n'\ldots\int_{-\infty}^{\infty}dt_1'\langle g|T\{V(t_1')\ldots V(t_n')\,a_{\mathbf{k}}(t_2)\,a_{\mathbf{k}}^{\dagger}(t_1)\}|g\rangle}{\sum_{n=0}^{\infty}\frac{(-i)^n}{n!}\int_{-\infty}^{\infty}dt_n'\ldots\int_{-\infty}^{\infty}dt_1'\langle g|T\{V(t_1')\ldots V(t_n')\}|g\rangle} \\ &\equiv \frac{\mathfrak{N}}{\mathfrak{D}}. \end{aligned} \tag{10.4.18}$$

In this form, the Green function is amenable to graphical analysis.

10.5. Graphical analysis of single-particle Green function

The denominator $\mathfrak{D}$ in (10.4.18) has been dealt with fully in Chapter 4 (the different range of the time integrations not being an essential feature of the graphical analysis). Indeed, the rules given there are sufficient to enable the numerator $\mathfrak{N}$ also to be calculated. While the full details are given in Appendix 10 A, and are certainly necessary for a proper understanding of the following sections, the essential argument may be summarized as follows:

(i) A general diagram corresponding to a term in $\mathfrak{N}$ has an external incoming line (arising from $a_{\mathbf{k}}^{\dagger}(t)$) and an external outgoing line (arising from $a_{\mathbf{k}}(t)$) together with a number (possibly zero) of unlinked (disconnected) vacuum-vacuum parts.

(ii) Contributions from $\mathfrak{D}$, much as is shown in section 4.16, exactly cancel the terms in $\mathfrak{N}$ from the vacuum-vacuum parts.

Therefore we are left with the simple result that G is formed only from contributions of linked diagrams having an external incoming and external outgoing line, drawn in all possible ways.

Thus we are now in a position to state a linked cluster theorem analogous to that of Chapter 4. The essential content of this is that $G(\mathbf{k}, t_2 - t_1)$ as defined by (10.4.18) can be rewritten

$$G(\mathbf{k}, t_2 - t_1) = \sum_{n=0}^{\infty} (-i)^{n-1} \int_{-\infty}^{\infty} dt_{2n-12n} \cdots \int_{-\infty}^{\infty} dt_{12} \sum_{\substack{KLMN \\ PQRS \\ ----\\ XYZT}}$$

$$\times \langle KL|v|NM\rangle \langle PQ|v|SR\rangle \ldots \langle XY|v|TZ\rangle \langle g|T\{a_K^{\dagger}(t_{12})$$

$$\times a_L^{\dagger}(t_{12}) \ldots a_T(t_{2n-12n})\, a_{\mathbf{k}}(t_2)\, a_{\mathbf{k}}^{\dagger}(t_1)\}|g\rangle_L, \quad (10.5.1)$$

where the final suffix L means that only linked distinct unlabelled diagrams are to be taken.

While the concept is explained fully at the end of Appendix 10A, perhaps we ought to indicate here what an unlabelled diagram means. On developing the theory rather as in Chapter 4, with labelled vertices $(1, 2)$, $(3, 4)$, ..., $(2n-1, 2n)$, it is found that any two graphs which have the same topological structure give the same contribution. Topologically equivalent graphs are grouped into families of size $2^n n!$. Thus we drop the usual $2^n n!$ which would ordinarily occur in the denominator of (10.5.1) and compute using contributions from only one member per family. This member is usually left unlabelled.

Symbolically (10.5.1) can be written

$G(\mathbf{k}, t_2 - t_1)$

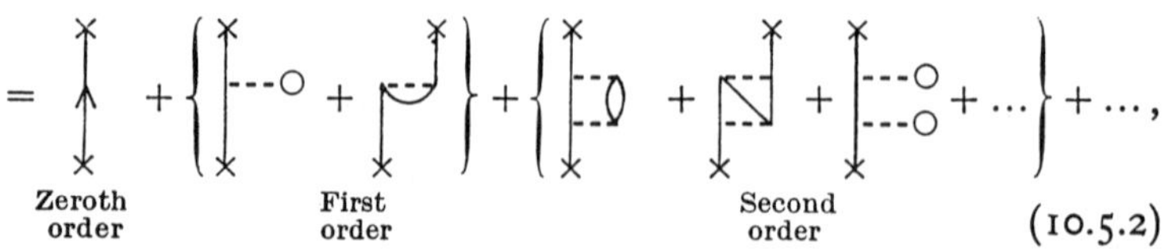
Zeroth
order
First
order
Second
order

(10.5.2)

where (cf. section 10.3) the terms represent pictorially the various physical processes an injected particle can undergo before it is recovered from the many-body system.

10.6. Fourier transform of Green function

Having obtained a convenient perturbation series for $G(\mathbf{k}, t)$, we are now in a position to Fourier transform, term by term, to obtain $G(\mathbf{k}, \omega)$, the quantity in which we are ultimately interested. We will see, below, that this greatly simplifies the mathematics of the perturbation series because the coupled arguments of the unperturbed propagators in t-space become uncoupled in ω-space. Again we discuss the simpler examples before going on to the general case.

10.6.1. *Zeroth- and first-order contributions*

The zeroth-order term gives a contribution of $-iG_0(\mathbf{k}, t_2 - t_1)$ (see (10A.4)), which on Fourier transforming obviously yields $G_0(\mathbf{k}, \omega)$. The first-order terms are of two types:

(*a*) *Direct.* This corresponds to the first of the two first-order terms displayed in (10.5.2). Its contribution (see Fig. 10.1), according to the rules spelled out in Appendix 10A is

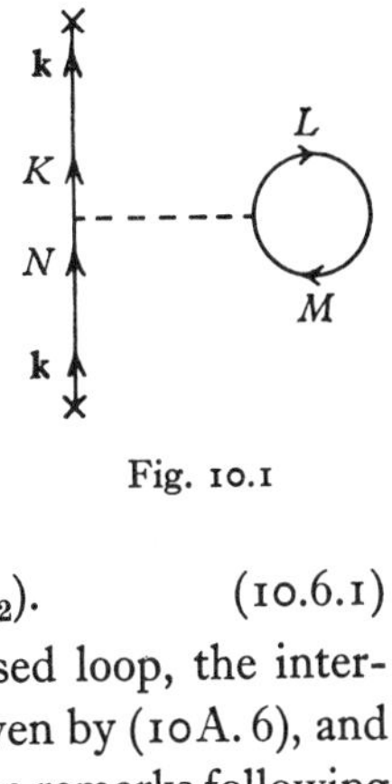

Fig. 10.1

$$G_{1d}(\mathbf{k}, t_2 - t_1) = \int_{-\infty}^{\infty} dt_{12} \sum_{KLMN} \times \langle KL|v|NM\rangle (-1)^1 (-i)^3 \delta_{\mathbf{k}K} \delta_{\mathbf{k}N} \delta_{LM} \times G_0(\mathbf{k}, t_2 - t_{12})\, G_0(\mathbf{k}, t_{12} - t_1)\, G_0(L, t_{12} - t_{12}). \tag{10.6.1}$$

The $(-1)^1$ arises since the graph has one closed loop, the interpretation of the zero-time Green function is given by (10A.6), and the complete independence of $\mathbf{k}$ and K (recall the remarks following (10.2.4)) should, at this stage, be stressed.

Transforming (10.6.1) now gives

$$G_{1d}(\mathbf{k}, \omega) = \int_{-\infty}^{\infty} d(t_2 - t_1)\, e^{i\omega(t_2 - t_1)} \int_{-\infty}^{\infty} dt_{12} \sum_{KLMN} \langle KL|v|NM\rangle \times (-1)^1 (-i)^3 \delta_{\mathbf{k}K} \delta_{\mathbf{k}N} \delta_{LM} \int_{-\infty}^{\infty} \frac{d\omega_1}{2\pi} \int_{-\infty}^{\infty} \frac{d\omega_2}{2\pi} \int_{-\infty}^{\infty} \frac{d\omega_3}{2\pi} \times G_0(\mathbf{k}, \omega_1)\, G_0(\mathbf{k}, \omega_2)\, G_0(L, \omega_3)\, e^{-i\omega_1(t_2 - t_{12})} e^{-i\omega_2(t_{12} - t_1)} e^{-i\omega_3(t_{12} - t_{12})}. \tag{10.6.2}$$

Using the result
$$\int_{-\infty}^{\infty} dt\, e^{i\omega t} = 2\pi\delta(\omega) \tag{10.6.3}$$
one can immediately perform the time integrations. Integrating over t_{12} gives a factor $2\pi\delta(\omega_1-\omega_2)$, while integration over t_2-t_1 gives either $2\pi\delta(\omega-\omega_2)$ or $2\pi\delta(\omega_1-\omega)$, depending on the way one does the integration. If one formally puts $t_1 = 0$ and integrates over t_2, one obtains the former result, whereas putting $t_2 = 0$ and integrating over t_1 gives the latter. Of course there is no paradox here, since the three arguments of the delta functions are linearly dependent, so that any two of the three will do.

Thus (10.6.2) reduces to
$$G_{1d}(\mathbf{k},\omega) = \sum_{KLMN} \langle KL|v|NM\rangle (-1)^1 (-i)^3 \delta_{\mathbf{k}K}\,\delta_{\mathbf{k}N}\,\delta_{LM} \times \int_{-\infty}^{\infty}\frac{d\omega_1}{2\pi}\int_{-\infty}^{\infty}\frac{d\omega_2}{2\pi}\int_{-\infty}^{\infty}\frac{d\omega_3}{2\pi}$$
$$\times G_0(\mathbf{k},\omega_1)\,G_0(\mathbf{k},\omega_2)\,G_0(L,\omega_3)\,2\pi\delta(\omega-\omega_1)\,2\pi\delta(\omega-\omega_2). \tag{10.6.4}$$
The ω_1 and ω_2 integrals are trivial, while that over ω_3 is $-i$ if $l < k_f$ and is zero otherwise if we note that it is simply a zero-time unperturbed Green function. Such an expression is defined by (10A.6), thus enabling us to apply (10.2.11) unambiguously. Hence, without any further trouble, we evaluate (10.6.4) to be
$$G_{1d}(\mathbf{k},\omega) = \Big(-\sum_{l<k_f}\langle \mathbf{k}L|v|\mathbf{k}L\rangle\Big)\,G_0^2(\mathbf{k},\omega)$$
$$\equiv M_{1d}\,G_0^2. \tag{10.6.5}$$
This expression has been used to define M_{1d}. The reason for doing this will emerge below.

(*b*) *Exchange.* This corresponds to the second of the two first-order terms displayed in (10.5.2). Its contribution is (see Fig. 10.2)

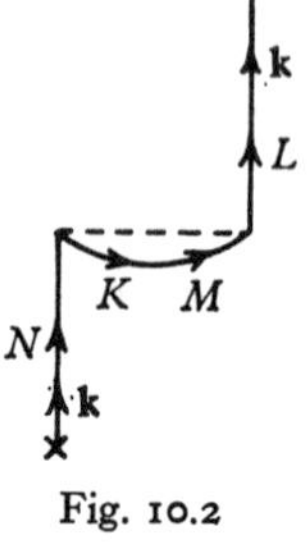

Fig. 10.2

$$G_{1e}(\mathbf{k},t_2-t_1) = \int_{-\infty}^{\infty} dt_{12} \sum_{KLMN} \langle KL|v|NM\rangle$$
$$\times (-1)^0 (-i)^3 \delta_{\mathbf{k}N}\,\delta_{KM}\,\delta_{L\mathbf{k}}\,G_0(\mathbf{k},t_{12}-t_1)$$
$$\times G_0(K,t_{12}-t_{12})\,G_0(\mathbf{k},t_2-t_{12}). \tag{10.6.6}$$
This time, we have a factor $(-1)^0$ since there are no closed loops.

This transforms much as in case (*a*). The analogue of (10.6.4), for instance, is

$$G_{1e}(\mathbf{k},\omega) = \sum_{KLMN} \langle KL|v|NM\rangle(-1)^0(-i)^3\,\delta_{\mathbf{k}N}\,\delta_{KM}\,\delta_{L\mathbf{k}}\int_{-\infty}^{\infty}\frac{d\omega_1}{2\pi}$$

$$\times\int_{-\infty}^{\infty}\frac{d\omega_2}{2\pi}\int_{-\infty}^{\infty}\frac{d\omega_3}{2\pi}G_0(\mathbf{k},\omega_1)\,G_0(\mathbf{k},\omega_2)\,G_0(K,\omega_3)\,\delta(\omega-\omega_1)\,\delta(\omega-\omega_2) \tag{10.6.7}$$

and this simplifies to

$$G_{1e}(\mathbf{k},\omega) = \left(\sum_{m<k_f}\langle M\mathbf{k}|v|\mathbf{k}M\rangle\right)G_0^2(\mathbf{k},\omega) \equiv M_{1e}\,G_0^2. \tag{10.6.8}$$

The significance of M_{1e} thus defined, will become apparent shortly.

10.6.2. *Second-order terms*

(*a*) *Direct polarization.* This corresponds to the first second-order term in (10.5.2). Its contribution is (see Fig. 10.3)

$$G_{2d}(\mathbf{k},t_2-t_1)$$
$$= (-i)\int_{-\infty}^{\infty}dt_{12}\int_{-\infty}^{\infty}dt_{34}\sum_{\substack{KLMN\\PQRS}}$$
$$\times\langle KL|v|NM\rangle\langle PQ|v|SR\rangle$$
$$\times(-1)^1(-i)^5\,\delta_{\mathbf{k}N}\,\delta_{KS}\,\delta_{\mathbf{k}P}\,\delta_{LR}\,\delta_{MQ}$$
$$\times G_0(\mathbf{k},t_{12}-t_1)\,G_0(K,t_{34}-t_{12})$$
$$\times G_0(\mathbf{k},t_2-t_{34})\,G_0(L,t_{34}-t_{12})$$
$$\times G_0(M,t_{12}-t_{34}). \tag{10.6.9}$$

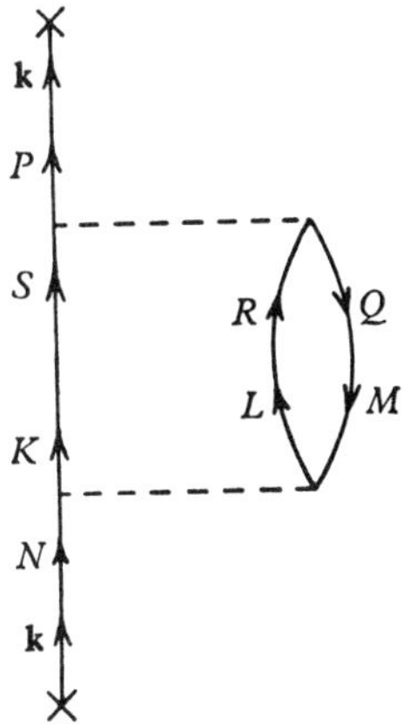

Fig. 10.3

This time there is one loop. Converting into ω-space, (10.6.9) becomes

$$G_{2d}(\mathbf{k},\omega) = (-i)\sum_{\substack{KLMN\\PQRS}}\langle KL|v|NM\rangle\langle PQ|v|SR\rangle(-1)^1(-i)^5$$

$$\times\,\delta_{\mathbf{k}N}\,\delta_{\mathbf{k}P}\,\delta_{KS}\,\delta_{LR}\,\delta_{MQ}\int_{-\infty}^{\infty}\frac{d\omega_1}{2\pi}\int_{-\infty}^{\infty}\frac{d\omega_2}{2\pi}\int_{-\infty}^{\infty}\frac{d\omega_3}{2\pi}\int_{-\infty}^{\infty}\frac{d\omega_4}{2\pi}\int_{-\infty}^{\infty}\frac{d\omega_5}{2\pi}$$

$$\times\,G_0(\mathbf{k},\omega_1)\,G_0(\mathbf{k},\omega_2)\,G_0(K,\omega_3)\,G_0(L,\omega_4)\,G_0(M,\omega_5)$$

$$\times\,2\pi\delta(\omega_1+\omega_5-\omega_3-\omega_4)\,2\pi\delta(\omega_3+\omega_4-\omega_2-\omega_5)\,2\pi\delta(\omega-\omega_1). \tag{10.6.10}$$

As previously (see comments after (10.6.3)), $\delta(\omega-\omega_2)$ could equally well replace any of the Dirac delta functions in (10.6.10). Thus, without any difficulty, (10.6.10) reduces to

$$G_{2d}(\mathbf{k},\omega) = \left(\sum_{KLM} \langle KL|v|\mathbf{k}M\rangle\langle\mathbf{k}M|v|KL\rangle \int_{-\infty}^{\infty}\frac{d\omega_3}{2\pi}\int_{-\infty}^{\infty}\frac{d\omega_4}{2\pi}\right.$$
$$\left.\times\, G_0(K,\omega_3)\,G_0(L,\omega_4)\,G_0(M,\omega_3+\omega_4-\omega)\right) G_0^2(\mathbf{k},\omega). \quad (10.6.11)$$

Now this result can be simplified further. In particular, the ω_4 integration can be done since we have the identity

$$i\int_{-\infty}^{\infty}\frac{d\omega'}{2\pi}G_0(L,\omega')\,G_0(M,\omega+\omega') = \begin{cases} \dfrac{-1}{\epsilon_M-\epsilon_L-\omega-i0} & (l<k_f<m), \\ \dfrac{1}{\epsilon_M-\epsilon_L-\omega+i0} & (m<k_f<l), \\ 0 & (\text{otherwise}). \end{cases}$$
$$(10.6.12)$$

This latter result is directly proved by using the explicit form (10.2.12) for G_0.

Further simplification can be achieved by employing (1.3.6) of Chapter 1, thus expressing our matrix elements in terms of the momentum transfer $\mathbf{q}$. In this way, without difficulty, (10.6.11) reduces to

$$\begin{aligned} G_{2d}(\mathbf{k},\omega) &= \left(\sum_q v_\mathbf{q}^2\int_{-\infty}^{\infty}\frac{d\omega''}{2\pi}G_0(\mathbf{k}_m-\mathbf{q},\omega-\omega'')\,F^{(0)}(\mathbf{q},\omega'')\right) G_0^2(\mathbf{k},\omega) \\ &= M_{2d}(\mathbf{k},\omega)\,G_0^2(\mathbf{k},\omega), \end{aligned} \quad (10.6.13)$$

where $\mathbf{k}_m$ denotes the momentum part of the momentum-spin index $\mathbf{k}$ and

$$F^{(0)}(\mathbf{k},\omega) = \sum_{|\mathbf{k}+\mathbf{q}|>k_f>|\mathbf{q}|}\frac{1}{\epsilon_\mathbf{q}-\epsilon_{\mathbf{k}+\mathbf{q}}-\omega+i0} - \sum_{|\mathbf{k}+\mathbf{q}|<k_f<|\mathbf{q}|}\frac{1}{\epsilon_\mathbf{q}-\epsilon_{\mathbf{k}+\mathbf{q}}-\omega-i0}, \quad (10.6.14)$$

the $\mathbf{k}$ in the latter equation and in (10.6.15) representing, in the usual way, the momentum part of the intermediate momentum-spin index K. For the G_0 inside the integral in (10.6.13), we have used, for the first time, the convention $G_0(L,\omega)\equiv G_0(\mathbf{l},\omega)$ since G_0 is spin-independent, and, once more, we use the convention of

(8.3.79) whereby $v_{\mathbf{q}} = v(\mathbf{q})/\Omega, v(\mathbf{q})$ (cf. Chapter 5) being of order unity.

Now the value of the second sum in (10.6.14) depends only on k, and thus is invariant if $k \to -k$. Furthermore, since q runs over all space, we may make the replacement $\mathbf{q} \to \mathbf{k}+\mathbf{q}$. In this way, the restrictions on the two sums in (10.6.14) become the same. Thus, summing (trivially) over spin, (10.6.14) becomes

$$F^{(0)}(\mathbf{k}, \omega) = 2 \sum_{|\mathbf{k}+\mathbf{q}|>k_f>|\mathbf{q}|} \left\{ \frac{1}{\omega - (\epsilon_{\mathbf{k}+\mathbf{q}} - \epsilon_{\mathbf{q}} - i0)} - \frac{1}{\omega + \epsilon_{\mathbf{k}+\mathbf{q}} - \epsilon_{\mathbf{q}} - i0} \right\}. \tag{10.6.15}$$

The factor $F^{(0)}(\mathbf{k}, \omega)$ of (10.6.15) is always associated with a 'bubble' of the kind shown in Fig. 10.3 and has an importance beyond that of the present application, as we will see later.

(*b*) *Exchange polarization.* This corresponds to the second member of the second-order terms displayed in (10.5.2). By writing down the time-dependent expressions (see Fig. 10.4) and transforming much as above, we obtain the analogue of (10.6.10) for the present case, namely,

$$\begin{aligned} G_{2e}(\mathbf{k}, \omega) &= (-i) \sum_{\substack{KLMN \\ PQRS}} \langle KL|v|NM\rangle \langle PQ|v|SR\rangle \\ &\times (-1)^0 (-i)^5 \delta_{\mathbf{k}N} \delta_{\mathbf{k}Q} \delta_{KS} \delta_{LR} \delta_{PM} \int_{-\infty}^{\infty} \frac{d\omega_1}{2\pi} \\ &\times \int_{-\infty}^{\infty} \frac{d\omega_2}{2\pi} \int_{-\infty}^{\infty} \frac{d\omega_3}{2\pi} \int_{-\infty}^{\infty} \frac{d\omega_4}{2\pi} \int_{-\infty}^{\infty} \frac{d\omega_5}{2\pi} \\ &\times G_0(\mathbf{k}, \omega_1)\, G_0(\mathbf{k}, \omega_2)\, G_0(K, \omega_3)\, G_0(L, \omega_4) \\ &\times G_0(M, \omega_5)\, 2\pi\delta(\omega_1 + \omega_5 - \omega_3 - \omega_4) \\ &\times 2\pi\delta(\omega_3 + \omega_4 - \omega_2 - \omega_5)\, 2\pi\delta(\omega - \omega_1). \end{aligned} \tag{10.6.16}$$

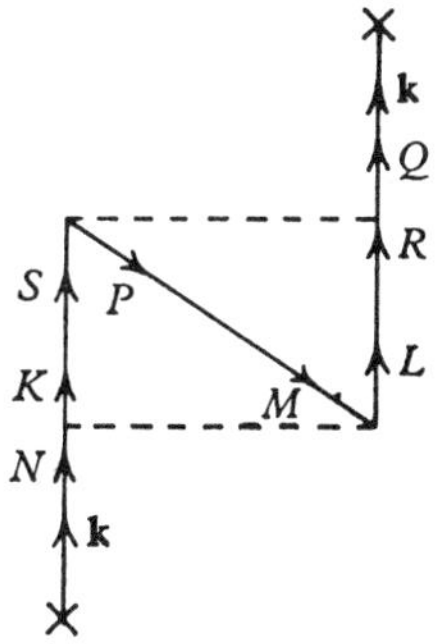

Fig. 10.4

As previously, $\delta(\omega - \omega_2)$ could equally well replace any of the Dirac delta functions in the latter expression. If one replaces, say, the first one in this way and integrates over ω_1 and ω_2, an expression analogous to (10.6.11), in which a $G_0^2(\mathbf{k}, \omega)$ term factorizes out, is obtained,

$$G_{2e}(\mathbf{k}, \omega) = M_{2e}(\mathbf{k}, \omega)\, G_0^2(\mathbf{k}, \omega), \tag{10.6.17}$$

where $M_{2e}(\mathbf{k}, \omega)$ is easily written down using (10.6.16).

We now proceed to a discussion of the general case.

10.6.3. *nth-order contributions*

Fig. 10.5 is meant to represent a general nth-order diagram, the topological structure of which is characterized by the symbol α. Of course, our drawing of the end vertices, as shown, make it to some degree special, but our arguments do not depend on these special features.

We now give each line (i.e. contraction of either propagating or non-propagating type) an ω_i-variable as shown. In this way the graph readily suggests the mathematical expression for the calculation of $G(\mathbf{k}, \omega)$ obtained by writing down $G(\mathbf{k}, t_2 - t_1)$ and transforming. We find in the usual way

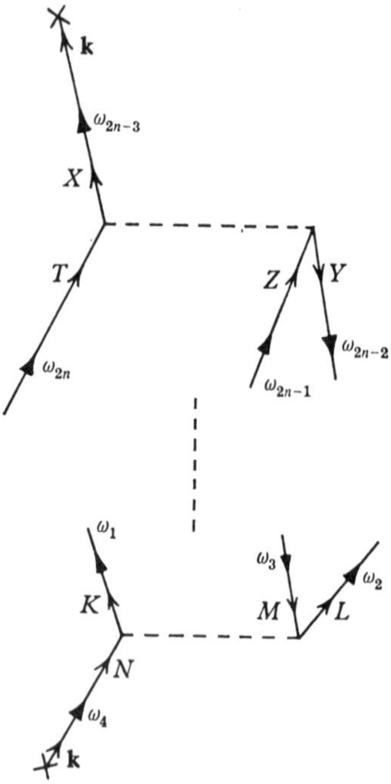

Fig. 10.5

$$
\begin{aligned}
G_{n\alpha}(\mathbf{k}, \omega) &= i^n(-1)^l \sum_{\substack{KLMN\\PQRS\\ \cdots \\ XYZT}} \langle KL|v|NM\rangle \\
&\times \langle PQ|v|SR\rangle \ldots \langle XY|v|TZ\rangle \\
&\times \delta_{\mathbf{k}N} \ldots \delta_{\mathbf{k}X} \int_{-\infty}^{\infty} \frac{d\omega_1}{2\pi} \int_{-\infty}^{\infty} \frac{d\omega_2}{2\pi} \ldots \\
&\times \int_{-\infty}^{\infty} \frac{d\omega_{2n}}{2\pi} G_0(K, \omega_1)\, G_0(L, \omega_2) \ldots G_0(Z, \omega_{2n-1})\, G_0(T, \omega_{2n}) \\
&\times 2\pi\delta(\omega_1 + \omega_2 - \omega_3 - \omega_4) \ldots 2\pi\delta(\omega_{2n-3} + \omega_{2n-2} - \omega_{2n-1} - \omega_{2n}) \\
&\times 2\pi\delta(\omega_{2n-3} - \omega). \qquad (10.6.18)
\end{aligned}
$$

The factor i^n arises from the product of the factor $(-1)^{n-1}$ shown in (10.5.1) with $(-i)^{2n+1}$ arising from $2n+1$ contractions. Once more, l is the number of closed loops and, because of the linear dependence of the energy arguments of the Dirac delta functions, we can write in place of any one of them the factor $\delta(\omega - \omega_4)$. By integrating over the two energy variables associated with the terminal lines (that is, as drawn, ω_4 and ω_{2n-3}), we can write (10.6.18) in the form

$$G_{n\alpha}(\mathbf{k}, \omega) = M_{n\alpha}(\mathbf{k}, \omega)\, G_0^2(\mathbf{k}, \omega). \tag{10.6.19}$$

10.7. Dyson's irreducible self-energy operator

We see from (10.6.18) and Fig. 10.5 how to write down the contribution to $G(\mathbf{k}, \omega)$ from any given diagram. It is important to remember that energy and momentum are conserved and each line carries an unperturbed Green function. For this reason, we write

$$G(\mathbf{k}, \omega) = \sum_{n, \alpha} G_{n\alpha}(\mathbf{k}, \omega)$$

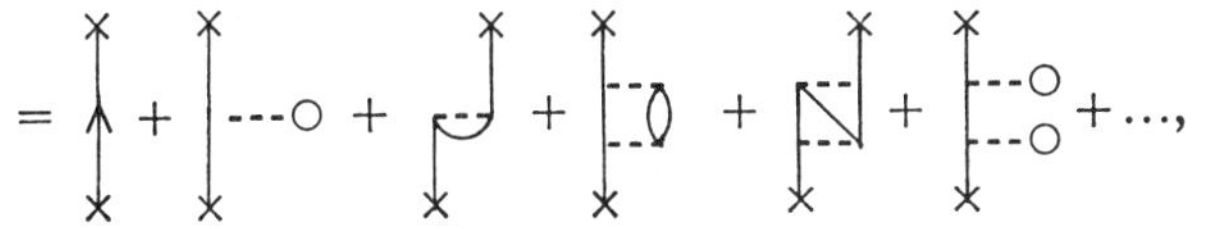

(10.7.1)

where now the diagrams on the right (in contrast with those of (10.5.2)) are used to denote contributions to $G(\mathbf{k}, \omega)$.

In evaluating contributions from diagrams, one soon learns to write these down with an economy of symbolism using the conservation results embodied in (10.6.18), together with that of (1.3.6) in Chapter 1. Thus, for example, if we were *now* to evaluate the contribution $G_{2d}(\mathbf{k}, \omega)$, we might replace Fig. 10.3 by Fig. 10.6, where the labelling automatically ensures energy and momentum conservation. Some emphasis is thus placed on momentum and energy transfer. One writes down the matrix elements appropriate to the interactions ($v(q)$ in each case), the various Green functions and multiplicative constants and then sums over the internal variables. In this way, we can reproduce (10.6.13).

Equation (10.7.1) turns out to have a very simple structure, as we go on to show. We have seen above that the zeroth-order contribution to $G(\mathbf{k}, \omega)$ is $G_0(\mathbf{k}, \omega)$. All others correspond to such graphs as that shown in Fig. 10.7, where the central part includes $n\,(\geqslant 1)$ interaction lines. The contribution from this graph is given by (10.6.19) and thus we can, if we wish, rewrite (10.7.1) as

$$G(\mathbf{k}, \omega) = G_0(\mathbf{k}, \omega) + G_0^2(\mathbf{k}, \omega)\, M_R(\mathbf{k}, \omega), \tag{10.7.2}$$

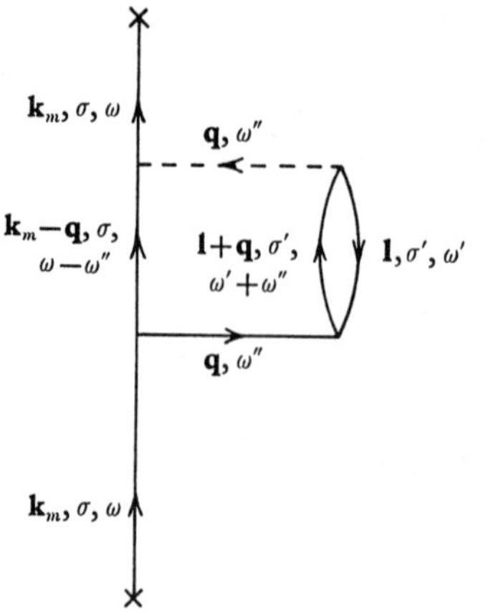

Fig. 10.6

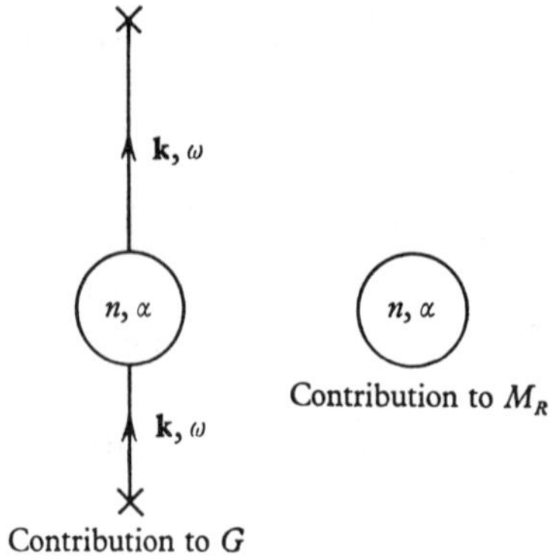

Fig. 10.7

where M_R, the reducible self-energy, is defined by

$$M_R(\mathbf{k}, \omega) = \sum_{n, \alpha} M_{n\alpha}(\mathbf{k}, \omega). \qquad (10.7.3)$$

The form (10.7.2) is not particularly fruitful *per se*, but it does serve to focus attention on the self-energy parts (bubbles (n, α)), representing contributions $M_{n\alpha}$ to (10.7.3). By analysing these bubbles in more detail, we will find, below, that we are able to write G in a form which is convenient both for its computation and application. We begin this analysis with some more definitions.

It may or may not be possible to draw the graph shown in Fig. 10.7 in the form indicated in Fig. 10.8. (The intermediate line shown necessarily carries the same **k** and ω indices as the end pair, since momentum, spin and energy are conserved at each vertex.) If such a redrawing is possible, the original graph (or self-energy part) is called reducible; if not, it is called irreducible. In the reducible case, when the description is as indicated, we have

$$M_{n\alpha}(\mathbf{k}, \omega) = G_0(\mathbf{k}, \omega)\, M_{n'\alpha'}(\mathbf{k}, \omega)\, M_{n''\alpha''}(\mathbf{k}, \omega). \qquad (10.7.4)$$

It is clear that we can keep on reducing any given graph or any given self-energy part until we arrive at the situation shown in Fig. 10.9 where we have strings of irreducible self-energy bubbles, each connecting line carrying indices **k** and ω. If Fig. 10.9 is simply

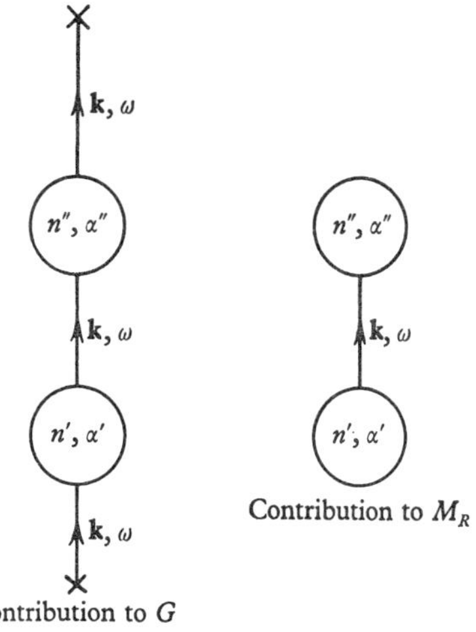

Fig. 10.8

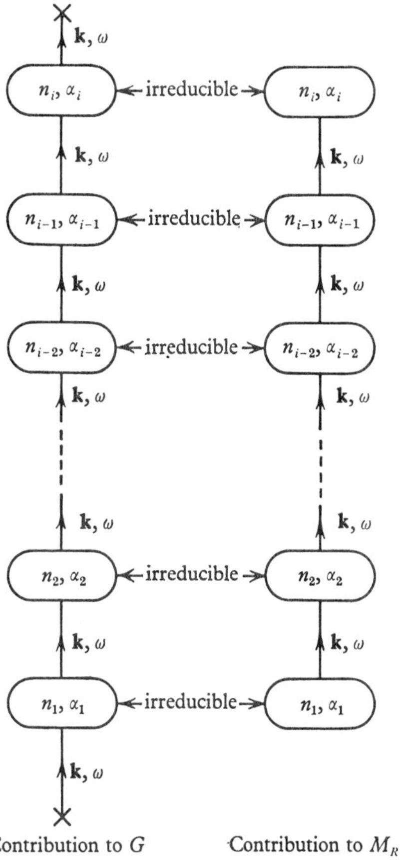

Fig. 10.9

a more detailed equivalent of Fig. 10.7 we have, by straightforward generalization of (10.7.4),

$$M_{n\alpha} = G_0^{i-1} M_{n_1\alpha_1} \dots M_{n_i\alpha_i}. \qquad (10.7.5)$$

To illustrate these points, we give, in Fig. 10.10, all graphs in first and second order and a few in third order, together with their contributions to $G(\mathbf{k}, \omega)$. The reducible and irreducible graphs are marked R and I respectively. The second-order terms correspond to the uncircled diagrams of Fig. 10 A. 7 in unlettered form. The quantities M_{1d}, M_{1e}, M_{2d}, M_{2e} have already been defined by (10.6.5), (10.6.8), (10.6.13), (10.6.17) respectively; the quantities M_{3p}, M_{2q}, M_{3r} are defined by the diagrams shown.

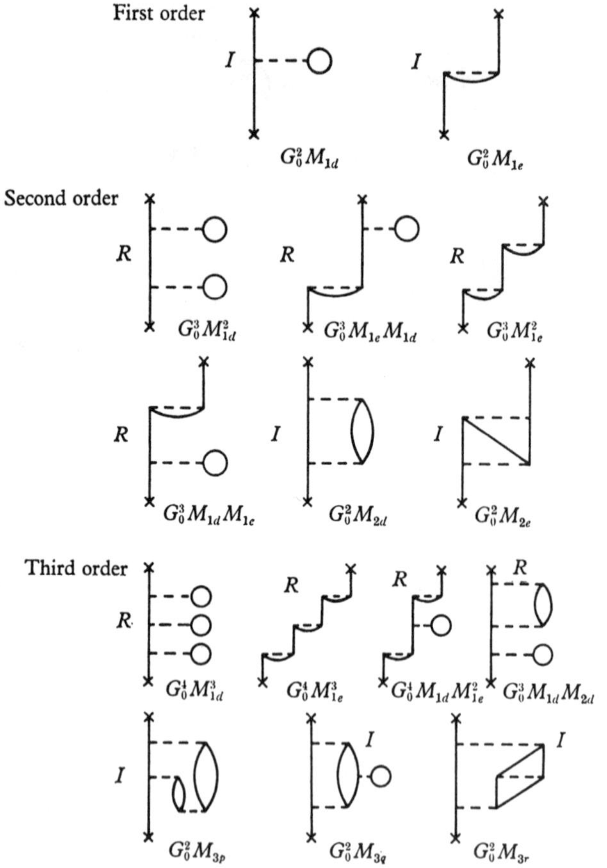

Fig. 10.10

We are now in a position to prove the important Dyson self-energy equation which, in section 3, we anticipated (recall (10.3.2)) and used. Let us begin by writing down the series for $G(\mathbf{k}, \omega)$, summing by vertex order according to (10.7.1), and then rearranging into a power series in G_0. We have (see Fig. 10.10)

$$
\begin{aligned}
G = {} & G_0 + (G_0^2 M_{1d} + G_0^2 M_{1e}) \\
& + (G_0^3 M_{1d}^2 + 2G_0^3 M_{1d} M_{1e} + G_0^3 M_{1e}^2 + G_0^2 M_{2d} + G_0^2 M_{2e}) \\
& + (G_0^4 M_{1d}^3 + G_0^4 M_{1e}^3 + \ldots) + \ldots \\
= {} & G_0 + G_0^2 (M_{1d} + M_{1e} + M_{2d} + M_{2e} + M_{3p} + \ldots) \\
& + G_0^3 (M_{1d} + M_{1e} + M_{2d} + \ldots)^2 + \ldots . \qquad (10.7.6)
\end{aligned}
$$

If we now define the sum of all irreducible self-energy parts by

$$M = M_{1d}+M_{1e}+M_{2d}+M_{2e}+M_{3p}+\ldots, \qquad (10.7.7)$$

then (10.7.6) becomes

$$G = G_0+G_0^2M+G_0^3M^2+\ldots = \frac{1}{G_0^{-1}-M}. \qquad (10.7.8)$$

The quantity M is the Dyson irreducible self-energy of section 10.3, and if one uses (10.2.12) to replace G_0^{-1} by $\epsilon_{\mathbf{k}}-\omega$, one obtains the fundamental form (10.3.2) for G. Actually, our 'proof' of (10.7.6) is only a low-order test, but a full proof follows without difficulty, for a general term of (10.7.6) can be written

$$G_0(M_{1d}+M_{1e}+\ldots)\,G_0(M_{1d}+M_{1e}+\ldots)\,G_0\ldots G_0(M_{1d}+M_{1e}+\ldots)\,G_0$$

$$=\Sigma G_0\,M_{n_1\alpha_1}\,G_0\,M_{n_2\alpha_2}\ldots G_0\,M_{n_j\alpha_j}\,G_0. \qquad (10.7.9)$$

If we recollect the result (10.7.5) applied to Fig. 10.9 we see that each ordered term of (10.7.9) corresponds to one and only one term of the original series. Thus, by summing (10.7.9) over all powers we generate the series (10.7.1) for G.

In applications, the main task is the calculation of M. Then the techniques of section 10.4 are applied for finding quasi-particle energies and life-times. Inevitably, in practice, M cannot be calculated exactly and we are usually content to isolate and calculate the dominant terms of (10.7.7). The strength of the Green function method lies in the fact that use of an approximate M in (10.7.8) amounts to having summed an infinite subset of terms in the perturbation series for G. Thus, we correct our original 'mistake' of having expanded at all in a perturbation series. Furthermore, by resumming, far from complicating the situation, we actually simplify it. The method of approximate solution is illustrated by the following cases.

10.7.1. *Hartree approximation*

Here, we write

$$M(\mathbf{k},\omega) \approx M_{1d}(\mathbf{k},\omega) = -\sum_{l<k_f} \langle \mathbf{k}L|v|\mathbf{k}L\rangle \qquad (10.7.10)$$

by (10.6.5). This quantity is real and independent of ω. Thus, application of (10.3.11) gives a quasi-particle energy of

$$\epsilon_{\mathbf{k}} = \mathcal{E}_{\mathbf{k}} + \sum_{l<k_f} \langle \mathbf{k}L|v|\mathbf{k}L\rangle, \tag{10.7.11}$$

while the lifetime is infinite.

We see that the Hartree method amounts to writing G in the approximate form

$$\frac{1}{G_0^{-1} - M_{1d}} = G_0 + G_0^2 M_{1d} + G_0^3 M_{1d}^2 + \ldots =$$

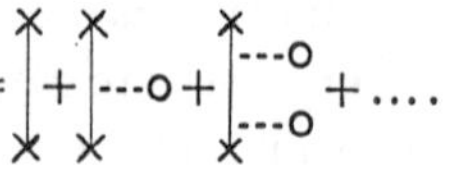

(10.7.12)

10.7.2. *Hartree–Fock approximation*

In this case, we have

$$M(\mathbf{k},\omega) \approx M_{1d}(\mathbf{k},\omega) + M_{1e}(\mathbf{k},\omega)$$
$$= -\sum_{l<k_f} \langle \mathbf{k}L|v|\mathbf{k}L\rangle + \sum_{l<k_f} \langle L\mathbf{k}|v|\mathbf{k}L\rangle \tag{10.7.13}$$

by (10.6.5) and (10.6.8). Once more, $M(\mathbf{k},\omega)$ is real and independent of ω, thus making the solutions of (10.3.11) and (10.3.12) trivial. We obtain undamped quasi-particle energies

$$\epsilon_{\mathbf{k}} = \mathcal{E}_{\mathbf{k}} + \sum_{l<k_f} \langle \mathbf{k}L|v|\mathbf{k}L\rangle - \sum_{l<k_f} \langle L\mathbf{k}|v|\mathbf{k}L\rangle. \tag{10.7.14}$$

The Hartree–Fock method corresponds to the approximate propagator

$$\frac{1}{G_0^{-1} - (M_{1d} + M_{1e})} = G_0 + G_0^2(M_{1d} + M_{1e}) + G_0^3(M_{1d} + M_{1e})^2 + \ldots$$

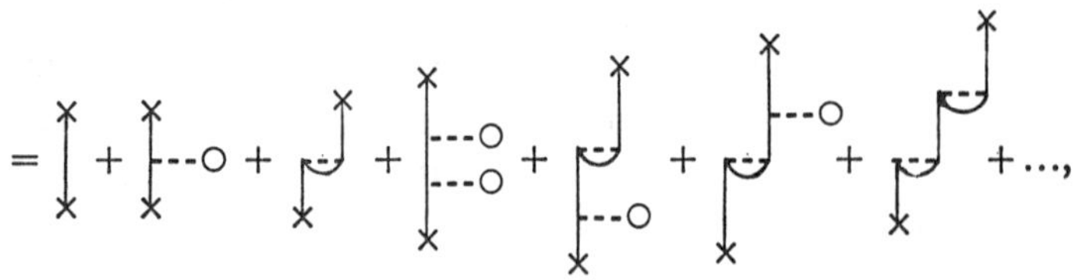

(10.7.15)

which means that every possible combination of direct and exchange interactions with passive particles, during the history of the introduced particle, is taken into account.

10.7.3. *Higher approximations*

In the case of a weakly coupled system with finite matrix elements, the way to improve the Hartree–Fock approximation is clear. We simply include the terms quadratic in the coupling by writing

$$M(\mathbf{k}, \omega) \approx M_{1d}+M_{1e}+M_{2d}+M_{2e}. \qquad (10.7.16)$$

The final two terms, given by (10.6.13) and (10.6.17) respectively, are explicitly ω-dependent and furthermore have imaginary parts. Thus, in this case, the solution of the equations (10.3.11) and (10.3.12) is non-trivial. For details of this and the more general problem of a low density short-range Fermion system, we refer the reader to Galitskii (1958), Klein (1962) and Pines (1961).

In the Coulomb case, we have an additional complication, arising, as usual, from the special nature of the potential. We see, from (10.6.13), that for $v(q) \propto q^{-2}$, M_{2d} is divergent. This is clearly another case of the 'piling up' of $v(q)$ factors of the kind encountered in Chapter 5, and it will be essential to sum all orders of selected graphs to remove the divergence. This is done (in the random phase approximation) by adding to the right-hand side of (10.7.16), all ring-type insertions of the kind shown in Fig. 10.11.

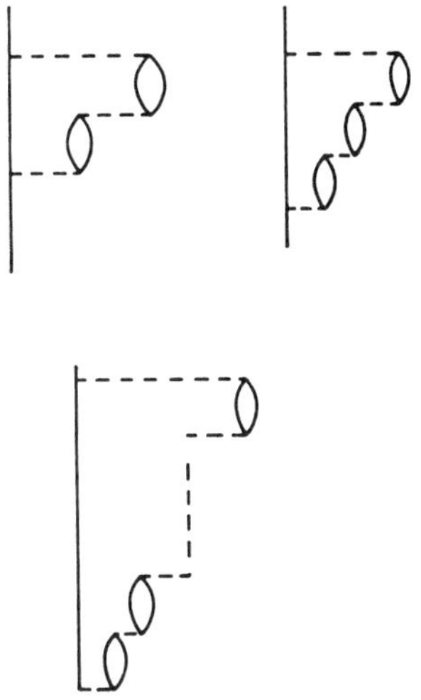

Fig. 10.11

Furthermore, it can be rigorously shown that, in this way, we obtain the most important terms at high densities. This leads back to the main results of Chapter 5 and we shall therefore not pursue the details further, the reader being referred to Dubois (1959) (see also, Ninham, 1964 for numerical corrections to Dubois's work) and Quinn & Ferrell (1958).

10.8. Collective motion. Preliminaries

We saw previously (Chapters 5 and 8) that on introducing interactions into an ideal system, new collective modes, detectable by coupling to external probes (say, charged particles as in the case of plasmons, or neutrons in the case of phonons), become possible. Our aim, below, is to describe these. Mathematically, this will amount to the discussion of a special form of the two-particle Green function and the techniques involved are similar to those used above. We will thus follow the one-particle propagator work rather closely, but, because of the similarities, our analysis below will be rather more condensed.

In our previous discussion of quasi-particles in section 10.3, we set up the problem by introducing a bare particle (or hole) of index $\mathbf{k}$ into the system at time t_1, and asking what the amplitude $G(\mathbf{k}, t_2-t_1)$ was for its recovery at time t_2. Under certain conditions, namely when the excitation energy was suitably chosen as a function of $\mathbf{k}$, it was found that G decayed comparatively slowly with time. In this way, the quasi-particle energies were found.

In our present problem, the same technique is used. One introduces a density fluctuation of wave index $\mathbf{k}$ into the interacting N-body system at time t_1 and asks for the probability of finding this same mode at time t_2. The appropriate amplitude is

$$F(\mathbf{k}, t_2-t_1) = i\langle\Psi_n(0)|\,T\{\rho_{\mathbf{k}t_2}\rho^\dagger_{\mathbf{k}t_1}\}|\Psi_n(0)\rangle, \tag{10.8.1}$$

where (cf. (10.2.1))

$$\rho_{\mathbf{k}t} = e^{iHt}\rho_{\mathbf{k}}e^{-iHt} \tag{10.8.2}$$

and recalling (1.3.4) and the classical form of $\rho_{\mathbf{k}}$ as given by (8.2.68)

$$\rho_{\mathbf{k}} = \sum_{\mathbf{q},\sigma} a^\dagger_{\mathbf{k}+\mathbf{q}\sigma}a_{\mathbf{q}\sigma}. \tag{10.8.3}$$

Note that as used in the density fluctuation theory, $\mathbf{k}$ represents momentum and not momentum and spin as in the single-particle Green function work.

For our immediate purposes, it is convenient to use (10.8.3) to rewrite (10.8.2) and its conjugate in the forms

$$\rho_{\mathbf{k}t} = \sum_{\mathbf{q},\sigma} a^\dagger_{\mathbf{k}+\mathbf{q}\sigma t}a_{\mathbf{q}\sigma t}; \quad \rho^\dagger_{\mathbf{k}t} = (\rho_{\mathbf{k}t})^\dagger = \rho_{-\mathbf{k}t}. \tag{10.8.4}$$

The latter may then be used to express (10.8.1) in the form

$$F(\mathbf{k}, t_2-t_1) = \sum_{\substack{\mathbf{q}_1\sigma_1\\ \mathbf{q}_2\sigma_2}} F_{\substack{\mathbf{q}_1\sigma_1\\ \mathbf{q}_2\sigma_2}}(\mathbf{k}, t_2-t_1), \tag{10.8.5}$$

where

$$F_{\substack{\mathbf{q}_1\sigma_1\\ \mathbf{q}_2\sigma_2}}(\mathbf{k}, t_2-t_1) = i\langle\Psi_n(0)|\,T(a^\dagger_{\mathbf{k}+\mathbf{q}_2\sigma_2t_2}\,a_{\mathbf{q}_2\sigma_2t_2}\,a^\dagger_{\mathbf{q}_1\sigma_1t_1}\,a_{\mathbf{k}+\mathbf{q}_1\sigma_1t_1})|\Psi_n(0)\rangle. \tag{10.8.6}$$

It should be noted that in (10.8.6) T is defined by its operation on a's and $a^{\dagger}$'s. Thus, in interpreting (10.8.1), we must think of it in a form such as (10.8.6). We thus find, because of the biquadratic form of the operand, that

$$T(\rho_{\mathbf{k}t_2}\rho^{\dagger}_{\mathbf{k}t_1}) = \begin{cases} \rho_{\mathbf{k}t_2}\rho^{\dagger}_{\mathbf{k}t_1} & (t_2 > t_1), \\ \rho^{\dagger}_{\mathbf{k}t_1}\rho_{\mathbf{k}t_2} & (t_1 > t_2). \end{cases} \tag{10.8.7}$$

In the spirit of section 10.3, we inquire what choice of excitation energies will lead to only slow attenuation in (10.8.5). The answer, once more, is determined by the poles of the Fourier transform of (10.8.5) into ω-space, namely,

$$F(\mathbf{k}, \omega) = \sum_{\substack{\mathbf{q}_1\sigma_1 \\ \mathbf{q}_2\sigma_2}} F_{\substack{\mathbf{q}_1\sigma_1 \\ \mathbf{q}_2\sigma_2}}(\mathbf{k}, \omega), \tag{10.8.8}$$

where
$$F_{\substack{\mathbf{q}_1\sigma_1 \\ \mathbf{q}_2\sigma_2}}(\mathbf{k}, t) = \int_{-\infty}^{\infty} \frac{d\omega}{2\pi} F_{\substack{\mathbf{q}_1\sigma_1 \\ \mathbf{q}_2\sigma_2}}(\mathbf{k}, \omega)\, e^{-i\omega t}. \tag{10.8.9}$$

Thus, we eventually become concerned with the graphical analysis of (10.8.6), which is a special case of a two-body Green function.

With these preliminaries and definitions, we now enter upon an exposition of the theory.

10.9 Spectral representation and collective modes

We could follow section 10.3 very closely here, beginning with the computed form of $F(\mathbf{k}, \omega)$ given by (10.9.9) below, this being the analogue of (10.3.2). Since, however, the spectral representation method is easier here than in the previous theory, we will use this approach in the present section.

By introducing the set of intermediate states defined by

$$H|r\rangle = E_r|r\rangle, \quad \langle r|r\rangle = 1, \tag{10.9.1}$$

the ground-state member of which is again given by (10.2.4), we obtain from (10.8.1), (10.8.2) and (10.8.7) the equation

$$F(\mathbf{k}, t) = \left\{\begin{array}{ll} i\sum\limits_r |\langle\Psi_n(0)|\rho_{\mathbf{k}}|r\rangle|^2\, e^{-i(E_r-E)t} & (t>0), \\ i\sum\limits_r |\langle\Psi_n(0)|\rho^{\dagger}_{\mathbf{k}}|r\rangle|^2\, e^{i(E_r-E)t} & (t<0), \end{array}\right\} \tag{10.9.2}$$

the former being the amplitude for finding at time t later the mode inserted at time 0 and the latter being the amplitude that the mode found at time 0 was present at time t earlier. In view of this interpretation, $F(\mathbf{k}, t) = F(\mathbf{k}, -t)$ and thus

$$\langle \Psi_n(0)|\rho_{\mathbf{k}}|r\rangle = \langle \Psi_n(0)|\rho_{\mathbf{k}}^{\dagger}|r\rangle. \tag{10.9.3}$$

Alternatively, if we remember (see (10.8.4)) that $\rho_{\mathbf{k}}^{\dagger} = \rho_{-\mathbf{k}}$, (10.9.3) implies that the reversal of the spatial direction of a wave does not alter its matrix elements.

A density-of-states function

$$S(\mathbf{k}, \omega) = \sum_r |\langle \Psi_n(0)|\rho_{\mathbf{k}}|r\rangle|^2 \delta(\omega - (E_r - E)) \tag{10.9.4}$$

is now introduced, in terms of which we have

$$F(\mathbf{k}, \omega) = \int_0^{\infty} d\omega' \, S(\mathbf{k}, \omega') \left(\frac{1}{\omega - \omega' + i\eta} - \frac{1}{\omega + \omega' - i\eta} \right). \tag{10.9.5}$$

It may be directly verified that (10.9.5) and (10.9.2) are Fourier transforms of each other.

If we were able to choose (cf. Chapter 8, especially (8.3.42))

$$S(\mathbf{k}, \omega) = S(k)\, \delta(\omega - \omega_{\mathbf{k}}), \tag{10.9.6}$$

where $\omega_{\mathbf{k}}$ is real ($\omega_{\mathbf{k}} \equiv \epsilon_{\mathbf{k}}$, with $\hbar = 1$), (10.9.5) would lead to

$$F(\mathbf{k}, \omega) = \delta(\mathbf{k}) \left(\frac{1}{\omega - \epsilon_{\mathbf{k}} + i\eta} - \frac{1}{\omega + \epsilon_{\mathbf{k}} - i\eta} \right), \quad F(\mathbf{k}, t) = iS(k)\, e^{i\omega_{\mathbf{k}} t}. \tag{10.9.7}$$

We obtain a pole in $F(\mathbf{k}, \omega)$ at $\omega = \omega_{\mathbf{k}}$ and undamped oscillations of frequency $\omega_{\mathbf{k}}$. In general, of course, this extreme situation is never realized exactly, but if for certain k, (10.9.5) has poles $\xi_{\mathbf{k}} = \epsilon_{\mathbf{k}} - i\Gamma_{\mathbf{k}}$ with $0 < \Gamma_{\mathbf{k}} \ll \epsilon_{\mathbf{k}}$, then $F(\mathbf{k}, t)$ takes on a slightly oscillatory form with excitation energy $\epsilon_{\mathbf{k}}$ and lifetime $1/\Gamma_{\mathbf{k}}$. This situation corresponds to an $S(\mathbf{k}, \omega)$ peaked around $\omega = \epsilon_{\mathbf{k}}$ with half-width $\Gamma_{\mathbf{k}}$. We see that our task, mathematically, is to find the poles of $F(\mathbf{k}, \omega)$.

Before commencing the graphical analysis of $F(\mathbf{k}, t)$ using (10.8.5) and (10.8.6), it is convenient to introduce a few definitions at this stage. For reasons which will be explained, we define a quantity $\epsilon(\mathbf{k}, \omega)$ (cf. Chapter 5) called the propagating dielectric function by the equation

$$1 - \frac{1}{\epsilon(\mathbf{k}, \omega)} = v_{\mathbf{k}} F(\mathbf{k}, \omega). \tag{10.9.8}$$

Also, in the same way as we anticipated the Dyson form of G (see (10.3.2)) we will now state the eventual form in which we will calculate F. It is

$$F(\mathbf{k},\omega) = \frac{Q(\mathbf{k},\omega)}{1+v_{\mathbf{k}}Q(\mathbf{k},\omega)}. \tag{10.9.9}$$

Q is called the irreducible or proper polarization and is the analogue of the Dyson self-energy M of single-propagator theory. By combining (10.9.8) and (10.9.9), we find

$$\epsilon(\mathbf{k},\omega) = 1+v_{\mathbf{k}}Q(\mathbf{k},\omega) \tag{10.9.10}$$

and the poles of F are found from

$$\epsilon(\mathbf{k},\omega) = 0. \tag{10.9.11}$$

Much as in our previous case, we will see that the computational effort, in any given application, goes into calculating Q in some approximate fashion. The procedure thereafter is clear in view of section 10.3.

10.10. Linked cluster theorem for particle-hole pairs

In this section we follow closely the methods used for the single-particle Green function in sections 10.4 and 10.5. Our procedure will be to calculate (10.8.6), then its Fourier transform and hence $F(\mathbf{k},\omega)$ from (10.8.8).

Turning, then, to (10.8.6), we rewrite the $a_{\mathbf{k}t}$'s using (10.4.5), invoke the composition law (10.4.6), recall (10.4.13) which says that the $a_{\mathbf{k}}(t)$'s and the U's commute and arrive at the result

$$F_{\substack{\mathbf{q}_1\sigma_1\\ \mathbf{q}_2\sigma_2}}(\mathbf{k},t_2-t_1) = \frac{\begin{aligned}i\sum_{n=0}^{\infty}\frac{(-1)^n}{n!}\int_{-\infty}^{\infty}dt'_n\ldots\int_{-\infty}^{\infty}dt'_1\,\langle g|T\{V(t'_1)\ldots V(t'_n)\\ \times a^{\dagger}_{\mathbf{k}+\mathbf{q}_2\sigma_2}(t_2)\,a_{\mathbf{q}_2\sigma_2}(t_2)\,a^{\dagger}_{\mathbf{q}_1\sigma_1}(t_1)\,a_{\mathbf{k}+\mathbf{q}_1\sigma_1}(t_1)\}|g\rangle\end{aligned}}{\sum_{0}^{\infty}\frac{(-i)^n}{n!}\int_{-\infty}^{\infty}dt'_n\ldots\int_{-\infty}^{\infty}dt'_1\langle g|T\{V(t'_1)\ldots V(t'_n)\}|g\rangle}. \tag{10.10.1}$$

We can examine the numerator of (10.10.1) graphically by

methods similar to those used in Appendix 10 A. The zeroth-order term is i times

$$
\begin{aligned}
&\langle g|T\{a^{\dagger}_{\mathbf{k}+\mathbf{q}_2\,\sigma_2}(t_2)\,a_{\mathbf{q}_2\,\sigma_2}(t_2)\,a^{\dagger}_{\mathbf{q}_1\,\sigma_1}(t_1)\,a_{\mathbf{k}+\mathbf{q}_1\,\sigma_1}(t_1)\}|g\rangle \\
&= \{a^{\dagger}_{\mathbf{k}+\mathbf{q}_2\,\sigma_2}(t_2)\,a_{\mathbf{q}_2\,\sigma_2}(t_2)\,a^{\dagger}_{\mathbf{q}_1\,\sigma_1}(t_1)\,a_{\mathbf{k}+\mathbf{q}_1\,\sigma_1}(t_1) \\
&\qquad + a^{\dagger}_{\mathbf{k}+\mathbf{q}_2\,\sigma_2}(t_2)\,a_{\mathbf{q}_2\,\sigma_2}(t_2)\,a^{\dagger}_{\mathbf{q}_1\,\sigma_1}(t_1)\,a_{\mathbf{k}+\mathbf{q}_1\,\sigma_1}(t_1)\} \\
&= \delta_{\mathbf{q}_1\,\sigma_1,\mathbf{q}_2\,\sigma_2}\,G_0(\mathbf{k}+\mathbf{q}_1,t_1-t_2)\,G_0(\mathbf{q}_1,t_2-t_1), \qquad (10.10.2)
\end{aligned}
$$

the second contracted term vanishing since $\mathbf{k} \neq \mathbf{0}$. The graph for this term is shown in Fig. 10.12. In general, the first graph is sufficient, but if we wish to emphasize that momentum was transferred to the system in order to create the particle-hole pair and removed from it at the moment of annihilation, we can add arms as shown. These arms do not necessarily constitute vertices, though clearly one way of providing, and later absorbing, momentum is via another particle.

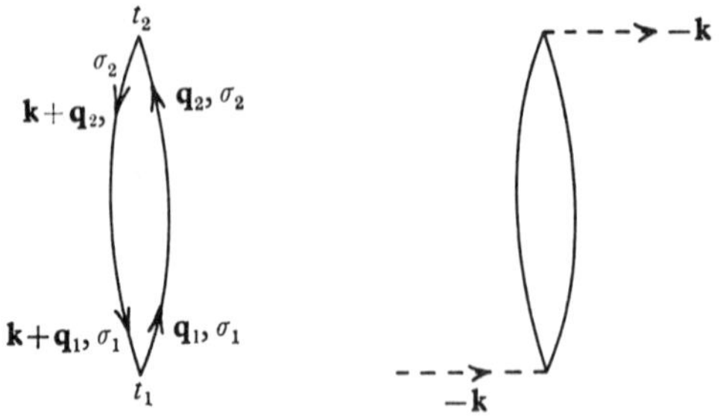

Fig. 10.12

Much as in Appendix 10 A, the t_1 and t_2 cusps can vary, taking either time order, as long as the topological structure is preserved. The zeroth-order term for a general two-particle Green function consists of two distinct lines. In the present special case, however, the ends are joined as we are discussing the creation of a particle $\mathbf{q}_1, \sigma_1$ and a hole $\mathbf{k}+\mathbf{q}_1, \sigma_1$ and their subsequent annihilation.

The graphical rules developed previously apply unchanged. Even the loop rule remains unaltered in the new situation. For, although we need a minus sign to allow for the fact, that to form from the $a^{\dagger}_{\mathbf{k}+\mathbf{q}_2\sigma_2}(t_2)\,a_{\mathbf{q}_2\sigma_2}(t_2)\,a^{\dagger}_{\mathbf{q}_1\sigma_1}(t_1)\,a_{\mathbf{k}+\mathbf{q}_1\sigma_1}(t_1)$, a standard 4-product of type $a^{\dagger}a^{\dagger}aa$ found always among the $V(t)$'s, the use of cusps rather than open ends always introduces an extra loop to rectify this. Thus, (10.10.2) can be directly written down from Fig. 10.12 in the form

$$(-1)^1\{-i\delta_{\mathbf{k}+\mathbf{q}_1\sigma_1,\,\mathbf{k}+\mathbf{q}_2\sigma_2}\,G_0(\mathbf{k}+\mathbf{q}_1,t_1-t_2)\}\\ \times\{-i\delta_{\mathbf{q}_1\sigma_1,\,\mathbf{q}_2\sigma_2}\,G_0(\mathbf{q}_1,t_2-t_1)\}.$$

10.10.1. *General particle-hole pairs*

One can continue to analyse the numerator of (10.10.1) in higher order. To write down all nth-order graphs, one draws (Fig. 10.13) a cusp at t_1 indicating the creation of a particle-hole pair, and another cusp at t_2 indicating such a pair being destroyed. Then the labelled interaction lines are drawn and particle and hole lines supplied in all possible ways. Each graph then unequivocally represents an nth-order contribution.

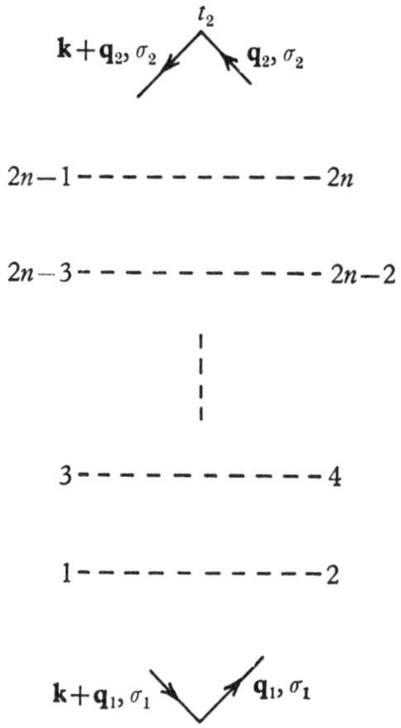

Fig. 10.13

We find there are two kinds of diagrams, linked (or connected) and unlinked (or disconnected), the definition of these terms being, once more, the obvious geometrical ones. As before, this results in a factorization of the numerator, one of the factors cancelling against the denominator. The remaining term involves linked graphs only and with these we can conveniently go over to unlabelled graphs (recall Appendix 10A and the relevant comments concerning unlabelled graphs on p. 354). In this way, we arrive at the formula

$$F_{\substack{\mathbf{q}_1\sigma_1\\ \mathbf{q}_2\sigma_2}}(\mathbf{k},t_2-t_1) = \sum_{n=0}^{\infty}(-i)^{n-1}\int_{-\infty}^{\infty}dt_{2n-12n}\cdots\int_{-\infty}^{\infty}dt_{12}\sum_{\substack{KLMN\\ PQRS\\ \cdots\\ XYZT}}$$
$$\times\langle KL|v|NM\rangle\langle PQ|v|SR\rangle\ldots\langle XY|v|TZ\rangle\langle g|\,T\{a_K^\dagger(t_{12})\,a_L^\dagger(t_{12})\ldots$$
$$\times\,a_T(t_{2n-12n})\,a^\dagger_{\mathbf{k}+\mathbf{q}_2\sigma_2}(t_2)\,a_{\mathbf{q}_2\sigma_2}(t_2)\,a^\dagger_{\mathbf{q}_1\sigma_1}(t_1)\,a_{\mathbf{k}+\mathbf{q}_1\sigma_1}(t_1)\}|g\rangle_L, \quad (10.10.3)$$

where the final suffix L means only distinct unlettered linked graphs are to be taken. Equation (10.10.3) is our present linked cluster theorem and is the counterpart of (10.5.1) in the single-particle Green function development.

10.11. Graphical analysis of propagator for particle-hole pairs

We now embark on an analysis of the various linked components of the term

$$\langle g|T\{a_K^\dagger(t_{12})\,a_L^\dagger(t_{12})\ldots a_T^\dagger(t_{2n-12n})\,a^\dagger_{\mathbf{k}+\mathbf{q}_2\sigma_2}(t_2)\,a_{\mathbf{q}_2\sigma_2}(t_2) \times a^\dagger_{\mathbf{q}_1\sigma_1}(t_1)\,a_{\mathbf{k}+\mathbf{q}_1\sigma_1}(t_1)\}|g\rangle. \quad (10.11.1)$$

(*a*) *Zeroth order.* We have already discussed this case. The appropriate graph is as in Fig. 10.12 and the corresponding mathematical expression is as in (10.10.2).

(*b*) *First order.* The six distinct diagrams which contribute in this order are shown in Fig. 10.14. Our rules can be used to write down the respective mathematical expressions. We have, for example, that graph (i) contributes

$$(-1)^2(-i)^4\,\delta_{K\mathbf{k}+\mathbf{q}_1\sigma_1}\delta_{L\mathbf{q}_2\sigma_2}\delta_{M\mathbf{k}+\mathbf{q}_2\sigma_2}\delta_{N\mathbf{q}_1\sigma_1}G_0(\mathbf{k}+\mathbf{q}_1, t_1-t_{12}) \times G_0(\mathbf{q}_2, t_2-t_{12})\,G_0(\mathbf{k}+\mathbf{q}_2, t_{12}-t_2)\,G_0(\mathbf{q}_1, t_{12}-t_1), \quad (10.11.2)$$

while graph (ii) contributes

$$(-1)^2(-i)^4\,\delta_{K\mathbf{k}+\mathbf{q}_1\sigma_1}\delta_{L\mathbf{q}_2\sigma_2}\delta_{M\mathbf{q}_1\sigma_1}\delta_{N\mathbf{k}+\mathbf{q}_2\sigma_2}G_0(\mathbf{k}+\mathbf{q}_1, t_1-t_{12}) \; G_0(\mathbf{q}_2, t_2-t_{12})\,G_0(\mathbf{q}_1, t_{12}-t_1)\,G_0(\mathbf{k}+\mathbf{q}_2, t_{12}-t_2). \quad (10.11.3)$$

The contributions from the other graphs can also be written down without difficulty.

(*c*) *Second order.* In this order there are already very many graphs, a selection of which are shown in Fig. 10.15. The set (i) consists of members of lower-order diagrams (Figs. 10.12, 10.14) joined at their cusps by interaction lines. These fall into the class of graphs later defined to be reducible. The first two members have been labelled because, below, we wish to write out explicitly their contributions. The remaining graphs (all irreducible) can be subdivided further for descriptive purposes. The set (ii) consists of zeroth- and first-order graphs with an appropriate number of self-energy insertions. Finally, there are also a number of graphs giving non-zero contributions and having entirely novel structures not describable in terms of lower-order diagrams. Some of these are

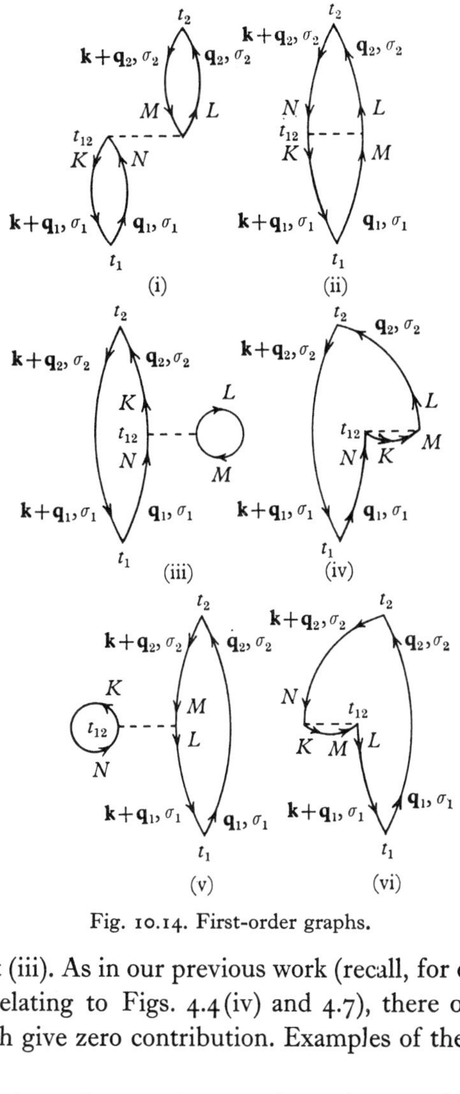

Fig. 10.14. First-order graphs.

shown in set (iii). As in our previous work (recall, for example, the discussion relating to Figs. 4.4 (iv) and 4.7), there occur certain graphs which give zero contribution. Examples of these are given by set (iv).

We can again apply our rules to evaluate the contributions of the various graphs. Thus, for example, the first graph of Fig. 10.15 contributes

$$\begin{aligned}(-1)^3(-i)^6\,&\delta_{K\mathbf{k}+\mathbf{q}_1\sigma_1}\,\delta_{LS}\,\delta_{MP}\,\delta_{N\mathbf{q}_1\sigma_1}\,\delta_{Q\mathbf{q}_2\sigma_2}\,\delta_{R\mathbf{k}+\mathbf{q}_2\sigma_2}\\ &\times G_0(\mathbf{k}+\mathbf{q}_1, t_1-t_{12})\,G_0(L, t_{34}-t_{12})\,G_0(M, t_{12}-t_{34})\\ &\times G_0(\mathbf{q}_1, t_{12}-t_1)\,G_0(\mathbf{q}_2, t_2-t_{34})\,G_0(\mathbf{k}+\mathbf{q}_2, t_{34}-t_2),\quad (10.11.4)\end{aligned}$$

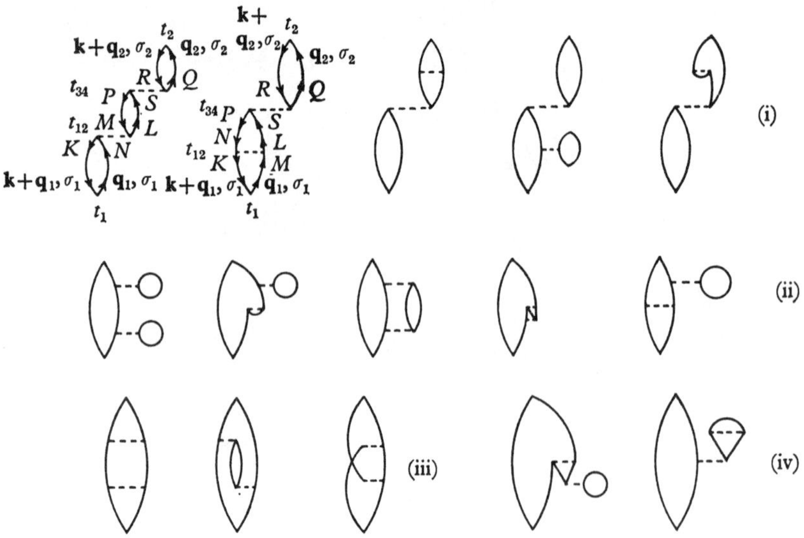

Fig. 10.15. Selection of second-order graphs.

while the second gives

$$(-1)^2(-i)^6\,\delta_{K\mathbf{k}+\mathbf{q}_1\sigma_1}\,\delta_{LS}\,\delta_{M\mathbf{q}_1\sigma_1}\,\delta_{NP}\,\delta_{Q\mathbf{q}_2\sigma_2}\,\delta_{R\mathbf{k}+\mathbf{q}_2\sigma_2}$$
$$\times G_0(\mathbf{k}+\mathbf{q}_1, t_1-t_{12})\,G_0(L, t_{34}-t_{12})\,G_0(\mathbf{q}_1, t_{12}-t_1)$$
$$\times G_0(N, t_{12}-t_{34})\,G_0(\mathbf{q}_2, t_2-t_{34})\,G_0(\mathbf{k}+\mathbf{q}_2, t_{34}-t_2). \quad (10.11.5)$$

Thus, we go up the series. Just as we wrote (10.5.1) in the form (10.5.2), we now represent (10.10.3) symbolically as

$$F_{\substack{\mathbf{q}_1\sigma_1\\ \mathbf{q}_2\sigma_2}}(\mathbf{k}, t_2-t_1)$$

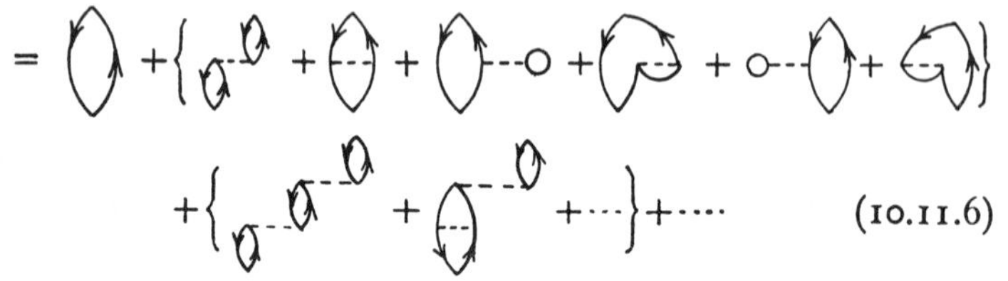

(10.11.6)

10.12. Poles of propagator and collective excitations

Our basic interest, it will be remembered, is in the poles of $F(\mathbf{k}, \omega)$. Thus we proceed to the calculation of the separate terms of the series (10.8.8) which give this function, by transforming (10.10.3). As usual, we begin with the simplest cases.

(*a*) *Zeroth order.* The contribution of this term to (10.10.3) is determined by (10.10.2). Thus, we have a zeroth-order value given by

$$F^{(0)}_{\substack{\mathbf{q}_1\sigma_1\\ \mathbf{q}_2\sigma_2}}(\mathbf{k}, t_2-t_1) = i\delta_{\mathbf{q}_1\sigma_1,\,\mathbf{q}_2\sigma_2} G_0(\mathbf{k}+\mathbf{q}_1, t_1-t_2)\, G_0(\mathbf{q}_1, t_2-t_1). \tag{10.12.1}$$

If these G_0's are replaced using (10.2.10), and F is transformed into ω-space, we find

$$\begin{aligned} F^{(0)}_{\substack{\mathbf{q}_1\sigma_1\\ \mathbf{q}_2\sigma_2}}(\mathbf{k}, \omega) &= i\delta_{\mathbf{q}_1\sigma_1,\,\mathbf{q}_2\sigma_2}\int_{-\infty}^{\infty}\frac{d\omega_1}{2\pi}\int_{-\infty}^{\infty}\frac{d\omega_2}{2\pi} G_0(\mathbf{k}+\mathbf{q}_1, \omega_1)\, G_0(\mathbf{q}_1, \omega_2) \\ &\qquad\times\int_{-\infty}^{\infty} dt\, e^{-i(\omega_2-\omega_1-\omega)t} \\ &= i\delta_{\mathbf{q}_1\sigma_1,\,\mathbf{q}_2\sigma_2}\int_{-\infty}^{\infty}\frac{d\omega_1}{2\pi}\int_{-\infty}^{\infty}\frac{d\omega_2}{2\pi} G_0(\mathbf{k}+\mathbf{q}_1, \omega_1)\, G_0(\mathbf{q}_1, \omega_2) \\ &\qquad\times 2\pi\,\delta(\omega_2-\omega_1-\omega). \end{aligned} \tag{10.12.2}$$

On performing the ω_2-integration, we obtain

$$F^{(0)}_{\substack{\mathbf{q}_1\sigma_1\\ \mathbf{q}_2\sigma_2}}(\mathbf{k}, \omega) = i\delta_{\mathbf{q}_1\sigma_1,\,\mathbf{q}_2\sigma_2}\int_{-\infty}^{\infty}\frac{d\omega_1}{2\pi} G_0(\mathbf{k}+\mathbf{q}_1, \omega_1)\, G_0(\mathbf{q}_1, \omega+\omega_1). \tag{10.12.3}$$

This integral has already been evaluated. It is given by (10.6.12). By (10.8.8), all one has to do, now, to calculate the zeroth-order contribution to $F(\mathbf{k}, \omega)$, is to sum (10.12.3) over $\mathbf{q}_1, \sigma_1, \mathbf{q}_2, \sigma_2$. In this way we find this contribution to be given by (10.6.15). (The occurrence of this same function in both situations was no accident, of course. They both arose from similar 'bubbles'. (Cf. the remarks following (10.6.15).)

It is obvious that, as in the single-particle propagator theory, we are going to find that each particle line will carry a characteristic energy ω_i. The alternative way of drawing diagrams illustrated in Fig. 10.12 will be useful for indicating energy as well as momentum conservation. Thus Fig. 10.16 represents the conservation conditions expressed in (10.12.2).

(*b*) *First order.* As we have seen (Fig. 10.14) there are six diagrams to consider. We will perform the calculation only for graph (i). The others can be dealt with along somewhat similar lines.

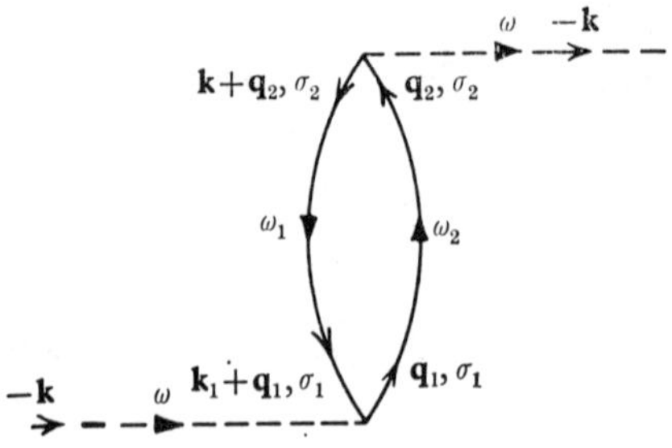

Fig. 10.16

Graph (i) contributes to the Wick expansion of the first-order form of (10.11.1), a term given by (10.11.2). By inserting this appropriately in (10.10.3) and transforming in a fashion similar to that whereby (10.12.2) was obtained from (10.12.1), we obtain a contribution to $F(\mathbf{k}, \omega)$ of

$$F^{(1a)}_{\substack{\mathbf{q}_1\sigma_1\\ \mathbf{q}_2\sigma_2}}(\mathbf{k}, \omega) = \int_{-\infty}^{\infty} dt_{12} \int_{-\infty}^{\infty} d(t_2-t_1) \sum_{KLMN} \langle KL|v|NM\rangle(-1)^2(-i)^4$$

$$\times \delta_{K\mathbf{k}+\mathbf{q}_1\sigma_1}\delta_{L\mathbf{q}_2\sigma_2}\delta_{M\mathbf{k}+\mathbf{q}_2\sigma_2}\delta_{N\mathbf{q}_1\sigma_1}\int_{-\infty}^{\infty}\frac{d\omega_1}{2\pi}\int_{-\infty}^{\infty}\frac{d\omega_2}{2\pi}\int_{-\infty}^{\infty}\frac{d\omega_3}{2\pi}\int_{-\infty}^{\infty}\frac{d\omega_4}{2\pi}$$

$$\times G_0(\mathbf{k}+\mathbf{q}_1, \omega_1)\, G_0(\mathbf{q}_2, \omega_2)\, G_0(\mathbf{k}+\mathbf{q}_2, \omega_3)\, G_0(\mathbf{q}_1, \omega_4)\, e^{-i\omega_1(t_1-t_{12})}$$

$$\times e^{-i\omega_2(t_2-t_{12})}\, e^{-i\omega_3(t_{12}-t_2)}\, e^{-i\omega_4(t_{12}-t_1)}\, e^{i\omega(t_2-t_1)}. \tag{10.12.4}$$

We met a situation similar to this previously (see (10.6.2)). Integration over t_{12} gives an energy-conserving factor,

$$2\pi\delta(\omega_1+\omega_2-\omega_3-\omega_4),$$

while integration over t_2-t_1 gives (equivalently)

$$2\pi\delta(\omega_2-\omega_3+\omega) \quad \text{or} \quad 2\pi\delta(\omega_1-\omega_4+\omega)$$

depending on the way the calculation is performed. The appropriate diagram (Fig. 10.17) shows how a certain momentum $-\mathbf{k}$ and energy ω is required to create a particle-hole pair. The latter propagates and subsequently annihilates to create a second particle-hole pair, energy and momentum being conserved during the process. Finally, the second pair propagates and later annihilates,

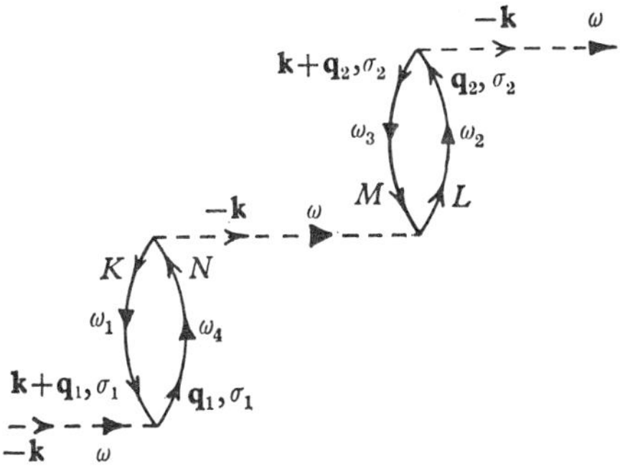

Fig. 10.17

yielding up once more the momentum $-\mathbf{k}$ and energy ω originally injected into the system.

The formula (10.12.4) can now be factorized as follows:

$$F^{(1a)}_{\substack{\mathbf{q}_1\sigma_1\\ \mathbf{q}_2\sigma_2}}(\mathbf{k},\omega) = \sum_{KLMN} \langle KL|v|NM\rangle \int_{-\infty}^{\infty}\frac{d\omega_1}{2\pi}\int_{-\infty}^{\infty}\frac{d\omega_4}{2\pi}\{\delta_{K\mathbf{k}+\mathbf{q}_1\sigma_1}\,\delta_{N\mathbf{q}_1\sigma_1}$$

$$\times G_0(\mathbf{k}+\mathbf{q}_1,\omega_1)\,G_0(\mathbf{q}_1,\omega_4)\,2\pi\delta(\omega_1-\omega_4+\omega)\}\int_{-\infty}^{\infty}\frac{d\omega_2}{2\pi}\int_{-\infty}^{\infty}\frac{d\omega_3}{2\pi}$$

$$\times\{\delta_{L\mathbf{q}_2\sigma_2}\,\delta_{M\mathbf{k}+\mathbf{q}_2\sigma_2}\,G_0(\mathbf{q}_2,\omega_2)\,G_0(\mathbf{k}+\mathbf{q}_2,\omega_3)\,2\pi\delta(\omega_2-\omega_3-\omega)\}. \tag{10.12.5}$$

Now, use of (10.12.2) in (10.12.15) yields

$$F^{(1a)}_{\substack{\mathbf{q}_1\sigma_1\\ \mathbf{q}_2\sigma_2}}(\mathbf{k},\omega) = -v_{\mathbf{k}}\,F^{(0)}_{\substack{\mathbf{q}_1\sigma_1\\ \mathbf{q}_1\sigma_1}}(\mathbf{k},\omega)\,F^{(0)}_{\substack{\mathbf{q}_2\sigma_2\\ \mathbf{q}_2\sigma_2}}(\mathbf{k},\omega). \tag{10.12.6}$$

By summing this expression over the $\mathbf{q}$'s and σ's, we see that the total contribution of the graph to $F(\mathbf{k},\omega)$ is

$$F^{(1a)}(\mathbf{k},\omega) = -v_{\mathbf{k}}\{F^{(0)}(\mathbf{k},\omega)\}^2. \tag{10.12.7}$$

(*c*) *nth order*. We see that the whole procedure is very like that used to calculate the contributions to the single-particle propagator. In section 10.10, we learned how to draw the appropriate momentum time graphs. These are readily converted into momentum-energy graphs by giving each particle line an energy index ω_i (see Fig. 10.18). The graphs then readily suggest their contributions to $F(\mathbf{k},\omega)$. Thus,

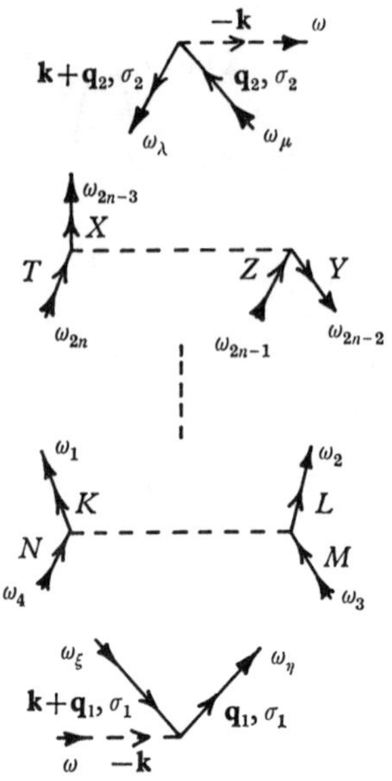

Fig. 10.18

for example, the nth-order graph characterized by structure α, say, in Fig. 10.18 contributes

$$F^{(n\alpha)}_{\substack{\mathbf{q}_1\sigma_1\\ \mathbf{q}_2\sigma_2}}(\mathbf{k},\omega) = i^{n-1}(-1)^l \sum_{\substack{KLMN\\ \text{- - - - -}\\ XYZT}} \langle KL|v|NM\rangle \ldots \langle XY|v|TZ\rangle\, \delta_{K*} \ldots \delta_{T*}$$

$$\int_{-\infty}^{\infty} \cdots \int_{-\infty}^{\infty} \frac{d\omega_1 \ldots d\omega_{2n}}{(2\pi)^{2n}} G_0(K,\omega_1)\, G_0(L,\omega_2) \ldots G_0(T,\omega_{2n})$$

$$\times\, 2\pi\delta(\omega_1+\omega_2-\omega_3-\omega_4)\, 2\pi \ldots 2\pi\delta(\omega_{2n-3}+\omega_{2n-2}-\omega_{2n-1}-\omega_{2n})$$

$$\times\, 2\pi\delta(\omega+\omega_\lambda-\omega_\mu) \tag{10.12.8}$$

and, once more, because of the linear dependence among the arguments, any one of the Dirac delta functions can be replaced by $\delta(\omega_\xi-\omega_\eta+\omega)$.

This expression may be compared with the corresponding formula (10.6.18) of one-body propagator theory. As in that case (cf. Fig. 10.6 and (10.6.13)), we soon learn to label diagrams with momenta, spins and energies with a minimum of symbolism, in such a way that all conservation requirements are satisfied. One is then able to write down the appropriate contribution.

10.12.1. *Irreducible or proper polarization*

The structure of the $F(\mathbf{k}, \omega)$ series, like that of $G(\mathbf{k}, \omega)$ turns out to have simplifying features and the theory is reminiscent of the previous case. The difference is that, whereas the basic reduction of diagrams contributing to G was into strings of subdiagrams connected together by single propagating lines, here the basic reduction is into strings of subdiagrams joined by single vertex lines.

As before, we begin with some definitions. Any graph of the kind shown in Fig. 10.18 ($n \geqslant 0$) is called a polarization diagram. It is represented symbolically as in Fig. 10.19 (i). It may or may

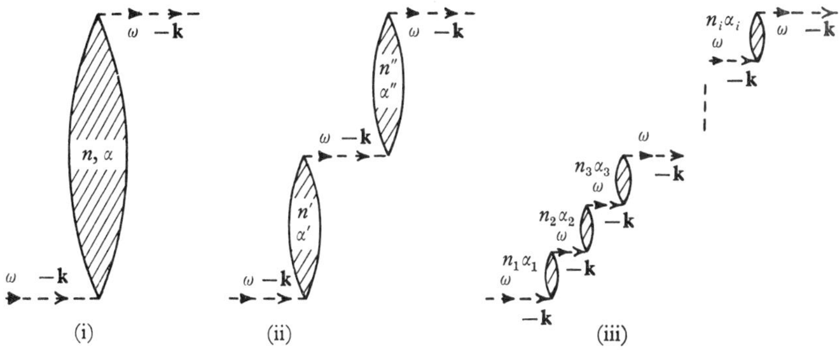

Fig. 10.19

not be possible to draw such a polarization graph in the form (ii) shown in Fig. 10.19. (The intermediate vertex shown necessarily carries the same momentum and energy indices as the lines at the end, because of the conservation requirements.) If such a redrawing is possible, the graph is called reducible, and if not, it is called irreducible. Thus, for example, in Fig. 10.14: (i) is reducible while the others are irreducible; in Fig. 10.15, the graphs of set (i) are reducible but the others are not (see comments thereon in the text). In the reducible case, we can refer to the component subdiagrams as polarization parts. Clearly, we can keep on reducing a graph in this way until we arrive at the situation (iii) of Fig. 10.19, when no further reduction is possible. We then have a series of irreducible polarization parts, each two joined by a single vertex and each such vertex carrying the same momentum and energy transfers.

We now wish to establish the present analogue of the result (10.7.4), namely, that if the graph shown in Fig. 10.19(ii) is re-drawn somewhat more explicitly as in Fig. 10.20, we have

$$F^{(n\alpha)}_{\substack{\mathbf{q}_1\sigma_1\\ \mathbf{q}_2\sigma_2}}(\mathbf{k},\omega) = -v_k \sum_{\mathbf{q}_2'\sigma_2'} F^{(n'\alpha')}_{\substack{\mathbf{q}_1\sigma_1\\ \mathbf{q}_2'\sigma_2'}}(\mathbf{k},\omega)$$

$$\times \sum_{\mathbf{q}_1'\sigma_1'} F^{(n''\alpha'')}_{\substack{\mathbf{q}_1'\sigma_1'\\ \mathbf{q}_2\sigma_2}}(\mathbf{k},\omega). \quad (10.12.9)$$

By summing over the $\mathbf{q}$ and σ variables (recall (10.8.8)) an immediate extension of this result is

$$F^{(n\alpha)}(\mathbf{k},\omega) = -v_{\mathbf{k}} F^{(n'\alpha')}(\mathbf{k},\omega)$$

$$\times F^{(n''\alpha'')}(\mathbf{k},\omega). \quad (10.12.10)$$

These results are generalizations of (10.12.6) and (10.12.7) respectively and are the analogues of (10.7.4) of one-particle Green function theory.

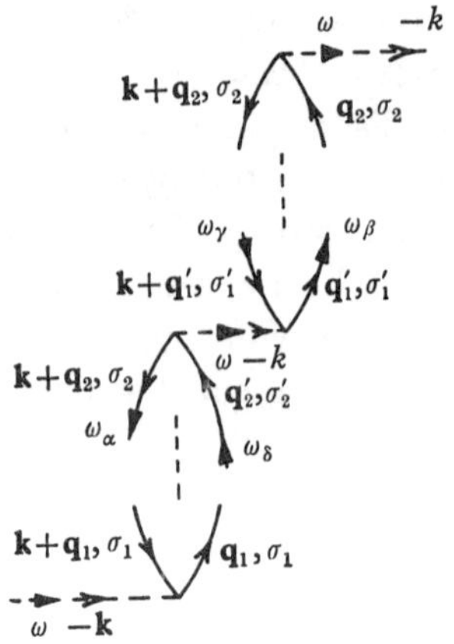

Fig. 10.20

The simplicity of the proof of (10.12.9) should not be obscured by the amount of symbolism we will use to write it down. Broadly speaking, the proof rests on the fact that the interaction line and the two polarization parts contribute separately the factors shown in (10.12.9). More explicitly (10.12.8) can be written

$$F^{(n\alpha)}_{\substack{\mathbf{q}_1\sigma_1\\ \mathbf{q}_2\sigma_2}} = i^{n-1}(-1)^l \sum_{\substack{KLMN\\ \cdots\\ \mathbf{q}_1'\sigma_1'\mathbf{q}_2'\sigma_2'\\ \cdots\\ XYZT}} \begin{array}{l} \langle KL|v|NM\rangle\ldots \\ \langle \mathbf{k}+\mathbf{q}_2'\,\sigma_2'\,\mathbf{q}_1'\,\sigma_1'\,|v|\,\mathbf{q}_2'\,\sigma_2'\,\mathbf{k}+\mathbf{q}_1'\,\sigma_1'\rangle\ldots \\ \langle XY|v|TZ\rangle \end{array}$$

$$\times\,\delta_{K*}\ldots\delta_{\mathbf{k}+\mathbf{q}_2'\sigma_2'*}\,\delta_{\mathbf{q}_2'\sigma_2'*}\,\delta_{\mathbf{q}_1'\sigma_1'*}\,\delta_{\mathbf{k}+\mathbf{q}_1'\sigma_1'*}\ldots$$

$$\times\int_{-\infty}^{\infty}\frac{d\omega_1}{2\pi}\ldots\int_{-\infty}^{\infty}\frac{d\omega_\alpha}{2\pi}\int_{-\infty}^{\infty}\frac{d\omega_\delta}{2\pi}\int_{-\infty}^{\infty}\frac{d\omega_\beta}{2\pi}\int_{-\infty}^{\infty}\frac{d\omega_\gamma}{2\pi}\ldots\int_{-\infty}^{\infty}\frac{d\omega_{2n}}{2\pi}$$

$$\times\,G_0(K,\omega_1)\ldots G_0(\mathbf{k}+\mathbf{q}_2',\omega_\alpha)\,G_0(\mathbf{q}_2',\omega_\delta)\,G_0(\mathbf{q}_1',\omega_\beta)$$

$$\times\,G_0(\mathbf{k}+\mathbf{q}_1',\omega_\gamma)\ldots G_0(T,\omega_{2n})\,2\pi\delta(\omega_\xi-\omega_\eta-\omega)$$

$$\times\,2\pi\delta(\omega_1+\omega_2-\omega_3-\omega_4)\ldots\cancel{2\pi\delta(\omega_\alpha+\omega_\beta-\omega_\gamma-\omega_\delta)}\,2\pi\ldots$$

$$2\pi\delta(\omega+\omega_\lambda-\omega_\mu). \quad (10.12.11)$$

In writing out this expression, we have exercised the option open to us (see comments after (10.12.8) and (10.12.4), etc.) of removing

$2\pi\,\delta(\omega_\alpha-\omega_\beta-\omega_\gamma-\omega_\delta)$ and inserting, at the head of the Dirac delta-function product, the factor $2\pi\,\delta(\omega_\xi-\omega_\eta+\omega)$. The matrix element for our special vertex reduces to $v_{\mathbf{k}}$ and can be removed from the summation. Then one can rewrite (10.12.11) as

$$
\begin{aligned}
F^{(n\alpha)}_{\substack{\mathbf{q}_1\sigma_1\\ \mathbf{q}_2\sigma_2}}(\mathbf{k},\omega) = i^{n-1}v_{\mathbf{k}}(-)^{l'}\Bigg\{ & \sum_{\mathbf{q}_2'\sigma_2'}\ \sum_{\substack{KLMN\\ \dots\\ \cancel{\mathbf{q}_1'\sigma_1'\mathbf{q}_2'\sigma_2'}}} \langle KL|v|NM\rangle \\
& \times \cancel{\langle \mathbf{k}+\mathbf{q}_2'\sigma_2'\mathbf{q}_1'\sigma_1'|v|\mathbf{q}_2'\sigma_2'\mathbf{k}+\mathbf{q}_1'\sigma_1'\rangle}\,\delta_{K*}\dots\delta_{\mathbf{k}+\mathbf{q}_2'\sigma_2'*}\,\delta_{\mathbf{q}_2'\sigma_2'*} \\
& \times \int_{-\infty}^{\infty}\frac{d\omega_1}{2\pi}\dots\int_{-\infty}^{\infty}\frac{d\omega_\alpha}{2\pi}\int_{-\infty}^{\infty}\frac{d\omega_\delta}{2\pi}\,G_0(K,\omega_1)\dots \\
& \times G_0(\mathbf{k}+\mathbf{q}_2',\omega_\alpha)\,G_0(\mathbf{q}_2',\omega_\delta)\,2\pi\delta(\omega_\xi-\omega_\eta+\omega) \\
& \times 2\pi\delta(\omega_1+\omega_2-\omega_3-\omega_4)\dots\cancel{2\pi\delta(\omega_\alpha+\omega_\beta-\omega_\gamma-\omega_\delta)}\Bigg\} \\
& \times(-1)^{l''}\Bigg\{\sum_{\mathbf{q}_1'\sigma_1'}\ \sum_{\substack{\cancel{\mathbf{q}_1'\sigma_1'\mathbf{q}_2'\sigma_2'}\\ \dots\\ XYZT}} \cancel{\langle \mathbf{k}+\mathbf{q}_2'\sigma_2'\mathbf{q}_1'\sigma_1'|v|\mathbf{q}_2'\sigma_2'\mathbf{k}+\mathbf{q}_1'\sigma_1'\rangle}\dots \\
& \qquad \dots\langle XY|v|TZ\rangle\,\delta_{\mathbf{q}_1'\sigma_1'*}\,\delta_{\mathbf{k}+\mathbf{q}_1'\sigma_1'*}\dots\delta_{T*} \\
& \times\int_{-\infty}^{\infty}\frac{d\omega_\beta}{2\pi}\int_{-\infty}^{\infty}\frac{d\omega_\gamma}{2\pi}\dots\int_{-\infty}^{\infty}\frac{d\omega_{2n}}{2\pi}\,G_0(\mathbf{q}_1',\omega_\beta) \\
& \times G_0(\mathbf{k}+\mathbf{q}_1',\omega_\gamma)\dots G_0(T,\omega_{2n})\,\cancel{2\pi\delta(\omega_\alpha+\omega_\beta-\omega_\gamma-\omega_\delta)} \\
& \times 2\pi\dots2\pi\,\delta(\omega+\omega_\lambda-\omega_\mu)\Bigg\}.
\end{aligned}
\tag{10.12.12}
$$

There are two comments we would like to make about this expression. First of all, the device of writing down expressions and then crossing them out is simply to indicate where the various sequences and products terminate. Secondly, l' and l'' denote the number of loops associated with the lower and upper polarization parts respectively. Because of the special nature of the diagram under consideration, $l=l'+l''$, and this fact was used in going from (10.12.11) to (10.12.12). Applying (10.12.11), now, to the constituent parts of (10.12.12) we have

$$
F^{(n\alpha)}_{\substack{\mathbf{q}_1\sigma_1\\ \mathbf{q}_2\sigma_2}}(\mathbf{k},\omega) = i^{n-1}v_{\mathbf{k}}\left\{\frac{1}{i^{n'-1}}\sum_{\mathbf{q}_2'\sigma_2'}F^{(n'\alpha')}_{\substack{\mathbf{q}_1\sigma_1\\ \mathbf{q}_2'\sigma_2'}}(\mathbf{k},\omega)\right\}\left\{\frac{1}{i^{n''-1}}\sum_{\mathbf{q}_1'\sigma_1'}F^{(n''\alpha'')}_{\substack{\mathbf{q}_1'\sigma_1'\\ \mathbf{q}_2\sigma_2}}(\mathbf{k},\omega)\right\}.
\tag{10.12.13}
$$

If we now use the result $n'+n''+1=n$, obtained by counting the vertices, (10.12.13) reduces to (10.12.9) as required.

Now, the reason for reducing any given graph into its irreducible parts becomes more apparent in view of our analysis in section 10.7. By carrying the reduction process from (ii) to (iii) in Fig. 10.19 and repeatedly using (10.12.10), we obtain

$$F^{(n\alpha)}(\mathbf{k},\omega) = (-v_{\mathbf{k}})^{i-1}F^{(n_1\alpha_1)}(\mathbf{k},\omega)\,F^{(n_2\alpha_2)}(\mathbf{k},\omega)\ldots F^{(n_i\alpha_i)}(\mathbf{k},\omega), \tag{10.12.14}$$

which may be compared with (10.7.5).

By analogy with the irreducible self-energy expression (10.7.7), we define a quantity $Q(\mathbf{k},\omega)$, called the irreducible (or proper) polarization, by the equation

$$Q = \sum_{n,\,\alpha} F^{(n\alpha)}. \tag{10.12.15}$$

This will turn out to be the function appearing in (10.9.9). The leading member of the series is $F^{(0)}$, as given by (10.6.15) and described by Fig. 10.16; the first-order terms are calculated from the graphs of Fig. 10.14 except the first, etc. Now, by a process similar to that used in obtaining the Dyson result (10.7.8), namely, by writing down all (reducible and irreducible) contributions to F in a series arranged in ascending vertex order, and then rearranging into a power series in the explicit v_k function, we find that

$$F = Q - v_{\mathbf{k}}Q^2 + v_{\mathbf{k}}^2Q^3 - \ldots = \frac{Q}{1+v_{\mathbf{k}}Q}. \tag{10.12.16}$$

Thus, we arrive at (10.9.9). This equation, together with (10.12.15), (10.9.10) and (10.9.11), is fundamentally important since it provides a method of systematically calculating the dielectric function and the collective modes.

10.12.2. *Frequency and wave number-dependent dielectric function*

In practice, one approximates to Q by taking only the most important terms of (10.12.15). For sufficiently weak potentials, for example, the only $v_{\mathbf{k}}$-independent term, $F^{(0)}$, will dominate. Even for stronger coupling, one may have to be content with this approximation because of the mathematical difficulties inherent in trying to do better. Thus, let us consider the example,

$$Q \approx F^{(0)} = 2 \sum_{|\mathbf{k}+\mathbf{q}|>k_f>|\mathbf{q}|} \left\{ \frac{1}{\omega-(\epsilon_{\mathbf{k}+\mathbf{q}}-\epsilon_{\mathbf{q}}-io)} - \frac{1}{\omega+(\epsilon_{\mathbf{k}+\mathbf{q}}-\epsilon_{\mathbf{q}}-io)} \right\}. \tag{10.12.17}$$

Then, by (10.12.16), we have

$$F \approx F^{(0)} - v_{\mathbf{k}} F^{(0)2} + v_{\mathbf{k}}^2 F^{(0)3} - \dots, \qquad (10.12.18)$$

which amounts to having summed, in (10.11.6), the infinite subseries

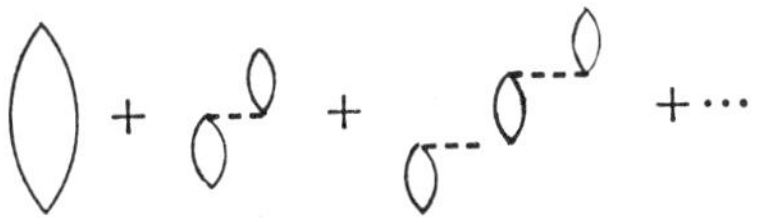

(10.12.19)

By (10.9.10), the dielectric constant is given by

$$\epsilon(\mathbf{k}, \omega) = 1 + v_{\mathbf{k}} F^{(0)}(\mathbf{k}, \omega) = \epsilon_{\text{R.P.A.}}(\mathbf{k}, \omega), \qquad (10.12.20)$$

the last step following by equation (5.4.19) of Chapter 5. We are thus back at the R.P.A. situation of that chapter, where the solutions of $\epsilon_{\text{R.P.A.}}(\mathbf{k}, \omega) = 0$ were examined in detail. We simply reiterate here, that these solutions provide the simplest mathematical descriptions of plasmons ($v_{\mathbf{k}} = 4\pi/\Omega k^2$) and zero sound ($v_{\mathbf{k}}$ finite as $k \to 0$).

For examples of approximations when more than $F^{(0)}$ has been included in the series (10.12.15), the reader is referred to Dubois (1959) (plasma oscillations) and Gottfried & Picman (1960) (zero sound).

10.13. Temperature-dependent Green functions

In the earlier sections of this chapter, we saw how we could reproduce the results of the earlier chapters of the book on zero-temperature problems in terms of Green functions, while, at the same time, making available a more powerful framework for tackling many-body problems.

In this section we wish to generalize the concept of Green functions to temperature-dependent problems and to establish the connexion between the approach of Chapter 9 and that of this chapter.

10.13.1. *Definitions*

We first recall the notation $\langle\langle \theta \rangle\rangle$ for denoting the expectation value of an operator θ, for a system with Hamiltonian H in thermodynamic equilibrium. As in Chapter 9, we consider the formulation in terms of the grand canonical ensemble. The defining equation is

$$\langle\langle\theta\rangle\rangle = \frac{\mathrm{Tr}\,[e^{-\beta(H-\mu N)}\theta]}{\mathrm{Tr}\,[e^{-\beta(H-\mu N)}]}, \tag{10.13.1}$$

where $\beta = 1/k_B T$, μ is the chemical potential and N is the number operator, defined by

$$N = \int d\mathbf{r}\,\psi^\dagger(\mathbf{r}t)\,\psi(\mathbf{r}t), \tag{10.13.2}$$

in terms of the Heisenberg field operators. Now we are in a position to define one- and two-particle Green functions by the equations

$$G(x_1, x_2) = i\langle\langle T(\psi(x_1)\,\psi^\dagger(x_2))\rangle\rangle \tag{10.13.3}$$

and

$$G(x_1, x_2;\, x_3, x_4) = i\langle\langle T(\psi(x_1)\,\psi(x_2)\,\psi^\dagger(x_3)\,\psi^\dagger(x_4)\rangle\rangle, \tag{10.13.4}$$

where we have written x for $\mathbf{r}$, t and T is the usual time-ordering operator. These definitions are obvious generalizations of those of section 10.2. We can define many-particle Green functions also in an analogous manner. As in section 10.2 we may split $G(x_1, x_2)$ into the retarded and advanced parts

$$G(x_1, x_2) = G^{(r)}(x_1, x_2) + G^{(a)}(x_1, x_2), \tag{10.13.5}$$

where

$$\left.\begin{aligned} G^{(r)}(x_1, x_2) &= i\langle\langle\psi(x_1)\,\psi^\dagger(x_2)\rangle\rangle \quad &&\text{for} \quad t_1 > t_2, \\ &= 0 && \qquad\;\; t_2 > t_1, \end{aligned}\right\} \tag{10.13.6}$$

and

$$\left.\begin{aligned} G^{(a)}(x_1, x_2) &= -i\langle\langle\psi^\dagger(x_2)\,\psi(x_1)\rangle\rangle \quad &&\text{for} \quad t_2 > t_1, \\ &= 0 && \text{for} \quad t_1 > t_2. \end{aligned}\right\} \tag{10.13.7}$$

The physical interpretation of $G(x_1, x_2)$ is the same as before. When $t_1 > t_2$, $G(x_1, x_2)$ describes the creation of a particle at $\mathbf{r}_2$ at time t_2 and its annihilation at $\mathbf{r}_1$ at later time t_1 when the system returns to its state of thermodynamic equilibrium. When $t_2 > t_1$, $G(x_1, x_2)$ describes the disturbance of the system from its state of thermodynamic equilibrium by the removal of a particle from point $\mathbf{r}_1$ and then its restoration to its former state by the later addition of the particle at point $\mathbf{r}_2$. Physical interpretation of $G(x_1, x_2;\, x_3, x_4)$ is similar.

For a Hamiltonian which is independent of time and also invariant under rotations and translations of the space axes, $G(x_1, x_2)$ is a function only of $|\mathbf{r}_1 - \mathbf{r}_2|$ and $t_1 - t_2$.

As for the zero-temperature case, it will often be convenient to use $\mathbf{k}$ and ω arguments, and as before we shall use G for all types of Green functions to avoid a proliferation of symbols. It is hoped the context and the arguments of the Green functions will be sufficient to identify any Green function that may occur in any equation. In addition, it should be noted that in this section, the Green functions depend also on β, though this is not explicitly displayed.

10.13.2. *Basic properties and a sum rule*

First we extend the definitions of $G^{(a)}$ and $G^{(r)}$ to imaginary values of the time variable.

We note that the definition of these Green functions for real time implies that the factor $e^{-\beta(H-\mu N)}$ is sufficient to give absolutely convergent values for the trace.

Since $G^{(a)}(x_1, x_2)$ may be written

$$G^{(a)}(x_1, x_2) = -\frac{i\,\mathrm{Tr}\,[e^{-\beta(H-\mu N)}e^{iHt_2}\psi^\dagger(\mathbf{r}_2, 0)\,e^{-iH(t_2-t_1)}\psi(\mathbf{r}_1, 0)\,e^{-iHt_1}]}{\mathrm{Tr}\,[e^{-\beta(H-\mu N)}]}, \tag{10.13.8}$$

it therefore follows that $G^{(a)}(x_1, x_2)$ will be an analytic function of the complex variable for $0 < \mathrm{Im}(t_1 - t_2) < \beta$. Similarly, $G^{(r)}(x_1, x_2)$ is analytic for $0 > \mathrm{Im}\,(t_1 - t_2) > -\beta$.

These are the desired analytic continuations. Next, we obtain a relation between the advanced and retarded one-particle Green functions.

Consider

$$\begin{aligned}
\{G^{(a)}(x_1, x_2)\}_{t_1=0} &= -i\langle\langle\psi^\dagger(x_2)\,\psi(x_1)\rangle\rangle \\
&= -i\,\frac{\mathrm{Tr}\,[e^{-\beta(H-\mu N)}\psi^\dagger(x_2)\,\psi(\mathbf{r}_1, 0)]}{\mathrm{Tr}\,[e^{-\beta(H-\mu N)}]} \\
&= -\frac{i\,\mathrm{Tr}\,[e^{-\beta(H-\mu N)}e^{\beta(H-\mu N)}\psi(\mathbf{r}_1, 0)\,e^{-\beta(H-\mu N)}\psi^\dagger(x_2)]}{\mathrm{Tr}\,[e^{-\beta(H-\mu N)}]} \\
&= -i\langle\langle e^{\beta(H-\mu N)}\psi(\mathbf{r}_1, 0)\,e^{-\beta(H-\mu N)}\psi^\dagger(x_2)\rangle\rangle.
\end{aligned} \tag{10.13.9}$$

In arriving at the third line, we have made use of the property that the trace of a product of operators is not altered if the products are cyclically interchanged.

The commutation relation between the number operator and $\psi(\mathbf{r}, 0)$ is

$$[N, \psi] = -\psi,$$

that is

$$N\psi(\mathbf{r}, 0) - \psi(\mathbf{r}, 0) N = -\psi.$$

Hence

$$\psi(\mathbf{r}, 0) N = (N+1)\psi(\mathbf{r}, 0), \tag{10.13.10}$$

and we conclude that

$$\psi(\mathbf{r}_1, 0) e^{\beta\mu N} = e^{\beta\mu(N+1)}\psi(\mathbf{r}_1, 0), \tag{10.13.11}$$

or

$$e^{-\beta\mu N}\psi(\mathbf{r}_1, 0) e^{\beta\mu N} = e^{\beta\mu}\psi(\mathbf{r}_1, 0). \tag{10.13.12}$$

Also we may write

$$e^{\beta H}\psi(\mathbf{r}_1, 0) e^{-\beta H} = \psi(\mathbf{r}_1, -i\beta), \tag{10.13.13}$$

thereby introducing Heisenberg operators with imaginary time arguments.

Substituting (10.13.11) and (10.13.13) into (10.13.9) then gives

$$\begin{aligned}\{G^{(a)}(x_1, x_2)\}_{t_1=0} &= -i\langle\langle e^{\beta\mu}\psi(\mathbf{r}_1, -i\beta)\psi^\dagger(x_2)\rangle\rangle \\ &= -e^{\beta\mu}\{G^{(r)}(x_1, x_2)\}_{t_1=-i\beta}\end{aligned}$$

that is

$$G^{(a)}(\mathbf{r}, t) = -e^{\beta\mu} G^{(r)}(\mathbf{r}, t - i\beta). \tag{10.13.14}$$

If we define Fourier transforms of these Green functions by the following equations

$$\left.\begin{aligned} G^{(a)}(\mathbf{k}, \omega) &= \int d\mathbf{r} \int_{-\infty}^{\infty} dt\, e^{-i\mathbf{k}.\mathbf{r}} e^{i\omega t} G^{(a)}(\mathbf{r}, t), \\ \text{and} \qquad G^{(r)}(\mathbf{k}, \omega) &= \int d\mathbf{r} \int_{-\infty}^{\infty} dt\, e^{-i\mathbf{k}.\mathbf{r}} e^{i\omega t} G^{(r)}(\mathbf{r}, t). \end{aligned}\right\} \tag{10.13.15}$$

Then (10.13.14) may be written as a relation connecting these transforms:

$$G^{(a)}(\mathbf{k}, \omega) = -e^{-\beta(\omega-\mu)} G^{(r)}(\mathbf{k}, \omega). \tag{10.13.16}$$

The spectral function $A(\mathbf{k}, \omega)$ is defined as

$$A(\mathbf{k}, \omega) = G^{(r)}(\mathbf{k}, \omega) - G^{(a)}(\mathbf{k}, \omega). \tag{10.13.17}$$

Introducing

$$f(\omega) = \frac{1}{e^{\beta(\omega-\mu)} + 1}, \tag{10.13.18}$$

we can write $$G^{(a)}(\mathbf{k}, \omega) = -f(\omega) A(\mathbf{k}, \omega) \tag{10.13.19}$$

and $$G^{(r)}(\mathbf{k}, \omega) = [1 - f(\omega)] A(\mathbf{k}, \omega). \tag{10.13.20}$$

In terms of the field operators

$$A(\mathbf{k}, \omega) = i \int d\mathbf{r} \int_{-\infty}^{\infty} dt\, e^{-i\mathbf{k}.\mathbf{r}} e^{i\omega t} [\langle\langle\{\psi(x)\psi^{\dagger}(0) + \psi^{\dagger}(0)\psi(x)\}\rangle\rangle]$$

and hence

$$\int \frac{d\omega}{2\pi} A(\mathbf{k}, \omega) = i \int d\mathbf{r}\, e^{-i\mathbf{k}.\mathbf{r}} [\langle\langle\{\psi(\mathbf{r}, 0)\psi^{\dagger}(0) + \psi^{\dagger}(0)\psi(\mathbf{r}, 0)\}\rangle\rangle]$$

$$= i \int d\mathbf{r}\, e^{-i\mathbf{k}.\mathbf{r}} \delta(\mathbf{r} - 0) = i,$$

that is $$\int A(\mathbf{k}, \omega)\, d\omega = 2\pi i. \tag{10.13.21}$$

This is an example of a 'sum rule'. We shall not give an exhaustive list of these but simply refer the interested reader to Martin & Schwinger (1959) where other references are also given.

10.13.3. *Non-interacting Fermions*

In order to illustrate the new concepts introduced we shall consider the case of non-interacting particles. Then we find almost immediately,

$$G_0^{(a)}(\mathbf{k}, \omega) = \int d\mathbf{r} \int dt\, e^{-i\mathbf{k}.\mathbf{r}} e^{i\omega t} G^{(a)}(\mathbf{r}, t)$$

$$= -i \iint dt\, e^{i\omega t} e^{-i\mathbf{k}.\mathbf{r}} \langle\langle\psi^{\dagger}(\mathbf{r}, t_2)\psi(0, t_1)\rangle\rangle\, d\mathbf{r}$$

$$= -i \int dt\, e^{i\omega t} \langle\langle a_{\mathbf{k}0}^{\dagger} a_{\mathbf{k}t}\rangle\rangle \tag{10.13.22}$$

$$= -i2\pi\, \delta(\omega - \epsilon_{\mathbf{k}}) \langle\langle a_{\mathbf{k}}^{\dagger} a_{\mathbf{k}}\rangle\rangle. \tag{10.13.23}$$

Similarly, we obtain

$$G_0^{(r)}(\mathbf{k}, \omega) = i2\pi\, \delta(\omega - \epsilon_{\mathbf{k}}) \langle\langle a_{\mathbf{k}} a_{\mathbf{k}}^{\dagger}\rangle\rangle. \tag{10.13.24}$$

Hence from equation (10.13.19)

$$A_0(\mathbf{k}, \omega) = i2\pi\, \delta(\omega - \epsilon_{\mathbf{k}}), \tag{10.13.25}$$

consistent with the sum rule (10.13.21). The average number of particles in state **k** is, from equation (10.13.20),

$$\langle\langle n(\mathbf{k})\rangle\rangle = \langle\langle a_{\mathbf{k}}^{\dagger} a_{\mathbf{k}}\rangle\rangle = f(\epsilon_{\mathbf{k}})$$

$$= \frac{1}{\exp\{\beta(\epsilon_{\mathbf{k}}-\mu)\}+1}, \qquad (10.13.26)$$

which is the usual result of statistical mechanics.

10.13.4. *Grand partition function*

In this section we derive expressions for the grand partition function Z_G in terms of Green functions. We first recall the definition of the grand partition function from Chapter 9 namely,

$$Z_G = \mathrm{Tr}\,[e^{-\beta(H^{(g)}-\mu N)}], \qquad (10.13.27)$$

where

$$H^{(g)} = H_0 + gV. \qquad (10.13.28)$$

Then, as we showed in Chapter 9, (9.5.4), we can write

$$\ln Z_G = \ln Z_G^0 - \beta \int_0^1 \frac{dg}{g} \langle\langle gV\rangle\rangle, \qquad (10.13.29)$$

where Z_G^0 is the grand partition function for non-interacting particles. We must still get an expression for $\langle\langle gV\rangle\rangle$ in terms of the Green functions.

From the equation of motion of the field operator $\psi(\mathbf{r}, t)$, namely

$$i\frac{\partial\psi}{\partial t} = [\psi, H], \qquad (10.13.30)$$

and the commutation relations of ψ and $\psi^{\dagger}$, we obtain

$$\left(i\frac{\partial}{\partial t}+\frac{\nabla^2}{2m}\right)\psi(\mathbf{r}, t) = g\int dx'\, v(x-x')\,\psi^{\dagger}(x')\,\psi(x')\,\psi(\mathbf{r}, t). \qquad (10.13.31)$$

Hence

$$\psi^{\dagger}(\mathbf{r}', t')\, i\frac{\partial}{\partial t}\psi(\mathbf{r}, t) = \psi^{\dagger}(\mathbf{r}', t') - \frac{\nabla^2}{2m}\psi(\mathbf{r}, t)$$

$$+ g\int dx''\,\psi^{\dagger}(\mathbf{r}', t')\,\psi^{\dagger}(x'')\,v(x-x'')\,\psi(x'')\,\psi(\mathbf{r}, t). \qquad (10.13.32)$$

Taking the expectation value in the grand canonical ensemble after putting $x = x'$, and integrating over x we obtain

$$\left[\pm\frac{\partial}{\partial t}G^{(a)}(x_1, x_2)\right]_{x_1=x_2} = \langle\langle H_0\rangle\rangle + 2\langle\langle gV\rangle\rangle. \quad (10.13.33)$$

Now
$$\langle\langle H_0\rangle\rangle = \left[-\frac{\nabla_1^2}{2m}G^{(a)}(x_1, x_2)\right]_{x_1=x_2}, \quad (10.13.34)$$

and hence

$$\langle\langle gV\rangle\rangle = \left\{\frac{1}{2}\left[-\frac{\partial}{\partial t}+i\frac{\nabla_1^2}{2m}\right]G^{(a)}(x_1, x_2)\right\}_{x_1=x_2} \quad (10.13.35)$$

$$= \frac{1}{2}\left\{-\frac{\partial}{\partial t}+\frac{\nabla^2}{2m}\int G^{(a)}(\mathbf{k}, \omega)\, e^{i\mathbf{k}.\mathbf{r}}\, e^{-i\omega t}\frac{d\mathbf{k}\, d\omega}{(2\pi)^4}\right\}_{\substack{\mathbf{r}=0\\ t=0}}$$

$$= \int \frac{1}{2}\left\{i\omega - i\frac{k^2}{2m}\right\} G^{(a)}(\mathbf{k}, \omega)\frac{d\omega\, d\mathbf{k}}{(2\pi)^4}$$

$$= \int \frac{d\omega\, d\mathbf{k}}{(2\pi)^4}\left(-\frac{i}{2}\right)\left(-\frac{k^2}{2m}+\omega\right) f(\omega)\, A(\mathbf{k}, \omega). \quad (10.13.36)$$

Combining (10.13.29) and (10.13.36) we obtain

$$\ln Z_G = \ln Z_G^0 + \beta\int_0^1 \frac{dg}{g}\frac{d\omega\, d\mathbf{k}}{(2\pi)^4}\frac{i}{2}\left(-\frac{k^2}{2m}+\omega\right) f(\omega)\, A(\mathbf{k}, \omega). \quad (10.13.37)$$

Thus, knowledge of the Green functions (or $A(\mathbf{k}, \omega)$) will enable us to determine Z_G and hence all thermodynamic properties.

10.13.5. *Physical interpretation of spectral function*

In section 10.13.3, for a system with Hamiltonian

$$H_0 = \Sigma \epsilon_{\mathbf{k}}\, a_{\mathbf{k}t}^{\dagger}\, a_{\mathbf{k}t},$$

we showed that $f(\epsilon_{\mathbf{k}})$ was the average (in the grand canonical ensemble) occupation number for state with energy $\epsilon_{\mathbf{k}}$. This result can be generalized as follows. If the Hamiltonian for a system can be reduced to the form

$$H = \Sigma \epsilon_l\, b_l^{\dagger}\, b_l,$$

where $b_l, b_l^{\dagger}$ are second quantized annihilation and creation operators referring to state of energy ϵ_l, then the average occupation number for state with label 'l' is $f(\epsilon_l)$.

Let $a_{\mathbf{k}}(\omega)$ be the Fourier transform of $a_{\mathbf{k}t}$ when $\phi_{\mathbf{k}}(\mathbf{r})$ is a plane-wave state $\frac{1}{\Omega^{\frac{1}{2}}} e^{i\mathbf{k}.\mathbf{r}}$. Then we may write

$$G^{(a)}(x_1, x_2) = \int -i\langle\langle a^{\dagger}_{\mathbf{k}'}(\omega)\, a_{\mathbf{k}''}(\omega)\rangle\rangle \frac{e^{-i\mathbf{k}'.\mathbf{r}_2}}{(2\pi)^3} e^{i\omega' t_2}\, d\mathbf{k}'\, d\omega'$$

$$\times \frac{e^{-\mathbf{k}''.\mathbf{r}_1}}{(2\pi)^3} e^{i\omega'' t_1}\, d\mathbf{k}''\, d\omega''.$$

Hence we deduce that

$$\begin{aligned} G^{(a)}(\mathbf{k}, \omega) &= -i\langle\langle a^{\dagger}_{\mathbf{k}}(\omega)\, a_{\mathbf{k}}(\omega)\rangle\rangle \\ &= -f(\omega)\, A(\mathbf{k}, \omega). \end{aligned} \tag{10.13.38}$$

Hence, in general, the average occupation number of any state with momentum $\mathbf{k}$ and energy ω is

$$-\frac{1}{i} G^{(a)}(\mathbf{k}, \omega) = \frac{1}{i} f(\omega)\, A(\mathbf{k}, \omega).$$

Since $f(\omega)$ is the average occupation number for a state of energy ω, $A(\mathbf{k}, \omega)/i$ is a 'weighting function' which defines the possible energies ω for state with momentum $\mathbf{k}$. Seen in this light the sum rule (10.13.21) is merely the statement that the total weight must be 1.

The density of particles at $\mathbf{r}, t$ is

$$\langle\langle \psi^{\dagger}(\mathbf{r}, t)\, \psi(\mathbf{r}, t)\rangle\rangle$$

and is independent of $\mathbf{r}$ and t.

Denoting it by n we obtain

$$n = \int \frac{d\mathbf{k}\, d\omega}{(2\pi)^4} \frac{G^{(a)}(\mathbf{k}, \omega)}{(-i)}. \tag{10.13.39}$$

For a system of free particles, $A_0(\mathbf{k}, \omega) = 2\pi i \delta(\omega - \epsilon_{\mathbf{k}})$. The total density of particles with momentum $\mathbf{k}$ is therefore

$$\begin{aligned} \langle\langle n(\mathbf{k})\rangle\rangle &= \int \frac{d\omega}{2\pi} \langle\langle n(\mathbf{k}, \omega)\rangle\rangle \\ &= f(\epsilon_{\mathbf{k}}). \end{aligned} \tag{10.13.40}$$

10.13.6. *Equations of motion*

In the general case, having reduced the statistical mechanical problem to the determination of the Green function, one generally appeals to some form of perturbation theory to calculate these Green functions.

In the earlier sections of this chapter, the diagrammatic method has been explored at great length. Therefore, in this section, we shall confine our attention to the writing down of the equations satisfied by these Green functions. The method of solutions, of course, must depend on the nature of the problem under consideration.

We naturally start out again from the Heisenberg equations of motion for the field operators:

$$i\hbar\frac{\partial}{\partial t}\psi = [\psi, H]. \tag{10.13.41}$$

Explicitly

$$[H_0, \psi(x)] = \frac{\hbar^2}{2m}\nabla^2\psi(x), \tag{10.13.42}$$

$$[H_0, \psi^\dagger(x)] = -\frac{\hbar^2}{2m}\nabla^2\psi^\dagger(x), \tag{10.13.43}$$

and

$$[V, \psi(x)] = -g\int v(x-x')\,\psi^\dagger(x')\,\psi(x')\,dx'\,\psi(x), \tag{10.13.44}$$

$$[V, \psi^\dagger(x)] = g\int \psi^\dagger(x')\,v(x-x')\,\psi(x')\,dx'\psi^\dagger(x), \tag{10.13.45}$$

where

$$v(x-x') = v(\mathbf{r}-\mathbf{r}')\,\delta(t-t'),$$

$$dx' = d\mathbf{r}'\,dt'.$$

Therefore

$$\begin{aligned}
i\frac{\partial}{\partial t_1}G(x_1,x_2) &= -\frac{\partial}{\partial t_1}[\theta(t_1-t_2)\langle\langle\psi(x_1)\,\psi^\dagger(x_2)\rangle\rangle] \\
&\quad +\frac{\partial}{\partial t_1}[\theta(t_2-t_1)\langle\langle\psi^\dagger(x_2)\,\psi(x_1)\rangle\rangle] \\
&= -\delta(t_1-t_2)\langle\langle\psi(x_1)\,\psi^\dagger(x_2)+\psi^\dagger(x_2)\,\psi(x_1)\rangle\rangle \\
&= i\langle\langle T([\psi(x_1),H],\psi^\dagger(x_2)\rangle\rangle-\delta^4(x_1-x_2).
\end{aligned} \tag{10.13.46}$$

Substituting the value of the commutator and rearranging we obtain

$$\begin{aligned}
&\left(i\frac{\partial}{\partial t_1}+\frac{\hbar^2\nabla_1^2}{2m}\right)G(x_1,x_2) \\
&= -\delta^4(x_1-x_2)+g\int d\mathbf{x}'\,v(x_1-x')\,G(\mathbf{r}',t',x_1;\mathbf{r}',t',x_2).
\end{aligned} \tag{10.13.47}$$

We note that the equation for the one-particle Green function involves a two-particle Green function.

Equation (10.13.47) may be rewritten in terms of $G_0(x_1, x_2)$ which satisfies

$$\left(i\frac{\partial}{\partial t_1}+\frac{\hbar^2\nabla_1^2}{2m}\right) G_0(x_1, x_2) = -\delta^4(x_1-x_2). \qquad (10.13.48)$$

This is therefore the single-particle Green function for a system of non-interacting particles.

Next, we write (10.13.47) as

$$\left(i\frac{\partial}{\partial t_1}+\frac{\hbar^2}{2m}\nabla_1^2\right) G(x_1, x_2) = -\delta^4(x_1-x_2)+\int dx'\, M(x_1, x')\, G(x', x_2), \qquad (10.13.49)$$

where M is the irreducible self-energy operator for elevated temperatures.

Then

$$G(x_1, x_2) = G_0(x_1, x_2)+\int G_0(x_1, x')\, M(x', x'')\, G(x'', x_2)\, dx'\, dx''. \qquad (10.13.50)$$

Formally we can write

$$G = G_0+G_0 MG, \qquad (10.13.51)$$

and proceed to expand in powers of the coupling constant just as for the zero-temperature case.

The equation satisfied by the two-particle Green function $G(x_1, x_2;\, x_3, x_4)$ is obtained in an analogous manner.

Thus

$$\begin{aligned} i\frac{\partial}{\partial t_1} & G(x_1, x_2;\, x_3, x_4) \\ &= i\langle\langle[T([\psi(x_1), H]\,\psi(x_2)\,\psi^\dagger(x_3)\,\psi^\dagger(x_4)]\rangle\rangle \\ &\quad +\delta^3(x_1-x_3)\,\langle\langle T(\psi(x_2)\,\psi^\dagger(x_4))\rangle\rangle-\delta^3(x_1-x_4)\,\langle\langle T(\psi(x_2)\,\psi^\dagger(x_3))\rangle\rangle \\ &= i\delta^3(x_1-x_3)\, G(x_2, x_4)-iG(x_2, x_3)\,\delta^3(x_1-x_4) \\ &\quad -i\langle\langle T\left\{\frac{\hbar^2}{2m}\nabla_1^2\psi(x_1)-g\int dx'\, v(\mathbf{r}-\mathbf{r}')\,\psi^\dagger(r')\,\psi(r')\,\psi(x_1)\right\}\psi(x_2) \\ &\qquad \psi^\dagger(x_3)\,\psi^\dagger(x_4)\rangle\rangle. \qquad (10.13.52) \end{aligned}$$

It then follows that

$$\begin{aligned} \left(i\frac{\partial}{\partial t}+\frac{\hbar^2}{2m}\nabla_1^2\right) & G(x_1, x_2;\, x_3, x_4) \\ &= -i\delta^3(x_1-x_3)\, G(x_2, x_4)+i\delta^3(x_1-x_4)\, G(x_2, x_3) \\ &\quad -g\int dx'\, v(x_1-x')\, G(\mathbf{r}', t', x_1, x_2;\, \mathbf{r}', t', x_3, x_4). \qquad (10.13.53) \end{aligned}$$

The equation for the two-particle Green function involves a three-particle Green function. It is clear that, in general, the equation for an n-particle Green function will involve not only an $n-1$-particle Green function but also an $n+1$-particle Green function. Thus, to obtain accurate solutions to any Green function equation we will have to solve an infinite set of coupled equations. This is clearly not possible. Progress is made for the two-particle Green function (the method is similar for higher-particle Green functions also) by writing its equation in terms of one- and two-particle Green functions only, and rewriting the last term of (10.13.53) as

$$\int \Gamma(x_1, x_2; x_5, x_6)\, G(x_5, x_6; x_3, x_4)\, dx_5\, dx_6.$$

Thus we introduce a new operator Γ in analogy with M for the one-particle Green function.

In terms of the unperturbed as well as the perturbed single-particle Green function, the equation for the two-particle function may now be written as

$$\begin{aligned} G(x_1, x_2; x_3, x_4) = {} & iG_0(x_1, x_3)\, G(x_2, x_4) - iG_0(x_1, x_4)\, G(x_2, x_3) \\ & + \int G_0(x_1, x_5)\, G_0(x_2, x_6)\, \Gamma(x_5, x_6; x_7, x_8)\, G(x_7, x_8; x_5, x_6) \\ & \times dx_5\, dx_6\, dx_7\, dx_8. \qquad (10.13.54) \end{aligned}$$

Once again a perturbation expansion is possible, or alternatively some other approximation, based on intuition, for Γ. We shall not pursue these methods here, however, but shall proceed to consider some important physical applications of single-particle Green functions.

10.13.7. *Boundary conditions*

Equation (10.13.47) is a first-order differential equation, and to determine a unique solution we require of course a boundary condition.

Equation (10.13.14) of section 10.13.2 can be written in the following form and used as the boundary condition

$$\{G(x_1, x_2)\}_{t_1=0} = -e^{\beta\mu} G(x_1, x_2)_{t_1=-i\beta}, \qquad (10.13.55)$$

where we have analytically continued the Green functions to imaginary value of the time variable. We now expand $G(\mathbf{k}, t-t')$ as a Fourier series in the form

$$G(\mathbf{k}, t-t') = \frac{1}{i\beta}\sum_{\nu} e^{-iZ_\nu(t-t')}\, G(\mathbf{k}, Z_\nu) \quad \text{for} \quad 0 \leqslant it, it' \leqslant \beta, \qquad (10.13.56)$$

where $$Z_\nu = (\pi\nu/-i\beta)+\mu \qquad (10.13.57)$$

and where ν is odd.

The inverse of (10.13.56) is

$$G(\mathbf{k}, Z_\nu) = \int_0^{-i\beta} dt\, e^{Z_\nu(t-t')}\, G(\mathbf{k}, t-t'), \qquad (10.13.58)$$

the integral being independent of t'. Since

$$G(\mathbf{k}, t) = G^{(r)}(\mathbf{k}, t) = \int \frac{d\omega}{2\pi}[1-f(\omega)\,A(\mathbf{k}, \omega)]\, e^{i\omega t}, \qquad (10.13.59)$$

$$G(\mathbf{k}, Z_\nu) = \int \frac{d\omega}{2\pi}\frac{A(\mathbf{k}, \omega)}{1+e^{-\beta(\omega-\mu)}}\int_0^{-\beta} dt\, e^{i\omega t}\, e^{i([\pi\nu/-i\beta]+\mu)t}$$

$$= \int \frac{d\omega}{2\pi}\frac{iA(\mathbf{k}, \omega)}{(Z_\nu - \omega)} \qquad (10.13.60)$$

gives the Fourier coefficients.

If we analytically continue $G(\mathbf{k}, Z)$ for values other than

$$Z = Z_\nu = \pi\nu/-i\beta+\mu$$

we can determine $A(\mathbf{k}, \omega)$ by measuring the discontinuity of $G(\mathbf{k}, Z)$ across the real axis

$$A(\mathbf{k}, \omega) = [G(\mathbf{k}, \omega+i\epsilon)-G(\mathbf{k}, \omega-i\epsilon)]. \qquad (10.13.61)$$

As an example consider the equation

$$\left(i\frac{\partial}{\partial t_1}+\frac{\nabla_1^2}{2m}\right) G_0(x_1, 0) = -\delta^4(x_1). \qquad (10.13.62)$$

Multiplying by $\exp\left\{-i\mathbf{k}.\mathbf{r}_1+i\left(\frac{\pi\nu}{-i\beta}+\mu\right)t_1\right\}$

and integrating over x_1 we obtain

$$\left[-\left(\frac{\pi\nu}{-i\beta}+\mu\right)-\frac{\mu^2}{2m}\right] G_0(\mathbf{k}, Z_\nu) = -1, \qquad (10.13.63)$$

and thus
$$G_0(\mathbf{k}, Z_\nu) = \frac{-1}{Z_\nu - k^2/2m}.$$

The analytically continued function is

$$G_0(\mathbf{k}, Z) = \frac{-1}{Z - k^2/2m}$$

and hence
$$A_0(\mathbf{k}, \omega) = 2\pi i \delta(\omega - k^2/2m)$$

the result we obtained earlier.

We have given the briefest introduction to equations of motion, to the boundary conditions and to the analytic continuation, of Green functions in the complex time or energy plane. A rapidly expanding literature exists on the applications of the Green functions method to physical problems. The reader may profitably consult Zubarev (1960), Aleckseev (1960) and Kadanoff & Baym (1962) for further details.

10.14. Superfluids

The Green function theories of superfluids form too large a subject to treat comprehensively here. Instead, it is our intention to indicate the essential modifications of the previous theory which must be made, before referring the reader to the more advanced accounts of, for example, Nozières (1963) and Abrikosov, Gorkov & Dzyaloshinski (1963). For this purpose we discuss (though in less detail than hitherto) the theory of the single-particle energy spectrum and will draw heavily on the contributions of Beliaev (1958), Hugenholtz & Pines (1959), Larkin & Migdal (1963).

10.14.1. *Bosons*

We consider, here, the model He II problem of Chapter 8, though clearly the techniques can also be applied to, say, phonons and magnons. The presence of the macroscopically occupied condensate of zero-momentum particles introduces a difficulty which must be considered at this point. To understand it, let us recollect the treatment of normal Fermion systems expounded in detail in previous sections of this chapter. There, a single-particle Green function was defined, such that its poles gave the one-particle energy spectrum. For computational purposes, the propagator was then converted into sums of time-ordered products, the expectation value of which was taken over the particle-hole vacuum state. The next step was to invoke Wick's theorem which, because of the vacuum-vacuum property, expressed immediately after (4.14.1), allowed us to replace each T-product average by the appropriate set of fully contracted terms. After that, it was necessary to sum the resulting series.

In the Boson case, we do not have to worry about particles and holes, of course, but there is no ready-made vacuum in the zero-momentum state. For, $a_{\mathbf{k}}|0\rangle = 0$ only for $\mathbf{k} \neq 0$, because of the existence of the condensate. Indeed $\langle 0|a_0^\dagger a_0|0\rangle$ represents the total particle number in the condensate. Thus, without further modification to the problem, such a state could not be a successful vacuum in the sense of the previous paragraph, because of the non-vanishing of some N-product averages containing $a_0^\dagger$ and a_0.

(*a*) *Revised Hamiltonian.* The above difficulty can be eliminated if we think in terms of excitations from the condensate. We define new operators (cf. Chapter 8, (8.3.59)) by

$$A_{\mathbf{k}} = a_{\mathbf{k}} a_0^\dagger/\surd N_0, \quad A_{\mathbf{k}}^\dagger = a_{\mathbf{k}}^\dagger a_0/\surd N_0 \quad (\mathbf{k} \neq 0), \qquad (10.14.1)$$

where N_0 is the average number of particles in the condensate. Then (cf. Chapter 8)

$$A_{\mathbf{k}} A_{\mathbf{l}} - A_{\mathbf{l}} A_{\mathbf{k}} = 0 = A_{\mathbf{k}}^\dagger A_{\mathbf{l}}^\dagger - A_{\mathbf{l}}^\dagger A_{\mathbf{k}}^\dagger \quad (\mathbf{l}, \mathbf{k} \neq 0), \qquad (10.14.2)$$

while

$$A_{\mathbf{k}} A_{\mathbf{l}}^\dagger - A_{\mathbf{l}}^\dagger A_{\mathbf{k}} = (a_0^\dagger a_0 \delta_{\mathbf{kl}} - a_{\mathbf{l}}^\dagger a_{\mathbf{k}})/N_0. \qquad (10.14.3)$$

We now replace the right-hand side of this equation by $\delta_{\mathbf{kl}}$ and thus obtain a set of Boson operators. Our reasons for doing this are that we are only concerned with operators on the interacting ground state or minor deviations therefrom (say by introducing an extra particle as in one-particle Green function theory). Thus we may evaluate (10.14.3) as though it were operating on the ground state, and, since $a_0^\dagger a_0$ and H are easily verified to commute (to order $1/N$), we may replace $a_0^\dagger a_0$ by N_0. To this order, the $a_{\mathbf{l}}^\dagger a_{\mathbf{k}}$ term is negligible.

For the Boson operators thus defined by (10.14.1), $|0\rangle$ is an appropriate vacuum, while the Hamiltonian

$$H = \Sigma \epsilon_{\mathbf{k}} a_{\mathbf{k}}^\dagger a_{\mathbf{k}} + \tfrac{1}{2}\Sigma \langle \mathbf{k}_1 \mathbf{k}_2|V|\mathbf{k}_3 \mathbf{k}_4\rangle a_{\mathbf{k}_1}^\dagger a_{\mathbf{k}_2}^\dagger a_{\mathbf{k}_3} a_{\mathbf{k}_4} \qquad (10.14.4)$$

becomes

$$H = T+V; \quad V = \sum_{i=1}^{8} V_i; \tag{10.14.5}$$

where, using momentum conservation,

$$T = \Sigma \epsilon_{\mathbf{k}} a_{\mathbf{k}}^{\dagger} a_{\mathbf{k}}, \tag{10.14.6}$$

$$V_1 = \tfrac{1}{2}\Sigma V_{\mathbf{q}} A_{\mathbf{k}_1}^{\dagger} A_{\mathbf{k}_2}^{\dagger} A_{\mathbf{k}_1-\mathbf{q}} A_{\mathbf{k}_2+\mathbf{q}}, \tag{10.14.7}$$

$$V_2 = \sqrt{N_0}\Sigma V_{\mathbf{k}_1} A_{\mathbf{k}_1}^{\dagger} A_{\mathbf{k}_2}^{\dagger} A_{\mathbf{k}_1+\mathbf{k}_2}, \tag{10.14.8}$$

$$V_3 = \sqrt{N_0}\Sigma V_{\mathbf{k}_1} A_{\mathbf{k}_1+\mathbf{k}_2}^{\dagger} A_{\mathbf{k}_1} A_{\mathbf{k}_2}, \tag{10.14.9}$$

$$V_4 = \frac{N_0}{2}\Sigma V_{\mathbf{k}} A_{\mathbf{k}}^{\dagger} A_{-\mathbf{k}}^{\dagger}, \tag{10.14.10}$$

$$V_5 = \frac{N_0}{2}\Sigma V_{\mathbf{k}} A_{-\mathbf{k}} A_{\mathbf{k}}, \tag{10.14.11}$$

$$V_6 = N_0\Sigma V_0 A_{\mathbf{k}}^{\dagger} A_{\mathbf{k}}, \tag{10.14.12}$$

$$V_7 = N_0\Sigma V_{\mathbf{k}} A_{\mathbf{k}}^{\dagger} A_{\mathbf{k}}, \tag{10.14.13}$$

$$V_8 = \frac{N_0^2}{2} V_0, \tag{10.14.14}$$

the above summations being over all $\mathbf{k}_i$ such that the subscripts on the operators do not vanish. This result is simply obtained by writing down all terms of (8.3.54) of Chapter 8 and then replacing, as above, $a_0^{\dagger} a_0$ by N_0.

Equations (10.14.5)–(10.14.14), together with the Boson commutation rules, now represent a well-defined problem. Particle number, of course, is not strictly conserved, but we can conserve it on the average using the Lagrange multiplier technique. Writing the number operator $N'_{\text{op.}} = \sum_{k \neq 0} A_{\mathbf{k}}^{\dagger} A_{\mathbf{k}}$ appropriate to particles not in the condensate, we thus become interested in the Hamiltonian

$$H' = H - \mu N'_{\text{op.}} = T' + V. \tag{10.14.15}$$

A proper procedure is now (i) to solve the problem associated with (10.14.15) for fixed N_0, μ, then (ii) to impose the normalization condition that $\langle N'_{\text{op.}}\rangle = N - N_0$ to relate N_0 and μ. Finally, (iii) one must decide (on energy grounds) what actually is the optimum N_0. Below, for considerations of space, we will sketch the solution of (i), referring the reader to Hugenholtz & Pines (1959) and Beliaev (1958) for the remainder. We will, however, recover the Bogoliubov result of Chapter 8 when the conditions are such that (ii) and (iii) are trivially solved.

In order to discuss the excitation spectrum for (i), we must assemble the appropriate machinery rather along the lines used for Fermions earlier in this chapter and in Chapter 4. Briefly, the procedure is as follows. We define N-products by the formula

$$N(AAA^{\dagger}A\ldots A^{\dagger}AA^{\dagger}) = A^{\dagger}A^{\dagger}\ldots A^{\dagger}AA\ldots A$$

(there being now no signature term). Then the usual property of vacuum-vacuum averages of N-products (see comment (iv) following (4.8.2)) follows. Next, T-products are defined in the usual way, except for the omission of signature:

$$T[A(t_1)\ldots A(t_n)] = A(t_{\alpha_1})\ldots A(t_{\alpha n}) \quad (t_{\alpha_1} > \ldots > t_{\alpha n}).$$

The contraction of A_1 and A_2 is then precisely as before. We have

$$\overbrace{A_1 A_2} \equiv \langle 0|T(A_1 A_2)|0\rangle, \qquad (10.14.16)$$

where time t_i is associated with the operator A_i. Thus, we obtain the usual form of Wick's theorem. Next, one defines a single-particle Green function, just as in (10.2.3), by

$$G(\mathbf{k}, t_2 - t_1) = i\langle\Psi_n(0)|T(A_{\mathbf{k}t_2} A_{\mathbf{k}t_1})|\Psi_n(0)\rangle. \quad (10.14.17)$$

The interpretation is exactly as before and we thus search for the poles of $G(\mathbf{k}, \omega)$ in the complex ω-plane, as a function of k.

(*b*) *Anomalous Green functions*. This is the appropriate stage to introduce two extraordinary Green functions

$$\tilde{G}(\mathbf{k}, t_2 - t_1) = i\langle\Psi_n(0)|T(A_{\mathbf{k}t_2} A_{-\mathbf{k}t_1})|\Psi_n(0)\rangle, \quad (10.14.18)$$

$$\bar{G}(\mathbf{k}, t_2 - t_1) = i\langle\Psi_n(0)|T(A^{\dagger}_{\mathbf{k}t_2} A^{\dagger}_{-\mathbf{k}t_1})|\Psi_n(0)\rangle, \quad (10.14.19)$$

the non-vanishing of which is characteristic of superfluid systems. Equation (10.14.19), for example, may be interpreted as the probability amplitude for creating a pair of particles which subsequently annihilate each other. Provision for this possibility is provided by the occurrence of the term (10.14.11) in the Hamiltonian. In the computation of (10.14.17), which is our chief concern, use may be made of (10.14.18) and (10.14.19) as auxiliary functions.

The method of calculation is, once more, via perturbation theory and we obtain (10.4.18) except for a's replaced by A's and $|g\rangle$ by the present vacuum $|0\rangle$. In the analysis of the perturbation series, we need the unperturbed form of (10.14.17), which is simply evaluated to be

$$G_0(\mathbf{k}, t_2 - t_1) = \begin{cases} i \exp\{-i\omega_{\mathbf{k}}(t_2 - t_1)\} & (t_2 > t_1) \\ 0 & (t_1 > t_2) \end{cases}, \tag{10.14.20}$$

where $\omega_{\mathbf{k}} = \epsilon_{\mathbf{k}} - \mu$, since we are working with the modified Hamiltonian (10.14.15). The Fourier transform of (10.14.20) is

$$G_0(\mathbf{k}, \omega) = (\omega_{\mathbf{k}} - \omega - i0)^{-1}. \tag{10.14.21}$$

We see that it is as though we put $k_f = 0$ in (10.2.11) and (10.2.12). Thus, when we come to draw diagrams by an adaptation of the rules for Fermion graphs, we omit all contractions travelling backwards in time and therefore (in view of the convention embodied in (10A.6)) all equal-time contractions. Composite diagrams are then put together from the elementary structures of Fig. 10.21 which represent the various terms of (10.14.7)–(10.14.14).

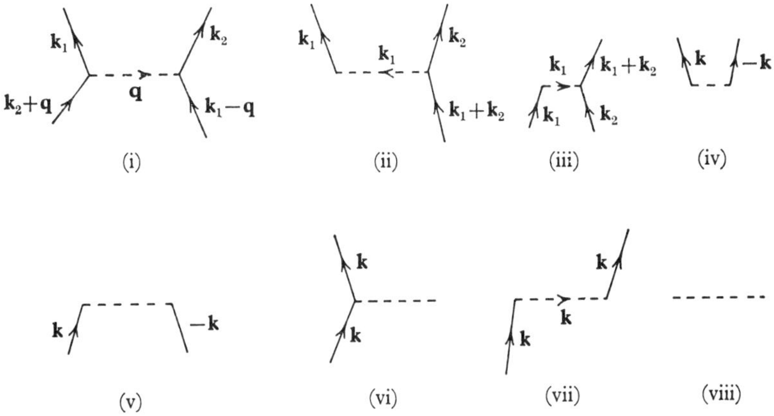

Fig. 10.21

Let us, to begin with, examine the $\mathfrak{D}$ terms (cf. (10.4.18)), which, in view of (10.14.5), may be written

$$\mathfrak{D} = \sum_{n=0}^{\infty} \frac{(-i)^n}{n!} \int_{-\infty}^{\infty} dt_n' \ldots \int_{-\infty}^{\infty} dt_1' \sum_{i_1, i_2 \ldots i_n = 1}^{8} \langle 0 | T V_{i_1}(t_1') \ldots V_{i_n}(t_n') | 0 \rangle. \tag{10.14.22}$$

As usual, we examine the integrand.

In first order, this takes the form

$$\sum_{i=1}^{8} \langle 0 | T V_i(t_1') | 0 \rangle$$

and, in this sum, only the 8th contribution, arising from (10.14.14) is non-zero. The second-order terms arise from

$$\sum_{i_1, i_2=1}^{8} \langle 0|TV_{i_1}(t_1')\, V_{i_2}(t_2')|0\rangle.$$

There are 64 pairs (i_1, i_2) to consider, but it is clear from Fig. 10.21 that only (iv, v) and (viii, viii) can contribute. Proceeding up the series in this way, we may write (10.14.22) symbolically as

$\mathfrak{D} = 1 +$ 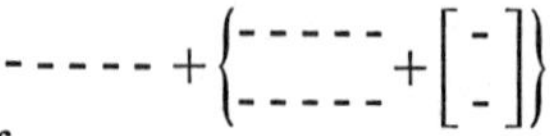 $+ \ldots.$ (10.14.23)

A similar analysis of

$$\mathfrak{N} = i \sum_{n=0}^{\infty} \frac{(-i)^n}{n!} \int_{-\infty}^{\infty} dt_n' \ldots \int_{-\infty}^{\infty} dt_1' \sum_{i_1, \ldots i_n=1}^{8} \langle 0|TV_{i_1}(t_1')\ldots V_{i_n}(t_n')$$
$$\times A_{\mathbf{k}}(t_2)\, A_{\mathbf{k}}^{\dagger}(t_1)|0\rangle \quad (10.14.24)$$

enables us to rewrite it as

$\mathfrak{N} =$

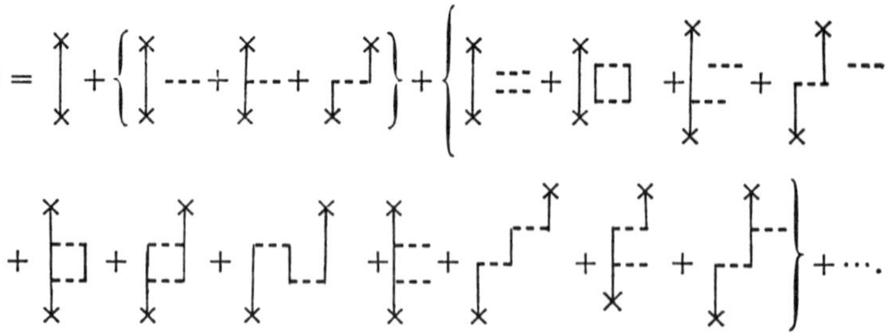

(10.14.25)

Once more, we are able to factorize, cancel a factor $\mathfrak{D}$, and establish a linked cluster theorem

$$G = \frac{\mathfrak{N}}{\mathfrak{D}}$$

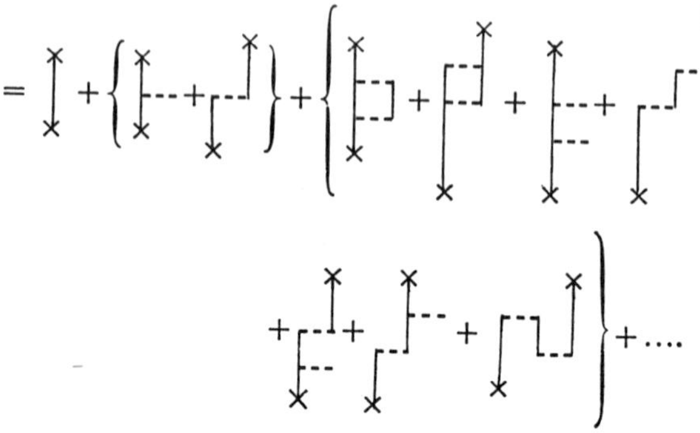

(10.14.26)

Terms in such series as (10.14.26) may be interpreted in energy as well as in time variables.

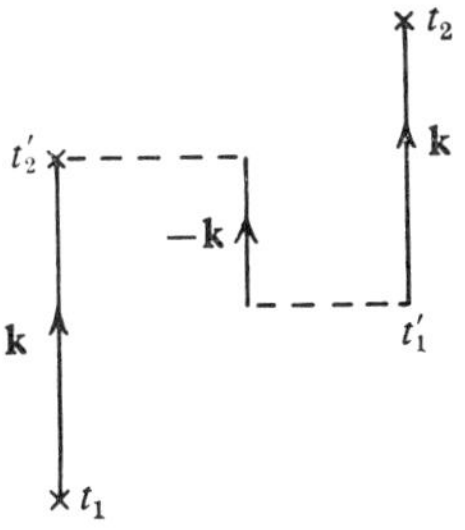

Fig. 10.22

EXAMPLE. Let us consider the 10th term of (10.14.26) (see also Fig. 10.22). Its contribution to G arises from the term

$$i\frac{(-i)^2}{2!}\int_{-\infty}^{\infty} dt_2' \int_{-\infty}^{\infty} \times dt_1' \langle 0|TV_4(t_1')\,V_5(t_2')\,A_{\mathbf{k}}(t_2)\,A_{\mathbf{k}}^{\dagger}(t_1)|0\rangle$$

plus an equal contribution from $V_5 V_4$: which readily yields

$$-iN_0^2 V_{\mathbf{k}}^2 \int_{-\infty}^{\infty} dt_2' \int_{-\infty}^{\infty} dt_1'\, G_0(\mathbf{k}, t_2'-t_1)\, G_0(\mathbf{k}, t_2-t_1')\, G_0(-\mathbf{k}, t_2'-t_1').$$

On applying the usual techniques of replacing $G_0(\mathbf{k}, t)$'s by $G_0(\mathbf{k}, \omega)$'s, the contribution to G in ω-space becomes

$$-iN_0^2 V_{\mathbf{k}}^2 \int_{-\infty}^{\infty} \exp\{i\omega(t_2-t_1)\}\, d(t_2-t_1) \int_{-\infty}^{\infty} dt_2' \int_{-\infty}^{\infty} dt_1' \int_{-\infty}^{\infty} \frac{d\omega_1}{2\pi}$$
$$\times \int_{-\infty}^{\infty} \frac{d\omega_2}{2\pi} \int_{-\infty}^{\infty} \frac{d\omega_3}{2\pi}\, G_0(\mathbf{k}, \omega_1) \exp\{-i\omega_1(t_2'-t_1)\}\, G_0(\mathbf{k}, \omega_2)$$
$$\times \exp\{-i\omega_2(t_2-t_1')\}\, G_0(-\mathbf{k}, \omega_3) \exp\{-i\omega_3(t_2'-t_1')\}$$
$$= -iN_0^2 V_{\mathbf{k}}^2\, G_0(-\mathbf{k}, -\omega)\, G_0^2(\mathbf{k}, \omega).$$

It should be noted that, when energy ω is associated with momentum $\mathbf{k}$, energy $-\omega$ is associated with $-\mathbf{k}$, just as one would intuitively expect on applying energy conservation to pair creation from the vacuum.

Similar developments enable us to write (10.14.18) and (10.14.19) in forms analogous to (10.14.26). For example,

$\tilde{G} =$ 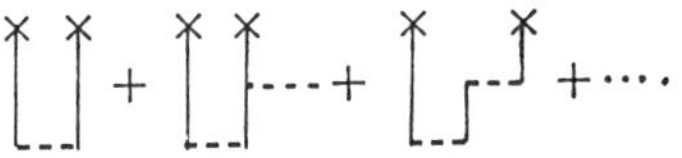 (10.14.27)

(c) *Generalized Dyson equation.* We would now like to distinguish between various kinds of self-energy parts. These fall into the three types indicated in Fig. 10.23. In (i) we have a type analogous to that obtained in the normal Fermion case, where an incoming particle takes part in some polarization process and afterwards emerges.

In (ii) we have a pair undergoing some (perhaps complicated) process which results in their mutual annihilation. The simplest example of this kind is shown in the left portion of Fig. 10.22. The component (iii) is similarly interpreted. For type (i) structures, we can associate

Fig. 10.23

self-energy parts using (10.6.19). Similar self-energy parts can be defined for structures of types (ii) and (iii) by using (10.6.19) with the final G^2 replaced by $G(\mathbf{k}, \omega)\, G(-\mathbf{k}, -\omega)$.

If we have

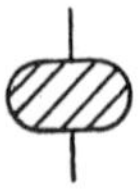 = either 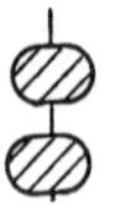 or  , (10.14.28)

then the left-hand side is said to be reducible. Similarly, with an analogous definition applied to (iii), if

 = either 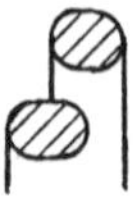 or 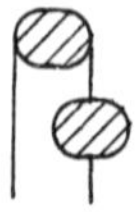 , (10.14.29)

then the left side is reducible. Otherwise, such components are said to be irreducible. Thus, for example, the component shown in Fig. 10.22 is itself reducible, consisting of two irreducible parts  and . Our next step is to define three irreducible self-energies M, $\bar{M}$ and $\tilde{M}$ obtained by summing the irreducible self-energy parts associated with subdiagrams of types (i), (ii) and (iii) respectively. (This is the analogue, in the present case, of (10.7.7).) In the Boson case, the symmetry of the statistics means that $\bar{M} = \tilde{M}$.

We must now generalize the simple Dyson equation

$$G = G_0 + G_0 MG, \qquad (10.14.30)$$

which is, of course, one way of writing (10.7.8). Clearly (10.14.30) will not do under present circumstances, for iteration will not produce any of the graphs of (10.14.26) which contain self-energy parts of the type shown in (10.14.27). From this remark, the generalization is clear. One replaces (10.14.30) by

$$G = G_0 + \tilde{G}\bar{M}G_0 + G_0 MG, \qquad (10.14.31)$$

which now iterates correctly. A somewhat similar examination of (10.14.27) reveals that

$$\tilde{G} = G_0(-\mathbf{k}, -\omega)\, M(-\mathbf{k}, -\omega)\, \tilde{G} + G_0(-\mathbf{k}, -\omega)\, \tilde{M}G. \quad (10.14.32)$$

Simple algebraic solution of (10.14.31) and (10.14.32) yields

$$G(\mathbf{k}, \omega) = \frac{\omega + \omega_{\mathbf{k}} + M(-\mathbf{k}, -\omega)}{[\omega + \omega_{\mathbf{k}} + M(-\mathbf{k}, -\omega)]\,[\omega - \omega_{\mathbf{k}} - M(\mathbf{k}, \omega)] + \bar{M}\tilde{M}} \qquad (10.14.33)$$

and

$$\tilde{G}(\mathbf{k}, \omega) = \frac{-\bar{M}}{[\omega + \omega_{\mathbf{k}} + M(-\mathbf{k}, -\omega)]\,[\omega - \omega_{\mathbf{k}} - M(\mathbf{k}, \omega)] + \bar{M}\tilde{M}}. \qquad (10.14.34)$$

The single-particle energy spectrum is determined by the vanishing of the denominator of (10.14.33), that is, by

$$\{\omega - \tfrac{1}{2}[M(\mathbf{k}, \omega) - M(-\mathbf{k}, -\omega)]\}^2 - \{\omega_{\mathbf{k}} + \tfrac{1}{2}[M(\mathbf{k}, \omega) + M(-\mathbf{k}, -\omega)]\}^2 + \bar{M}\tilde{M} = 0. \quad (10.14.35)$$

For general consideration of this equation, we refer the reader to the original papers. In particular, steps (ii) and (iii) following (10.14.15) now require consideration. We will content ourselves here with establishing contact with Chapter 8 by recovering the Bogoliubov spectrum (8.3.74) when (ii) and (iii) become trivial considerations.

In this approximation (cf. Chapter 8) the coupling is taken to be so weak that the number of particles excited out of the condensate is small in comparison with N, and we work only to first order. The leading term in the energy (cf. equation (8.3.76) of Chapter 8 and (10.14.14) above) is just due to the static interaction of N

particles. Thus, μ, the chemical potential, is obtained by adding an extra particle and finding the energy change. This is a trivial calculation using the above mechanism and we have (remembering we may replace N_0 by N)

$$\mu = NV_0. \tag{10.14.36}$$

Further, to this order (cf. (10.14.12) and (10.14.13), and (10.14.11) respectively) one finds

$$G_0^2(\mathbf{k}, \omega)\, M(\mathbf{k}, \omega) =$$ 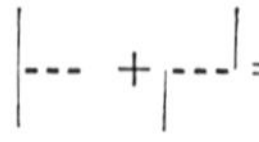 $$= G_0^2(\mathbf{k}, \omega)\, N(V_0 + V_{\mathbf{k}}), \tag{10.14.37}$$

and

$$G_0(\mathbf{k}, \omega)\, G_0(-\mathbf{k}, -\omega)\, \tilde{M} =$$  $$= G_0(\mathbf{k}, \omega)\, G_0(-\mathbf{k}, -\omega)\, NV_{\mathbf{k}}. \tag{10.14.38}$$

On substituting these results into (10.14.35), we obtain the Bogoliubov result (8.3.74) of Chapter 8.

The Bogoliubov result contains the characteristic phonon part, of course. Actually, the requirement that this should occur in general has its implications for (10.14.35). If $\omega = 0$ when $\mathbf{k} = 0$, (10.14.35) yields

$$\mu = M(0, 0) - \tilde{M}(0, 0) \tag{10.14.39}$$

(the ambiguity of sign being resolved on examining the weak coupling results (10.14.36), (10.14.37) and (10.14.38)). Hugenholtz & Pines prove this result using perturbation theory and the assumption of a macroscopically occupied condensate. By straightforward Taylor series expansion of (10.14.35), one finds a sound velocity specified by

$$\omega'(0) = -[(\partial\tilde{M}/\partial k)/(\partial\tilde{M}/\partial\omega)]_{k=\omega=0},$$

which emphasizes once more the crucial role of pair creation and annihilation in superfluids.

10.14.2. *Fermions*

In the Boson case, by exploiting the macroscopic occupancy of the condensate, we were able to arrive in a systematic way at a natural correlation between pairs $\mathbf{k}$, $-\mathbf{k}$, and a natural method of computation which conserves particle number on the average only. The Fermion case is not so clearcut. For the usual vacuum, which we will write $|g\rangle_N$ to emphasize it is that appropriate to an N-particle system, the Hamiltonian will conserve particle number. Yet we know from Chapter 7 that it is mathematically desirable to violate this strict conservation condition. The way out of this dilemma is to make the hypothesis that $|g\rangle_N$ can be replaced by a new vacuum

$$|0\rangle = \Sigma C_N |g\rangle_N; \quad \Sigma |C_N|^2 = 1, \tag{10.14.40}$$

the states of this series being obtained from each other by adding or subtracting pairs K, $-K$, no other states being allowed. The precise values of the weighting coefficients are, at this stage, unspecified.

The vacuum (10.14.40) bears some resemblance to the previous one, but there is the radical difference that

$$\overline{a_K^\dagger \, a_{-K}^\dagger} = (\overline{a_{-K} \, a_K})^\dagger \neq 0. \tag{10.14.41}$$

In fact, because of the way our vacuum was constructed, we have

$$\langle 0| T a_{\mathbf{k}}(t_2)\, a_{-\mathbf{l}}(t_1)|0\rangle = \delta_{\mathbf{kl}} \Sigma C_N^* \, C_{N+2} [{}_N\langle g| T a_{\mathbf{k}}(t_2)\, a_{-\mathbf{k}}(t_1)|g\rangle_{N+2}], \tag{10.14.42}$$

where (recall the convention stated after (10.2.4)), we take **k** and **l** to be momentum-spin indices in this subsection. We cannot proceed further in simplifying (10.14.42) until the C's are known. On the other hand, we can evaluate the ordinary unperturbed Green function. For

$$\langle 0| T a_{\mathbf{k}}(t_2)\, a_{\mathbf{l}}^\dagger(t_1)|0\rangle = \delta_{\mathbf{kl}} \Sigma |C_N|^2 \, {}_N\langle g| T a_{\mathbf{k}}(t_2)\, a_{\mathbf{k}}^\dagger(t_1)|g\rangle_N. \tag{10.14.43}$$

This is just a weighted sum of normal free-particle propagators, the Fourier transforms of which are given by (10.2.12). It should be noted that the only dependence of these terms on k_f, or equivalently N, is through the $\pm i0$ part. If, as happens in computations, we need not distinguish these limits, the dependence on N drops out and (10.14.43) transforms to give

$$\delta_{\mathbf{kl}}(\omega_{\mathbf{k}} - \omega)^{-1} \Sigma C_N^2 = \delta_{\mathbf{kl}}(\omega_{\mathbf{k}} - \omega)^{-1}. \tag{10.14.44}$$

Thus, in computations, we treat the ordinary unperturbed propagator in the usual way.

The method now develops as in the normal theory, except that we must always take care to give special treatment to the terms which give rise to the reduced Hamiltonian defined by (7.4.3) of Chapter 7.

The result is a formalism closely following that of the Boson treatment above. The analogue of (10.14.26), for example, is

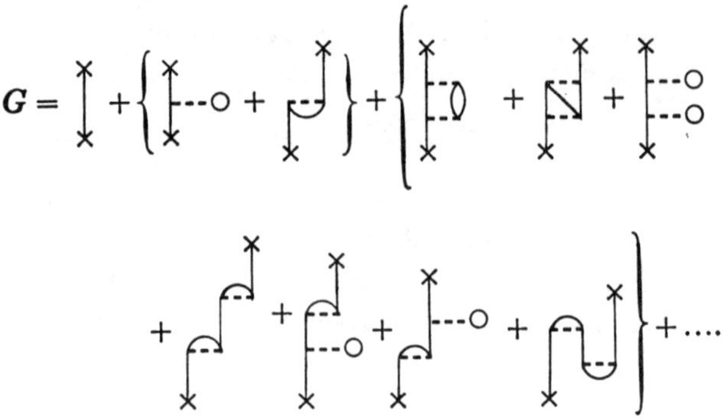

(10.14.45)

To the order shown, it differs from (10.5.2) only by the inclusion of the 10th term, which arises from (see Fig. 10.24 (i)) terms of the kind

$$a_K^\dagger(t_{12})\, a_L^\dagger(t_{12})\, a_M(t_{12})\, a_N(t_{12})\, a_P^\dagger(t_{34})\, a_Q^\dagger(t_{34})\, a_R(t_{34})\, a_S(t_{34})\, a_k(t_2)\, a_k^\dagger(t_1).$$

This should not be confused with the vanishing graph (ii) of Fig. 10.24.

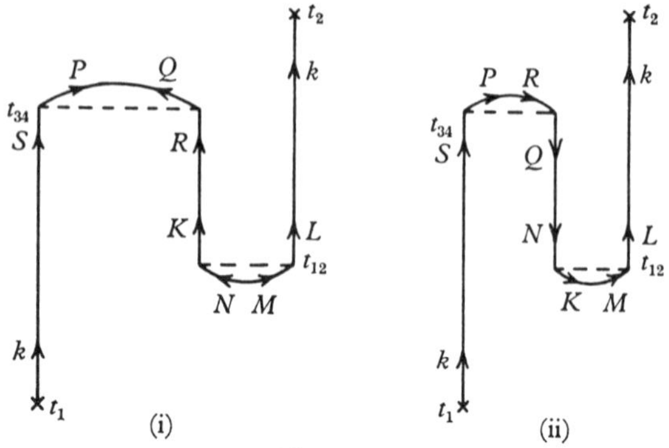

Fig. 10.24

Thus, we arrive at (10.14.31)–(10.14.35) once more, with the one difference that $\tilde{M} = -\overline{M}$ for Fermions.

To establish contact with the B.C.S. theory of Chapter 7, let us examine the case of weak coupling. The first order of M is

$$M = M_{1d} + M_{1e}, \qquad (10.14.46)$$

these two quantities being specified by (10.6.5) and (10.6.8) respectively. In this approximation M is independent of ω. Also, for reasonable potentials, one may assume $M(\mathbf{k}) = M(-\mathbf{k})$. Equation (10.14.46) is, of course, the analogue of (10.14.37); the corresponding analogue of (10.14.38) is readily calculated to be

$$\tilde{M}(\mathbf{k}, \omega) = -i\sum_{k'} \langle -\mathbf{k}\mathbf{k}|v|\mathbf{k}'-\mathbf{k}'\rangle \int_{-\infty}^{\infty} \frac{d\omega'}{2\pi} \tilde{G}_0(\mathbf{k}', \omega')\, e^{i\omega' 0}, \tag{10.14.47}$$

which is thus, to this approximation, independent of ω. Hence, (10.14.35) may be solved explicitly to give an energy spectrum

$$E_k = \sqrt{[(\omega_{\mathbf{k}} + M(\mathbf{k}))^2 + (\tilde{M}(\mathbf{k}))^2]}. \tag{10.14.48}$$

Unlike (10.14.38) for the Boson case, (10.14.47) does not immediately give $\tilde{M}$, since the final integral depends, through (10.14.42), say, on the rather inaccessible C_N coefficients. But, as it happens, this difficulty can be circumvented rather conveniently for weak coupling. For then $\tilde{G} \sim \tilde{G}_0$, and thus (10.14.34) may be rewritten with the aid of (10.14.48) as

$$\tilde{G}_0(\mathbf{k}, \omega) = \frac{\tilde{M}(\mathbf{k})}{\omega^2 - E_{\mathbf{k}}^2}. \tag{10.14.49}$$

Substitution of this function into (10.14.47) leads immediately to the B.C.S. integral equation. Thus we establish contact with the B.C.S. results with a formalism which enables one, in principle at least, to obtain further refinements of that theory.

Problems

P. 10(i). The single-particle energy spectrum for the Hamiltonian

$$H = \Sigma\omega_K a_K^\dagger a_K - \tfrac{1}{2}\Sigma\Delta_K(a_K^\dagger a_{-K}^\dagger + a_{-K} a_K)$$

is most easily solved by diagonalization, using the Bogoliubov transformations of Chapters 7 and 8. One finds the exact result

$$E_K = \sqrt{[\omega_K^2 - \Delta_K \Delta_{-K}]},$$

where Δ_K is even for Bosons and odd for Fermions.

Obtain this result for Bosons using the method of section 10.14.1, that is, by evaluating the Green function as an infinite series which can be exactly summed. (It is as though we drop all potential energy terms in (10.14.5) except V_4 and V_5.) For a more complicated example, see Mattuck (1964).

P. 10 (ii). In the Fermion case, the formalism of section 10.14.2, because it always involves potential energy terms of the type $a^\dagger a^\dagger aa$, is not immediately applicable to the above problem. How, then, can one obtain the correct result by this kind of technique?

APPENDIX 3A

SECOND QUANTIZED FORM OF ONE-PARTICLE OPERATORS

Let the one-particle operator part of an N-particle Hamiltonian be

$$U_T = \sum_{i=1}^{N} U(\mathbf{r}_i). \qquad (3\text{A}.1)$$

The result of $U(\mathbf{r}_i)$ operating on a single-particle basis function $\phi_l(\mathbf{r}_i)$ may be expressed as a linear combination of the complete single-particle set $\phi_m(\mathbf{r}_i)$, through the equation

$$U(\mathbf{r}_i)\,\phi_l(\mathbf{r}_i) = \sum_m \langle m|U|l\rangle\,\phi_m(\mathbf{r}_i). \qquad (3\text{A}.2)$$

Now we use the fact that a general N-body ket $|N\rangle$ may be constructed by allowing N creation operators $c^\dagger$ to act on the vacuum state. The ket may therefore be written

$$|N\rangle = \Pi c^\dagger|0\rangle. \qquad (3\text{A}.3)$$

For Bosons, for example, this is a symmetrized product of N of the ϕ_l's and hence from (3 A. 2) we have

$$U(\mathbf{r}_i)\prod_j \phi_{l_j}(\mathbf{r}_j) = \sum_m \langle m|U|l_i\rangle\,\phi_m(\mathbf{r}_i)\prod_{j\neq i}\phi_{l_j}(\mathbf{r}_j). \qquad (3\text{A}.4)$$

But U_T is symmetric in the interchange of particle co-ordinates, and hence commutes with the symmetrizing operator. Hence

$$U_T\,\Pi c^\dagger|0\rangle = \sum_i\sum_m \langle m|U|l_i\rangle\, c_m^\dagger\,\Pi'^{l_i}c^\dagger|0\rangle$$

$$= \sum_{imn}\langle m|U|n\rangle\,\delta_{nl_i}\,c_m^\dagger\,\Pi'^{l_i}c^\dagger|0\rangle, \qquad (3\text{A}.5)$$

where Π'^{l_i} implies that, in the product of $c^\dagger$'s, we must omit $c_{l_i}^\dagger$. Further, we can show that

$$c_n\,\Pi c^\dagger|0\rangle = \sum_{l_i}\delta_{l_i n}\,\Pi'^{l_i}c^\dagger|0\rangle, \qquad (3\text{A}.6)$$

and hence (3 A. 5) becomes

$$U_T\,\Pi c^\dagger|0\rangle = \sum_{mn}\langle m|U|n\rangle\,c_m^\dagger c_n\,\Pi c^\dagger|0\rangle. \qquad (3\text{A}.7)$$

Therefore, we may write for the operator U_T

$$U_T = \sum_{mn} \langle m|U|n\rangle c_m^\dagger c_n. \tag{3 A.8}$$

This argument has been given for Bosons, but the same result can be obtained by replacing symmetrized products by determinants for Fermions.

In particular, when U is the kinetic energy operator, we can obviously bring (3 A. 8) to diagonal form by choosing $|m\rangle$ and $|n\rangle$ as plane waves. The result (3.6.3) then follows.

APPENDIX 4A.1

WICK'S THEOREM

The major part of the work, below, is to establish the theorem for ordinary products in the form (4.10.1). Once this is done, the result (4.14.1) for time-ordered products follows without great difficulty.

1. Wick's theorem for ordinary products

This depends on the following result:

$$N(A_1 A_2 \dots A_n) B = \sum_{r=1}^{n} N(A_1 A_2 \dots \underbrace{A_r \dots A_n B}) + N(A_1 \dots A_n B), \tag{4A.1.1}$$

the proof of which is as follows. (i) If B is a destruction operator, then all $\underbrace{A_r B}$ vanish (see example following equation (4.9.2)) and (4A.1.1) is trivially satisfied. (ii) If B is a creation operator and the A's are all creation operators, again all the $\underbrace{A_r B}$ vanish and (4A.1.1) is true.

Thus, it is sufficient to prove (4A.1.1) for the case when B is a creation operator and the A's are all destruction operators. Then, from (i) and (ii), the general result will follow. We prove the result by induction.

For the special circumstances under consideration, equation (4A.1.1) asserts that

$$A_1 \dots A_n B = \sum_{r=1}^{n} (-1)^{r+n} \underbrace{A_r B} A_1 A_2 \dots \not{A}_r \dots A_n + (-1)^n B A_1 \dots A_n, \tag{4A.1.2}$$

the signature $(-1)^{r+n}$ resulting from $(r-1)$ interchanges of A_r and then $(n-1)$ of B. This equation is certainly true for $n = 1$ as is readily verified using (4.9.1). Our aim now is to assume that (4A.1.2) is true and show this implies the same result with n replaced by $n+1$. Let us, therefore, premultiply (4A.1.2) by the destruction operator A_0 to give

$$A_0 A_1 \dots A_n B = \sum_{r=1}^{n} (-1)^{r+n} \underbrace{A_r B} A_0 A_1 A_2 \dots \not{A}_r \dots A_n + (-1)^n A_0 B A_1 \dots A_n. \tag{4A.1.3}$$

Now we use, in the final term, the result that

$$A_0 B = N(A_0 B) + \underbracket{A_0 B} = -BA_0 + \underbracket{A_0 B}.$$

This latter pairing gives rise to a term which can be merged into the summation by summing from $r = 0$. Thus, (4 A. 1.3) may be written

$$A_0 A_1 \ldots A_n B = \sum_{r=0}^{n} (-1)^{r+n} \underbracket{A_r B} A_0 A_1 A_2 \ldots \not{A}_r \ldots A_n + (-1)^{n+1} B A_0 A_1 \ldots A_n$$

or, with a slight change of notation,

$$A_1 A_2 \ldots A_{n+1} B = \sum_{r=1}^{(n+1)} (-1)^{r+(n+1)} \underbracket{A_r B} A_1 A_2 \ldots \not{A}_r \ldots A_{n+1} + (-1)^{n+1} B A_1 A_2 \ldots A_{n+1}.$$

The result (4 A. 1.1) now follows by induction.

We wish to extend (4 A. 1.1) for N-products containing pairings. Let

$$N(A_1 A_2 A_3 \ldots A_x \ldots A_y \ldots A_z \ldots A_n),$$

with an arbitrary number of pairings, denote such product. Then

$$\begin{aligned} &N(A_1 A_2 A_3 \ldots A_x \ldots A_y \ldots A_z \ldots A_n) B \\ &\quad = \Sigma N(A_1 A_2 A_3 \ldots A_x \ldots A_r \ldots A_y \ldots A_z \ldots A_n B) \\ &\qquad + N(A_1 A_2 A_3 \ldots A_x \ldots A_y \ldots A_z \ldots A_n B), \quad (4\,\mathrm{A}.\ 1.4) \end{aligned}$$

where the sum is over all hitherto unpaired A_r's. The proof of (4 A. 1.4) is achieved by removing the pairings from the N-product, applying (4 A. 1.1) and then re-inserting the pairings. Specifically, the left-hand side of (4 A. 1.4) is

$$\begin{aligned} &\pm \underbracket{A_2 A_y}\, \underbracket{A_x A_z} \ldots N(A_1 \not{A}_2 A_3 \ldots \not{A}_x \ldots \not{A}_y \ldots \not{A}_z A_n) B \\ &\quad = \pm \underbracket{A_2 A_y}\, \underbracket{A_x A_z} \ldots \{\Sigma N(A_1 \not{A}_2 \ldots \not{A}_x \ldots \underbracket{A_r \ldots \not{A}_y \ldots \not{A}_z \ldots A_n B}) \\ &\qquad + N(A_1 \not{A}_2 \ldots \not{A}_x \ldots \not{A}_y \ldots \not{A}_z \ldots A_n B)\}. \end{aligned}$$

On re-inserting the pairings we obtain the right-hand side of (4 A. 1.4).

We now turn to the proof of Wick's theorem for ordinary products. Once more we use the method of induction. As we have already remarked, the theorem reduces to (4.9.1) when $n = 2$. Let us, then, suppose (4.10.1) holds. We have to prove the same result with n replaced by $n+1$. To this end, we multiply (4.10.1) on the right by A_{n+1} and use the result (4 A. 1.1) and its corollary (4 A. 1.4).

The first term on the right-hand side is

$$N(A_1 A_2 \ldots A_n) A_{n+1} = N(A_1 \ldots A_n A_{n+1}) + \sum_r N(A_1 \ldots \underbrace{A_r \ldots A_n A_{n+1}}). \tag{4A.1.5}$$

The next class of terms arises from N-products containing one pairing only. The total contributions from these terms is

$$\sum_{1 \leqslant x < y \leqslant n} N(A_1 \ldots \underbrace{A_x \ldots A_y} \ldots A_n) A_{n+1} = \sum_{1 \leqslant x < y \leqslant n} N(A_1 A_2 \ldots \underbrace{A_x \ldots A_y} \ldots$$
$$\ldots A_n A_{n+1}) + \sum_{1 \leqslant x < y \leqslant n} \sum_{r \neq x, y} N(A_1 \ldots \underbrace{A_x \ldots A_r} \ldots \underbrace{A_y \ldots A_n A_{n+1}}). \tag{4A.1.6}$$

Now we add (4A.1.5) and (4A.1.6). We then obtain a term $N(A_1 \ldots A_n A_{n+1})$ without pairings. Also the total contribution of terms with one pairing is found by adding the second term on the right-hand side of (4A.1.1) to the first series on the right of (4A.1.2) to give

$$\sum_{1 \leqslant x < y \leqslant n+1} N(A_1 A_2 \ldots \underbrace{A_x \ldots A_y} \ldots A_n A_{n+1}).$$

By adding to the sum of (4A.1.5) and (4A.1.6) all other contributions, one clearly generates the right-hand side of (4.10.1) with n replaced by $n+1$. The theorem is thus proved.

The theorem continues to hold if the A_i's of (4.10.1) are now supposed to be linear combinations of creation and destruction operators (in particular, if they are ψ's, $\psi^\dagger$'s, $\psi_\pm$'s and $\psi^\dagger_\pm$'s). For, suppose A_i^k denote a (generally mixed) set of creation and destruction operators.

Then

$$\left(\sum_k A_1^k \sum_l A_2^l \ldots\right) = \sum_{k, l, \ldots} (A_1^k A_2^l \ldots)$$
$$= \sum_{k, l, \ldots} \{N(A_1^k A_2^l \ldots) + N(\underbrace{A_1^k A_2^l} \ldots) + \ldots + \text{fully paired terms}\} \quad \text{(using (4.10.1))}.$$
$$= \sum_{k, l, \ldots} N(A_1^k A_2^l \ldots) + \sum_{k, l, \ldots} N(\underbrace{A_1^k A_2^l} \ldots) + \ldots + \sum_{k, l, \ldots} \text{fully paired terms}$$
$$= N\left(\sum_k A_1^k \sum_l A_2^l \ldots\right) + N\left(\underbrace{\sum_k A_1^k \sum_l A_2^l} \ldots\right) + \ldots$$
$$\text{(using (4.9.10))}.$$

This establishes the desired result. Now we can proceed to discuss the theorem for time-ordered products.

2. Wick's theorem for time-ordered products

Equation (4.13.1) corresponds to (4.14.1) for two operators only. There is no need in the statement of Wick's theorem for T-products to treat first (as was done previously for ordinary products) the case when the A's represent simple a's, b's, a^+'s and b^+'s and then the case when they represent linear combinations thereof. The proof is as follows.

Let us suppose that the time-ordering of $t_1, t_2, \ldots, t_n$ is such that $t_{\alpha_1} > t_{\alpha_2} > \ldots > t_{\alpha_n}$. Then the T-product is easily converted to an ordinary product to which (4.10.1) may be applied.

We have

$$\begin{aligned} T(A_1 A_2 \ldots A_n) &= \pm A_{\alpha_1} A_{\alpha_2} \ldots A_{\alpha_n} \\ &= \pm \{N(A_{\alpha_1} A_{\alpha_2} \ldots A_{\alpha_n}) + N(\underbrace{A_{\alpha_1} A_{\alpha_2}} \ldots A_{\alpha_n}) \\ &\quad + N(\underbrace{A_{\alpha_1} A_{\alpha_2} A_{\alpha_3}} \ldots A_{\alpha_n}) + \ldots \\ &\quad + N(\underbrace{A_{\alpha_1} A_{\alpha_2}} \underbrace{A_{\alpha_3} A_{\alpha_4}} \ldots A_{\alpha_n}) + \ldots + \ldots \\ &\quad + \ldots + \text{all fully paired terms}\}. \end{aligned} \tag{4A.1.7}$$

Now let us recall (4.13.3), which tells us that, for our particular time sequence, pairings may be replaced everywhere by contractions. Doing this in (4A.1.7) and then switching every term back to its original time sequence, thereby absorbing the initial ± 1 factor, we arrive at the desired result (4.14.1).

APPENDIX 4A.2

THE LOOP THEOREM

The proof of this theorem is not hard in principle, though it is a little trouble to write down generally because of notational difficulties. Thus, we begin by seeing how the theorem works for a specific example.

Consider the special case

$$a_K^\dagger a_L^\dagger a_M a_N a_P^\dagger a_Q^\dagger a_R a_S a_X^\dagger a_Y^\dagger a_Z a_T, \tag{4A.2.1}$$

the graph for which is shown in Fig. 4A.2.1. The time indices are the usual ones but have been suppressed to avoid repetition. The demonstration of the theorem for (4A.2.1) is divided into two parts.

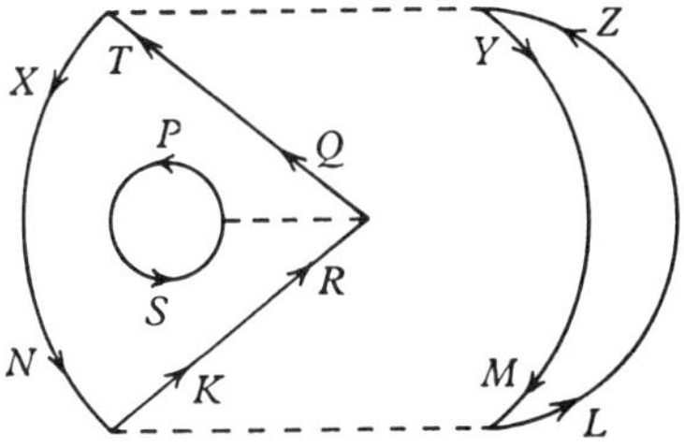

Fig. 4A.2.1

(i) Expression (4A.2.1) can be rewritten as the product of three factors suggested by the loops in the diagram, thus:

$$(+1)(a_K^\dagger a_N a_Q^\dagger a_R a_X^\dagger a_T)(a_P^\dagger a_S)(a_L^\dagger a_M a_Y^\dagger a_Z). \tag{4A.2.2}$$

The relative positions of the suffices within each factor are the same as in (4A.2.1) and the same contractions are preserved. By analogy with the graphical description, we will call the corresponding mathematical factors themselves loops. The initial signature in (4A.2.2) as a result of rearranging the various loop factors from (4A.2.1) is $+1$ as may be checked by direct counting. (We shall see below that this is a general result when resolving a fully contracted term into loops.)

(ii) Now we note that each loop can be rewritten as follows:

$$a_K^\dagger a_N a_Q^\dagger a_R a_X^\dagger a_T = -a_R a_K^\dagger a_N a_X^\dagger a_T a_Q^\dagger = -\{-iG_0(K, t_{34}-t_{12})\,\delta_{KR}\}$$
$$\times\{-iG_0(N, t_{12}-t_{56})\,\delta_{NX}\}\{-iG_0(T, t_{56}-t_{34})\,\delta_{TQ}\},$$

$$a_P^\dagger a_S = -a_S a_P^\dagger = -\{-iG_0(S, -0)\,\delta_{SP}\},$$

$$a_L^\dagger a_M a_Y^\dagger a_Z = -a_Z a_L^\dagger a_M a_Y^\dagger = -\{-iG_0(L, t_{56}-t_{12})\,\delta_{ZL}\}$$
$$\times\{-iG_0(M, t_{12}-t_{56})\,\delta_{MY}\}. \tag{4A.2.3}$$

Each loop contributes a factor of -1 times the appropriate product of $-iG_0$'s and Kronecker delta functions as indicated by rule (ii), p. 115. Thus the loop product (4A.2.2) is $(-1)^3$ times the appropriate contraction terms. This proves rule (ii) for the example (4A.2.1) for which $l = 3$. (In general, we shall show that any loop contributes a signature of -1 and this, taken together with the final remarks of part (i), above, proves rule (ii) for an arbitrary case).

We begin the general proof by considering the arbitrary fully contracted term

$$\overbrace{a_K^\dagger a_L^\dagger a_M a_N a_P^\dagger a_Q^\dagger a_R a_S \ldots a_X^\dagger a_Y^\dagger a_Z a_T}, \tag{4A.2.4}$$

the special symbol above the operator product denoting some general set of contractions in which every operator participates. For a convenient exposition of the proof, it is desirable to rewrite (4A.2.4) as

$$\overbrace{a_K^\dagger a_N a_L^\dagger a_M a_P^\dagger a_S a_Q^\dagger a_R \ldots a_X^\dagger a_T a_Y^\dagger a_Z}, \tag{4A.2.5}$$

where the same pairs have been contracted in (4A.2.5) as in (4A.2.4). There are an even number of sign changes involved (two interchanges within each 4-product) and so the initial signature is $+1$ as indicated. As an example (4A.2.1) may be rewritten as

$$a_K^\dagger a_N a_L^\dagger a_M a_P^\dagger a_S a_Q^\dagger a_R a_X^\dagger a_T a_Y^\dagger a_Z. \tag{4A.2.6}$$

We will refer to the pairs $a_K^\dagger a_N, a_L^\dagger a_M, \ldots, a_Y^\dagger a_Z$ as matched and will use the notation $\underbrace{a^\dagger a}$ if we wish to emphasize that the $a^\dagger$ and a concerned are matched.

(i) Now let us factorize (4A.2.5) into loops to obtain an expression analogous to (4A.2.2), above. In any loop, only matched pairs are involved. For example, if we include $a_K^\dagger$ which is represented by the line leaving the vertex end 1, then because we have a loop, we must include the matching a_N which is represented by the line entering 1. Thus, we may factorize any loop from (4A.2.5) by successively removing matched pairs from the total expression. In this process, each jump over an $a^\dagger$ or an a contributes a factor $(-1)^2 = +1$. Thus, we extract a loop factor from (4 A. 2. 5) which necessarily has the form

$$\overline{\underbrace{a^\dagger a}\,\underbrace{a^\dagger a}\ldots\underbrace{a^\dagger a}},\qquad (4A.2.7)$$

the contractions involved being the same as those in (4 A. 2.4) and (4 A. 2.5). Now we may observe that on factorizing (4 A. 2.7) from (4 A. 2.5), the remainder is a fully contracted product of adjacent matched pairs, just as (4 A. 2.5) was. Thus, we can repeat the process and factorize out further loops, again with positive signature, and finally rewrite (4 A. 2.4) as a product of l loops, say, each of the form indicated in (4 A. 2.7).

(ii) To complete the proof, what must now be done is to show that the general loop (4 A. 2.7) can be rewritten in the form

$$-\overline{aa}^\dagger\,\overline{aa}^\dagger\ldots\overline{aa}^\dagger.\qquad (4A.2.8)$$

Then, the product of l loops equivalent to (4 A. 2.4), will be equal to $(-1)^l$ times the factors $-iG_0(A, t_{\nu\sigma}-t_{\lambda\mu})\,\delta_{AB}$ of rule (ii), thus proving the theorem.

We begin by considering a general 'arc' defined by

$$\overline{\underbrace{a^\dagger a}\,\underbrace{a^\dagger a}\ldots\underbrace{a^\dagger_\gamma a_\delta}\ldots\underbrace{a^\dagger a}\,a^\dagger_\alpha a_\beta},\qquad (4A.2.9)$$

the final pair $a^\dagger_\alpha a_\beta$ not necessarily matching. If they do match, then the term (4 A. 2.9) becomes a loop. A typical schematic representation of (4 A. 2.9) is shown in Fig. 4 A. 2.2.

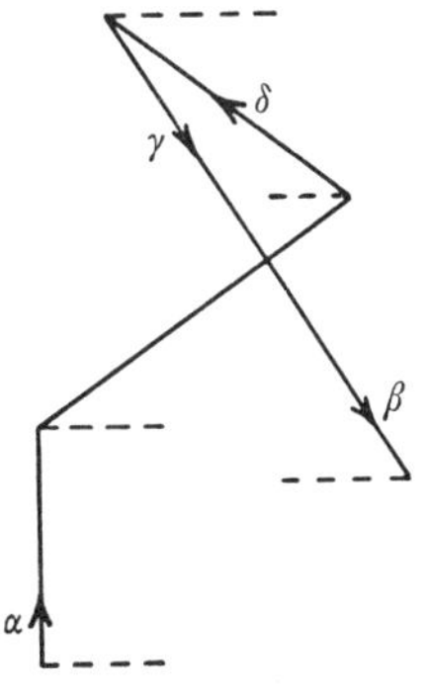

Fig. 4 A.2.2

Now let us suppose, as indicated in the Figure, that a_β is contracted with $a_\gamma^\dagger$, the partner of which in the matching is a_δ (see (4A.2.9)). Then, we may remove $\overbracket{a_\gamma^\dagger a_\beta}$ explicitly from (4A.2.9) by an even number of jumps and rewrite it as $-\overbracket{a_\beta a_\gamma^\dagger}$. Next, the minus sign thus obtained can be reabsorbed into the remainder by bringing a_δ, by an odd number of jumps, into the position to the right of $a_\alpha^\dagger$ previously occupied by a_β, the contraction partner of a_δ, of course, being kept the same. Thus, (4A.2.9) becomes

$$(\overbracket{\underbracket{a^\dagger a}\underbracket{a^\dagger a}\ldots\underbracket{a^\dagger a}\not{a}_\gamma^\dagger \not{a}_\delta \underbracket{a^\dagger a}\ldots\underbracket{a^\dagger a}a_\alpha^\dagger a_\delta})(\overbracket{a_\beta a_\gamma^\dagger}). \qquad (4\text{A}.2.10)$$

Graphically, what we have done is to remove the final arm from Fig. 4A.2.2. Now the first factor of (4A.2.10) simply defines another similar arc. Thus, the above process of extracting $\overbracket{aa^\dagger}$ terms can be repeated until we arrive at the form

$$\overbracket{a^\dagger a}\overbracket{aa^\dagger}\overbracket{aa^\dagger}\ldots\overbracket{aa^\dagger}\overbracket{a_\beta a_\gamma^\dagger}, \qquad (4\text{A}.2.11)$$

the first contraction being all that remains of the original arc (4A.2.9). On reversing the order and sign of this term, (4A.2.11) becomes identical in form with (4A.2.8) and the theorem is therefore proved.

APPENDIX 5A.1

SUMMATION OF RING DIAGRAMS

The third-order ring diagram shown in Fig. 5 A. 1.1 (i) contributes to the correlation energy an amount

$$E_R^{(3)} = \frac{\Omega}{2^3\pi^9}\int d\mathbf{q}\int d\mathbf{k}_1\int d\mathbf{k}_2\int d\mathbf{k}_3\left(\frac{e^2}{q^2}\right)^3\frac{m}{q^2+\mathbf{q}.(\mathbf{k}_1+\mathbf{k}_2)} \times\frac{m}{q^2+\mathbf{q}.(\mathbf{k}_1+\mathbf{k}_3)}.$$

Fig. 5 A.1.1

We can thus deduce that a typical nth-order ring diagram shown in Fig. 5 A. 1.1 (ii) contributes an amount

$$E_R^{(n)} = \sum_{\substack{\mathbf{k}_1\ldots\mathbf{k}_n\\ \mathbf{q}}}\left(\frac{4}{2}\frac{v(q)}{\Omega}\right)^n\frac{-m}{q^2+\mathbf{q}.(\mathbf{k}_1+\mathbf{k}_2)}\cdots\frac{-m}{q^2+\mathbf{q}.(\mathbf{k}_1+\mathbf{k}_n)}$$

$$= 2^n\left(\frac{\Omega}{8\pi^3}\right)^{n+1}(-m)^{n-1}\int d\mathbf{k}_1\ldots d\mathbf{k}_n\,d\mathbf{q}\left(\frac{v(q)}{\Omega}\right)^n\frac{1}{D_1+D_2} \times\frac{1}{D_1+D_3}\cdots\frac{1}{D_1+D_n}, \quad (5\text{A. }1.1)$$

where $$D_i(\mathbf{q},\mathbf{k}_i) = \frac{q^2}{2}+\mathbf{q}.\mathbf{k}_i.$$

Fig. 5 A.1.1 (ii) represents only one of the possible ring diagrams of order n—the one in which the hole state $\mathbf{k}_1$ is singled out. The contributions of all ring diagrams of order n will therefore be

$$E_{R,\text{ total}}^{(n)} = 2^n\left(\frac{\Omega}{(2\pi)^3}\right)^{n+1}(-m)^{n-1}\int d\mathbf{k}_1\ldots d\mathbf{k}_n\,d\mathbf{q}\left(\frac{v(q)}{\Omega}\right)^n$$

× (inverse product of energy denominators of (5 A. 1.1) plus sum of corresponding products arising from other diagrams). (5 A. 1.2)

In order to express the sum of integrals involving different products of energy denominators as a single integral over $\mathbf{q}$, we introduce a Feynman propagator $F(\mathbf{q}, t_i)$ defined by

$$F(\mathbf{q}, t_i) = \int d\mathbf{k}_i \, e^{-|t_i| D_i(\mathbf{q}, \mathbf{k}_i)}, \tag{5 A. 1.3}$$

and a further useful function $A_n(\mathbf{q})$ defined in terms of F as

$$A_n(\mathbf{q}) = \frac{1}{n}\int_{-\infty}^{\infty} dt_1 \dots dt_n \, F(\mathbf{q}, t_1) \dots F(\mathbf{q}, t_n) \, \delta(t_1 + t_2 + \dots + t_n). \tag{5 A. 1.4}$$

The following result can be proved by the method of induction

$$A_n(\mathbf{q}) = 2\int d\mathbf{k}_1 \dots d\mathbf{k}_n \, \{\Sigma \text{ energy denominators occurring in (5 A. 1.2)}\}. \tag{5 A. 1.5}$$

We shall not give the proof in the general case, but restrict ourselves to the case $n = 3$. From (5 A. 1.4)

$$A_3(\mathbf{q}) = \frac{1}{3}\int_{-\infty}^{\infty} dt_1 \, dt_2 \, dt_3 \, F(\mathbf{q}, t_1) \, F(\mathbf{q}, t_2) \, F(\mathbf{q}, t_3) \, \delta(t_1 + t_2 + t_3)$$

$$= \frac{1}{3}\int d\mathbf{k}_1 \, d\mathbf{k}_2 \, d\mathbf{k}_3 \int_{-\infty}^{\infty} dt_1 \, dt_2 \, e^{-D_1|t_1|} \, e^{-D_2|t_2|} \, e^{-D_3|t_1+t_2|}. \tag{5 A. 1.6}$$

The time integral occurring in (5 A. 1.6) can be written as twice the sum of the following three integrals:

$$\text{(i)} \quad \int_0^{\infty} dt_1 \int_0^{\infty} dt_2 \, e^{-D_1 t_1} \, e^{-D_2 t_2} \, e^{-D_3(t_1+t_2)} = \frac{1}{D_1 + D_3} \, \frac{1}{D_2 + D_3}; \tag{5 A. 1.7}$$

$$\text{(ii)} \quad \int_0^{\infty} dt_1 \int_{-t_1}^{0} dt_2 \, e^{-D_1 t_1} \, e^{D_2 t_2} \, e^{-D_3(t_1+t_2)} = \frac{1}{D_1 + D_3} \, \frac{1}{D_1 + D_2}; \tag{5 A. 1.8}$$

$$\text{(iii)} \quad \int_0^{\infty} dt_1 \int_{-\infty}^{-t_1} dt_2 \, e^{-D_1 t_1} \, e^{D_2 t_2} \, e^{D_3(t_1+t_2)} = \frac{1}{D_1 + D_2} \, \frac{1}{D_2 + D_3}. \tag{5 A. 1.9}$$

The energy denominators of (5 A. 1.7), (5 A. 1.8) and (5 A. 1.9) are those corresponding to the three ring diagrams of order 3 which are shown in Figs. 5 A. 1.1 (iii), 5 A. 1.1 (iv) and 5 A. 1.1 (v) respectively.

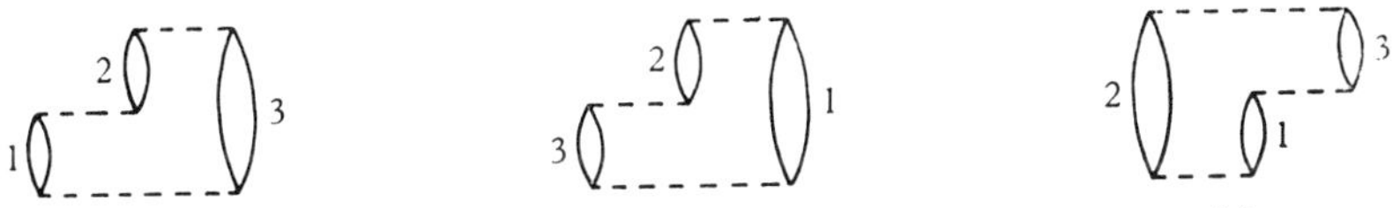

(iii) (iv) (v)

Fig. 5A.1.1 (*cont.*)

Thus, from (5 A. 1.6), we find

$$A_3(\mathbf{q}) = \frac{2}{3}\int d\mathbf{k}_1\, d\mathbf{k}_2\, d\mathbf{k}_3 \left\{\frac{1}{D_1+D_2}\frac{1}{D_1+D_3}+\frac{1}{D_1+D_2}\frac{1}{D_2+D_3}\right.$$

$$\left. +\frac{1}{D_1+D_3}\frac{1}{D_2+D_3}\right\}$$

$$= 2\int d\mathbf{k}_1\, d\mathbf{k}_2\, d\mathbf{k}_3 \frac{1}{(D_1+D_2)(D_1+D_3)}. \qquad (5\,\text{A. 1.10})$$

Using (5 A. 1.5) we may write (5 A. 1.2) as

$$E^{(n)}_{R,\,\text{total}} = 2^n \left(\frac{\Omega}{(2\pi)^3}\right)^{n+1} \int d\mathbf{q} \left(\frac{v(q)}{\Omega}\right)^n A_n(q)\,(-m)^{n-1}. \qquad (5\,\text{A. 1.11})$$

Since $$\delta(t_1+\ldots+t_n) = \frac{q}{2\pi}\int_{-\infty}^{\infty} du\, e^{iq(t_1+\ldots t_n)u}$$

we may write $$A_n = \frac{q}{2\pi n}\int_{-\infty}^{\infty} [Q_q(u)]^n\, du, \qquad (5\,\text{A. 1.12})$$

where

$$Q_q(u) = \int_{\substack{k<k_f \\ |\mathbf{k}+\mathbf{q}|>k_f}} d\mathbf{k} \int_{-\infty}^{\infty} dt \exp\left\{-|t|\left(\frac{q^2}{2}+\mathbf{q}.\mathbf{k}\right)+iqtu\right\}$$

$$= 2\pi\left[1+\frac{1}{2q}(1-\tfrac{1}{4}q^2+u^2)\ln\frac{(1+\frac{1}{2}q)^2+u^2}{(1-\frac{1}{2}q)^2+u^2}\right.$$

$$\left. -u\tan^{-1}\frac{1+\frac{1}{2}q}{u} - u\tan^{-1}\frac{1-\frac{1}{2}q}{u}\right]. \qquad (5\,\text{A. 1.13})$$

It is easy to show that

$$\operatorname*{Lt}_{q\to 0} Q_q(u) = Q_0(u) = 4\pi(1-u\tan^{-1}u) = 4\pi R. \qquad (5\,\text{A. 1.14})$$

The total contribution of all second- and higher-order ring diagrams to the correlation energy is therefore

$$\sum_{n=2}^{\infty} E_{R,\,\text{total}}^{(n)} = -\frac{\Omega}{(2\pi)^3}\frac{1}{m}\int d\mathbf{q}\,\frac{q}{2\pi}\int_{-\infty}^{\infty} du \sum_{n=2}^{\infty}\frac{(-1)^n}{2n}[Q_q(u)]^n\left[\frac{2v(q)\,m}{8\pi^3}\right]^n$$

or

$$\left(\frac{E_{\text{corr.}}\,2a_0}{e^2 N}\right) = \frac{3}{16\pi^6}\int_{-\infty}^{\infty} du\int\frac{d\mathbf{q}}{q^3}\left(\frac{\pi^2q^2}{\alpha r_s}\right)^2\left\{\ln\left(1+\frac{\alpha r_s\,Q_q(u)}{\pi^2q^2}\right) - \frac{\alpha r_s\,Q_q(u)}{\pi^2 q^2}\right\}, \quad (5\,\text{A}.\,1.15)$$

where the last integral is written in dimensionless form in which q is measured in units of k_f and $\alpha^3 = 4/9\pi$.

When the integration in (5 A. 1.15) is carried out using Q_0 for Q, we obtain the result quoted in the text.

APPENDIX 5A.2

TIME-DEPENDENT HARTREE–FOCK THEORY OF ELECTRON GAS

In this Appendix, we show how the main results on the screening of electrons in an electron gas can be derived from the time-dependent Hartree–Fock theory. We shall follow rather closely the treatment of Ehrenreich & Cohen (1959).

The equation of motion for the Dirac density matrix γ in the time-independent formulation of Chapter 1 is

$$H\gamma - \gamma H = 0. \quad (5\,\text{A}.\,2.1)$$

In the time-dependent theory, this is generalized to

$$i\hbar\frac{\partial\gamma}{\partial t} = H\gamma - \gamma H. \quad (5\,\text{A}.\,2.2)$$

We now consider a single-particle Hamiltonian in which the potential is a function of time, i.e.

$$H = T + V(\mathbf{r}, t), \quad (5\,\text{A}.\,2.3)$$

where T is the kinetic energy operator.

The procedure is to expand the Dirac matrix γ in the form

$$\gamma = \gamma_0 + \gamma_1 + \dots, \tag{5 A. 2.4}$$

and to write
$$V(\mathbf{r}, t) = \int V(\mathbf{k}, t)\, e^{i\mathbf{k}.\mathbf{r}}\, d\mathbf{k}. \tag{5 A. 2.5}$$

We substitute (5 A. 2.4) and (5 A. 2.5) into (5 A. 2.2), and linearize by neglecting terms involving products of γ_1 and V, since γ_1 is expected to be first order in V.

As in the time-independent theory of Chapter 5, section 5.2, the potential V consists of an unscreened potential V_0 plus the potential V_e due to the electronic screening cloud. This latter potential, as usual, is related to the change $\Delta\rho$ in the electron density through the Poisson equation
$$\nabla^2 V_e = -4\pi e^2 \Delta\rho. \tag{5 A. 2.6}$$

In terms of the density matrix $\Delta\rho$ is given by

$$\Delta\rho(\mathbf{r}) = \mathrm{Tr}\,(\delta(\mathbf{r}_e - \mathbf{r})\,\gamma_1). \tag{5 A. 2.7}$$

If we now calculate matrix elements of γ_1 between states $\mathbf{k}$ and $\mathbf{k}+\mathbf{q}$, we then find from (5 A. 2.2)

$$\begin{aligned} i\hbar \frac{\partial}{\partial t} \langle \mathbf{k}|\gamma_1|\mathbf{k}+\mathbf{q}\rangle & \\ &= \langle \mathbf{k}|[H_0, \gamma_1]|\mathbf{k}+\mathbf{q}\rangle + \langle \mathbf{k}|[V, \gamma_0]|\mathbf{k}+\mathbf{q}\rangle \\ &= (E_{\mathbf{k}} - E_{\mathbf{k}+\mathbf{q}})\,\langle \mathbf{k}|\gamma_1|\mathbf{k}+\mathbf{q}\rangle + [f_0(E_{\mathbf{k}+\mathbf{q}}) - f_0(E_{\mathbf{k}})]\,V(\mathbf{q}, t). \end{aligned} \tag{5 A. 2.8}$$

Here, we have introduced a slight generalization by retaining the Fermi function $f_0(E)$; clearly at absolute zero this is unity for E less than the Fermi energy, and zero otherwise. The Poisson equation (5 A. 2.6) gives us, as in section 5.2, a relation between $V_e(\mathbf{q}, t)$ and γ_1, which is explicitly

$$V_e(\mathbf{q}, t) = \frac{4\pi e^2}{q^2\Omega} \sum_{\mathbf{k}'} \langle \mathbf{k}'|\gamma_1|\,\mathbf{k}'+\mathbf{q}\rangle. \tag{5 A. 2.9}$$

Thus, substituting this into equation (5 A. 2.8) we obtain

$$\begin{aligned} i\hbar \frac{\partial}{\partial t} \langle \mathbf{k}|\gamma_1|\mathbf{k}+\mathbf{q}\rangle &= (E_{\mathbf{k}} - E_{\mathbf{k}+\mathbf{q}})\,\langle \mathbf{k}|\gamma_1|\,\mathbf{k}+\mathbf{q}\rangle \\ &\quad + \frac{4\pi e^2}{q^2\Omega}[f_0(E_{\mathbf{k}+\mathbf{q}}) - f_0(E_{\mathbf{k}})] \sum_{\mathbf{k}'} \langle \mathbf{k}'\,|\gamma_1|\,\mathbf{k}'+\mathbf{q}\rangle. \end{aligned} \tag{5 A. 2.10}$$

We can use this time-dependent self-consistent field method to make a rather simple calculation of the frequency and wave number-dependent dielectric constant $\epsilon(\mathbf{q}, \omega)$, by making an assumption about the time dependence of $V(\mathbf{q}, t)$. Thus, we suppose that the external potential $V_0(\mathbf{q}, t)$ acts on the system with time dependence $e^{\eta t} e^{i\omega t}$, where $\eta \to 0$, as we have seen, corresponds to an adiabatic switching on of the perturbation. Obviously, the electron distribution is modified, and this is most readily described by the polarization $P(\mathbf{q}, t)$, related to the electric field $E(\mathbf{q}, t)$ and the dielectric constant $\epsilon(\mathbf{q}, \omega)$ by

$$P(\mathbf{q}, t) = \frac{1}{4\pi}[\epsilon(\mathbf{q}, \omega) - 1]\, E(\mathbf{q}, t), \qquad (5\,\text{A}.\,2.11)$$

the polarization $P(\mathbf{q}, t)$ is related to the change in electron density by

$$\operatorname{div} \mathbf{P} = e\Delta\rho \qquad (5\,\text{A}.\,2.12)$$

or

$$-iqP(\mathbf{q}, t) = e\Delta\rho(\mathbf{q}, t). \qquad (5\,\text{A}.\,2.13)$$

The electric field $E(\mathbf{q}, t)$ is given by

$$eE(\mathbf{q}, t) = -iqV(\mathbf{q}, t). \qquad (5\,\text{A}.\,2.14)$$

Equation (5 A. 2.8) is easily solved for $\langle \mathbf{k}|\gamma_1|\mathbf{k}+\mathbf{q}\rangle$ by assuming that $\langle \mathbf{k}|\gamma_1|\mathbf{k}+\mathbf{q}\rangle$ and $V_e(q, t)$ have the same time dependence as $V_0(q, t)$. The change in electron density $\Delta\rho(\mathbf{q}, t)$ may then be obtained from (5 A. 2.7), and the dielectric constant deduced from (5 A. 2.12)–(5 A. 2.14). The result is

$$\epsilon(\mathbf{q}, \omega) = 1 - \lim_{\eta \to 0} \frac{4\pi e^2}{q^2 \Omega} \sum_{\mathbf{k}} \frac{f_0(E_{\mathbf{k}+\mathbf{q}}) - f_0(E_{\mathbf{k}})}{E_{\mathbf{k}+\mathbf{q}} - E_{\mathbf{k}} - \hbar\omega + i\hbar\eta}, \qquad (5\,\text{A}.\,2.15)$$

as was first obtained by Lindhard (1954). We readily can show that, when $\omega = 0$, we regain the result (5.7.1) of the Hartree time-independent calculation.

APPENDIX 5A.3

COMMUTATOR OF SAWADA HAMILTONIAN AND $d_{\mathbf{q}}^{\dagger}(\mathbf{k}\sigma)$

To evaluate $[H_s, d_{\mathbf{q}}^{\dagger}(\mathbf{k}\sigma)]$, we note first that the d's and $d^{\dagger}$'s obey the following commutation rules (see especially Brout & Carruthers, 1963);

$$[d_{\mathbf{q}}(\mathbf{k}\sigma), d_{\mathbf{q}'}^{\dagger}(\mathbf{k}'\sigma')] = \delta_{\sigma\sigma'}(\delta_{\mathbf{k}+\mathbf{q}\mathbf{k}'+\mathbf{q}'}\, a_{\mathbf{k}\sigma}^{\dagger}\, a_{\mathbf{k}'\sigma} - \delta_{\mathbf{k}\mathbf{k}'}\, a_{\mathbf{k}'+\mathbf{q}'\sigma}^{\dagger}\, a_{\mathbf{k}+\mathbf{q}\sigma}), \tag{5 A. 3.1}$$

$$[d_{\mathbf{q}}(\mathbf{k}\sigma), d_{\mathbf{q}'}(\mathbf{k}'\sigma')] = \delta_{\sigma\sigma'}(\delta_{\mathbf{k}+\mathbf{q}\mathbf{k}'}\, a_{\mathbf{k}\sigma}^{\dagger}\, a_{\mathbf{k}'+\mathbf{q}'\sigma} - \delta_{\mathbf{k}'+\mathbf{q}'\mathbf{k}}\, a_{\mathbf{k}'\sigma}^{\dagger}\, a_{\mathbf{k}+\mathbf{q}\sigma}). \tag{5 A. 3.2}$$

Now we assert that for our purposes the latter may be replaced by Boson commutation rules, namely,

$$[d_{\mathbf{q}}(\mathbf{k}\sigma), d_{\mathbf{q}'}^{\dagger}(\mathbf{k}'\sigma')] = \delta_{\mathbf{q}\mathbf{q}'}\,\delta_{\mathbf{k}\mathbf{k}'}\,\delta_{\sigma\sigma'}, \tag{5 A. 3.3}$$

$$[d_{\mathbf{q}}(\mathbf{k}\sigma), d_{\mathbf{q}'}(\mathbf{k}'\sigma')] = 0. \tag{5 A. 3.4}$$

If we assume these rules for the moment, we have, using (5.4.5)

$$\begin{aligned}
[V_s, d_{\mathbf{Q}}^{\dagger}(\mathbf{K}\beta)] &= \frac{\lambda}{2N}\sum_{\mathbf{q}} v(\mathbf{q}) \sum_{\substack{\mathbf{k},\,\mathbf{k}'\\ \sigma,\,\sigma'}} [A_{\mathbf{q}}(\mathbf{k}\sigma)\, A_{\mathbf{q}}^{\dagger}(\mathbf{k}'\sigma'), d_{\mathbf{Q}}^{\dagger}(\mathbf{K}\beta)] \\
&= \frac{\lambda}{2N}\sum_{\mathbf{q}} v(\mathbf{q}) \sum_{\substack{\mathbf{k},\,\mathbf{k}'\\ \sigma,\,\sigma'}} \{[A_{\mathbf{q}}(\mathbf{k}\sigma), d_{\mathbf{Q}}^{\dagger}(\mathbf{K}\beta)]\, A_{\mathbf{q}}^{\dagger}(\mathbf{k}'\sigma') \\
&\qquad + A_{\mathbf{q}}(\mathbf{k},\sigma)\,[A_{\mathbf{q}}^{\dagger}(\mathbf{k}'\sigma'), d_{\mathbf{Q}}^{\dagger}(\mathbf{K}\beta)]\},
\end{aligned} \tag{5 A. 3.5}$$

where, for short, we have written

$$A_{\mathbf{q}}(\mathbf{k}\sigma) = d_{\mathbf{q}}(\mathbf{k}\sigma) + d_{-\mathbf{q}}^{\dagger}(-\mathbf{k}\sigma). \tag{5 A. 3.6}$$

At this stage we use the rules (5 A. 3.3) and (5 A. 3.4) to evaluate the commutators in (5 A. 3.5). They are respectively

$$\left.\begin{aligned}
[A_{\mathbf{q}}(\mathbf{k}\sigma), d_{\mathbf{Q}}^{\dagger}(\mathbf{K}\beta)] &= \delta_{\mathbf{q}\mathbf{Q}}\,\delta_{\mathbf{k}\mathbf{K}}\,\delta_{\sigma\beta}, \\
[A_{\mathbf{q}}^{\dagger}(\mathbf{k}\sigma), d_{\mathbf{Q}}^{\dagger}(\mathbf{K}\beta)] &= \delta_{-\mathbf{q}\mathbf{Q}}\,\delta_{-\mathbf{k}\mathbf{K}}\,\delta_{\sigma\beta};
\end{aligned}\right\} \tag{5 A. 3.7}$$

and so (5 A. 3.5) becomes

$$[V_s, d_{\mathbf{Q}}^{\dagger}(\mathbf{K})] = \frac{\lambda}{N} v(\mathbf{Q}) \sum_{\mathbf{k}'\sigma'} A_{\mathbf{Q}}^{\dagger}(\mathbf{k}'\sigma'). \tag{5 A. 3.8}$$

To justify the use of (5 A. 3.3) and (5 A. 3.4) instead of (5 A. 3.1) and (5 A. 3.2), one must trace through in the above analysis the consequence of using the exact relations. Nothing happens until (5 A. 3.7) is reached. These equations should be replaced by

$$[A_{\mathbf{q}}(\mathbf{k}\sigma), d^{\dagger}_{\mathbf{Q}}(\mathbf{K})] = [d_{\mathbf{q}}(\mathbf{k}\sigma), d^{\dagger}_{\mathbf{Q}}(\mathbf{K})] + [d^{\dagger}_{-\mathbf{q}}(-\mathbf{k}\sigma), d^{\dagger}_{\mathbf{Q}}(\mathbf{K})] \tag{5 A. 3.9}$$

and

$$[A^{\dagger}_{\mathbf{q}}(\mathbf{k}'\sigma'), d^{\dagger}_{\mathbf{Q}}(\mathbf{K})] = [d^{\dagger}_{\mathbf{q}}(\mathbf{k}'\sigma'), d^{\dagger}_{\mathbf{Q}}(\mathbf{K})] + [d_{-\mathbf{q}}(-\mathbf{k}\sigma), d^{\dagger}_{\mathbf{Q}}(\mathbf{K})], \tag{5 A. 3.10}$$

where the commutators on the right are given by (5 A. 3.1) and (5 A. 3.2). Let us now put (5 A. 3.9) and (5 A. 3.10) into (5 A. 3.5) and remember, from our initial discussion of this method, we need only consider the effect of (5 A. 3.5) acting on Ψ, the exact ground state of the system. Approximating Ψ for small r_s, by Ψ_0, the independent particle wave function, (5 A. 3.10) acting on Ψ differs from (5 A. 3.7) acting on Ψ only by terms which vanish as $r_s \to 0$ (remembering that momentum indices written as sums refer to particles and those written singly refer to holes). Similarly, when acting on Ψ, (5 A. 3.9) can be replaced by (5 A. 3.7). This justifies the use of Boson commutation rules for the second bracket in (5 A. 3.5) if we restrict our interest to leading-order terms. Turning to the first bracket in (5 A. 3.5) the same sort of reasoning can be applied. The effect of the $A^{\dagger}_{\mathbf{q}}(\mathbf{k}'\sigma')$ operator modifying Ψ has only the effect of introducing a few terms of order $(1/N)$ into the difference between the effects of the two sets of commutation rules, and these are, of course, negligible.

Combining (5.4.13) and (5 A. 3.8), then, we have

$$[H_s, d^{\dagger}_{\mathbf{q}}(\mathbf{k}\sigma)] = \omega_{\mathbf{q}}(\mathbf{k}\sigma)\, d^{\dagger}_{\mathbf{q}}(\mathbf{k}\sigma) + (\lambda/N)\, v(\mathbf{q}) \sum_{\mathbf{k}'\sigma'} \{d_{\mathbf{q}}(\mathbf{k}'\sigma') + d^{\dagger}_{-\mathbf{q}}(-\mathbf{k}'\sigma')\}, \tag{5 A. 3.11}$$

and, taking conjugates,

$$[H_s, d_{\mathbf{q}}(\mathbf{k}\sigma)] = -\omega_{\mathbf{q}}(\mathbf{k}\sigma)\, d_{\mathbf{q}}(\mathbf{k}\sigma) - (\lambda/N)\, v(\mathbf{q}) \sum_{\mathbf{k}'\sigma'} \{d^{\dagger}_{\mathbf{q}}(\mathbf{k}'\sigma') + d_{-\mathbf{q}}(-\mathbf{k}'\sigma')\} \tag{5 A. 3.12}$$

giving us the desired results.

APPENDIX 5 A.4

QUASI-PARTICLE EXCITATION ENERGY AT HIGH DENSITIES

We show here how the quasi-particle excitation energy may be evaluated at the Fermi limit in the random-phase approximation (Gell-Mann, 1957; see also Silverstein, 1962; Rice, 1965).

Following Gell-Mann we consider those states of the electron gas in which only a small number of particles ν are in excited states. Then, the energy E may be written

$$E = E_0 + \sum_{j=1}^{\nu} \{\epsilon(k_j) - \epsilon(p_j)\} + O(\nu/N). \qquad (5\text{A}.4.1)$$

For free electrons

$$\epsilon(p_f) = \frac{p_f^2}{\alpha^2 r_s^2} \quad : \quad \alpha = \left(\frac{4}{9\pi}\right)^{\frac{1}{3}} \qquad (5\text{A}.4.2)$$

and

$$\left(\frac{d\epsilon}{dp}\right)_{p=p_f} = \frac{2}{\alpha^2 r_s^2}. \qquad (5\text{A}.4.3)$$

We must therefore calculate $(d\epsilon/dp)_{p_f}$ at high density for interacting electrons. $\epsilon(p)$ is, in fact, the energy lost when a particle, with the system in its ground state, is then annihilated. Clearly $p \leqslant 1$ here. Similarly, starting again from the ground state and creating an electron of momentum $\mathbf{k}\,(k \geqslant 1)$, then $\epsilon(k)$ is the energy gain.

To see how to calculate these quantities, let us consider a second-order term in the ground-state energy, namely

$$-\frac{1}{\pi^4}\int \frac{d\mathbf{q}}{q^4} \sum_{\substack{p_1<1 \\ |\mathbf{p}_1+\mathbf{q}|>1}} \int_{\substack{p_2<1 \\ |\mathbf{p}_2+\mathbf{q}|>1}} d\mathbf{p}_2 \, \frac{1}{q^2 + \mathbf{q}\cdot(\mathbf{p}_1+\mathbf{p}_2)}. \qquad (5\text{A}.4.4)$$

Now remove a particle with spin up and momentum $\mathbf{p}$. There are four contributions to the loss of energy:

(*a*) The one-particle state with momentum $\mathbf{p}$ and spin is no longer occupied and must be omitted from the sum.

(*b*) For one of the spin states, a term $\mathbf{p}_1 + \mathbf{q} = \mathbf{p}$ is added to the sum.

(*c*) and (*d*). Corresponding contributions from the sum or integral over $\mathbf{p}_2$ rather than $\mathbf{p}_1$. These are equal to (*a*) and (*b*).

Thus, the contribution of this term to the energy lost by annihilating a particle is

$$-\frac{1}{\pi^4}\left\{\int_{|\mathbf{p}+\mathbf{q}|>1}\frac{d\mathbf{q}}{q^4}\int_{\substack{p_2<1\\ \mathbf{p}_2+\mathbf{q}>1}} d\mathbf{p}_2\,\frac{1}{q^2+\mathbf{q}\cdot(\mathbf{p}+\mathbf{p}_2)}-\int_{|\mathbf{p}-\mathbf{q}|<1}\frac{d\mathbf{q}}{q^4}\right.$$
$$\left.\times\int_{\substack{p_2<1\\ |\mathbf{p}_2+\mathbf{q}|>1}} d\mathbf{p}_2\,\frac{1}{\mathbf{q}\cdot(\mathbf{p}+\mathbf{p}_2)}\right\}. \quad (5\text{ A. }4.5)$$

We can consider similarly the contribution of each term in the ground-state energy. We deal now specifically with $(d\epsilon/dp)_{p_f}$, and we find for the exchange energy

$$\epsilon(p) = -\frac{1}{\pi^2\alpha r_s}\int_{p_2<1} d\mathbf{p}_2\,\frac{1}{(\mathbf{p}+\mathbf{p}_2)^2}. \quad (5\text{ A. }4.6)$$

We obtain

$$\left(\frac{d\epsilon}{dp}\right)_{p_f} = \frac{2}{\pi\alpha r_s}\int_{-1}^{1}\frac{x\,dx}{2(1-x)} \equiv \frac{1}{\pi\alpha r_s}\int_0^2 dq\left(\frac{1-q}{q}\right), \quad (5\text{A. }4.7)$$

which is logarithmically divergent, as stressed initially by Bardeen (1936).

The term in (5 A. 4.5), on differentiation, gives a contribution to $(d\epsilon/dp)_{p_f}$ of

$$-\frac{8}{\pi^2}\int_{-1}^{1}\frac{x\,dx}{[2(1-x)]^2} \quad (5\text{ A. }4.8)$$

and collecting together the results (5 A. 4.3), (5 A. 4.7) and (5 A.4.8) we have

$$\left(\frac{d\epsilon}{dp}\right)_{p_f} = \frac{2}{\alpha^2 r_s^2}+\frac{2}{\pi\alpha r_s}\int_{-1}^{1}\frac{x\,dx}{2(1-x)}\left\{1-\frac{4\alpha r_s}{\pi}\frac{1}{2(1-x)}+\dots\right\}.$$
$$(5\text{ A. }4.9)$$

Following the arguments used in calculating the ground-state energy, it can be shown that the quantity in the { } of (5 A. 4.9) is simply the first two terms in the expansion of

$$\left[1+\frac{4\alpha r_s}{\pi}\frac{1}{2(1-x)}\right]^{-1}. \quad (5\text{ A. }4.10)$$

Thus

$$\left(\frac{d\epsilon}{dp}\right)_{pf} = \frac{2}{\alpha^2 r_s^2} + \frac{1}{\pi\alpha r_s}\int_0^2 dq \left(\frac{1-q}{q}\right)\frac{1}{[1+(2\alpha r_s/\pi q)]}$$

$$= \frac{2}{\alpha^2 r_s^2} + \frac{1}{\pi\alpha r_s}\left[-2+\left(\frac{1+2\alpha r_s}{\pi}\right)\ln\left(\frac{(2\alpha r_s/\pi)+2}{2\alpha r_s/\pi}\right)\right], \tag{5 A. 4.11}$$

which is the desired result.

Thus, at low temperatures, we obtain the specific heat

$$\frac{C_v}{C_v^0} = \left[1+\frac{\alpha r_s}{2\pi}\left\{-\ln r_s + \ln\left(\frac{\pi}{\alpha}\right) - 2\right\} + \dots\right]. \tag{5 A. 4.12}$$

Numerically this is

$$\frac{C_v}{C_v^0} = [1+0{\cdot}083 r_s(-\ln r_s - 0{\cdot}203)+\dots]^{-1}. \tag{5 A. 4.13}$$

The term in $r_s \ln r_s$ was first given by Pines (1955). His method was not adequate for the calculation of the constant term.

APPENDIX 7A.1

SOME MATHEMATICS ASSOCIATED WITH THE LAWS OF SIMILARITY

Evaluation of critical temperature

We must consider (7.6.22), namely

$$\frac{1}{\rho V} = \int_0^{\delta} \frac{d\epsilon}{\epsilon} \tanh \tfrac{1}{2}\beta_c \epsilon \equiv \int_0^{\beta_c \delta} \frac{dx}{x} \tanh \tfrac{1}{2}x \equiv F(\beta_c \delta)$$

and solve for $\beta_c \delta$, supposing this quantity to be large.

If we show that

$$\exp\left[\int_0^{\beta_c \delta} \frac{dx}{x} \tanh \tfrac{1}{2}x\right] = 1{\cdot}14\beta_c \delta \quad (\text{large } \beta_c \delta)$$

then the required result $e^{1/\rho V} = 1{\cdot}14\beta_c \delta$ follows immediately.

It is therefore necessary to show that

$$\lim_{X\to\infty} \frac{\exp\left[\int_0^X \frac{dx}{x} \tanh \tfrac{1}{2}x\right]}{X} = 1{\cdot}14. \qquad (7\text{A}.1.1)$$

Proof. The left-hand side of (7 A. 1.1) is evidently equal to

$$\lim_{X\to\infty} \frac{\exp\left[\int_0^X \frac{dx}{x} \tanh \tfrac{1}{2}x\right]}{\exp\left[\int_1^X \frac{dx}{x}\right]}$$

$$= \lim_{X\to\infty} \left[\exp\left\{\int_0^1 \frac{dx}{x} \tanh \tfrac{1}{2}x + \int_1^X \frac{dx}{x} (\tanh \tfrac{1}{2}x - 1)\right\}\right]$$

$$= \exp\left[\int_0^1 \frac{dx}{x} \tanh \tfrac{1}{2}x\right] \exp\left[\int_1^{\infty} \frac{dx}{x} (\tanh \tfrac{1}{2}x - 1)\right]$$

$$= \text{finite number} = 1{\cdot}14.$$

Proof that $\Delta(\beta)/\Delta(\infty)$ against β_c/β yields an absolute curve. We first recall that

$$\Delta(\infty) = 2\delta e^{-1/\rho V}, \quad \frac{1}{\beta_c} = 1{\cdot}14\delta e^{-1/\rho V}.$$

Hence it follows that $\beta_c^2 = 3{\cdot}08/(\Delta(\infty))^2$. We must investigate the relationship between β and Δ through

$$\frac{1}{\rho V} = \int_0^{\delta} \frac{d\epsilon}{\sqrt{[\epsilon^2 + \Delta^2]}} \tanh \{\tfrac{1}{2}\beta \sqrt{[\epsilon^2 + \Delta^2]}\}.$$

We have

$$\frac{1}{\rho V}=\int_0^{\delta}\frac{\beta_c\,d\epsilon}{\sqrt{[\beta_c^2\epsilon^2+\beta_c^2\Delta^2]}}\tanh\frac{1}{2}\frac{\beta}{\beta_c}\sqrt{\{[\beta_c^2\epsilon^2+\beta_c^2\Delta^2]\}}$$

$$=\int_0^{\beta_c\delta}\frac{dx}{\sqrt{\left[x^2+3\cdot 08\left(\frac{\Delta(\beta)}{\Delta(\infty)}\right)^2\right]}}\tanh\left\{\frac{1}{2}\frac{\beta}{\beta_c}\sqrt{\left[x^2+3\cdot 08\left(\frac{\Delta(\beta)}{\Delta(\infty)}\right)^2\right]}\right\}.$$

Now let $\beta_c\delta\to\infty$, keeping β/β_c and $\Delta(\beta)/\Delta(\infty)$ fixed. The situation is analogous to that of the previous section, and we have

$$e^{1/\rho V}=\exp\left[\int_0^{\beta_c\delta}\frac{dx}{\sqrt{\left[x^2+3\cdot 08\left(\frac{\Delta(\beta)}{\Delta(\infty)}\right)^2\right]}}\tanh\left\{\frac{1}{2}\frac{\beta}{\beta_c}\times\sqrt{\left[x^2+3\cdot 08\left(\frac{\Delta(\beta)}{\Delta(\infty)}\right)^2\right]}\right\}\right]$$

$$\sim\beta_c\delta\exp\left[\int_0^{1}\frac{dx}{\sqrt{\left[x^2+3\cdot 08\left(\frac{\Delta(\beta)}{\Delta(\infty)}\right)^2\right]}}\tanh\left\{\frac{1}{2}\frac{\beta}{\beta_c}\times\sqrt{\left[x^2+3\cdot 08\left(\frac{\Delta(\beta)}{\Delta(\infty)}\right)^2\right]}\right\}\right]\exp\left[\int_1^{\infty}\frac{dx}{\sqrt{\left[x^2+3\cdot 08\left(\frac{\Delta(\beta)}{\Delta(\infty)}\right)^2\right]}}\times\left(\tanh\left\{\frac{1}{2}\frac{\beta}{\beta_c}\sqrt{\left[x^2+3\cdot 08\left(\frac{\Delta(\beta)}{\Delta(\infty)}\right)^2\right]}\right\}-1\right)\right].$$

If we replace $e^{1/\rho V}$ by $1\cdot 14\beta_c\delta$, we find

$$\ln 1\cdot 14=\int_0^{1}\frac{dx}{\sqrt{\left[x^2+3\cdot 08\left(\frac{\Delta(\beta)}{\Delta(\infty)}\right)^2\right]}}\tanh\left\{\frac{1}{2}\frac{\beta}{\beta_c}\sqrt{\left[x^2+3\cdot 08\left(\frac{\Delta(\beta)}{\Delta(\infty)}\right)^2\right]}\right\}+\int_1^{\infty}\frac{dx}{\sqrt{\left[x^2+3\cdot 08\left(\frac{\Delta(\beta)}{\Delta(\infty)}\right)^2\right]}}\left(\tanh\left\{\frac{1}{2}\frac{\beta}{\beta_c}\times\sqrt{\left[x^2+3\cdot 08\left(\frac{\Delta(\beta)}{\Delta(\infty)}\right)^2\right]}\right\}-1\right),$$

which shows immediately that $\Delta(\beta)/\Delta(\infty)$ is a function of β_c/β only. This is the desired result.

APPENDIX 7A.2

FRÖHLICH'S HAMILTONIAN FOR ELECTRON-PHONON COUPLING

We start from a Hamiltonian of the form

$$H' = H_{\text{el.}} + H_f + H_{\text{int.}}, \tag{7A.2.1}$$

where the three terms represent the Hamiltonian of the electrons, that of the field and that of the interaction respectively.

If $\mathbf{P}(\mathbf{r}, t)$ is a longitudinal displacement of the lattice, we may assume

$$H_f = \frac{1}{2}\int (M\dot{P}^2 + Ms'^2(\operatorname{div}\mathbf{P})^2)\,n\,d\mathbf{r}, \tag{7A.2.2}$$

with curl $\mathbf{P} = 0$, where M is the mass of an ion of the lattice, n is the number of ions per unit volume and s' is the velocity of sound when the electron-lattice interaction is switched off.

Following Fröhlich, we now introduce a complex function $\mathbf{B}$ related to $\mathbf{P}$ through

$$\operatorname{div}\mathbf{B} = \operatorname{div}\mathbf{P} + \frac{i\dot{P}}{s'}, \quad \operatorname{curl}\mathbf{B} = 0. \tag{7A.2.3}$$

As in the usual methods of field theory (see, for example, Heitler, 1954) we now develop $\mathbf{B}$ in plane waves, and write, with V the total volume,

$$\mathbf{B} = \frac{\mathbf{w}}{w}\sum_{\mathbf{w}}\left(\frac{2\hbar}{nVMws'}\right)^{\frac{1}{2}} b_{\mathbf{w}}\,e^{i\mathbf{w}.\mathbf{r}} \tag{7A.2.4}$$

from which one finds

$$H_f = \tfrac{1}{2}nMs'^2\int |\operatorname{div}\mathbf{B}|^2\,d\mathbf{r}$$

$$= \frac{1}{2}\sum \hbar ws'(b^{\dagger}_{\mathbf{w}}b_{\mathbf{w}} + b_{\mathbf{w}}b^{\dagger}_{\mathbf{w}}). \tag{7A.2.5}$$

Now, using the rules appropriate for Bosons, we have

$$[b_{\mathbf{w}}, b_{\mathbf{v}}] = b_{\mathbf{w}}b_{\mathbf{v}} - b_{\mathbf{v}}b_{\mathbf{w}} = 0; \quad [(b_{\mathbf{w}}, b^{\dagger}_{\mathbf{v}})] = \delta_{\mathbf{v},\mathbf{w}}. \tag{7A.2.6}$$

As we have seen on many occasions, the electronic part may be written:

$$H_{\text{el.}} = \sum_{\mathbf{k}} \epsilon_k\, a^{\dagger}_{\mathbf{k}}\, a_{\mathbf{k}}, \tag{7A.2.7}$$

with $\epsilon_k = \hbar^2 k^2/2m$, while the interaction Hamiltonian is given by

$$H_{\text{int.}} = C' \int \psi^\dagger \psi \,\text{div}\,\mathbf{P}\,d\mathbf{r}$$
$$= i \sum_{\mathbf{wk}} D_w (b_{\mathbf{w}} a_{\mathbf{k}}^\dagger a_{\mathbf{k}-\mathbf{w}} - b_{\mathbf{w}}^\dagger a_{\mathbf{k}-\mathbf{w}}^\dagger a_{\mathbf{k}}). \qquad (7\text{A}.2.8)$$

Thus, the total Hamiltonian H' becomes

$$H' = \sum_{\mathbf{k}} \epsilon_{\mathbf{k}} a_{\mathbf{k}}^\dagger a_{\mathbf{k}} + \sum_{\mathbf{w}} \hbar w s' (b_{\mathbf{w}}^\dagger b_{\mathbf{w}} + \tfrac{1}{2}) + i \sum_{\mathbf{w}} D_w (b_{\mathbf{w}} \rho_{\mathbf{w}}^\dagger - b_{\mathbf{w}}^\dagger \rho_{\mathbf{w}}), \qquad (7\text{A}.2.9)$$

where $\rho_{\mathbf{w}}$ represents a Fourier coefficient of the electronic density operator, and is given by

$$\rho_{\mathbf{w}} = \sum_{\mathbf{k}} a_{\mathbf{k}-\mathbf{w}}^\dagger a_{\mathbf{k}} = \rho_{-\mathbf{w}}^\dagger. \qquad (7\text{A}.2.10)$$

Canonical transformation of H'

To proceed, we seek a canonical transformation which removes the interaction term as completely as possible.

Fröhlich employs a unitary operator e^S, having the form

$$S = \sum_{\mathbf{w}} S_{\mathbf{w}}, \quad S_{\mathbf{w}} = -\gamma_{\mathbf{w}} b_{\mathbf{w}} + \gamma_{\mathbf{w}}^\dagger b_{\mathbf{w}}^\dagger = -S_{\mathbf{w}}^\dagger, \qquad (7\text{A}.2.11)$$

where

$$\gamma_{\mathbf{w}} = \sum_{\mathbf{k}} \phi(\mathbf{k},\mathbf{w})\, a_{\mathbf{k}}^\dagger a_{\mathbf{k}-\mathbf{w}}, \quad \gamma_{\mathbf{w}}^\dagger = \sum_{\mathbf{k}} \phi^*(\mathbf{k},\mathbf{w})\, a_{\mathbf{k}-\mathbf{w}}^\dagger a_{\mathbf{k}}. \qquad (7\text{A}.2.12)$$

Here, $\phi(\mathbf{k},\mathbf{w})$ is to be chosen so that the interaction term in the transformed Hamiltonian

$$H = e^{S\dagger} H' e^S \qquad (7\text{A}.2.13)$$

is as small as possible. The choice adopted by Fröhlich is

$$\phi(\mathbf{k},\mathbf{w}) = -\frac{-iD_w}{\epsilon_{\mathbf{k}-\mathbf{w}} - \epsilon_{\mathbf{k}} + \hbar w s}(1 - \Delta(\mathbf{k},\mathbf{w})), \qquad (7\text{A}.2.14)$$

where

$$\Delta(\mathbf{k},\mathbf{w}) \equiv \Delta^2(\mathbf{k},\mathbf{w}) = \begin{cases} 1, & \text{if } (\epsilon_{\mathbf{k}-\mathbf{w}} - \epsilon_{\mathbf{k}} + \hbar w s)^2 < \Gamma_w^2, \\ 0 & \text{if } (\epsilon_{\mathbf{k}-\mathbf{w}} - \epsilon_{\mathbf{k}} + \hbar w s)^2 > \Gamma_w^2, \end{cases} \qquad (7\text{A}.2.15)$$

and Γ_w has to be chosen to prevent divergencies. The final result turns out to be insensitive to the choice of the energy Γ_w. Here s is the renormalized velocity of sound.

The essential result which then emerges is that the transformed Hamiltonian may be written as

$$H = \sum_{\mathbf{k}} \epsilon_k a_{\mathbf{k}}^\dagger a_{\mathbf{k}} + \sum_{\mathbf{w}} \hbar w s b_{\mathbf{w}}^\dagger b_{\mathbf{w}} + \sum_{w} \tfrac{1}{2}\hbar w s'$$
$$+ i \sum_{\mathbf{w},\mathbf{k}} D_w (b_{\mathbf{w}} a_{\mathbf{k}}^\dagger a_{\mathbf{k}-\mathbf{w}} - b_w^\dagger a_{\mathbf{k}-\mathbf{w}}^\dagger a_{\mathbf{k}}) \Delta(\mathbf{k}, \mathbf{w}) + H_s, \quad (7\text{A}.2.16)$$

where

$$H_s = -\frac{1}{2} \sum_{\mathbf{w},\mathbf{k},} \frac{D_w^2 (1 + \Delta(\mathbf{k},\mathbf{w}))(1 - \Delta(\mathbf{q},\mathbf{w}))}{\epsilon_{\mathbf{q}-\mathbf{w}} - \epsilon_{\mathbf{q}} + \hbar w s} (a_{\mathbf{k}}^\dagger a_{\mathbf{k}-\mathbf{w}} a_{\mathbf{q}-\mathbf{w}}^\dagger a_{\mathbf{q}}$$
$$+ \text{complex conjugate}). \quad (7\text{A}.2.17)$$

We note that, while in (7 A. 2.16), the physical interpretation of $a_{\mathbf{k}}$ and $b_{\mathbf{w}}$ is different from that in the original Hamiltonian, since they describe respectively electrons plus lattice deformation and ionic oscillations plus electronic density oscillation, they satisfy the same commutation rules as before.

The last term H_s is the essential part of the Hamiltonian for the theory of superconductivity. It represents an effective interaction between 'electrons', via the medium of the lattice. It is this interaction which has been simulated in the treatment of Chapter 7.

APPENDIX 10A

DETAILED GRAPHICAL ANALYSIS OF SINGLE-PARTICLE GREEN FUNCTION

Let us write the numerator of (10.4.18) quite explicitly as

$$\mathfrak{N} = i\sum_0^\infty \frac{(-1)^n}{n!}\frac{1}{2^n}\int_{-\infty}^{\infty} dt_{2n-12n}\cdots\int_{-\infty}^{\infty} dt_{12} \sum_{\substack{KLMN\\ \cdots\cdots\\ XYZT}} \langle KL|v|NM\rangle$$

$$\times \langle PQ|v|SR\rangle\ldots\langle XY|v|TZ\rangle\langle g|T\{a_K^\dagger(t_{12})\,a_L^\dagger(t_{12})\,a_M(t_{12})$$

$$\times a_N(t_{12})\ldots a_T(t_{2n-1\,2n})\,a_{\mathbf{k}}(t_2)\,a_{\mathbf{k}}^\dagger(t_1)\}|g\rangle, \qquad (10\,\text{A}.\,1)$$

where now, of course, t_1, t_2 and t_{12} are independent times no longer related by the convention $t_1 = t_2 = t_{12}$ of Chapter 4. Our attention thus turns to the expression

$$\langle g|T\{a_K^\dagger(t_{12})\,a_L^\dagger(t_{12})\,a_M(t_{12})\,a_N(t_{12})\,a_P^\dagger(t_{34})\,a_Q^\dagger(t_{34})\ldots$$

$$\times \ldots a_X^\dagger(t_{2n-12n})\,a_Y^\dagger(t_{2n-12n})\,a_Z(t_{2n-12n})\,a_T(t_{2n-12n})$$

$$\times a_{\mathbf{k}}(t_2)\,a_{\mathbf{k}}^\dagger(t_1)\}|g\rangle. \quad (10\,\text{A}.\,2)$$

We saw that as a special case of Wick's theorem, the expectation value of any T-product is given by the sum of all fully contracted terms. This result will now be used to study (10 A. 2) systematically in all orders. In this connexion (4.13.16) of Chapter 4 should be recalled. Combining this with (4.13.5) we then obtain

$$\overline{a_K(t)\,a_L^\dagger(t')} = -iG_0(K, t-t')\,\delta_{KL} = -\overline{a_L^\dagger(t')\,a_K(t)}, \quad (10\text{A}.3)$$

a result which is extensively used below.

Following the precedent set earlier, we examine low orders before taking on the general term.

Zeroth order. Here, the only term is

$$\overline{a_{\mathbf{k}}(t_2)\,a_{\mathbf{k}}^\dagger(t_1)} = -iG_0(\mathbf{k}, t_2 - t_1). \qquad (10\text{A}.4)$$

Using the conventions of Fig. 4.2, Chapter 4, we represent (10 A. 4) pictorially as in Fig. 10 A. 1.

To emphasize that the line has definite ends, i.e. begins at time t_1 and ends at time t_2, the end points are marked by crosses. This is in contrast with such graphs as are shown in, say, Fig. 4.3 of Chapter 4,

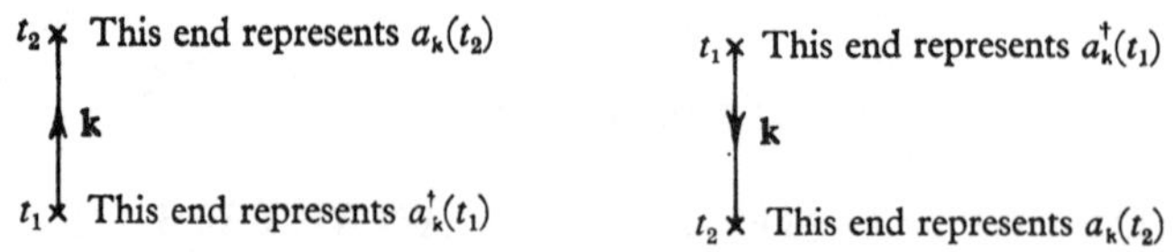

Fig. 10 A.1. The joining of the ends pictorially denotes the contraction. Particles travel forward in time since if $t_2 > t_1$, for non-vanishing G_0 we require $k > k_f$ (recall (10.2.11)). Similarly, holes move backwards in time.

where the lines can continue indefinitely. All graphs of the latter kind arise from uncontracted terms and do not occur in the vacuum-vacuum graphs of Chapter 10.

The rules for representing (10 A.4) graphically may be stated as follows. One marks (by crosses) two times t_1 and t_2 in any order and draws an arrowed line bearing index **k** from t_1 to t_2. When one allows t_1 and t_2 to vary, one obtains both time sequences as illustrated in Fig. 10 A.1.

First order. Here, there are $3! = 6$ possible sets of contractions, namely,

(i) $$a_K^\dagger(t_{12})\, a_L^\dagger(t_{12})\, a_M(t_{12})\, a_N(t_{12})\, a_{\mathbf{k}}(t_2)\, a_{\mathbf{k}}^\dagger(t_1) = (-i)^3\, \delta_{KN}\, \delta_{LM}\, G_0(K,0)\, G_0(M,0)\, G_0(\mathbf{k}, t_2 - t_1),$$

(ii) $$a_K^\dagger(t_{12})\, a_L^\dagger(t_{12})\, a_M(t_{12})\, a_N(t_{12})\, a_{\mathbf{k}}(t_2)\, a_{\mathbf{k}}^\dagger(t_1) = -(-i)^3\, \delta_{KM}\, \delta_{LN}\, G_0(K,0)\, G_0(L,0)\, G_0(\mathbf{k}, t_2 - t_1),$$

(iii) $$a_K^\dagger(t_{12})\, a_L^\dagger(t_{12})\, a_M(t_{12})\, a_N(t_{12})\, a_{\mathbf{k}}(t_2)\, a_{\mathbf{k}}^\dagger(t_1) = -(-i)^3\, \delta_{K\mathbf{k}}\, \delta_{LM}\, \delta_{N\mathbf{k}}\, G_0(\mathbf{k}, t_2 - t_{12})\, G_0(L,0)\, G_0(\mathbf{k}, t_{12} - t_1),$$

(iv) $$a_K^\dagger(t_{12})\, a_L^\dagger(t_{12})\, a_M(t_{12})\, a_N(t_{12})\, a_{\mathbf{k}}(t_2)\, a_{\mathbf{k}}^\dagger(t_1) = -(-i)^3\, \delta_{KN}\, \delta_{L\mathbf{k}}\, \delta_{M\mathbf{k}}\, G_0(K,0)\, G_0(\mathbf{k}, t_2 - t_{12})\, G_0(\mathbf{k}, t_{12} - t_1),$$

(v) $$a_K^\dagger(t_{12})\, a_L^\dagger(t_{12})\, a_M(t_{12})\, a_N(t_{12})\, a_{\mathbf{k}}(t_2)\, a_{\mathbf{k}}^\dagger(t_1) = (-i)^3\, \delta_{KM}\, \delta_{L\mathbf{k}}\, \delta_{N\mathbf{k}}\, G_0(K,0)\, G_0(\mathbf{k}, t_2 - t_{12})\, G_0(\mathbf{k}, t_{12} - t_1),$$

(vi) $$a_K^\dagger(t_{12})\, a_L^\dagger(t_{12})\, a_M(t_{12})\, a_N(t_{12})\, a_{\mathbf{k}}(t_2)\, a_{\mathbf{k}}^\dagger(t_1) = (-i)^3\, \delta_{K\mathbf{k}}\, \delta_{LN}\, \delta_{M\mathbf{k}}\, G_0(\mathbf{k}, t_2 - t_{12})\, G_0(L,0)\, G_0(\mathbf{k}, t_{12} - t_1),$$

(10A.5)

where, consistent with the discussion in the early part of section 15, Chapter 4, the zero-time propagator is defined by writing $t' = t+0$ in (10A.3). Thus we always have

$$G(\mathbf{k}, 0) \equiv G(\mathbf{k}, -0). \qquad (10\text{A}.6)$$

The graphs corresponding to (10A.5) (with $t_2 > t_1$) are shown in Fig. 10A.2.

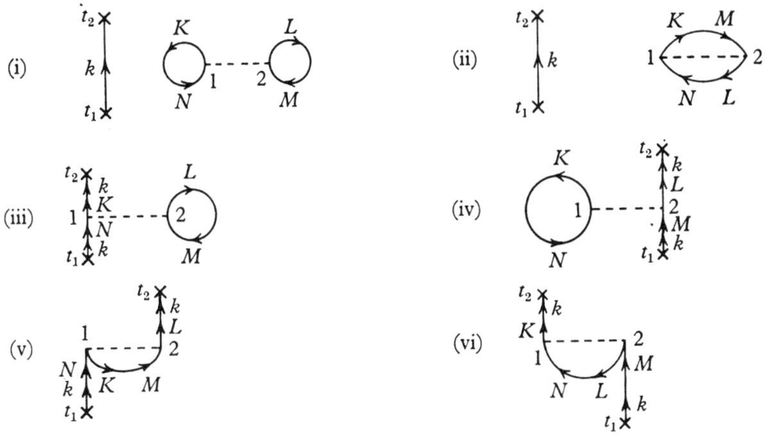

Fig. 10A.2

The following points should be noted:

(*a*) The prescription for obtaining all first-order diagrams is this. We draw an appropriately labelled dotted line (1 on the left, 2 on the right) at time t_{12} and two crosses, one labelled t_1 and the other t_2. Then, every distinct graph is drawn such that one line leaves t_1, one enters t_2 and one enters and leaves each of 1 and 2. (Once more, the conventions of Chapter 4 are used and, in particular, no attention is paid to the direction of arrows in equal-time contractions.) Inserting labels (**k**'s entering t_2 and leaving t_1, K, L leaving 1 and 2 respectively and M, N entering 2, 1 respectively) then gives us the desired graphs.

To illustrate this, Fig. 10A.3 (0) shows the bare structure on which the particle lines are to be draped, while graphs (i)–(vi) show the result of doing this draping in all distinct ways. The diagrams so obtained are identical with those of Fig. 10A.2.

If, now, we allow t_1, t_2 and the vertex time t_{12} to take on all values so that the crosses and the horizontal vertex can be thought of as free to meander on the page, we obtain all possible time values of the terms of (10 A.5). We thus regard a graph of the kind shown in Fig. 10 A. 3 as representative of all time sequences. With this convention, in zeroth order, one only need draw the first of the

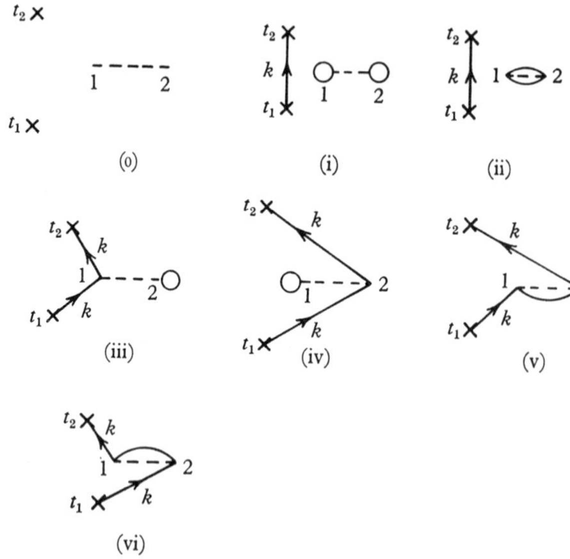

Fig. 10A. 3

two diagrams shown in Fig. 10 A. 1. Furthermore, since only the topology of graphs matter, they are sometimes 'straightened out' to present a neater appearance! An example of this is the way the graphs of Fig. 10 A. 2 are obtained from those of Fig. 10 A. 3.

(*b*) The converse of (*a*) is also true. From a labelled diagram, the appropriate member of (10 A. 5) can be identified. One may either proceed from first principles and write down the left-hand side, explicitly drawing in the contractions, or use the rules by which each doubly labelled line segment carries a Kronecker delta and an unperturbed (equal- or non-equal-) time Green function. The rule for automatically allocating the sign, depending on the number of closed loops, is the same as in Chapter 4, this following from our original choice of the order of $a_{\mathbf{k}t_2}$ and $a^{\dagger}_{\mathbf{k}t_1}$ in (10.2.3).

(*c*) In Green function theory, also, we have linked (connected) and unlinked (disconnected) graphs, the definitions being the obvious topological ones. Thus, in Fig. 10 A. 3, (i) and (ii) are unlinked while (iii)–(vi) are linked. Shortly, we will have another linked cluster theorem.

The linked graphs, once more, are the only ones which enter the theory in an essential way and these may be given simple physical interpretations. Thus, in Fig. 10 A. 3, (iii) and (iv) represent a particle freely propagating from t_1 to t_2, except for interaction with a passive particle at time t_{12}, while (v) and (vi) describe free particle propagation from t_1 to t_{12}, exchange interaction taking place, followed by free propagation to t_2.

(*d*) The linked terms vanish unless t_{12} is intermediate between t_1 and t_2. This is because the product $G_0(\mathbf{k}, t_2 - t_{12})\, G_0(\mathbf{k}, t_{12} - t_1)$ always appears in the contributions from the linked terms in (10 A. 5), and from (10.2.11) such a product vanishes if one of the time arguments is positive and the other negative. Thus,

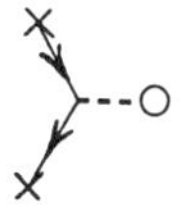 and 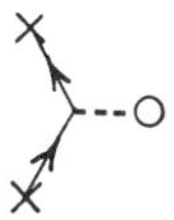

contribute but 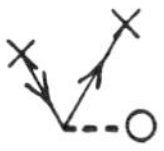 and 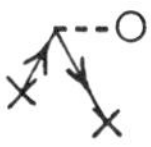 do not. Of course this is no more than a necessary criterion for the success of the Feynman method, for diagrams of the latter kind violate the Feynman rule that particles always travel forwards in time and holes always backwards.

Second order. Here, there are $5! = 120$ terms arising from the expansion of

$$\langle g|\, T\{a_K^\dagger(t_{12})\, a_L^\dagger(t_{12})\, a_M(t_{12})\, a_N(t_{12})\, a_P^\dagger(t_{34})\, a_Q^\dagger(t_{34})\, a_R(t_{34})\, a_S(t_{34}) \times a_{\mathbf{k}}(t_2)\, a_{\mathbf{k}}^\dagger(t_1)\}\,|g\rangle. \quad (10\,A.7)$$

Clearly, it is not convenient to write down explicitly all these terms to obtain the analogue of the first-order expressions (10 A. 5). Nevertheless, in view of our unequivocal method of drawing contractions graphically, it is not difficult to give a prescription for drawing all possible graphs.

One draws the structure shown in Fig. 10A.4 and supplies directed lines in all possible way such that one line leaves t_1, one enters t_2 and one enters and leaves each end of each vertex in the usual way. Thus one obtains 40 disconnected and 80 connected graphs. Now one can imagine t_1, t_2, t_{12} and t_{34} to take on any time sequence but always, however, keeping the original topology.

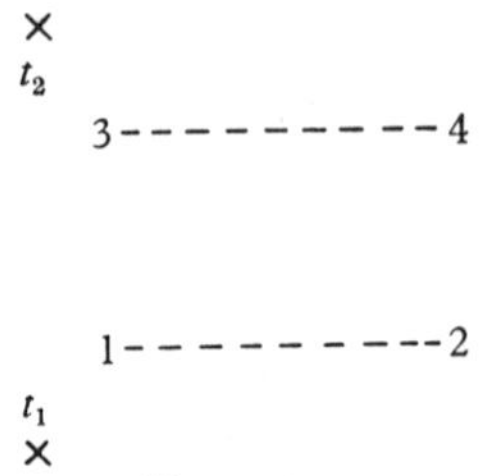

Fig. 10A.4

The disconnected graphs fall into two categories. There are 24 of the kind 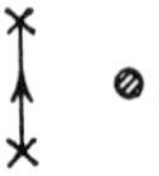 , and these are shown in Fig. 10A.5. Then there are 16 of the kind 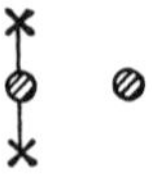 and these are displayed in Fig. 10A.6. The 80 linked graphs have the general form  and are given in Fig. 10A.7.

Having drawn the graphs (linked and unlinked), one allocates the same index **k** to the line leaving t_1 and to that entering t_2. Then one supplies the appropriate labels to the lines entering and leaving the vertices (K, L leaving 1, 2 respectively, M, N entering 2, 1 respectively, P, Q leaving 3, 4 respectively, R, S entering 4, 3 respectively).

We note the following points:

(*a*) Many of these graphs can be ignored for momentum- and spin-conserving interactions since they are associated then with vanishing matrix elements. We know, for example, from Chapter 4 (cf. discussions relating to Figs. 4.7 and 4.4 (iv); also that on Fig. 4.15), that the circled disconnected graphs of Fig. 10A.5 are of this kind, as, for identical reasons, are the circled linked graphs of Fig. 10A.7.

(*b*) For the same reason as in first order, the time sequence for a non-vanishing contribution must be such that lines entering t_2 and leaving t_1 must be arrowed in the same direction in time. For, by momentum-conservation, each of these lines carries the same symbol **k** and thus the product $G_0(\mathbf{k}, t_{12}-t_1)\, G_0(\mathbf{k}, t_2-t_{34})$ always occurs. (Recall note (*d*) of the first-order discussion.)

(*c*) As in first order, the linked graphs can be interpreted

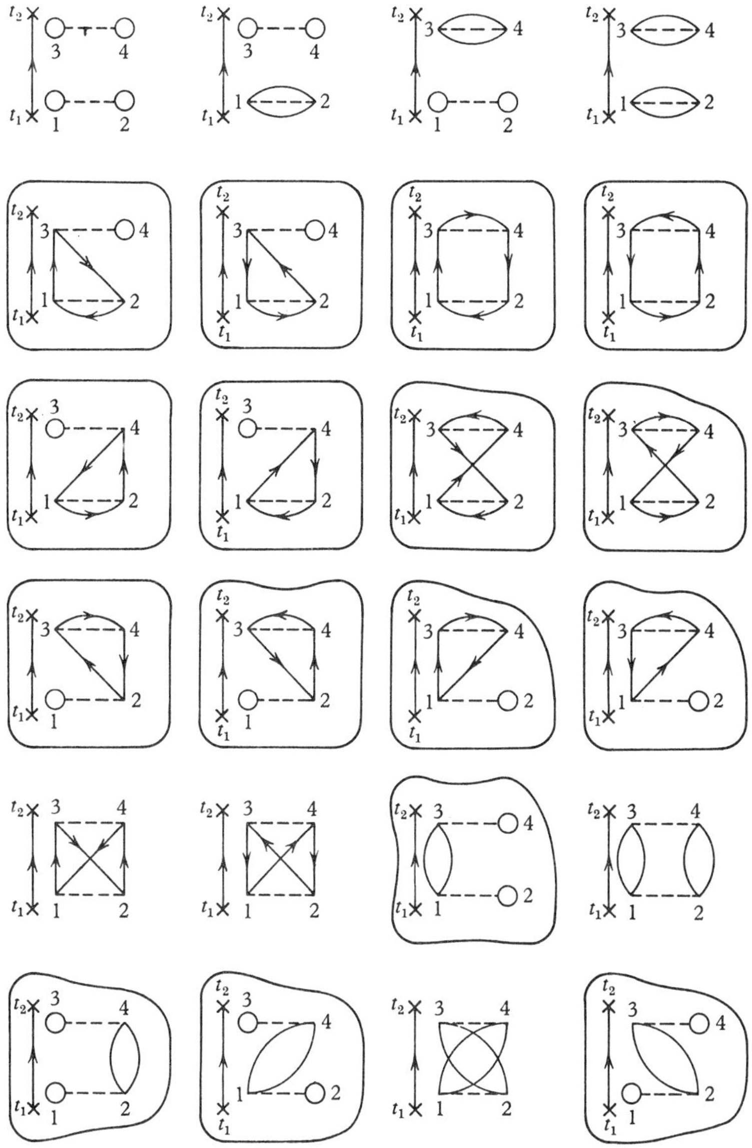

Fig. 10A.5

physically. Thus, for example, in Fig. 10A.8, (i) represents a particle exciting from the Fermi sea a particle-hole pair which subsequently annihilate each other. Diagram (ii) shows a particle exciting a particle-hole pair, where, subsequently the original particle annihilates with the hole.

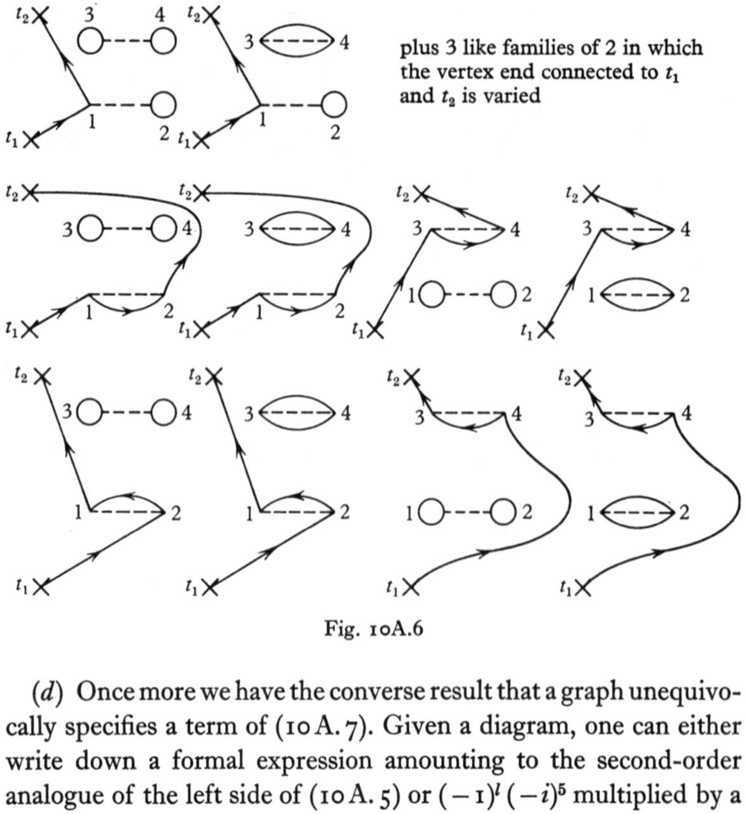

Fig. 10A.6

(*d*) Once more we have the converse result that a graph unequivocally specifies a term of (10A.7). Given a diagram, one can either write down a formal expression amounting to the second-order analogue of the left side of (10A.5) or $(-1)^l(-i)^5$ multiplied by a product of Kronecker deltas and unperturbed propagators specified by the diagram. Once more, l denotes the number of closed loops in the graph.

nth order. With the experience gained above, it is now clear how to discuss the general case. There are $(2n+1)!$ terms in the expansion of

$$\langle g|T\{a_K^\dagger(t_{12})\,a_L^\dagger(t_{12})\,a_M(t_{12})\,a_N(t_{12})\,a_P^\dagger(t_{34})\ldots a_X^\dagger(t_{2n-12n})\,a_Y^\dagger(t_{2n-12n}) \\ \times a_Z(t_{2n-12n})\,a_T(t_{2n-12n})\,a_{\mathbf{k}}(t_2)\,a_{\mathbf{k}}^\dagger(t_1)|g\rangle, \quad (10\text{A}.8)$$

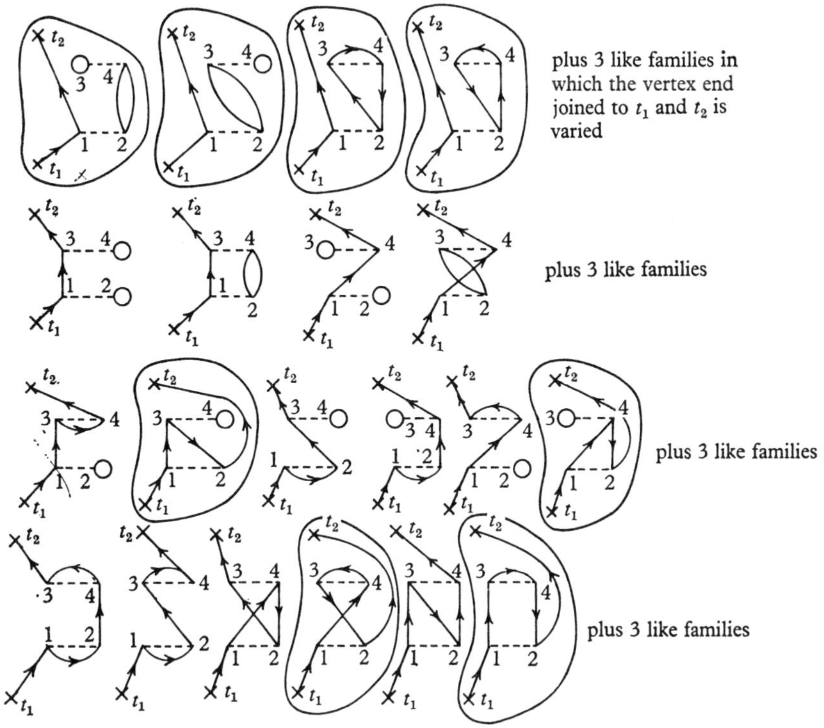

Fig. 10A.7

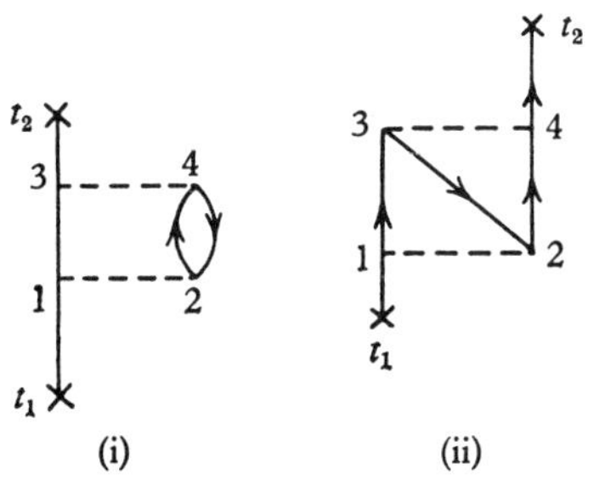

Fig. 10A.8

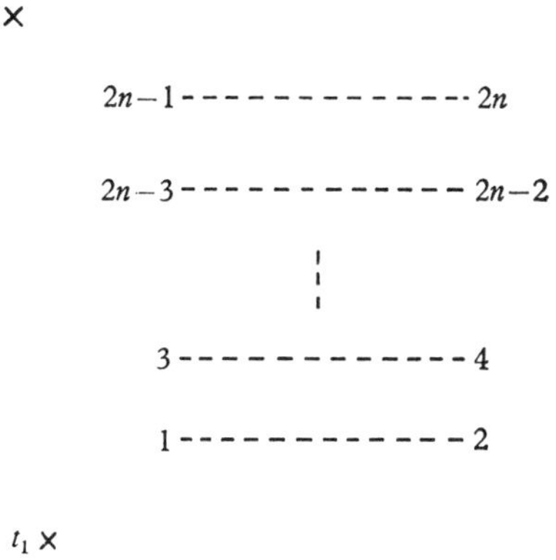

Fig. 10A.9

and these are denoted by graphs constructed as follows. First, the structure shown in Fig. 10A.9 is drawn. Then, directed lines are supplied in all possible ways, one entering t_2, one leaving t_1 and one entering and leaving each vertex end. Then line labels are attached at each vertex, K, L, M, N to 1, 2, P, Q, R, S to 3, 4, ..., X, Y, Z, T to $2n-1, 2n$ in the proper order (so that K, L leave 1, 2 respectively, M, N enter 2, 1 respectively, etc.). The indices at t_1 and t_2 are both $\mathbf{k}$ on account of momentum and spin conservation.

Once more, we may note (*a*) many graphs give zero contributions on account of momentum conservation, (*b*) the lines leaving t_1 and entering t_2 must be in the same time sense for a non-zero contribution, (*c*) the graphs admit of simple physical interpretations and finally, and most importantly, (*d*) each graph, conversely has a mathematical equivalent which is readily written down. Each line gives a Kronecker delta times a free-particle Green function. One multiplies all these, together with another factor of $(-1)^l(-i)^{2n+1}$, where l is the number of closed loops.

We have learned, above, how to graphically analyse (10A.1), and thereby (10.4.18). The diagrams which arose fell into two classes, linked and unlinked. It is now possible to eliminate the unlinked graphs entirely from the theory much as discussed in Chapter 4. For the remaining linked graphs, one may simplify matters slightly by introducing unlabelled diagrams. These arise conveniently by noting that nth-order labelled linked graphs always occur in families of size $2^n n!$ obtained by permuting the pairs $(1, 2), (3, 4), \ldots, (2n-1, 2n)$ with each other and the individual members of each pair internally. Furthermore, there is often no ambiguity in dropping the arrows and the t_1 and t_2 time labels. As examples, we may quote 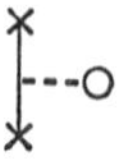

representing the $2^1 1!$ graphs 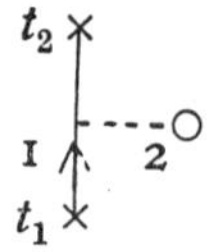 and 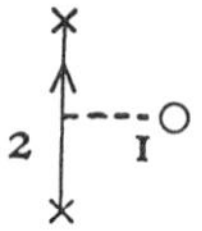 and 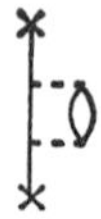

representing the $2^2 2!$ graphs

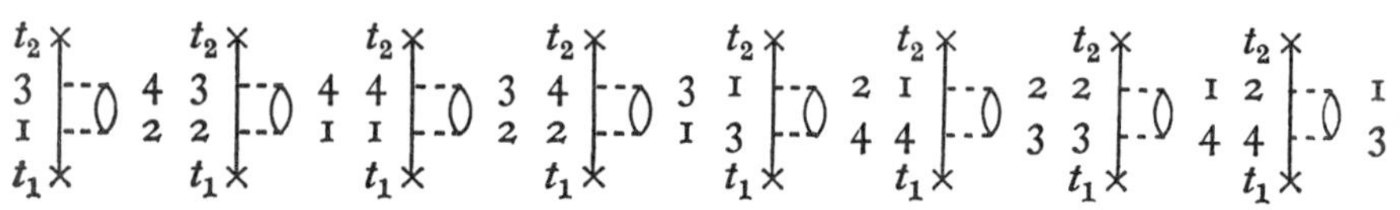

The latter graphs are more easily identified in Fig. 10A.7 and associated families in the forms

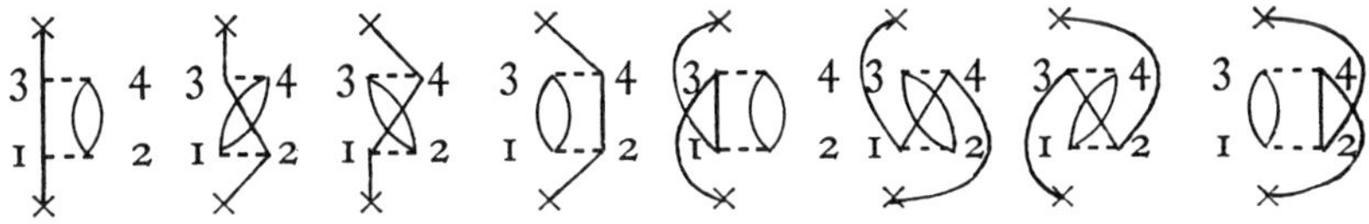

respectively.

Collecting the contributions discussed fully to all orders in this Appendix, the basic result (10.5.1) and its symbolic form (10.5.2) are readily understood.

Additional Problems for Chapter 7

P.7(ii). In (9) above, choose

$$V_{KK'} = \begin{cases} \frac{1}{\Omega}\sum_{l=0}^{\infty}\sum_{m=-l}^{l} 2\pi V_l Y_{lm}(\theta,\phi) Y_{lm}(\theta',\phi') & \text{when } |\epsilon_K|, |\epsilon_{K'}| < \delta, \\ 0 & \text{otherwise.} \end{cases} \tag{10}$$

where θ, ϕ are the angular parts of **k**. (All we have done here is to expand the plane waves involved in the definition of

$$V(K, -K, K', -K')$$

in terms of spherical harmonics and neglected the k-dependence over the thin energy shell. If we choose to write $V_l = 0$ $(l \neq 0)$ and $2\pi V_0 = -V$, we would return to (7.5.19).) Show, following closely the techniques used in arriving at (7.4.24), that for any l for which $V_l < 0$ there is a solution of (9) given by

$$\Delta(\theta, \phi) = 2\delta\Gamma e^{-1/\rho V_l} Y_{lm}(\theta, \phi), \tag{11}$$

where

$$\ln \Gamma = -\iint |Y_{lm}(\theta, \phi)|^2 \ln |Y_{lm}(\theta, \phi)| \sin\theta \, d\theta \, d\phi. \tag{12}$$

(Thus, even if V_0 is repulsive, if at least one V_l is attractive, pairing should result, though not necessarily with the gap parameter (11), which may be only one of many solutions of (9). The physical significance of the above analysis is that at sufficiently low temperatures ^{3}He may be a superfluid, for while the repulsive core ensures that $V_0 > 0$, the longer range attractive van der Waals force should ensure that at higher l, $V_l < 0$. The above example is discussed in P. W. Anderson & P. Morel, *Phys. Rev.* **123**, 1911 (1961). So far such superfluid behaviour has not been observed.)

P.7(iii). Recall the discussion of section 1.11 in Chapter 1 on quasi-particles. The B.C.S. theory takes the interaction of quasi-particles into account through the final term of (7.6.8). Obtain the analogue of (7.6.16) when this term is omitted and test the validity of the assertion in section 1.11 that as many quasi-particles are created, the effect of their interactions is to build up to destroy the independent quasi-particle picture.

REFERENCES

Numbers in parentheses following each reference indicate the text pages where the reference has been cited

Abel, W. R., Anderson, A. C. & Wheatley, J. C. (1966). *Phys. Rev. Letters*, **17**, 74. (180)
Abrikosov, A. A. (1957). *Sov. Phys. J.E.T.P.* **5**, 1174; *J. Phys. Chem. Solids*, **2**, 199. (253)
Abrikosov, A. A. & Khalatnikov, I. M. (1959). *Rep. Prog. Phys.* **22**, 329. (182)
Abrikosov, A. A., Gorkov, L. P. & Dzyaloshinski, I. E. (1963). *Methods of Quantum Field Theory in Statistical Physics*. New Jersey: Prentice Hall. (348, 397)
Aleckseev, A. I. (1960). *Sov. Phys. Uspekhi*, **4**, 23. (397)
Anderson, P. W. (1958). *Phys. Rev.* **112**, 1900. (249)
Anderson, P. W. (1959). *J. Phys. Chem. Solids*, **11**, 26. (251)
Baker, G. A. (1964). *Phys. Rev.* **131**, 1869. (206)
Bardeen, J. (1936). *Phys. Rev.* **50**, 1098. (430)
Bardeen, J. (1937). *Phys. Rev.* **52**, 688. (139)
Bardeen, J., Cooper, L. N. & Schrieffer, J. R. (1957). *Phys. Rev.* **108**, 1175. (215, 226)
Bardeen, J. & Pines, D. (1955). *Phys. Rev.* **99**, 1140. (217)
Barnett, M. P., Birss, F. W. & Coulson, C. A. (1958). *Mol. Phys.* **1**, 44. (49)
Bell, J. S. & Squires, E. J. (1961). *Adv. Phys.* **10**, 211. (204, 206)
Beliaev, S. T. (1959). *The Many Body Problem*. Paris: Dunod. (296)
Beliaev, S. T. (1958). *Sov. Phys. J.E.T.P.* **7**, 289, 299. (397, 399)
Bethe, H. A. (1965). *Phys. Rev.* **138**, B 804. (207)
Bethe, H. A., Brandow, B. H. & Petschek, A. G. (1963). *Phys. Rev.* **129**, 225. (202)
Blandin, A., Daniel, E. & Freidel, J. (1959). *Phil. Mag.* **4**, 180. (139)
Blatt, J. M. (1964). *Theory of Superconductivity*. New York: Academic Press. (228, 320)
Beaumont, C. F. A. & Reekie, J. (1955). *Proc. Roy. Soc.* A, **228**, 363. (306)
Bloch, C. (1963). *Studies in Statistical Mechanics*, vol. III. Amsterdam: North Holland. (333)
Bogoliubov, N. N. (1947). *J. Phys. (U.S.S.R.)*, **11**, 23. (298)
Bogoliubov, N. N. (1958). *Sov. Phys. J.E.T.P.* **7**, 41. (235)
Bogoliubov, N. N. & Zubarev, D. N. (1955). *Sov. Phys. J.E.T.P.* **1**, 83. (298)
Bohm, D. & Pines, D. (1953). *Phys. Rev.* **92**, 609. (279)
Bohm, D. & Pines, D. (1952). *Phys. Rev.* **85**, 338. (148)
Bohr, A., Mottelson, B. R. & Pines, D. (1958). *Phys. Rev.* **110**, 936. (204)
Born, M. & Huang, K. (1954). *Dynamical Theory of Crystal Lattices*. Oxford University Press. (263)
Breit, G. (1947). *Phys. Rev.* **71**, 215. (270)

Brout, R. (1963). *Phys. Rev.* **131**, 899. (292)
Brout, R. & Carruthers, P. (1963). *Lectures on the Many Electron Problem.* New York: Interscience. (90, 172, 427)
Brueckner, K. A. (1955). *Phys. Rev.* **100**, 36. (67)
Brueckner, K. A. (1959). *The Many Body Problem.* Paris: Dunod. (144)
Brueckner, K. A. & Gammel, J. L. (1958). *Phys. Rev.* **109**, 1023. (195, 203)
Brueckner, K. A. & Goldman, D. T. (1960). *Phys. Rev.* **117**, 207. (209)
Brueckner, K. A. & Masterson, K. S. Jr. (1962). *Phys. Rev.* **128**, 2267. (203)
Byers, N. & Yang, C. N. (1961). *Phys. Rev. Lett.* **7**, 46. (250)
Carr, W. J. Jr. (1961). *Phys. Rev.* **122**, 1437. (162, 170, 173)
Carr, W. J. Jr. & Maradudin, A. (1964). *Phys. Rev.* **133**, A 371. (164)
Carr, W. J. Jr., Coldwell-Horsfall, R. A. & Fein, A. E. (1961). *Phys. Rev.* **124**, 747. (163)
Carruthers, P. (1961). *Rev. Mod. Phys.* **33**, 921. (278)
Cochran, W. (1963). *Rep. Prog. Phys.* **26**, 1. (265)
Cohen, M. & Feynman, R. P. (1957). *Phys. Rev.* **107**, 13. (295)
Coldwell-Horsfall, R. A. & Maradudin, A. A. (1960). *J. Math. Phys.* **1**, 395. (171)
Coldwell-Horsfall, R. A. & Maradudin, A. A. (1963). *J. Math. Phys.* **4**, 582. (163)
Cooper, L. N. (1956). *Phys. Rev.* **104**, 1189. (225)
Coulson, C. A. (1938). *Proc. Camb. Phil. Soc.* **34**, 204. (46)
Coulson, C. A. & Neilson, A. H. (1961). *Proc. Phys. Soc.* **78**, 831. (39)
Daniel, E. & Vosko, S. H. (1960). *Phys. Rev.* **120**, 2041. (166)
de Gennes, P. G. (1959). *Physica*, **25**, 825. (283)
Dickinson, W. C. (1950). *Phys. Rev.* **80**, 563. (34)
Dirac, P. A. M. (1930). *Proc. Camb. Phil. Soc.* **26**, 376. (10)
Dirac, P. A. M. (1958). *The Principles of Quantum Mechanics.* Oxford: Clarendon Press. (268)
Douglass, D. H. Jr. & Falicov, L. M. (1964). *Progress in Low Temperature Physics*, vol. 4. Amsterdam: North Holland. (255)
Dubois, D. F. (1959). *Ann. Phys.* **7**, 174; *ibid.* **8**, 24. (367, 385)
Egelstaff, P. A. (1963). *Symposium on Inelastic Scattering of Neutrons in Solids and Liquids.* Chalk River. (273)
Ehrenreich, H. & Cohen, M. H. (1959). *Phys. Rev.* **115**, 786. (137, 424)
Enderby, J. E., Gaskell, T. & March, N. H. (1965). *Proc. Phys. Soc.* **85**, 217. (283, 309)
Emery, V. J. (1959). *Nucl. Phys.* **12**, 69. (205)
Emery, V. J. & Sessler, A. M. (1960). *Phys. Rev.* **119**, 248. (206)
Faddeev, L. D. (1961). *Sov. Phys. J.E.T.P.* **12**, 1014. (207)
Feynman, R. P. (1939). *Phys. Rev.* **56**, 340. (34)
Feynman, R. P. (1953). *Phys. Rev.* **91**, 1291. (292)
Feynman, R. P. (1955). *Progress in Low Temperature Physics*, vol. 1. Ed. C. J. Gorter. Amsterdam: North Holland. (292, 294, 298)
Feynman, R. P. & Cohen, M. (1956). *Phys. Rev.* **102**, 1189. (284, 294)
Foldy, L. L. (1951). *Phys. Rev.* **83**, 397. (34)
Foldy, L. L. (1961). *Phys. Rev.* **124**, 649. (321)

Fröhlich, H. (1950). *Phys. Rev.* **79**, 845. (217)
Fuchs, K. (1935). *Proc. Roy. Soc.* A, **151**, 585. (162)
Galitskii, V. M. (1958). *Sov. Phys. J.E.T.P.* **7**, 104. (367)
Galitskii, V. M. & Migdal, A. M. (1958). *Sov. Phys. J.E.T.P.* **7**, 96. (349)
Gammel, J. L. & Thaler, R. M. (1960). *Progress in Elementary Particle and Cosmic Ray Physics*, **5**, 99. Amsterdam: North Holland. (188)
Garland, J. W. Jr. (1963). *Phys. Rev. Lett.* **11**, 114. (217)
Gaskell, T. (1961). *Proc. Phys. Soc.* **77**, 1182. (163, 168)
Gaskell, T. (1962). *Proc. Phys. Soc.* **80**, 1091. (168, 321)
Gell-Mann, M. (1957). *Phys. Rev.* **106**, 369. (170, 429)
Gell-Mann, M. & Brueckner, K. A. (1957). *Phys. Rev.* **106**, 364. (68, 137)
Gell-Mann, M. & Low, F. E. (1951). *Phys. Rev.* **84**, 350. (80)
Ginzburg, V. L. & Landau, L. D. (1950). *Sov. Phys. J.E.T.P.* **20**, 1064. (252)
Girardeau, M. (1962). *Phys. Rev.* **127**, 1809. (321, 323)
Glauber, R. J. (1955). *Phys. Rev.* **98**, 1692. (274)
Goldstone, J. (1957). *Proc. Roy. Soc.* A, **239**, 267. (67)
Gombàs, P. (1949). *Die Statistische Theorie des Atoms und ihre Anwendungen.* Vienna: Springer. (32)
Goodman, B. B. (1962). *I.B.M. J. Res. Develop.* **63**. (255)
Gor'kov, L. P. (1959). *Sov. Phys. J.E.T.P.* **9**, 1364. (252)
Gottfried, K. & Picman, L. (1960). *Math.-fys. Meddr*, **32**, 13. (385)
Green, H. S. (1952). *J. Chem. Phys.* **20**, 1274. (13)
Gurnee, E. F. & Magee, J. L. (1950). *J. Chem. Phys.* **18**, 142. (46)
Hall, G. G. (1961). *Phil. Mag.* **6**, 249. (43)
Hall, G. G., Jones, L. L. & Rees, D. (1965). *Proc. Roy. Soc.* A, **283**, 1393. (42)
Hamada, T. & Johnston, I. D. (1962). *Nucl. Phys.* **34**, 383. (203)
Hartree, D. R. (1957). *The Calculation of Atomic Structures.* New York: Wiley. (28)
Heitler, W. (1954). *The Quantum Theory of Radiation.* Oxford: Clarendon Press. (263, 434)
Henshaw, D. G. & Woods, A. D. B. (1961). *Phys. Rev.* **121**, 1266. (295)
Herman, F. & Skillman, S. (1965). *Atomic Structure Calculations.* New Jersey: Prentice Hall. (28, 31)
Huang, K. (1960). *Phys. Rev.* **119**, 1129. (309)
Huang, K. (1963). *Statistical Mechanics.* New York and London: Wiley. (290, 316)
Hubbard, J. (1957). *Proc. Roy. Soc.* A, **240**, 539. (68)
Hubbard, J. (1958). *Proc. Roy. Soc.* A, **243**, 336. (155)
Hugenholtz, N. M. & Pines, D. (1959). *Phys. Rev.* **116**, 489. (397, 399)
Hugenholtz, N. M. & Van Hove, L. (1958). *Physica*, **24**, 363. (207)
Isihara, A. & Yee, D. D. H. (1964). *Phys. Rev.* **136**, A, 618. (309)
Inelastic Scattering of Neutrons from Solids and Liquids (1963). International Atomic Energy Agency, Vienna. (276)
James, H. M. & Coolidge, A. S. (1939). *Phys. Rev.* **55**, 873. (46)

Johnson, M. D., Hutchinson, P. & March, N. H. (1964). *Proc. Roy. Soc.* A, **282**, 283. (282)
Kadanoff, L. P. & Martin, P. C. (1963). *Ann. Phys.* (*N.Y.*), **24**, 419. (283)
Kadanoff, L. P. & Baym, G. (1962). *Quantum Statistical Mechanics.* New York: W. A. Benjamin Inc. (397)
Kanazawa, H. & Watabe, M. (1960). *Progr. Theor. Phys.* **23**, 408. (173)
Kanazawa, H., Misawa, S. & Fujita, E. (1960). *Progr. Theor. Phys.* **23**, 426. (173)
Kanazawa, H. & Matsudaira, N. (1960). *Progr. Theor. Phys.* **23**, 433. (173)
Kilby, G. E. (1961). *Proc. Phys. Soc.* **78**, 673. (48)
Kittel, C. (1963). *Quantum Theory of Solids.* New York: John Wiley. (269)
Klein, A. (1962). *Cargèse Lectures in Theoretical Physics.* New York: W. A. Benjamin Inc. (367)
Kohn, W. & Luttinger, J. M. (1960). *Phys. Rev.* **118**, 41. (68, 186, 333)
Kokkedee, J. J. J. (1962). *Physica*, **28**, 893. (278)
Kokkedee, J. J. J. (1963). *Phys. Lett.* **4**, 78. (278)
Landau, L. D. (1957). *Sov. Phys. J.E.T.P.* **3**, 920; *ibid.* (1957). **5**, 101. (175)
Landau, L. D. (1959). *Sov. Phys. J.E.T.P.* **8**, 70. (183, 343)
Landau, L. D. & Lifshitz, I. M. (1958). *Statistical Physics.* Oxford: Pergamon. (294, 325, 338)
Langer, J. S. & Vosko, S. H. (1959). *J. Phys. Chem. Solids*, **12**, 196. (139)
Larkin, A. I. & Migdal, A. B. (1963). *Sov. Phys. J.E.T.P.* **17**, 1146. (397)
Lassila, K. E., Hull, M. H. Jr., Ruppel, H. M., McDonald, F. A. & Breit, G. (1962). *Phys. Rev.* **126**, 881. (203)
Lee, T. D., Huang, K. & Yang, C. N. (1957). *Phys. Rev.* **106**, 1135. (309)
Lester, W. A. & Krauss, M. (1964). *J. Chem. Phys.* **41**, 1407. (41)
Lieb, E. H. (1963). *Phys. Rev.* **130**, 2518. (317)
Lieb, E. H. (1963). *Phys. Rev.* **130**, 1616. (324)
Lieb, E. H. & Liniger, W. (1963). *Phys. Rev.* **130**, 1605. (324)
Lieb, E. H. & Sakakura, A. Y. (1964). *Phys. Rev.* **133**, A, 899. (322)
Lighthill, M. J. (1958). *Fourier Analysis and Generalized Functions.* Cambridge University Press. (139, 309)
Lindhard, J. (1954). *Kgl. danske Mat. fys. Medd.* **28**, 8. (139, 426)
London, F. (1954). *Superfluids*, vol. II. New York: John Wiley. (255)
London, F. (1961). *Superfluids*, vol. I. New York: Dover. (292, 293)
Luban, M. (1962). *Phys. Rev.* **128**, 965. (290)
Luttinger, J. M. (1966). *Phys. Rev.* **150**, 202. (217)
Luttinger, J. M. & Ward, J. C. (1960). *Phys. Rev.* **118**, 1417. (333)
Lynton, E. A. (1959). *The Many Body Problem.* Paris: Dunod. (292)
Macke, W. (1950). *Z. Naturf.* **5a**, 192. (68, 137)
March, N. H. (1958). *Phys. Rev.* **110**, 604. (163)
March, N. H. & Donovan, B. (1954). *Proc. Phys. Soc.* **67**, 464. (173)
March, N. H. & Murray, A. M. (1960). *Phys. Rev.* **120**, 830. (11, 139)
March, N. H. & Murray, A. M. (1961). *Proc. Roy. Soc.* A, **261**, 119. (11, 13, 15, 139)

March, N. H. & Sampanthar, S. (1962). *Acta Phys. Hungarica*, **14**, 61. (166, 168)
March, N. H. & Young, W. H. (1959). *Phil. Mag.* **4**, 384. (173)
Martin, P. C. & Schwinger, J. (1959). *Phys. Rev.* **115**, 1342. (344, 389)
Maslen, V. W. (1956). *Proc. Phys. Soc.* A, **69**, 734. (41)
Mattis, D. C. (1965). *Theory of Magnetism.* New York: Harper & Row. (263)
Mattuck, R. D. (1964). *Ann. Phys.* **27**, 216. (410)
Mayer, J. E. (1955). *Phys. Rev.* **100**, 1579. (166)
Messiah, A. (1961). *Quantum Mechanics*, vol. 1. Amsterdam: North-Holland. (273)
Misawa, S. (1965). *Phys. Rev.* **140**, A 1645. (174)
Montroll, E. W. & Ward, J. C. (1958). *Phys. Fluids*, **1**, 55. (325)
Moravesik, M. J. (1963). *The Two Nucleon Interaction.* Oxford: Clarendon Press. (187)
Moszkowski, S. A. & Scott, B. L. (1960). *Ann. Phys., N.Y.*, **11**, 65. (202)
Nishen, B. W. (1964). *Ann. Phys. (N.Y.)*, **28**, 220. (367)
Nozières, P. (1963). *The Theory of Interacting Fermi Systems.* New York: W. A. Benjamin. (80, 183, 343, 349, 397)
Nozières, P. & Pines, D. (1958). *Phys. Rev.* **111**, 442. (155, 163)
Overhauser, A. W. (1962). *Phys. Rev.* **128**, 437. (174)
Pauling, L. & Podolsky, B. (1929). *Phys. Rev.* **34**, 109. (36)
Penrose, O. & Onsager, L. (1956). *Phys. Rev.* **104**, 576. (290)
Peshkov, V. P. (1946). *Sov. Phys. J.E.T.P.* **16**, 1000. (292)
Pines, D. (1955). *Solid State Phys.* **1**, 267. (431)
Pines, D. (1961). *The Many Body Problem.* New York: W. A. Benjamin Inc. (155, 159, 367)

Pippard, A. B. (1953). *Proc. Roy. Soc.* A, **216**, 547. (214)
Placzek, G. (1952). *Phys. Rev.* **86**, 377. (283)
Prange, R. E. (1963). *Phys. Rev.* **129**, 2495. (249)
Quinn, J. J. & Ferrell, R. A. (1958). *Phys. Rev.* **112**, 812. (367)
Raimes, S. (1961). *The Wave Mechanics of Electrons in Metals.* Amsterdam: North Holland. (148)
Randolph, P. D. (1964). *Phys. Rev.* **134**, A 1238. (283)
Razavy, M. (1963). *Phys. Rev.* **130**, 1091. (203)
Rice, T. M. (1965). *Ann. Phys., N.Y.*, **31**, 100. (163, 429)
Rickayzen, G. (1959). *Phys. Rev.* **115**, 795. (249)
Rickayzen, G. (1965). *Theory of Superconductivity.* New York: Interscience. (218)
Ritchie, R. H. (1957). *Phys. Rev.* **106**, 874. (175)
Roothaan, C. C., Sachs, L. M. & Weiss, A. W. (1960). *Rev. Mod. Phys.* **32**, 186. (39)
Rosenfeld, L. (1951). *Theory of Electrons.* Amsterdam: North Holland. (310)
Saint-James, D. & de Gennes P. G. (1963). *Phys. Lett.* **7**, 306. (256)
Sawada, K. (1957). *Phys. Rev.* **106**, 372. (137)
Sawada, K., Brueckner, K. A., Fukuda, N. & Brout, R. (1957). *Phys. Rev.* **108**, 507. (155)

Schafroth, M. R. (1955). *Phys. Rev.* **100**, 463. (320)

Schafroth, M. R. (1960). Theoretical aspects of superconductivity. *Solid State Physics*, **10**. (320)

Schiff, L. I. (1955). *Quantum Mechanics*. New York: McGraw Hill. (270)

Schrieffer, J. R. (1964). *Theory of Superconductivity*. New York: W. A. Benjamin. (218, 349)

Schwartz, C. (1959). *Ann. Phys.* (*N.Y.*), **6**, 156. (42, 43)

Schweber, S. S. (1961). *Introduction to Relativistic Quantum Field Theory*. Illinois: Row Reterson and Co. (59)

Scott, J. M. C. (1952). *Phil. Mag.* **43**, 859. (34)

Silverstein, S. D. (1962). *Phys. Rev.* **128**, 631. (429)

Slater, J. C. (1951). *Phys. Rev.* **81**, 385. (23)

Slater, J. C. & Kirkwood, J. G. (1931). *Phys. Rev.* **37**, 682. (290)

Sommerfeld, A. (1932). *Z. Phys.* **78**, 283. (32)

Sondheimer, E. H. & Wilson, A. H. (1951). *Proc. Roy. Soc.* A, **210**, 173. (14)

Stephen, M. J. (1962). *Proc. Phys. Soc.* **79**, 994. (321)

Thouless, D. J. (1960). *Ann. Phys., N.Y.*, **10**, 553. (335)

Thouless, D. J. (1961). *The Quantum Mechanics of Many Body Systems*. New York: Academic Press. (177)

Tredgold, R. H. (1957). *Phys. Rev.* **105**, 1421. (166)

Ueda, S. (1961). *Progr. Theor. Phys.* **26**, 45. (168)

Valatin, J. G. (1958). *Nuovo Cimento*, **7**, 843. (235)

Van Hove, L. (1954). *Phys. Rev.* **95**, 249, 1374. (269, 276)

Van Hove, L. (1960). *Physica*, **26**, S 200. (68)

Van Hove, L., Hugenholtz, N. M. & Howland, L. P. (1961). *Quantum Theory of Many-Particle Systems*. New York: Benjamin. (269, 276, 278)

Vinen, W. F. (1961). *Proc. Roy. Soc.* A, **260**, 218. (311)

Werner, E. (1959). *Nucl. Phys.* **10**, 688. (195)

Wigner, E. P. (1934). *Phys. Rev.* **46**, **1002**. (161)

Wigner, E. P. (1938). *Trans. Faraday Soc.* **34**, 678. (161)

Wilson, A. H. (1958). *The Theory of Metals*. Cambridge University Press. (244)

Wong, D. Y. (1964). *Nucl. Phys.* **56**, 213. (203)

Young, W. H. & March, N. H. (1958). *Phys. Rev.* **109**, 1854. (44)

Young, W. H. & March, N. H. (1960). *Proc. Roy. Soc.* A, **256**, 62. (166, 173)

Zener, C. (1933). *Nature, Lond.* **132**, 968. (184)

Ziman, J. M. (1960). *Electrons and Phonons*. Oxford: Clarendon Press. (263, 265, 275)

Ziman, J. M. (1964). *Principles of the Theory of Solids*. Cambridge University Press. (170, 278)

Zubarev, D. N. (1960). *Sov. Phys. Uspekhi*, **3**, 320. (397)

SUBJECT INDEX

A CATALOG OF SELECTED

DOVER BOOKS

IN SCIENCE AND MATHEMATICS

A CATALOG OF SELECTED

DOVER BOOKS

IN SCIENCE AND MATHEMATICS

QUALITATIVE THEORY OF DIFFERENTIAL EQUATIONS, V.V. Nemytskii and V.V. Stepanov. Classic graduate-level text by two prominent Soviet mathematicians covers classical differential equations as well as topological dynamics and ergodic theory. Bibliographies. 523pp. 5⅜ × 8½. 65954-2 Pa. $10.95

MATRICES AND LINEAR ALGEBRA, Hans Schneider and George Phillip Barker. Basic textbook covers theory of matrices and its applications to systems of linear equations and related topics such as determinants, eigenvalues and differential equations. Numerous exercises. 432pp. 5⅜ × 8½. 66014-1 Pa. $10.95

QUANTUM THEORY, David Bohm. This advanced undergraduate-level text presents the quantum theory in terms of qualitative and imaginative concepts, followed by specific applications worked out in mathematical detail. Preface. Index. 655pp. 5⅜ × 8½. 65969-0 Pa. $13.95

ATOMIC PHYSICS (8th edition), Max Born. Nobel laureate's lucid treatment of kinetic theory of gases, elementary particles, nuclear atom, wave-corpuscles, atomic structure and spectral lines, much more. Over 40 appendices, bibliography. 495pp. 5⅜ × 8½. 65984-4 Pa. $12.95

ELECTRONIC STRUCTURE AND THE PROPERTIES OF SOLIDS: The Physics of the Chemical Bond, Walter A. Harrison. Innovative text offers basic understanding of the electronic structure of covalent and ionic solids, simple metals, transition metals and their compounds. Problems. 1980 edition. 582pp. 6⅛ × 9¼. 66021-4 Pa. $15.95

BOUNDARY VALUE PROBLEMS OF HEAT CONDUCTION, M. Necati Özisik. Systematic, comprehensive treatment of modern mathematical methods of solving problems in heat conduction and diffusion. Numerous examples and problems. Selected references. Appendices. 505pp. 5⅜ × 8½. 65990-9 Pa. $12.95

A SHORT HISTORY OF CHEMISTRY (3rd edition), J.R. Partington. Classic exposition explores origins of chemistry, alchemy, early medical chemistry, nature of atmosphere, theory of valency, laws and structure of atomic theory, much more. 428pp. 5⅜ × 8½. (Available in U.S. only) 65977-1 Pa. $10.95

A HISTORY OF ASTRONOMY, A. Pannekoek. Well-balanced, carefully reasoned study covers such topics as Ptolemaic theory, work of Copernicus, Kepler, Newton, Eddington's work on stars, much more. Illustrated. References. 521pp. 5⅜ × 8½. 65994-1 Pa. $12.95

PRINCIPLES OF METEOROLOGICAL ANALYSIS, Walter J. Saucier. Highly respected, abundantly illustrated classic reviews atmospheric variables, hydrostatics, static stability, various analyses (scalar, cross-section, isobaric, isentropic, more). For intermediate meteorology students. 454pp. 6⅛ × 9¼. 65979-8 Pa. $14.95

GEOMETRY OF COMPLEX NUMBERS, Hans Schwerdtfeger. Illuminating, widely praised book on analytic geometry of circles, the Moebius transformation, and two-dimensional non-Euclidean geometries. 200pp. 5⅜ × 8¼.
63830-8 Pa. $8.95

MECHANICS, J.P. Den Hartog. A classic introductory text or refresher. Hundreds of applications and design problems illuminate fundamentals of trusses, loaded beams and cables, etc. 334 answered problems. 462pp. 5⅜ × 8½. 60754-2 Pa. $9.95

TOPOLOGY, John G. Hocking and Gail S. Young. Superb one-year course in classical topology. Topological spaces and functions, point-set topology, much more. Examples and problems. Bibliography. Index. 384pp. 5⅜ × 8¼.
65676-4 Pa. $9.95

STRENGTH OF MATERIALS, J.P. Den Hartog. Full, clear treatment of basic material (tension, torsion, bending, etc.) plus advanced material on engineering methods, applications. 350 answered problems. 323pp. 5⅜ × 8½. 60755-0 Pa. $8.95

ELEMENTARY CONCEPTS OF TOPOLOGY, Paul Alexandroff. Elegant, intuitive approach to topology from set-theoretic topology to Betti groups; how concepts of topology are useful in math and physics. 25 figures. 57pp. 5⅜ × 8½.
60747-X Pa. $3.50

ADVANCED STRENGTH OF MATERIALS, J.P. Den Hartog. Superbly written advanced text covers torsion, rotating disks, membrane stresses in shells, much more. Many problems and answers. 388pp. 5⅜ × 8½. 65407-9 Pa. $9.95

COMPUTABILITY AND UNSOLVABILITY, Martin Davis. Classic graduate-level introduction to theory of computability, usually referred to as theory of recurrent functions. New preface and appendix. 288pp. 5⅜ × 8½. 61471-9 Pa. $7.95

GENERAL CHEMISTRY, Linus Pauling. Revised 3rd edition of classic first-year text by Nobel laureate. Atomic and molecular structure, quantum mechanics, statistical mechanics, thermodynamics correlated with descriptive chemistry. Problems. 992pp. 5⅜ × 8½. 65622-5 Pa. $19.95

AN INTRODUCTION TO MATRICES, SETS AND GROUPS FOR SCIENCE STUDENTS, G. Stephenson. Concise, readable text introduces sets, groups, and most importantly, matrices to undergraduate students of physics, chemistry, and engineering. Problems. 164pp. 5⅜ × 8½. 65077-4 Pa. $6.95

THE HISTORICAL BACKGROUND OF CHEMISTRY, Henry M. Leicester. Evolution of ideas, not individual biography. Concentrates on formulation of a coherent set of chemical laws. 260pp. 5⅜ × 8½. 61053-5 Pa. $6.95

THE PHILOSOPHY OF MATHEMATICS: An Introductory Essay, Stephan Körner. Surveys the views of Plato, Aristotle, Leibniz & Kant concerning propositions and theories of applied and pure mathematics. Introduction. Two appendices. Index. 198pp. 5⅜ × 8½. 25048-2 Pa. $7.95

THE DEVELOPMENT OF MODERN CHEMISTRY, Aaron J. Ihde. Authoritative history of chemistry from ancient Greek theory to 20th-century innovation. Covers major chemists and their discoveries. 209 illustrations. 14 tables. Bibliographies. Indices. Appendices. 851pp. 5⅜ × 8½. 64235-6 Pa. $18.95